MAKING DATA SEXY

Annie CUSHING | ANNIELYTICS.COM

MAKING DATA SEXY

A STEP-BY-STEP VISUALIZATION GUIDE FOR MICROSOFT® EXCEL® 2016 WINDOWS USERS

CHART GRAPHICS
Annie Cushing

TECHNICAL LEAD
Destinee Cushing

ART DIRECTOR
Jessica Maldonado

COPY EDITORS
Wendy Notarnicola
Malory Wood

COVER PHOTO
©iStockphoto/goir

Copyright ©2019 by Annie Cushing

All rights reserved. No part of this book may be reproduced or transmitted in any form or by any means, electronic or mechanical, including photocopying, recording, or by any information storage and retrieval system, without written permission from the publisher, except for the inclusion of brief quotations in a review.

Trademarks
All terms mentioned in this book that are known to be trademarks or service marks have been appropriately capitalized. The publisher cannot attest to the accuracy of this information. Use of a term in the book should not be regarded as affecting the validity of any trademark or service mark.

Microsoft Excel is copyright Microsoft. Windows is copyright Microsoft.

Warning and Disclaimer
This book is designed to provide information about Microsoft Excel. Every effort has been made to make this book as complete and as accurate as possible, but no warranty of fitness is implied.

The information is provided on an as-is basis. The author shall have neither the liability nor responsibility to any person or entity with respect to any loss or damages arising from the information contained in this book or from the use of the discs, videos, or programs that may accompany it.

ISBN: 978-1-7330491-0-8

www.makingdatasexy.com
www.annielytics.com

I dedicate this book to my amazing kids,
Destinee, Victoria, Camilla, and Kian!
(I know, I know, I'm a breeder.)

Thank you for being patient with me as I invested so much of my life into this project! You endured many screenshots of excerpts from the book in our family chat, letting me know how funny or clever they were, whether you really believed that or not. That's amore!

You guys are my greatest inspiration and source of joy. You've been my confidants and sounding boards through all the exhilarating and painful moments that led up to this point. And I hope that you can draw a modicum of inspiration from just the feat of publishing this book. If I, being a perennial perfectionist, can overcome my history of crippling perfectionism enough to write a book like this, you can do anything your wild hearts dream of!

I would also be remiss to not give a shout out to my amazing dog, Finnegan (**@FinnCushing** on the 'gram)! No one loves people as enthusiastically as you do! You're always ready to greet total strangers and make them feel like the most important person on the sidewalk at that moment! You're the best Chief Ball Chaser any entrepreneur could hope to have!

ACKNOWLEDGMENTS

I would like to acknowledge a few people who have been especially instrumental in helping me through the arduous book-publishing process.

My Tribe
Shout out to my dragon boating besties, Karen Anderson and Dian Balanovich, for helping me stay sane(ish) through the whole book-writing process by dragging me out onto the water for our "chaddles," where we prioritized chatting (and snacking) over actually elevating our heart rates. You are the only ones I would trust to help me bury a body!

My Grammar Sage
Thank you, Michelle Lowery (michellelowery.com), for fielding my many, many grammar questions, most of which came on Saturdays! (Boundaries are overrated.) Your insights on the publishing process were also really helpful.

My Coffee Shop (But Seriously)
I'm not sure how many authors credit their coffee shop, but the hardest part of being an entrepreneur, for me, is the lack of interaction with people. (No offense, Finnegan!) When I lived in Philly, I started working mornings at the OCF Coffee House in Fairmount—and probably wrote at least 80% of this book there. Over time, the baristas became like family and cheered me on as I trudged through the book-writing process. Coe Burchfield and Josh Brechter get a special shout out because they knew how to disrupt everyone's quiet, peaceful morning routine with their nonstop shenanigans. Apologies to the introverts who got caught up in the friendly fire!

My NY Hosts
At some point I thought to myself, *This process of book writing is going way too smoothly. How can I move this car into the far-left lane while staying in third gear? Oh, I know! I'll move to New York!* I could not have transitioned as well as I did (jury still out on that one) without the assistance of Kathleen and David Whitehead. Thank you for stepping up to the plate and opening your home again and again. I may never be a legit New Yorker (smile way too much and offer money to total strangers to move heavy packages up to my apartment), but you guys have helped me at least blend in a little better.

ABOUT THE AUTHOR

Annie Cushing is a web analytics consultant in New York City and blogs at **Annielytics.com**. She lives for data. By day she gathers it and makes it runway ready. By night she dreams about her next data vis fix. The only thing that thrills her as much as getting knee deep in data is teaching others how to sexy up their data.

When she's not tethered to a computer, you are most likely to find her out on the water somewhere, preparing for her next race.

If you have extra marketing budget you don't know what to do with, you can find out how to lob it over the fence at her on her services page: **bit.ly/mdsepc-services**.

CONTENTS

Chapter 1
Getting Started

Some Preliminary Thoughts 2
What This Book Isn't 2
Sexy Data? What Does That Even Mean? 2
Tone 2
Links 3
Screenshots 3
Tutorial Breakdown 3
AKA 3
Uses 3
Tips 3
Category / Metric 4
Tutorial 4
Hundred-Foot View 5
Example File 5
    Optional Purchase 5
    What to Expect 6
    Don't Try This at Home, Kids 6
Template(s) 6
    Optional Purchase 6
    Why Only Some 6

Know the Lingo 7

Keyboard Shortcuts 7

Chapter 2
Massage Your Data   9

Handling Blanks 9
Tutorial 9
Charting Options 9

Formatted Tables 10
Benefits of Tables 10
Apply a Table to Your Data 11
Design a Table Style 12
    Tips from the Trenches 12
        Create Your Branded Theme First 12
        Scan Table Styles 12
    Two Approaches 13
        From Scratch 13
        Duplicate and Customize 17

CONTENTS

Number Formatting Pet Peeve 17
Pivot Tables 18
Tabular Data: Excel's Choking Hazard 18
Create a Pivot Table 19
    Requirements 19
    Tutorial 19
Design a Pivot Table Style 24
    Tips from the Trenches 24
        Create Your Branded Theme First 24
        Scan Pivot Table Styles 24
    Modify a Pivot Table Style 24

Unpivot Data 27
Example File 27
Tutorial 28
Avoid Security Warning 32
Example Visualization 33
Downloadable Files 33

Fix Dates 34
Functions You'll Need 34
    DATE 34
    LEFT / RIGHT 34
Formatting You'll Need 35

Chapter 3
Design with Elegance   37

Two Main Objectives 37
Alignment with Branding 37
Freedom from Clutter 38

A Tale of Two Organizations 38
DOJ's Disastrous Charts 38
The Makeover 39
Be Like Chevron 41
How to Analyze an Organization's Branding 42

How to Summon Formatting Options 43
Alternatives to Keyboard Shortcut 43
One and You're Done 43
Not Just for Charts 43
Tip for Hard-to-Reach Chart Elements 44

CONTENTS

Key Formatting Techniques **44**
Overall Aesthetics **44**
   Stick to Branded Colors 44
      Create Tints and Shades 45
      Design for Colorblind Users 47
   Point Out Actionable Insights 48
   Customize Number Formatting 49
      Use Cases 49
      Working with a TEXT function 50
      Simplify Large Numbers 50
      Reformat Negative Numbers 52
      Fix Alignment 52
   Colorize Icons 54
   Paint Formats 55
Chart Aesthetics **55**
   Add Data Labels 55
      How to Add 55
      Making Room 55
      Format Painter Trick 56
      Reassign Values 56
   Ditch Gridlines 57
   Modify Chart Title 57
   Give Chart Title Space 58
   Add Annotations 58
      How to Add 58
      Formatting Text Boxes 58
      Set Default Text Box Formatting 60
      Pull in Values 60
   Add Border to Touching Data Series 62
   Align Strategically 64
      Align Options 64
      Snap to Options 65
      Group to Preserve Your Work 66
   Resize Fonts 67
   Adjust Chart Series Spacing 67
   Thin Out Axes 67
      Date Axes 68
      Value Axes 69
      Text Axes 69
   Remove Unnecessary Axes 70
   Add an Axis Title 70
   Add Markers 70
   Hack Excel's Charting Engine 71
      Scariest Dialog in Excel? 71
      Anatomy of the Select Data Source Dialog 72
      Add Data Series to Chart 76

CONTENTS

Copy-and-Paste Tips 76
    Paste as Values 76
    Calculate on the Fly 77
    Copy Column Widths 77
    Transpose 77
    Paste Visible Cells Only 77
    Move without Pasting 78
    Paste Chart Formats 78
Bring Sexy Back 79

Chapter 4
Set It and Forget It   81

Good Data Gone Bad   81

Hidden Schmidden 81

$125 Million up in Smoke 81

Who Needs Automation Anyway? 81

Need for Codified Process   81

Templatize Everything   82

Create a Store of Chart Templates 82
    How to Create a Chart Template 83
    How to Apply a Template 84
    Tips 84

Create a Branded Template 85
    Broken but Hackable 85
    Steps to Creating a Template 85

Distribute Themes and Templates 96

Assign Someone to Manage Process   97

Chapter 5
Create Charts   99

Comparison   99

Bar Chart 99
    Uses 99
    Tips 100
    Category / Metric 100
    Tutorial 100
    Variations 101
        Highlight with Color *101*
        Add a Mean Line *102*
    Example File 106
    Templates 106

CONTENTS

Column Chart **107**
    Uses 107
    Tips 107
    Category / Metric 108
    Tutorial 108
    Variations 108
        Highlight with Color *108*
        Add a Mean Line *109*
    Example File 112
    Templates 112

Pictograph **113**
    AKA 113
    Uses 113
    Tips 113
    Category / Metric 113
    Tutorial 113
    Example File 117
    Template 117

Clustered Column Chart **118**
    AKA 118
    Uses 118
    Tips 118
    Category / Metric 119
    Tutorial 119
    Variations 119
        Add a Mean Line *119*
        Show % Change *123*
    Example File 127
    Template 127

Stacked Column Chart **128**
    Uses 128
    Tips 128
    Category / Metric 129
    Tutorial 129
    Variation 130
        Add Column Totals *130*
    Example File 133
    Templates 133

Clustered Stacked Column Chart **134**
    AKA 134
    Uses 134
    Tips 134
    Category / Metric 134
    Tutorial 134

CONTENTS

    Example File 137
    Template 137
Diverging Bar Chart **138**
    AKA 138
    Uses 138
    Tips 138
    Category / Metric 138
    Tutorial 138
    Variation 142
        Center Label *142*
    Example File 145
    Template 145
Diverging Stacked Bar Chart **146**
    AKA 146
    Uses 146
    Tips 146
    Category / Metric 146
    Tutorial 147
    Example File 149
    Template 149
Floating Bar Chart **150**
    AKA 150
    Uses 150
    Tips 151
    Category / Metric 151
    Tutorial 151
    Example File 157
    Template 157
Dot Plot **158**
    AKA 158
    Uses 158
    Tips 158
    Category / Metric 159
    Tutorial 159
    Variation 165
        Jitter Plot *165*
    Example File 167
    Template 167
Cycle Plot **168**
    AKA 168
    Uses 168
    Tips 168
    Category / Metric 168
    Tutorial 169

CONTENTS

    Example File.. 172
    Template 172
Slope Chart... **173**
    AKA 173
    Uses.. 173
    Tips 173
    Category / Metric 173
    Tutorial... 173
    Example File.. 178
    Template 178
Lollipop Graph... **179**
    Uses.. 179
    Tips 179
    Category / Metric 179
    Tutorial... 179
    Variation... 182
        Add Data Labels *182*
    Example File.. 182
    Templates 182
Dumbbell Plot.. **183**
    AKA 183
    Uses.. 183
    Tips 183
    Category / Metric 183
    Tutorial... 184
    Example File.. 189
    Template 189
Combination Chart **190**
    AKA 190
    Uses.. 190
    Tips 190
    Category / Metric 191
    Tutorial... 191
    Variation... 193
        Use Area Chart *193*
    Example File.. 193
    Template 193
Small Multiples.. **194**
    AKA 194
    Uses.. 194
    Tips 194
    Category / Metric 195
    Tutorial... 195

CONTENTS

Variations	198
Example File	198
Template	198
Sparklines	**199**
AKA	199
Uses	199
Tips	199
Category / Metric	200
Tutorial	200
Example File	201
Template	201
Trend	**202**
Line Chart	**202**
Uses	202
Tips	202
Category / Metric	203
Tutorial	203
Variations	204
Add Sizzle	*204*
Add a Target Line	*208*
Add a Trendline	*212*
Example File	212
Templates	212
Band Chart	**213**
AKA	213
Uses	213
Tips	213
Category / Metric	213
Tutorial	213
Variation	218
Add Single Band	*218*
Example File	218
Template	218
Area Chart	**219**
AKA	219
Uses	219
Tips	219
Category / Metric	220
Tutorial	220
Example File	221
Template	221
Baseline Delta Chart	**222**
Uses	222

CONTENTS

    Tips 222
    Category / Metric 222
    Tutorial 223
    Example File 225
    Template 225
Stacked Area Chart **226**
    Uses 226
    Tips 226
    Category / Metric 227
    Tutorial 227
    Example File 228
    Template 228

Composition **229**
Pie Chart **229**
    AKA 229
    Uses 229
    Tips 229
    Category / Metric 229
    Tutorial 230
    Variations 231
        Accent with Color *231*
        Exploding Pie Chart *231*
        Donut Chart *231*
        Exploding Donut Chart *231*
    Example File 231
    Templates 231
100% Stacked Bar Chart **232**
    AKA 232
    Uses 232
    Tips 232
    Category / Metric 233
    Tutorial 233
    Variation 236
        Add Data Labels *236*
    Example File 236
    Template 236
100% Stacked Area Chart **237**
    AKA 237
    Uses 237
    Tips 237
    Category / Metric 237
    Tutorial 237
    Template 238
Treemap **240**
    AKA 240

CONTENTS

Uses.	240
Tips	240
Category / Metric	240
Tutorial	241
Variation	244
Highlight with Color	*244*
Example File	245
Template	245
Sunburst Chart	**246**
AKA	246
Uses.	246
Tips	246
Category / Metric	247
Tutorial	247
Variation	248
Highlight with Color	*248*
Example File	248
Template	248
Distribution	**249**
Box & Whisker Chart	**249**
AKA	250
Uses.	250
Tips	250
Category / Metric	250
Tutorial	251
Example File	252
Template	252
Histogram	**253**
Uses.	253
Tips	253
Category / Metric	254
Tutorial	254
Example File	255
Template	255
Pareto Chart	**256**
Uses.	256
Tips	256
Category / Metric	257
Tutorial	257
Example File	257
Template	257
Stock Chart	**258**
AKA	258

CONTENTS

Uses.	258
Tips	258
Category / Metric	259
Tutorial	259
High-Low-Close	*259*
Open-High-Low-Close	*261*
Volume-High-Low-Close	*263*
Volume-Open-High-Low-Close	*264*
Example File	264
Templates	264
Relationship	**265**
Scatter Plot	**265**
AKA	265
Uses	265
Tips	266
Category / Metric	268
Tutorial	268
Variations	272
Apply Polynomial Trendline	*272*
Add a Category	*272*
Isolate Trendlines	*276*
Example File	276
Template	276
Bubble Chart	**277**
Uses	277
Tips	277
Category / Metric	277
Tutorial	277
Variation	279
Add a Category	*279*
Example File	279
Template	279
Heatmap	**280**
AKA	280
Uses	280
Tips	280
Category / Metric	281
Tutorial	281
Example File	284
Template	284
Radar Chart	**285**
AKA	285
Uses	285
Tips	285
Category / Metric	286

CONTENTS

Tutorial	286
Example File	289
Template	289
Progress	**290**
Gauge Chart	**290**
AKA	290
Uses	290
Tips	290
Category / Metric	290
Tutorial	291
Example File	296
Template	296
Waterfall Chart	**297**
AKA	297
Uses	297
Tips	297
Category / Metric	299
Tutorial	299
Example File	299
Template	299
Maps	**300**
AKA	300
Uses	300
Tips	300
Category / Metric	300
Tutorials	301
Option 1	*301*
Option 2	*302*
Example File	302
Template	302

Chapter 6
Parting Thoughts — 305

Go Strut Your Stuff — **305**

Join In the Fun — **305**

More Resources — **306**

Suggestions? Corrections? — **306**

Like It? Review It! — **306**

CHAPTER 1

Getting Started

Organizations have never had more access to data. We are inundated by it at every turn, from enterprise systems to our watches to our refrigerators. Most of us aren't in need of making data from our refrigerators sexy; however, just about everything else is fair game! That is what you'll learn to do in this book.

Making Data Sexy is a comprehensive guide on how to create compelling visualizations in Microsoft Excel—60+ total visualizations.

I place a heavy emphasis on elegance, branding, simplicity, and branding. In fact, to demonstrate that even an organization with the simplest branding can create sexy, branded visualizations, I used a palette of only three colors to create all the visualizations in the book **(Fig 1.1)**.

Fig 1.1: I used only three colors for all the visualizations in this book.

It may be tempting to jump right into the tutorials in the Create Charts chapter. But the Design with Elegance chapter will set the tone for the importance of aligning your visualizations with your branding, and the Set It and Forget It chapter will show you how to create branded themes and templates that will save you duplication of effort by making everything you create reusable—for the benefit of all who create data visualizations in your organization.

I have made a conscious effort to only include text that I think will be helpful. To that end, I will use lists over paragraph text wherever possible. My objective in this Spartan approach is to create a resource that you can keep by your desk and pull up for quick references when you need to learn how to create a particular visualization—or just need inspiration.

This book won't teach you everything you need to know about Excel, but it is basically everything I wish I had known when I started my sojourn into data visualization.

CHAPTER 1 • GETTING STARTED

SOME PRELIMINARY THOUGHTS

WHAT THIS BOOK ISN'T

Everyone who writes a data visualization book comes into it with certain opinions and prejudices. I try to keep mine to an absolute minimum. I will share some caveats from time to time, but the emphasis will be more on pitfalls that I've fallen into or issues I've witnessed in other people's visualizations.

That said, I draw the line at 3-D visualizations **(Fig 1.2)**. Excel offers them as an option with a number of its charts, but you're on your own if you choose to use them.

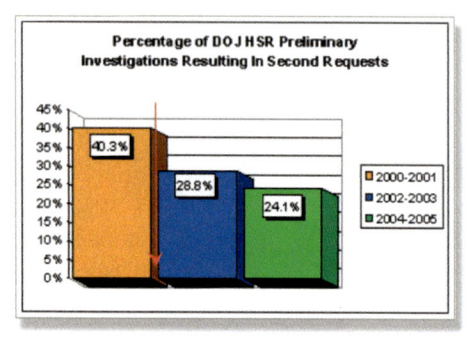

Fig 1.2: Don't be like the DOJ.

SEXY DATA?
WHAT DOES THAT EVEN MEAN?

My tagline for **Annielytics.com** is "I make data sexy." Hence the book title. But what I'm really talking about is making data both elegant and actionable—maximizing both form and function. You want your visualizations to be compelling—downright galvanizing at times. It's very difficult to do that when your data is trapped in tables, the de facto standard for many analysts, marketers, and managers. As a budding data artist, tables will rarely do the job as well as a beautifully formatted (and carefully structured) visualization.

Case in point: I once had a client who was running expensive paid search campaigns on Google and Bing. They hired me to transform their collection of tables into a dashboard. They were investing the majority of their budget in Google Ads and threw some pity money at Bing for a spattering of ads on their network. What they didn't realize—until they saw their data come to life in a dashboard I created—is that those tiny Bing ads were outperforming their Google ads 8:1. Their task at hand was clear: optimize their Google Ads campaigns better and/or reallocate more of their budget to Bing Ads. That's the power of sexy data. It would have been very difficult to see the story their data was telling them through the tables they were using.

TONE

I am insufferably visual and rely heavily on analogies and imagery. I also have a writing style that can be… How do I describe it?…On the lighter side. You will find references to everything from *The Breakfast Club* to *Stranger Things*. That is by design. Sometimes the issues and bugs that I'll expose in this book border on infuriating, and it's easier to laugh than cry.

My end goal is to make the training as painless as possible. I'm a coach by nature, and I recognize that Excel can be an intimidating program to work with. Therefore, I actively try to minimize that sense of intimidation by taking you by the hand and walking you through some of these techniques that have been at times a tremendous source of intimidation and frustration for me.

I literally spent five hours one Friday night going through very frustrating tutorials on how to create a gauge chart **(Fig 1.3)**. The well-meaning instructors provided

Fig 1.3: Don't let this unassuming chart fool you; it's the most complicated one in the whole book!

CHAPTER 1 · GETTING STARTED

instructions that either didn't work if your value was greater than 360 (the number of degrees in a circle) or suffered from leaps of logic. Nothing will frustrate me as much as telling me to do something without explaining why.

As a result, I strive to explain the why behind the how in every tutorial, sometimes even including a Hundred-Foot View section before more complex tutorials that will give you a breakdown of the overall strategy to minimize anxiety. Picture me next to third base yelling for you to keep going and press through the discomfort to take home!

LINKS

I use Bitly, a URL shortening service, to shorten all the links I share in the book. Each link starts with mdsepc, which is short for Making Data Sexy Excel for PC.

SCREENSHOTS

I used Microsoft Office 2016 to create the screenshots in this book, except for the map charts. I used Office 365, Microsoft's subscription service, for those. If you're using Office 365 there might be slight variations in workflow and screenshots. Updates to 2016 also come with slight variations. I will note where there are significant changes in the interface.

It's also worth noting that menus might vary slightly depending on your screen resolution, so if your menu items are more (or less) consolidated than what you see in the screenshots in this book, that is most likely why.

TUTORIAL BREAKDOWN

For each visualization I break down in the Create Charts chapter, I include a number of support sections to help you wrap your mind around each chart type. The primary emphasis is on the tutorial, obviously, but knowing how to physically create a chart is about half the battle. Knowing when to use it and how to extract actual insights from it takes you from impressive to imminently useful in an organization.

I provide an example of extracting insights from the data in the Box & Whisker Chart tutorial. I couldn't help myself.

AKA

If a chart has other names, I include them in this section. Knowing other terms for a particular chart can be particularly helpful when you're doing requirements gathering or searching for more information.

USES

I pulled together example uses for each chart type. These lists aren't exhaustive. You might find new uses I haven't thought of, but I included some common uses to give you ideas for how to use each of them. Think of them as more suggestive than prescriptive.

TIPS

I included tips for each of the chart types to alert you to common mistakes I've seen out in the wild, bugs I ran into when experimenting, and potential workarounds for said bugs. To create these sections, I performed searches for each of the chart types in Google Images and took note of common formatting faux pas I observed and recorded any issues I ran into when I was testing in my crazy Excel lab.

Admittedly, I was pushing Excel to its limits at times, so you might not run into some of the issues I experienced. Or you might be working with a more powerful desktop that doesn't crash and want a belly rub just because you ask it to sort a treemap **(Fig 1.4)**! #basedonatruestory

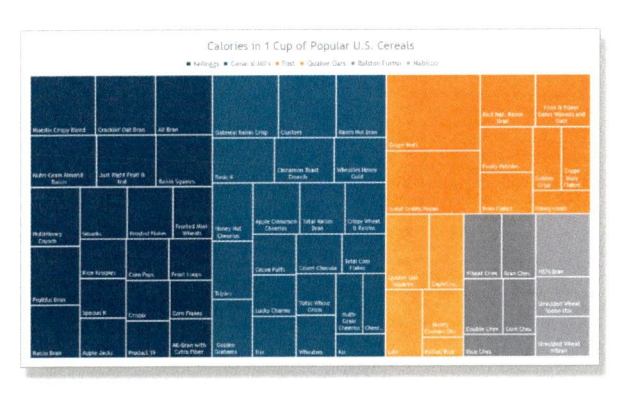

Fig 1.4: The treemap is sexy but volatile, but I warn you of volatilities you may experience.

3

CHAPTER 1 • GETTING STARTED

CATEGORY / METRIC

For each visualization, I provide guidance on what kind of dataset qualifies. There's no sense investing time into learning to create a sexy scatter plot, just to find out that your dataset doesn't meet the requirements of two metrics.

There are many different terms I could have used here: variables, measures, etc. I decided on the two that I thought would be easiest for neophytes to understand.

Think of categories as buckets for your data. A few that you'll run into in this book will be countries, marketing channels, gender, and weapons. Metrics are values you will aggregate and/or perform calculations on. A few metrics that you'll see in the visualizations in this book will be website visits, revenue, income, temperatures, Facebook reach, Google Trends search volume, stock prices, and deaths by police.

The tricky thing was deciding what to call the individual items inside a category (e.g., state names under the category State). I differentiated these by calling them groups or category items/elements.

I try to keep things as simple as possible, without degrading the nuanced nature of data visualization. To that end, I don't include highly customized Frankencharts (think Frankenstein, except in data form)

when tallying the number of categories and metrics each chart requires. For example, a scatter plot traditionally uses two metrics and no categories. However, you can hack Excel to add in a category that you can decorate with color and/or shapes. I show you how in both the Scatter Plot and the Bubble Chart tutorials. But since this is more of a hack, I don't include it in the Category / Metric section. I still list both charts as requiring 0 categories / 2 metrics **(Fig 1.5)**.

I did pretty extensive research to account for every use case I could think of or find out in the wild when calculating these, but if you can squeeze an extra category or metric into your chart, go for it!

TUTORIAL

For each visualization I cover, I try to include the steps that I believe are most essential. I tend to be a minimalist when it comes to data visualization. My job, as an analyst, is to find the data that will drive the most insight for its intended recipients. One time an instructor told me, with regard to visualizations I was adding to a dashboard, "Either make me care or take it out." I take that approach now with every dashboard I create. Once I decide what data I want to display, I seek to create visualizations that communicate those findings with clarity and finesse.

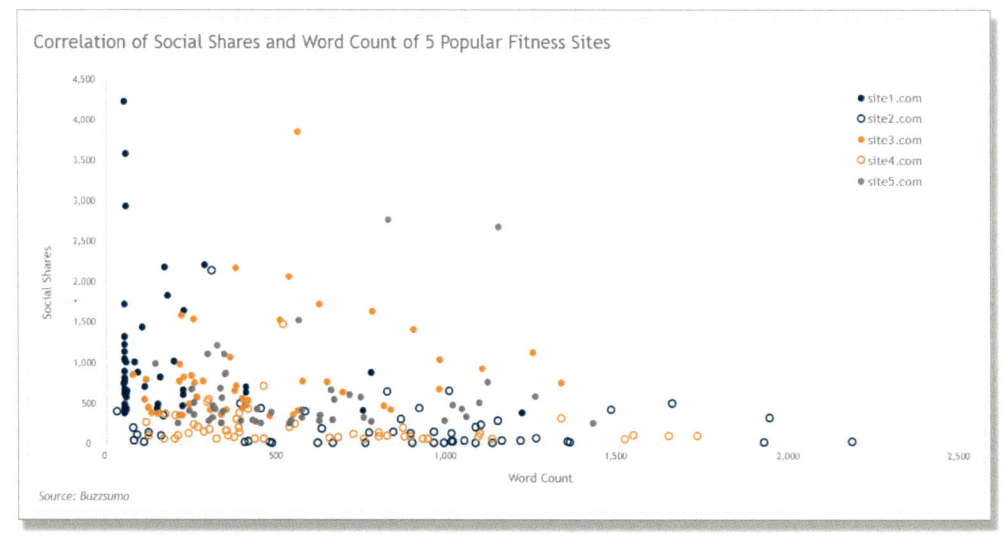

Fig 1.5: Hacked alternative charts aren't included in category/metric counts.

CHAPTER 1 • GETTING STARTED

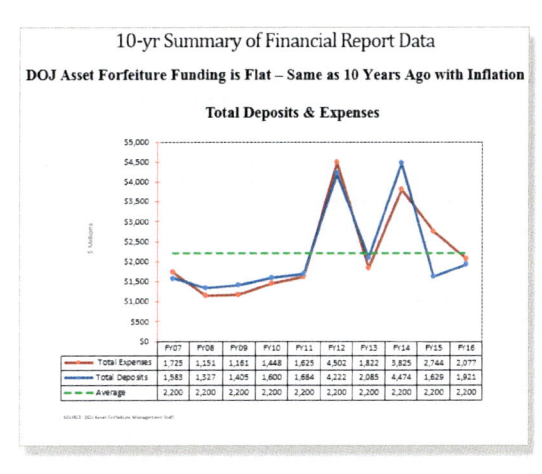

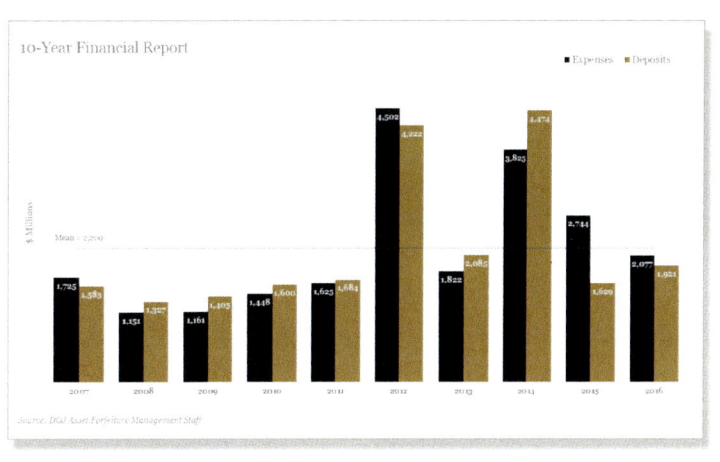

Fig 1.6: I demonstrate how to simplify (and brand!) your visualizations to make the data sing.

I remember when I was a junior reporter, my first editor told me that if a reader needs to reread a sentence in my article to understand it, I've failed as a writer. She gave me the goal of writing articles that are so clear and flow so effortlessly, the reader can smoothly wend from paragraph to paragraph without the need to backtrack. In a lot of ways, my responsibility as an analyst isn't that far off from that goal that helped establish my writing style early in my career. Only now I communicate with data more than words.

I also try to keep the visualizations in this book as distraction free as possible. Excel's default settings are fraught with unnecessary pixels that communicate more noise than signal. It is your job to tame Excel's wild proclivities—that is, if you want to become a data whisperer. If you choose to titrate some of the chart accoutrements back in, who am I to judge? My goal is more to make you mindful of what you allow into your charts **(Fig 1.6)**.

Finally, my intention with this book is to write a series of standalone tutorials so that you can use it as a desk top reference of sorts. Although a few diehards might read the book from cover to cover (you GO, data-hungry mavericks!), most will thumb through it and then pull it out when they need some quick guidance on how to create a chart. Therefore, there might be some duplication of effort from one tutorial to the next. That's because I won't assume in the 36th tutorial that you read the same steps in the 17th tutorial. That said, I included most formatting techniques I used more than once in the Design with Elegance chapter.

HUNDRED-FOOT VIEW

With the charts that require significant hacking to get what we want, I add a Hundred-Foot View section to the tutorial before the first step. I did this because I was often frustrated by being told to tweak obscure settings or add error bars or flip rows and columns (you get the idea) without explaining why. I don't want to create automatons; I want you to understand the reasoning for every step we take. Every move we make. Every bond we break….

EXAMPLE FILE

Optional Purchase

If you'd like to see how I massaged the data using tables, pivot tables, etc., you can purchase the example files from the **makingdatasexy.com** website: **bit.ly/mdsepc-purchase**. The free files that contain the data used in the book can be downloaded from **bit.ly/mdsepc-data**.

CHAPTER 1 · GETTING STARTED

What to Expect

Each tutorial has an example file you can download and explore. These files may have any of the following worksheets:

- *Raw Data:* This is where you can find the unadulterated data I worked with. These datasets are typically in the condition they were in when I downloaded them, and they are oftentimes quite ugly. I also include links to datasets that are publicly available. If there's private data, I sometimes anonymize it. *Note:* These are available for free download on the website.
- *Extracted Data:* Some of the data sources are robust. If I extracted a subset of the original data source to create a visualization, I put that extract to its own worksheet to minimize confusion.
- *Pivot Table:* If I needed to use a pivot table (usually to restructure my data into a crosstab format), it will be in a different worksheet from my raw data. I always place my pivot tables in a separate worksheet from my raw data. When I'm building out dashboards, I'll have the dashboard in one worksheet, my raw data in another, and then a third worksheet for my massaged data.
- *Formatted Data:* If I needed to massage my data in some way, but not with a pivot table, I labeled the worksheet this way. Those worksheets typically include formatted tables.
- *Final:* These tabs typically take on the name of the chart type, e.g., Dumbbell Plot.

Don't Try This at Home, Kids

I rarely, if ever, include my raw data in the same worksheet as my visualization because...ew. However, I did include it in the example files because I wanted it to be easy to reference the data that was used to create the chart **(Fig 1.7)**.

TEMPLATE(S)

Optional Purchase

Some may read this book and decide that they don't want to learn how to create all the charts contained within. For those of you who want to fast track by starting with a chart template, they can be purchased from the **makingdatasexy.com** website: **bit.ly/mdsepc-purchase**.

Why Only Some

Not all of the charts will allow you to save them as templates. There are several reasons for this:

- *It's one of the chart types new to Excel 2016.* At the time of writing, none of the new charts in Excel 2016 save as a chart template. These include the Histogram, Waterfall, Pareto, Box & Whisker, Treemap, and Sunburst charts. I believe this was an oversight on the team's part. Hopefully they will remedy this issue in a future update.
- *It's just too hacktastic.* Some of the charts in this book are too customized to save as templates. A few of these include the dot plot, slope chart, dumbbell plot, clustered column chart, and gauge chart. Excel gets all tangled up in itself if you try to apply the template to a new dataset. With these charts, I recommend creating a file that includes these visualizations with a sample dataset and then copying both into a new workbook when you want to use them. Then, all you'll need to do is point it to your new data using the Select Data Source dialog and customize it to taste.

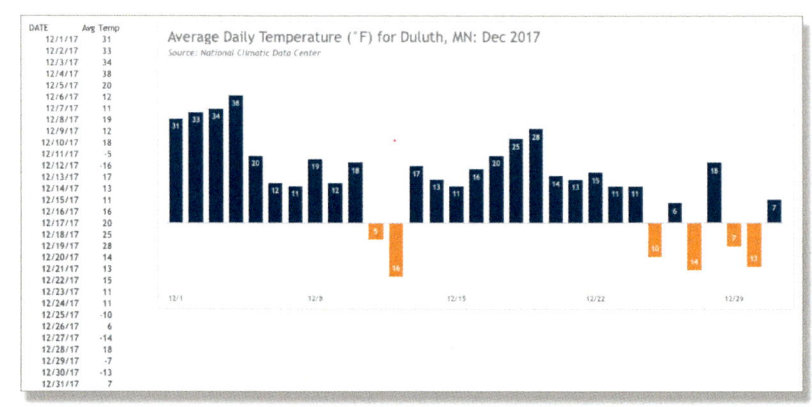

Fig 1.7: I'd never do this in real life.

💬 **LEARN MORE** See the Hack Excel's Charting Engine section of the Design with Elegance chapter to learn how to rock the Select Data Source dialog.

✦ *It's not actually a chart.* Most of visualizations in this book are charts. However, a couple aren't. The bouncer was at the door, but they were young and really cute and he got distracted. The guilty parties I'm outing here are sparklines and heatmaps **(Fig 1.8)**. Since they're not objects, they can't be saved as templates. Again, with these visualizations, I recommend saving them to your file of visualizations that can't be saved as charts for quick reference. Because I'm a systems person, I recommend including any necessary instructions you (or a team member) might need to customize the visualization. You can do this in a Read This file you store in the same folder. Not sure I've actually read one of those, but someone might! At minimum, you will be CYA compliant.

💬 **LEARN MORE** See the Create a Store of Chart Templates section of the Set It and Forget It chapter to learn how to work with templates.

KNOW THE LINGO

If you come across a term you're not familiar with, don't just keep reading, hoping it will become clear. I try to define everything as I go, but because this book is structured as more of a desktop reference, you may choose to jump into a tutorial without the benefit of learning the difference between a data series and data point, for example.

To help you along the learning curve, Microsoft has published a comprehensive dictionary of terms, which you can access from the **office.com** website: **bit.ly/mdsepc-glossary**. I recommend bookmarking it for future use.

KEYBOARD SHORTCUTS

Wherever possible, I try to share keyboard shortcuts. My favorite reference is Exceljet's cheatsheet, which includes shortcuts for Windows and Mac: **bit.ly/mdsepc-shortcuts**.

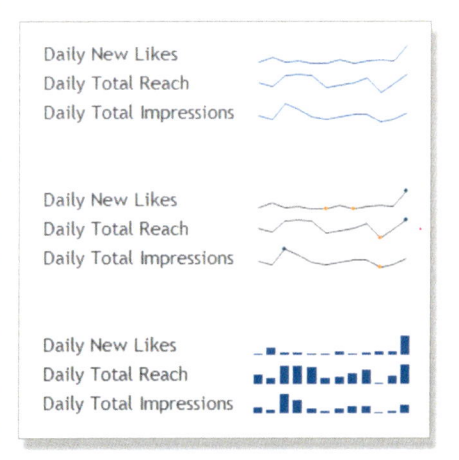

Fig 1.8: Heatmaps and Sparklines are visualizations but technically not charts.

CHAPTER 2

Massage Your Data

To quote the great philosopher, John Bender (*The Breakfast Club*), "Screws fall out all the time. The world is an imperfect place."

Similarly, data is the product of an imperfect world. Many times it comes to you poorly formatted, disorganized, and missing a few teeth. You will then need to decide how you'll overcome those deficiencies before you start plotting it.

Although this chapter will barely scrape the surface of the kinds of data cleaning—sometimes called "data munging"—you'll need to do at times, I do want to cover a few strategies and tools you will need to be comfortable with.

HANDLING BLANKS

One of the issues we sometimes need to make judgment calls on is how to handle blank cells in our data. In the example file for the Histogram tutorial, I include a dataset I downloaded from The World Bank that includes life expectancy at birth. I pulled out the 2015 variable (geek speak for column) to visualize it with a histogram. But the data contained multiple blank cells, so I had to decide what to do with the blanks. I chose to delete them.

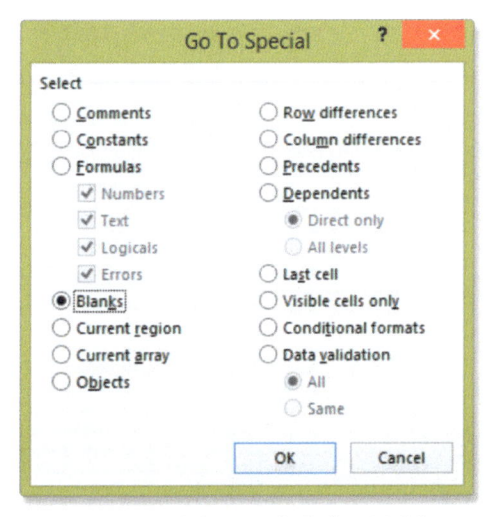

Fig 2.1: Select Blanks from the Go To Special dialog.

TUTORIAL

If you find yourself in a position in which you need to delete blank cells, take the following steps:

Step 1: Click-and-drag over your data range that contains blank cells, then navigate to **Home ▸ Editing ▸ Find & Select ▸ Go To Special**. (Alternatively, press the F5 key on your keyboard and select the **Special** button.)

Step 2: In the **Go To Special** dialog, select **Blanks** and click **OK (Fig 2.1)**.

Step 3: Right-click anywhere in the worksheet and choose **Delete** from the contextual menu.

CHARTING OPTIONS

Excel also gives you options on how to handle blanks, null values, and hidden data in charts. By default, hidden data isn't displayed in a chart, and empty cells or null values are displayed as gaps. For most visualizations, I'm satisfied with the default functionality, but if you want to modify how Excel treats these values, you can learn how on the **office.com** website: **bit.ly/mdsepc-display-blank**.

9

CHAPTER 2 • MASSAGE YOUR DATA

FORMATTED TABLES

I'm a fan—nay, evangelist—of formatted tables because of how easy Excel makes it to do analysis on the fly. Before we jump in, let's talk about why you want to start incorporating them into your workflow.

BENEFITS OF TABLES

These tables aren't merely decorative. They provide extra functionalities, just a few of which include:

- **Filters:** These alone make converting to a table worthwhile. Table filters have intelligence attached to them beyond your standard filter. For example, Excel serves up text filters if a column contains text and number filters if it contains numerical data.

- **Total Row:** Once you create your table, with any cell in it selected, you can navigate to **Table Tools ▶ Design ▶ Table Style Options ▶ Total Row** in the tab that appears when you're working in the table. If you have multiple columns with values, the total will only be applied to the far-right column. For the other columns, selecting the cell where you want the total will trigger a drop-down menu **(Fig 2.2)**. Use it to select how you want your data aggregated. If you want each of your columns to have totals, you'll need to first select the cell with the lone total, then drag the fill handle to the left **(Fig 2.3)**. It feels a little awkward—like you're reaching over someone's head to plug your laptop in at a coffee shop—but doing this will autofill those columns with a total. If your columns use different number formatting (e.g., whole numbers, decimals, and percentages) you may need to fix the formatting. You can always use the Format Painter (under **Home ▶ Clipboard**) to apply formatting.

- **Dynamic Ranges:** If you add data to your table, it will automatically expand to envelop the new row. The magic in this feature is that if you've created a chart from your table, it will update as well.

- **Dynamic Named Ranges:** If you need to reference a column from a table or the entire table in

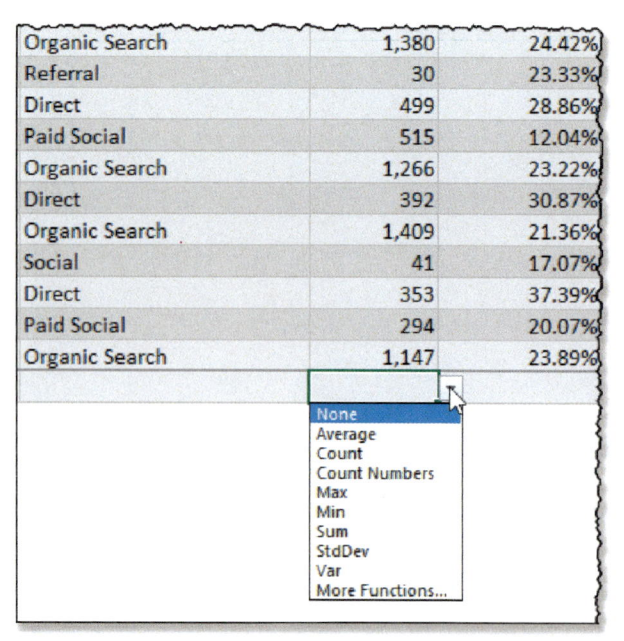

Fig 2.2: Choose how you want your table totals to be aggregated.

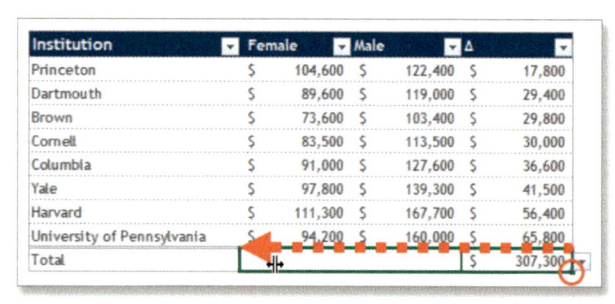

Fig 2.3: Add totals to each column by dragging the fill handle.

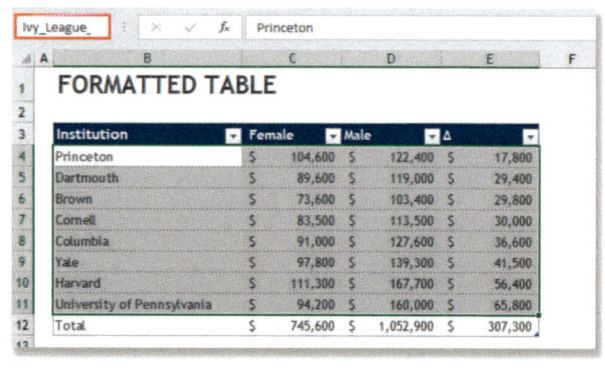

Fig 2.4: Excel allows you to assign your table a custom name.

CHAPTER 2 · MASSAGE YOUR DATA

a formula, you can use the table or column names Excel automatically assigns. Tables will be given the name Table + an index, e.g., Table1, Table2, etc. In a VLOOKUP formula, for example, instead of selecting the entire table, you can just start typing in the table name. Alternatively, you can assign your table an intuitive name. I recommend doing this if you'll reference it in a VLOOKUP because then you can just reference your table name. This helps keep larger workbooks organized. To name your table, click anywhere inside your table, press Ctrl-A to select the data (not the headers), click inside **Name Box** in the upper-left corner of your worksheet, and enter your custom table name **(Fig 2.4)**. Now you can reference it instead of Table[index #] or a range in formulas **(Fig 2.5)**. Learn more on the **office.com** website: **bit.ly/mdsepc-table-lookup**.

Fig 2.5: Assigning your table a custom name can make it easier to manage and reference in a formula.

APPLY A TABLE TO YOUR DATA

Excel provides numerous predefined table styles you can use to apply some structure to your data. To convert your data range to a table, select any cell in the range, then navigate to **Home ▸ Styles ▸ Format as Table** and select a table style from the drop-down menu **(Fig 2.6)**. Alternatively, press Ctrl-T to apply your default table style.

💬 LEARN MORE See the Design a Table Style section of this chapter to learn how to align your tables with your branding.

You'll be prompted with a **Format As Table** dialog, asking you to verify that Excel has selected your data range correctly and that your table has headers **(Fig 2.7)**. Once you've confirmed everything is copacetic, click **OK** to close the dialog.

Note: Your options might look different from mine, depending on what your default theme is.

💬 LEARN MORE See the Create a Branded Template section of the Set It and Forget It chapter to learn how to create branded themes.

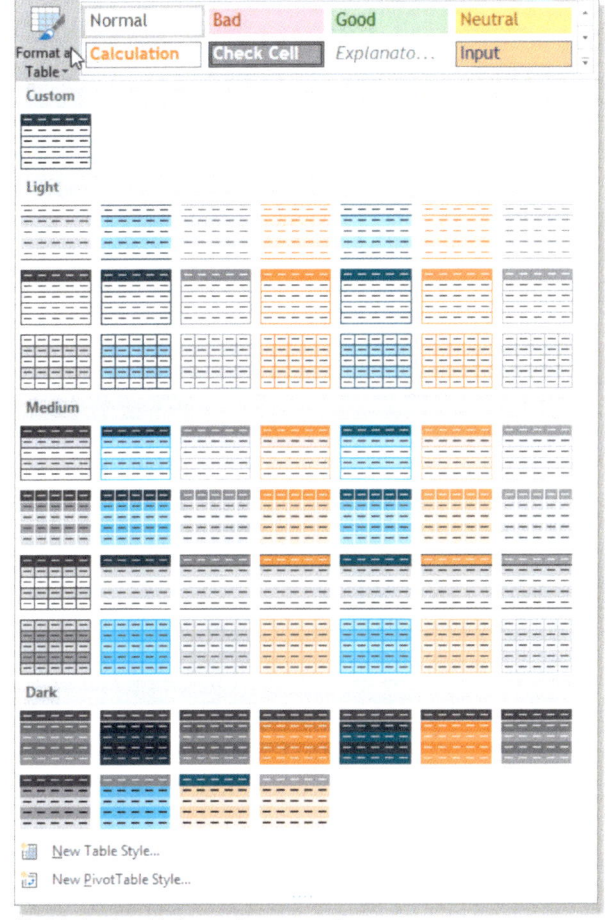

Fig 2.6: Apply table formatting to your dataset.

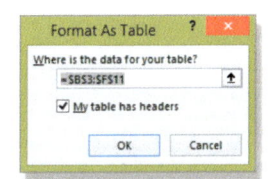

Fig 2.7: Just Excel checking to make sure your table is pointing to the right dataset.

11

CHAPTER 2 • MASSAGE YOUR DATA

If you just want to apply the default table style to your dataset, select any cell and press Ctrl-T, which will open the **Create Table** dialog. You'll learn how to create a custom default table for your documents in the next section.

You'll also learn a hack for how to add your custom default table to your global template in the Set It and Forget It chapter. This is essential because Excel only allows you to set a custom table as the default for the notebook you're currently working in, for some reason I can't wrap my mind around **(Fig 2.8)**.

Note: The **Create Table** dialog looks exactly like the **Format As Table** dialog but was given a different name. #twinning

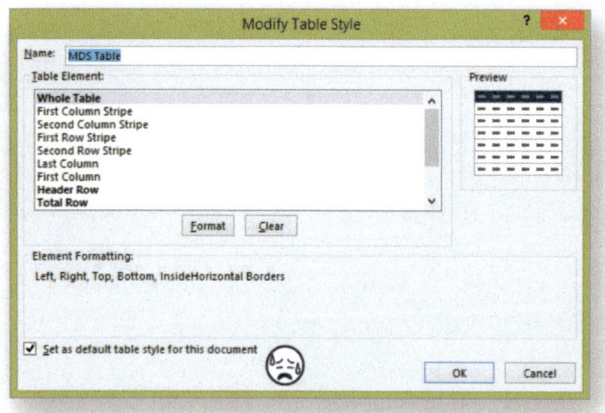

Fig 2.8: Wait…what…why?!?

DESIGN A TABLE STYLE

I rarely surface tables in a dashboard. They're usually tucked away on the worksheet where I do all my processing. I usually name it Formatted. I even hide this worksheet when I send the dashboard to a client. It's kinda similar to throwing dirty dishes in the oven when your home is on the market and your real estate agent gives you less than an hour to get it in show condition. From what I've read. *whistles*

Even though I ultimately hide the worksheet, I still use branded tables for client dashboards. They're easy to create, and they communicate finesse and diligence to detail, which translates into confidence. If an intern (especially one who reads this book!) comes along and needs to clean up your slovenly workbook, you may not be long for that contract…or job.

Tips from the Trenches
Before we jump in, I'll share a few tips that may save you some frustration and duplication of effort.

Create Your Branded Theme First
Table styles are built off your theme (**Page Layout ▸ Themes ▸ Themes**). As you'll learn in the Set It and Forget It chapter—arguably the most important chapter of this book—it's very important to build a theme using your brand colors and fonts. I can't underscore the importance of this enough. You can build the sexiest visualizations, but if they don't align with your brand, they will appear unfinished and amateur. Therefore, you might not want to create a custom table until you create a theme.

Scan Table Styles
To see a preview of a table style, just hover over it with a cell from your dataset selected. Your dataset will update on the fly with a preview of that style **(Fig 2.9)**.

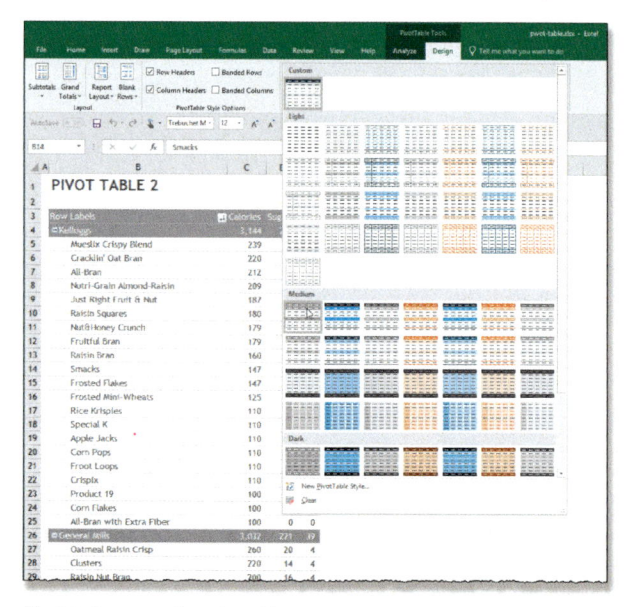

Fig 2.9: See a preview of a table style by hovering over it.

CHAPTER 2 · MASSAGE YOUR DATA

Two Approaches
From Scratch

It's not a terribly intuitive process to create a custom table from scratch, but even just customizing your borders and header row will go a long way toward making your tables sexy. I usually create a table style by modifying an existing style. I'm only demonstrating this method to get you comfortable with the more important settings.

I highly recommend using these techniques to tweak an existing style, but you do you!

The number of steps makes the process look intimidating, but that's more because I wanted to make sure each of the steps is crystal clear.

Step 1: Select any cell in your dataset and navigate to **Home** ▸ **Styles** ▸ **Format as Table** ▸ **New Table Style**. This will open the **New Table Style** dialog. Set **Name**. I set mine to MDS Table **(Fig 2.10)**.

Step 2: The first thing I want to do is set the outside and inside borders for my table. To set both, set **Table Element** to **Whole Table** and click the **Format** button to open the **Format Cells** dialog. Open the **Border** tab to access your border formatting options **(Fig 2.11)**.

Step 3: I use a bolder line for the outside border of my table than for the interior cell borders because this helps establish the outer perimeter of my table. I'll start with the outside border. In the **Line Style** field, set your style option. I kept the default line option. Next, set **Line Color**. I used the lightest gray from my branded palette. Finally, I chose the **Outline** option in the **Presets** field. The preview window will update with a very subtle (barely perceptible) view of your option **(Fig 2.12)**. Click **OK** to close the **Format Cells** dialog.

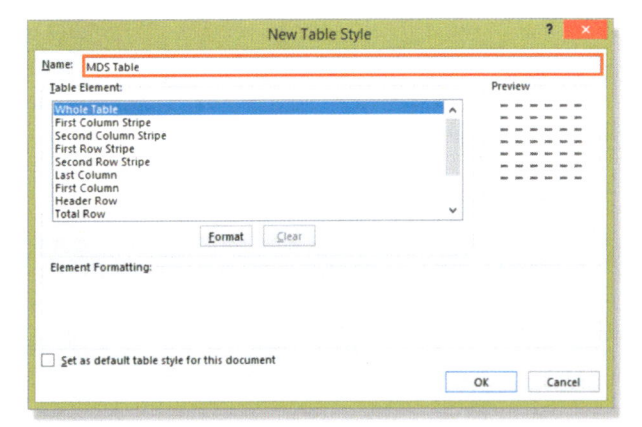

Fig 2.10: Assign your table style a name via the New Table Style dialog.

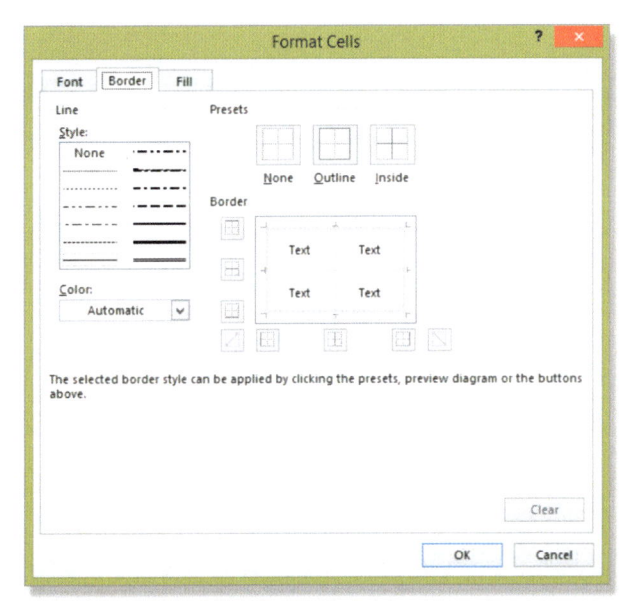

Fig 2.11: Open the Border tab of the Format Cells dialog to format your border.

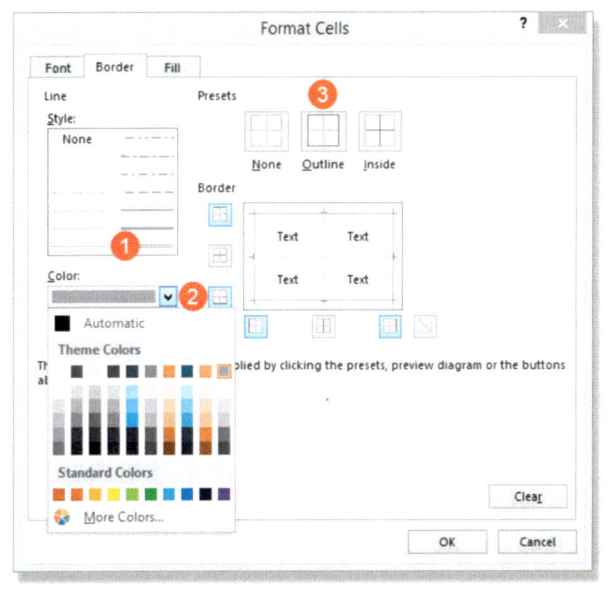

Fig 2.12: Set your outside border options.

13

CHAPTER 2 • MASSAGE YOUR DATA

Tip: If you decide to change your border color, you'll need to reselect your option in the **Presets** field or lines in the **Border** preview window.

Step 4: Next, we'll set the row borders. If Excel gave us the option to reduce the width of the lines more, I would. Its thinnest line (the default) is still a little obtrusive for my taste. The next best thing, in my opinion, is the dotted option. I like the second one down in the left column. I set **Line Color** to the same light gray. You'll notice that most of Excel's preset table options don't include a border for columns. I like that because they're unnecessary pixels, so I'll maintain that with my custom table. To apply this style to only row borders, click the middle line inside the **Border** preview window. Again, you'll see your preview update with your addition. Click **OK** to close the **Format Cells** dialog **(Fig 2.13)**.

Note: Unfortunately, there's no option to preview your actual table with your new formatting, so you need to back out of the **Format Cells** dialog and the **New Table Style** dialog if you want to preview your formatting options as you go. Because I'm a creature of habit and like what I like, I use the same formatting with different colors across clients, but when I was first figuring out my table preferences, my **Modify Table Style** dialog was a revolving door.

Step 5: Next, we'll format our header row. In the **New Table Style** dialog, select the **Header Row** table element, then click the **Format** button. Open the **Fill** tab to set **Background Color (Fig 2.14)**. I set mine to my darker branded blue color. At least Excel gives you

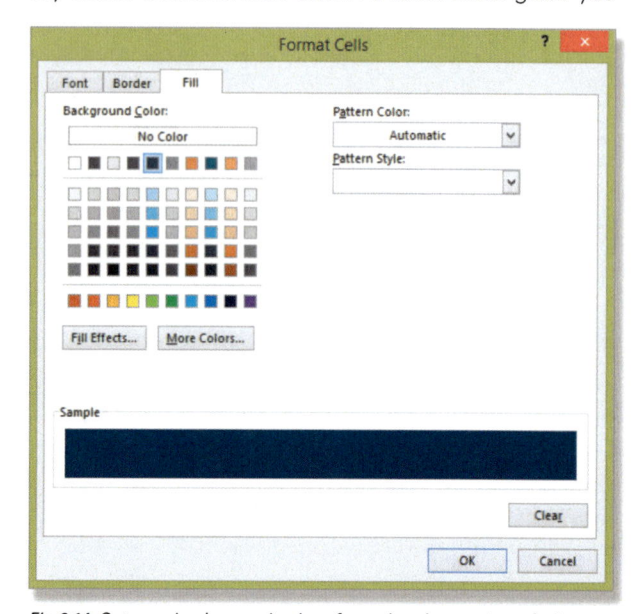

Fig 2.14: Set your background color of your header row to make it pop.

Fig 2.13: Set inside borders. I don't recommend using the Inside preset.

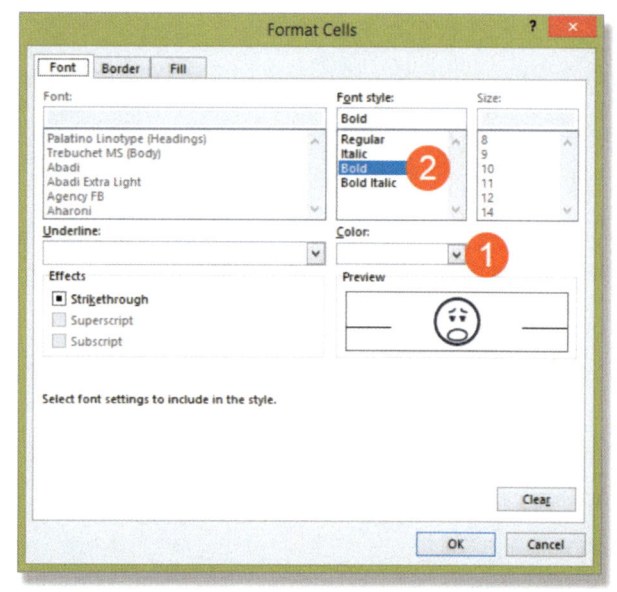

Fig 2.15: Previewing a white font is...tricky.

14

a decent preview under **Sample**. If you use a darker color, don't click **OK** yet because you'll need to set your header font to a lighter color, like white.

Step 6: Open the **Font** tab and set **Color**. If you chose a light color for your background, your default color might work. I set **Color** to white. I also set **Font style** to **Bold (Fig 2.15)**. It would be nice if Excel threw light colors on a dark background in the preview window. You choose to fly blind when you're working with white. Click **OK** to close the **Format Cells** dialog, then click **OK** again to close the **New Table Style** dialog.

Tip: If you build complex tables, you may want to experiment with some of the stripe options. As a minimalist, I try to avoid every unnecessary pixel. However, if formatting helps to process your data, format away! Function > form.

Note: You can't change your font in this dialog. That's because Excel uses the font from your theme.

💬 **LEARN MORE** See the Create a Branded Template section of the Set It and Forget It chapter to learn how to create branded themes.

Step 7: I like adding a double line to the top border of my total row. To set it for your table, select **Total Row** in the **New Table Style** dialog and click **Format**. In the **Format Cells** dialog, open the **Border** tab. Set **Line Style** to the double line. Set **Line Color**. I use the same color as the rest of my borders. In the **Border** preview window, click the top border. Click **OK** to close the **Format Cells** dialog **(Fig 2.16)**.

Step 8: To preview the total row formatting, select any cell in the table, then navigate to **Table Tools ▸ Design ▸ Table Style Options ▸ Total Row (Fig 2.17)**.

Step 9: If you want each of your columns to have totals, you'll need to first select the cell with the lone total, then drag the fill handle to the left **(Fig 2.18)**. Learn more from the Benefits of Tables section.

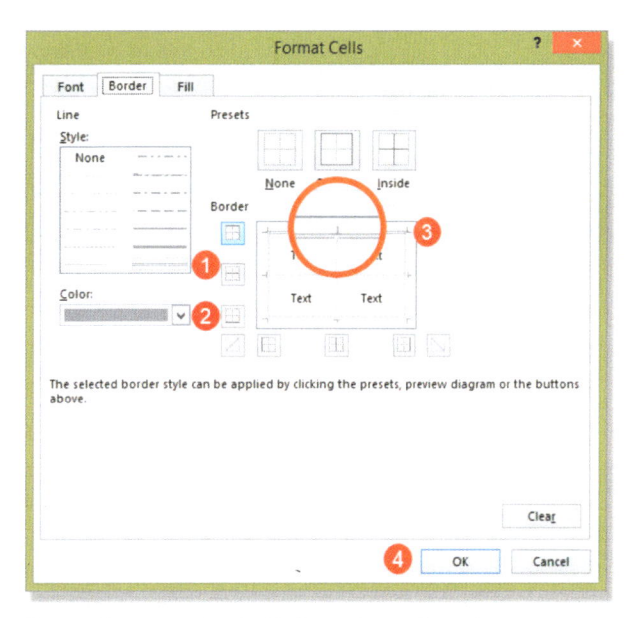

Fig 2.16: You can apply a double line to the top of a total row.

Fig 2.17: Preview your total row by selecting it as an option for your table.

Fig 2.18: Add totals to each of your columns by dragging the fill handle.

Warning: If you hover over the bottom-right corner of the total cell without first selecting it, you'll see a double-headed arrow instead of the crosshairs. Don't drag that! It will remove columns from your table. You don't need to know how I know that. *rolls eyes*

15

CHAPTER 2 · MASSAGE YOUR DATA

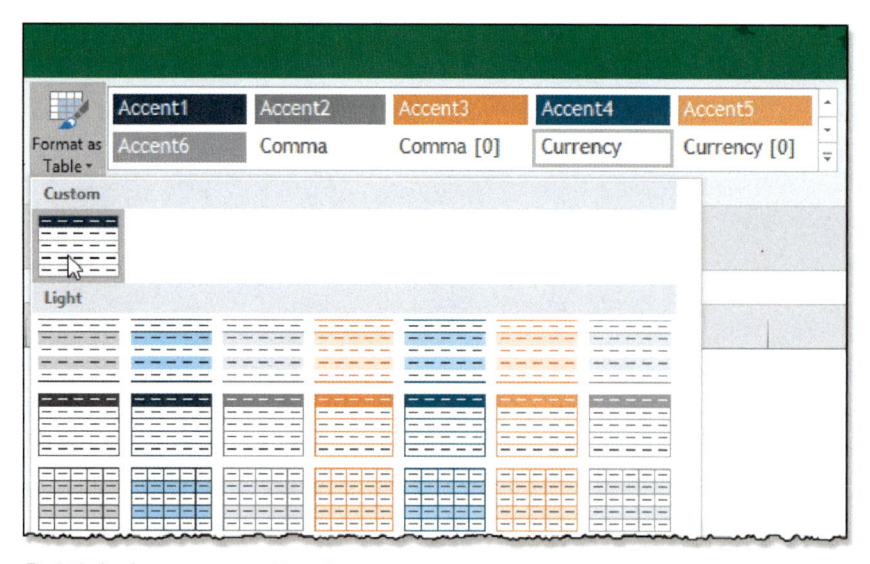

Fig 2.19: Apply your custom table style.

Fig 2.20: Verify your range to finalize your table.

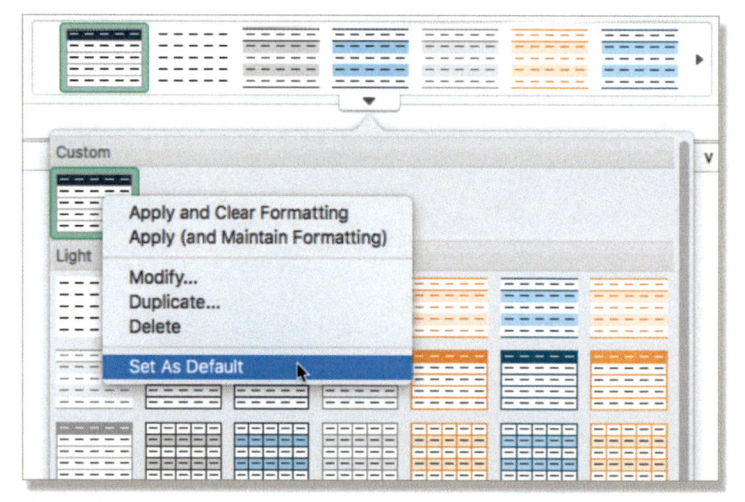

Fig 2.21: Set your branded table as the default style.

Step 10: Now your table style is ready for its first catwalk. To apply it to a dataset, select any cell in your table and navigate to **Home ▸ Styles ▸ Format as Table**, then select your custom style under **Custom (Fig 2.19)**.

Step 11: There's nothing to do in the **Format As Table** dialog. It's just asking you to verify the range and that it has a header row. Click **OK** to apply your custom table **(Fig 2.20)**.

Step 12: Once you're happy with your formatting options, you may want to set your custom formatted table as the default table style. If you create tables using menus, it won't make a difference to your workflow. However, if you use Excel's keyboard shortcut to create a table (Ctrl-T), it uses your default style. To set it, navigate to **Table Tools ▸ Design ▸ Table Styles**, then right-click on your custom style thumbnail and choose **Set As Default** from the contextual menu **(Fig 2.21)**. Just keep in mind, this will only apply it as the default for the document you're working in, which makes me stabby.

16

CHAPTER 2 • MASSAGE YOUR DATA

💬 **LEARN MORE** See the Templatize Everything section of the Set It and Forget It chapter to learn a hack that will help you get around this limitation.

Duplicate and Customize

A much easier alternative to creating a table from scratch is to duplicate one of the built-in styles and tweak it. To duplicate a style, right-click on any of the style thumbnails and choose **Duplicate** from the contextual menu **(Fig 2.22)**.

Now you can just modify the elements you want to by following the same steps in the tutorial above.

Note: Excel's built-in table styles are built from your theme, so creating a theme that aligns with your brand—using your brand's colors and fonts—will ensure the table styles you have to choose from will be on point with your branding.

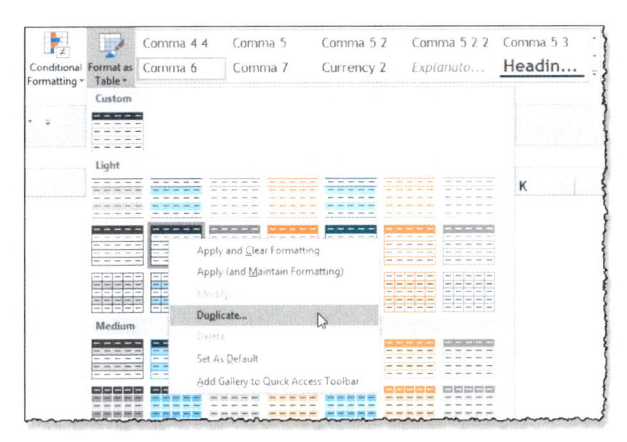

Fig 2.22: Create a table style by duplicating one of Excel's built-in styles.

💬 **LEARN MORE** See the Create a Branded Template section of the Set It and Forget It chapter to learn how to create branded themes.

There's much more you can do with custom tables, which you can learn more about from the **office.com** website: **bit.ly/mdsepc-table-format**.

NUMBER FORMATTING PET PEEVE

The only thing I don't like about Excel's formatted tables is they format 0 as a hyphen **(Fig 2.23)**. It's a matter of preference, but I don't like to see hyphens littering my tables and chart axes. I'd rather the number 0 be displayed as an actual 0.

If you share my hyphen aversion, you can replace them with 0s by selecting data columns in your table and pressing Ctrl-1 to open the **Format Cells** dialog. The code before the first semicolon controls the formatting for positive numbers, the code between the first and second semicolons controls the formatting

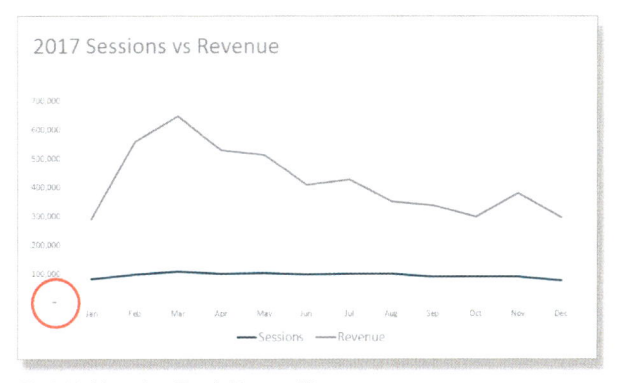

Fig 2.23: Not a fan, Excel. Not one bit.

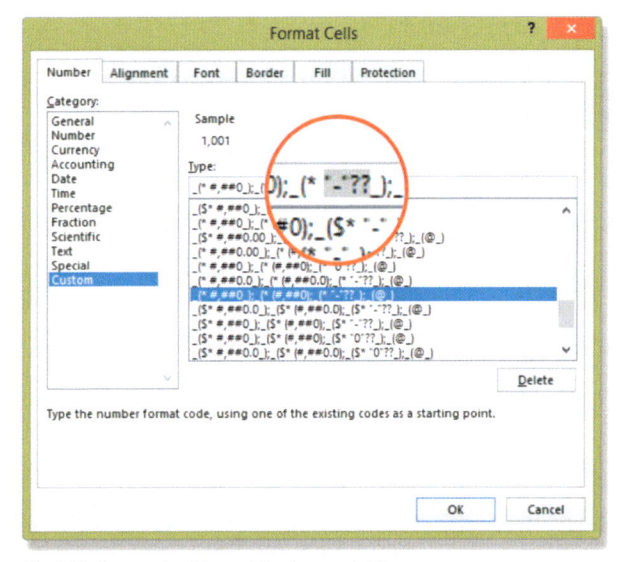

Fig 2.24: Customize 0 formatting in your tables.

17

CHAPTER 2 · MASSAGE YOUR DATA

for negative numbers, and the code after the second semicolon controls the formatting for 0. Just replace "–"?? (highlighted in the accompanying screenshot) with a 0, and you'll be golden **(Fig 2.24)**.

> 💬 **LEARN MORE** See the Customize Number Formatting section of the Design with Elegance chapter to learn how to finesse your number formatting.

Warning: Be careful not to run roughshod over the space that comes before the offending characters or the underscore that comes after, or your numbers won't align properly.

PIVOT TABLES

TABULAR DATA: EXCEL'S CHOKING HAZARD

There are times when the data you need to visualize is in a format that Excel can't process. Tabular data with multiple categories is especially problematic. An example might be a dataset that includes both date and marketing channel or cereal manufacturer and brand. (You will see both of these examples in this book.) If you try to feed Excel's charting engine a table like the one in the accompanying screenshot, it will choke and sputter—and basically create a scene **(Fig 2.25)**.

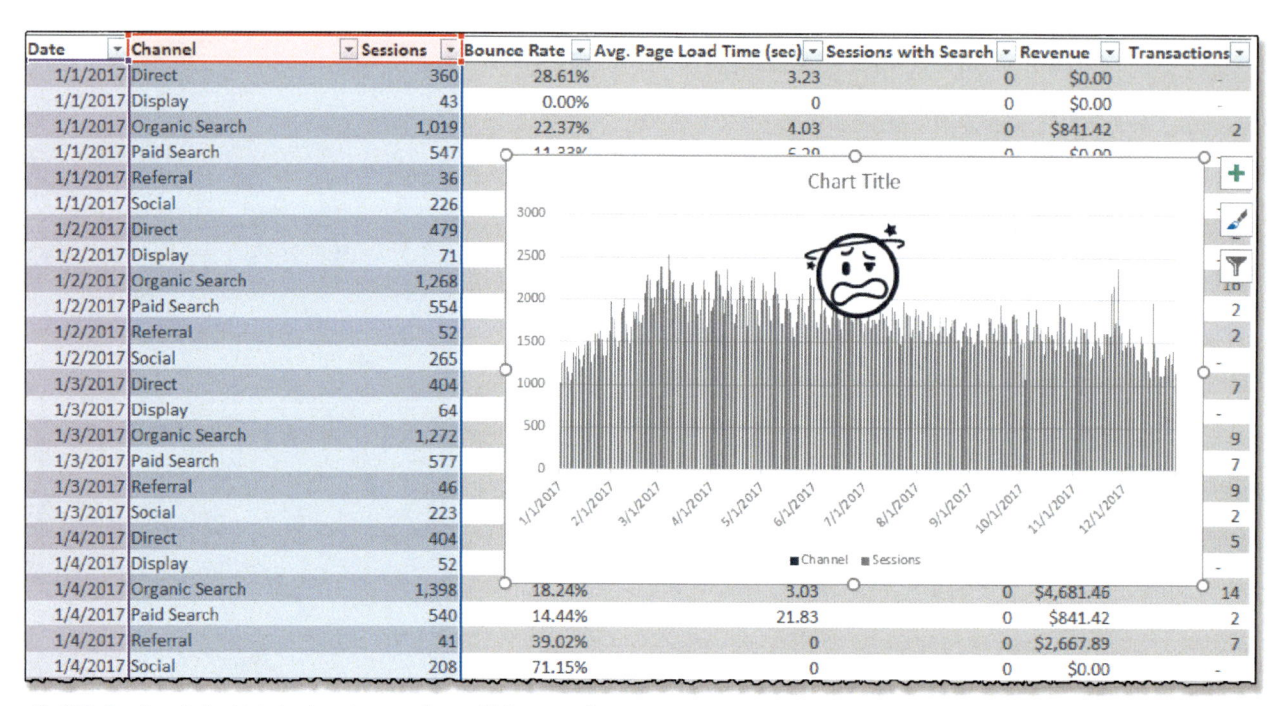

Fig 2.25: Excel can't chart tabular data that contains multiple categories (e.g., Date and Channel).

Fig 2.26: Category 1: Month; category 2: Marketing Channel; metric: Revenue.

CHAPTER 2 • MASSAGE YOUR DATA

Revenue	Column Labels							
Row Labels	Organic Search	Direct	Referral	Paid Search	Paid Social	Social	(Other)	Email	Grand Total
Jan	$ 119,190	$ 58,457	$ 75,020	$ 36,021		$ 2,406			$ 291,093
Feb	$ 249,250	$ 84,465	$ 155,173	$ 67,645		$ 3,746			$ 560,279
Mar	$ 323,941	$ 99,351	$ 153,818	$ 62,710		$ 8,588			$ 648,408
Apr	$ 269,134	$ 62,776	$ 142,934	$ 49,749		$ 5,778			$ 530,372
May	$ 236,117	$ 65,743	$ 144,651	$ 67,462		$ 1,178			$ 515,150
Jun	$ 203,595	$ 55,873	$ 84,331	$ 57,534		$ 6,895			$ 408,228
Jul	$ 290,314	$ 82,878	$ 4,618	$ 23,494	$ 22,899	$ 3,731	$ 864		$ 428,798
Aug	$ 206,488	$ 77,399	$ 3,700	$ 0	$ 59,740	$ 2,732	$ 0		$ 350,059
Sep	$ 211,928	$ 61,038	$ 6,560	$ 0	$ 54,818	$ 2,946	$ 978		$ 338,268
Oct	$ 179,541	$ 68,580	$ 9,617	$ 0	$ 39,017	$ 1,227	$ 1,138		$ 299,120
Nov	$ 204,713	$ 98,706	$ 3,605	$ 1,729	$ 57,531	$ 11,051	$ 1,774	$ 203	$ 379,313
Dec	$ 168,555	$ 70,824	$ 5,928	$ 864	$ 39,660	$ 4,283	$ 3,067	$ 1,216	$ 294,397
Grand Total	$ 2,662,766	$886,091	$789,954	$ 367,208	$ 273,665	$54,560	$7,822	$1,419	$ 5,043,485

Fig 2.27: Our final pivot table.

Therefore, if you have a dataset with two categories, before you can create a chart from it, you first need to pivot it. Prepare yourself for a gross simplification, but pivoting your data essentially puts one category in the first column and divides the other category into the remaining columns **(Fig 2.26)**. Everything in between is your one metric.

You create that pivot in Excel by creating a pivot table. Pivot tables accomplish many purposes, and entire resources have been published about the many uses of pivot tables. But we will mostly use them to massage datasets with two categories into a crosstab format (with one category along the top and the other along the left edge of the table). We'll also use them to optimize the structure of our treemap chart, which you'll see in the Treemap tutorial, and save ourselves a lot of manual effort in the Cycle Plot tutorial. However, they are both edge cases.

CREATE A PIVOT TABLE
Requirements
For one of the charts we'll be building later in this book, we'll need the following:

+ Date in rows, grouped by month
+ Marketing Channel in columns
+ Sessions in values

We'll need a pivot table to morph our data into a matrix format that Excel can process. All right, let's get started.

Tutorial
Step 1: Select any cell in your dataset and navigate to **Insert ▸ Tables ▸ PivotTable**.

Step 2: In the **Create PivotTable** dialog, set **Choose where you want the PivotTable report to be placed** to **New Worksheet (Fig 2.28)**. If you want to drop it on the same worksheet as your raw data, select the cell you want to drop it in. I don't recommend this though. Click **OK** to close the dialog.

Tip: I recommend naming the tabs for each of your worksheets. I named my pivot table worksheet Pivot Table. Getting into a habit of organizing your workbooks

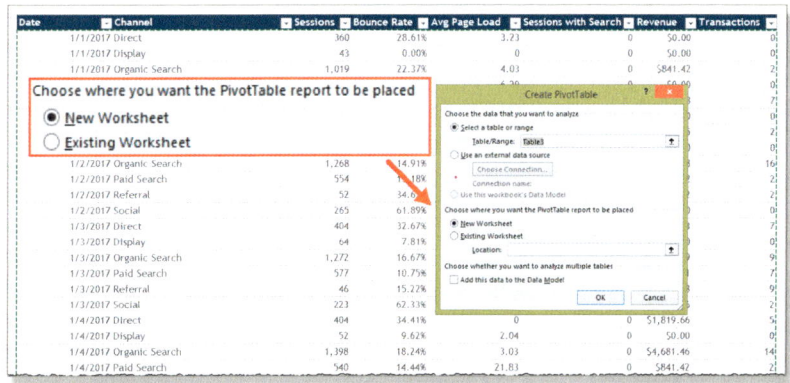

Fig 2.28: Decide where you want to place your pivot table.

19

CHAPTER 2 • MASSAGE YOUR DATA

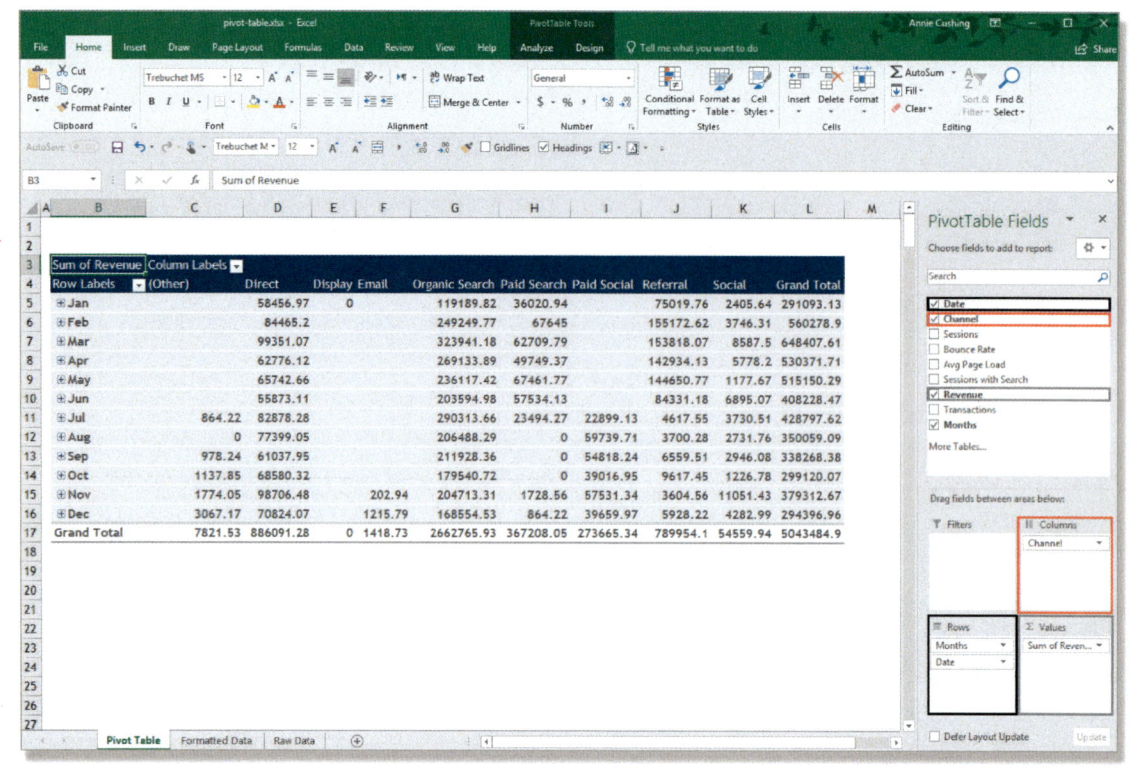

Fig 2.29: Creating pivot tables is a…drag. (Oh c'mon, that joke practically wrote itself!)

as you go can save you headaches later, especially with large workbooks. I also sort my worksheets so that my Raw Data worksheet is to the far right, and my worksheets get increasingly formatted/massaged as you move to the left. If I'm building a dashboard, it will be positioned to the far left.

Step 3: In the **PivotTable Fields List**, drag Date to **Rows**, Channel to **Columns**, and Revenue (or whatever metric you want) to **Values (Fig 2.29)**. When you place a category in the Rows and Columns fields, you'll only want to add one metric to the Values field. I've seen pivot tables with more than one value, but they are very difficult to interpret. All the charts in this book that use pivot tables use one metric.

Step 4: Excel separates Date into Date and Months in the Rows field. We can drag Date out of the Rows field, which will leave us with Months.

Step 5: Your pivot table may automatically aggregate the data into months. That's what we want in this example. If it doesn't, right-click any cell in the Date column of your pivot table (not the **PivotTable Fields** pane) and choose **Group** from the contextual menu, which will open the **Grouping** dialog. Set **By** to **Months (Fig 2.30)**.

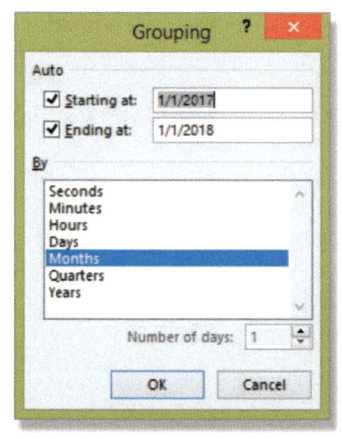

Fig 2.30: Aggregate by months using the Group option, if Excel doesn't do it automatically.

20

CHAPTER 2 • MASSAGE YOUR DATA

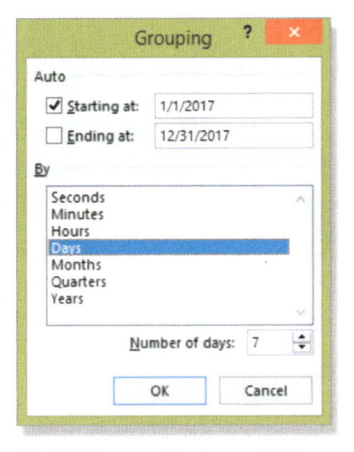

Fig 2.31: Aggregating by week takes a bit more tweaking.

Tip: If you want to group by week, Excel doesn't offer that as a default option in the Grouping dialog, but you can choose **Days** and then set **Number of days** to 7 **(Fig 2.31)**. You just might need to adjust the **Starting at** date setting if you want to start on a particular day of the week (like Sunday). If you don't like the labels (they're cumbersome), when you copy your data as values, you can drop something like Week 1 into the first row, then drag the fill handle down to increment your weeks by 1.

Step 6: I want to sort my columns in descending order, based on the total. The easiest way to do this is to right-click anywhere in the total row and choose **Sort ▸ Sort Largest to Smallest**. You can also use the drop-down filter you access from the **Column Labels** cell that sits in the header row of your pivot table. You can tell we have a sorted pivot table because there's now a tiny arrow that needs to share a room with the filter arrow in that cell **(Fig 2.32)**.

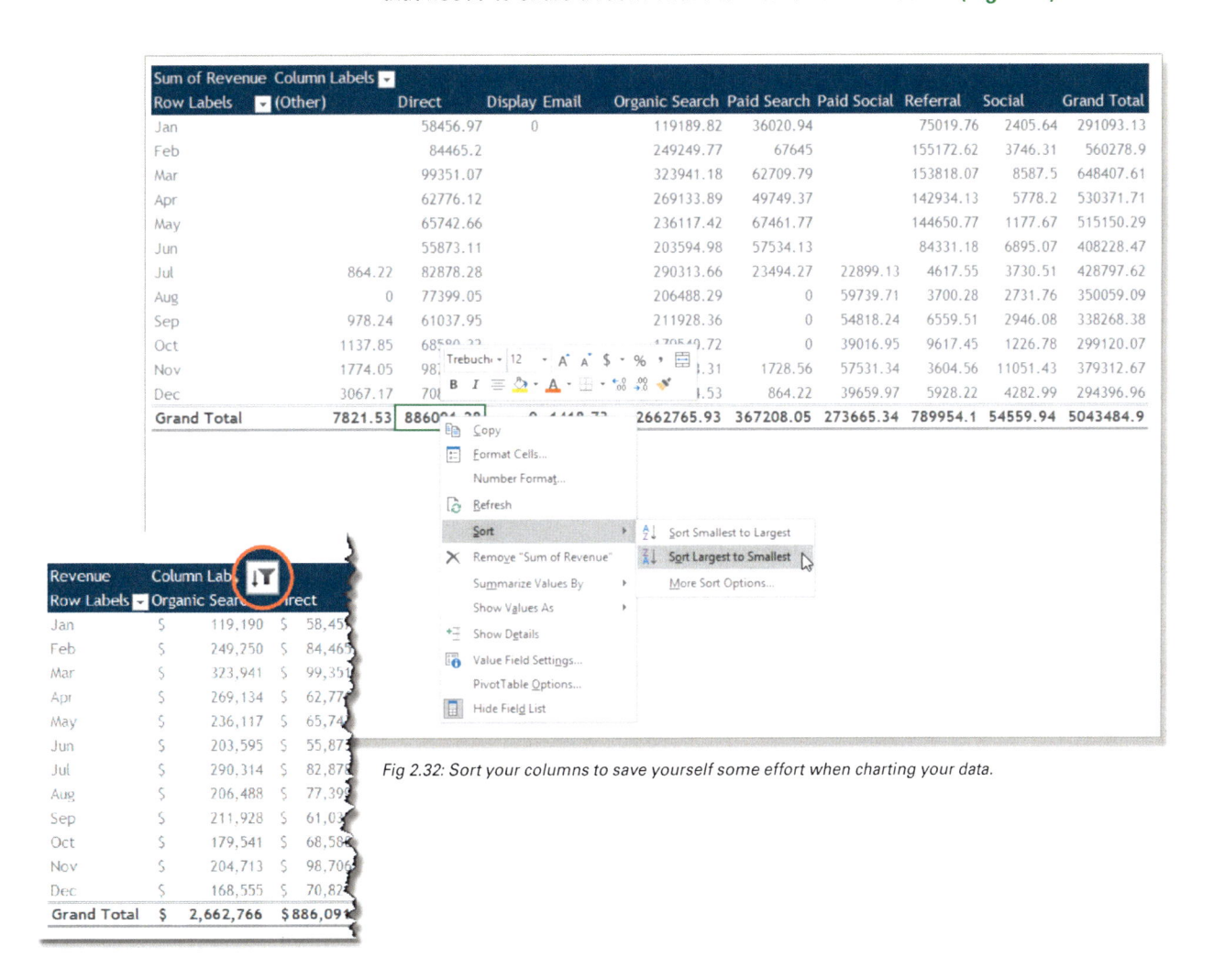

Fig 2.32: Sort your columns to save yourself some effort when charting your data.

21

CHAPTER 2 · MASSAGE YOUR DATA

Note: This trick works in any column, not just a total column. But the only row it works in is a total row.

Step 7: You can also filter out columns that are irrelevant, like the Display and (Other) columns in my dataset. You access this filter from the **Column Labels** cell. Deselect the ones you want to filter out from the filter list **(Fig 2.33)**.

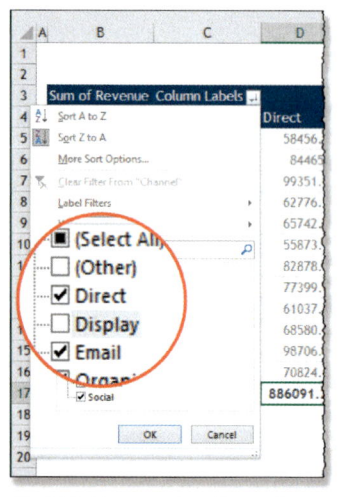

Fig 2.33: You can also filter columns out using the Column Labels filter.

Step 8: Select your table, sans the top row, and any total rows or columns. Copy-and-paste your massaged data from the pivot table as values from the **Paste Options** button: **Paste Values ▸ Values & Number Formatting (Fig 2.34)**.

Step 9: Now let's format our data. Since we pulled Revenue into our **Values** field, we'll navigate to **Home ▸**

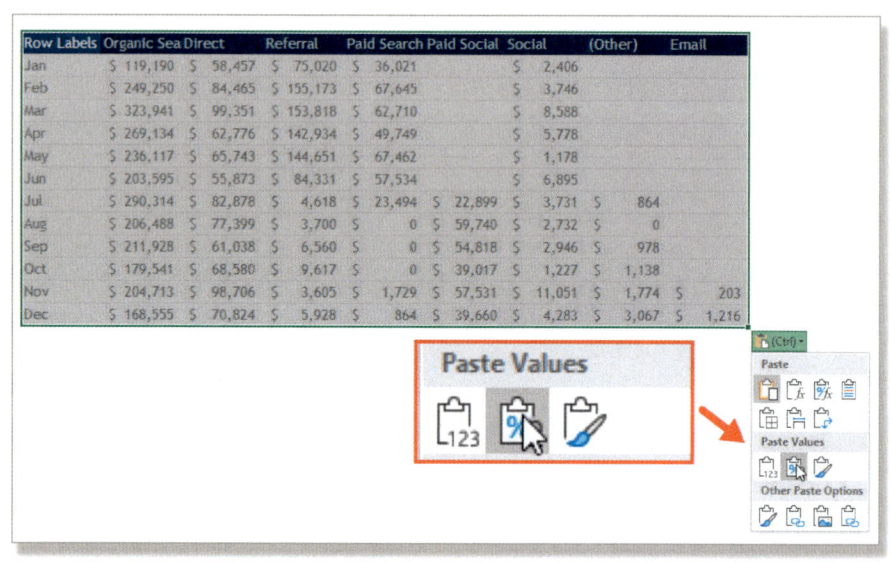

Fig 2.34: Copy data as values from pivot table.

Number and select **Accounting Number Format**. I don't need cents in my chart, so I typically reduce decimals by pressing the **Decrease Decimal** button twice. (It's located four buttons to the right of the **Accounting Number Format** button.) And I replace the hyphen in Excel's 0 formatting with a 0 to prevent hyphens in my tables and axis labels **(Fig 2.35)**.

💬 **LEARN MORE** See the Customize Number Formatting section of the Design with Elegance chapter to learn how to finesse your number formatting.

Tip: In this pivot table, we don't have metrics in our column headers because we pulled a category (Channel) into the Columns field. But when you don't have a

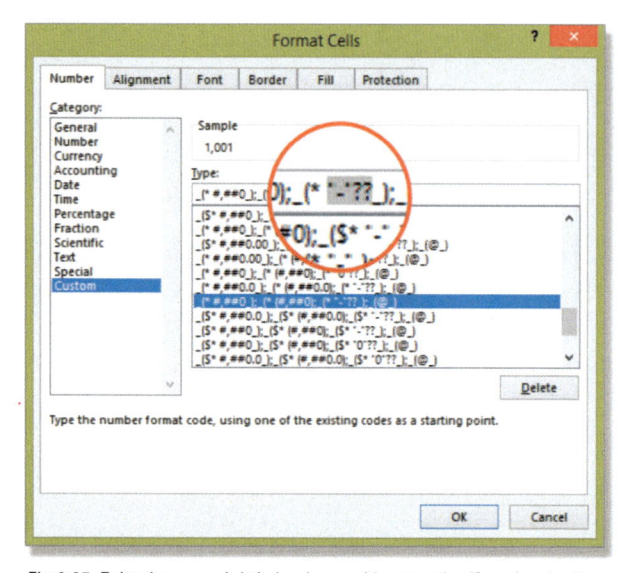

Fig 2.35: Evict those unsightly hyphens without notice if you're charting accounting data.

22

CHAPTER 2 · MASSAGE YOUR DATA

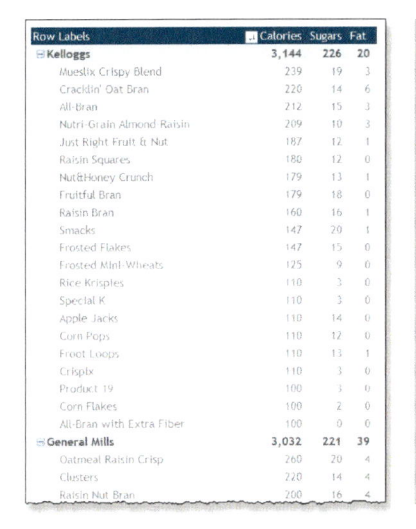

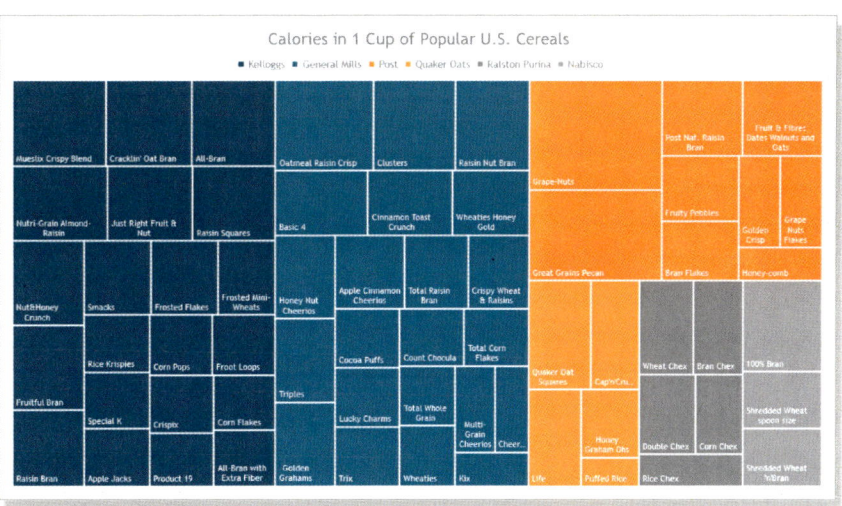

Fig 2.36: Get rid of aggregation type in column headers using this hack.

second category, your metric(s) will be in your headers (technically *PivotTable Field Name* but whatevs). Excel tries to be helpful by adding the aggregation type to the header, e.g., *Sum of Revenue*, *Average of Conversion Rate*, etc. If you want to remove the aggregation type (I usually do), you can't just delete it. Excel will throw shade your way in the form of an error message. However, if you add a space after the metric, Excel's like, "*Ohhh! Revenue with a spaaace.* Then, by all means, carry on!" You can modify the headers by double-clicking them to open the **Value Field Settings** dialog, using the **Formula Bar**, or very carefully clicking to select the text and overwrite it. I included another pivot table in the downloadable workbook to illustrate this, as well as a pivot table with subtotals **(Fig 2.36)**. We use this pivot table in the Treemap tutorial.

Step 10: Make any final formatting tweaks to your data before creating visualizations out of it **(Fig 2.37)**. It will make the chart-creation process smoother. You will get very comfortable with massaging your data into this format by the end of the book!

For all the datasets in this book with multiple categories, I created a pivot table to get the data in

Fig 2.37: Our data is finally in a crosstab format.

the right format (i.e., crosstab format), but then I copied and pasted it as values to remove the pivot table formatting. The reason is that when you create a chart from a pivot table, you get a pivot chart, which comes with a host of formatting issues. If you love vintage 90s Microsoft formatting, take full advantage of pivot charts. I don't prefer them and will build dynamic charts from scratch over using them. Dynamic charts fall outside the scope of this book. (I had to draw the line somewhere.) However, if you're so inclined, you can check out the video tutorial I did for Marketing Land, where I demonstrate how to update a chart using radio buttons: **bit.ly/ mdsepc-dynamic-chart**.

23

CHAPTER 2 • MASSAGE YOUR DATA

DESIGN A PIVOT TABLE STYLE

Tips from the Trenches

The design options for pivot tables are nearly identical to those we explored in the Design a Table Style section; you just have a few more to choose from. Therefore, I won't completely reinvent the wheel and explain the steps in the same level of detail.

Create Your Branded Theme First

Pivot table styles are built off your theme (**Page Layout ▸ Themes ▸ Themes**). As you'll learn in the Set It and Forget It chapter, it's very important to build a theme using your brand colors and fonts. I can't underscore the importance of this enough. You can build the sexiest visualizations, but if they don't align with your brand, they will appear unfinished and amateur. Therefore, you might not want to create a custom pivot table until you create a theme and global template.

Scan Pivot Table Styles

To test any of the pivot table styles, select your pivot table, then hover over a new style, which can be found by navigating to **PivotTable Tools ▸ Design ▸ PivotTable Styles**. Your dataset will update on the fly with a preview of that style **(Fig 2.38)**.

Modify a Pivot Table Style

Just as we saw with formatted tables, if a predefined pivot table style doesn't meet your needs, you can create and apply a custom pivot table style. It can be a more intimidating process than creating a custom table style because there are more options, and a format that works well for a pivot table with one structure may look weird when restructured, like when you add subtotals. As a result, I recommend tweaking a built-in style over building one from scratch. That's what we'll do here **(Fig 2.39)**.

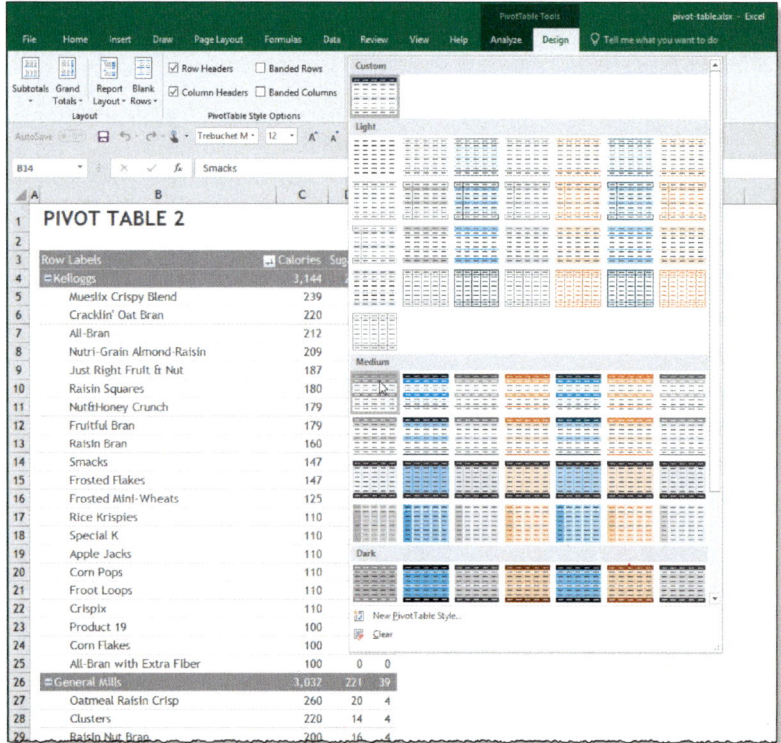

Fig 2.39: Our final pivot table style.

Fig 2.38: See a preview of a pivot table style by hovering over it.

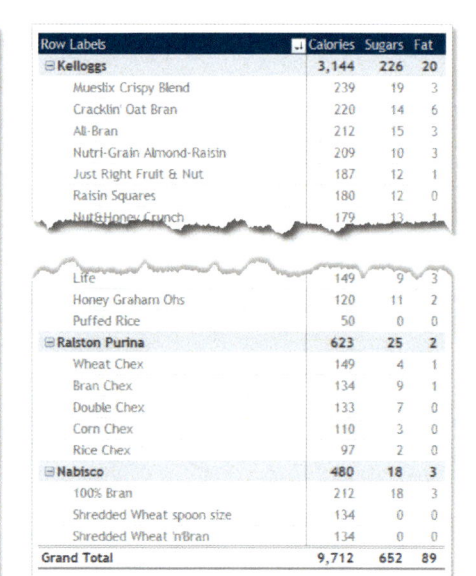

24

CHAPTER 2 · MASSAGE YOUR DATA

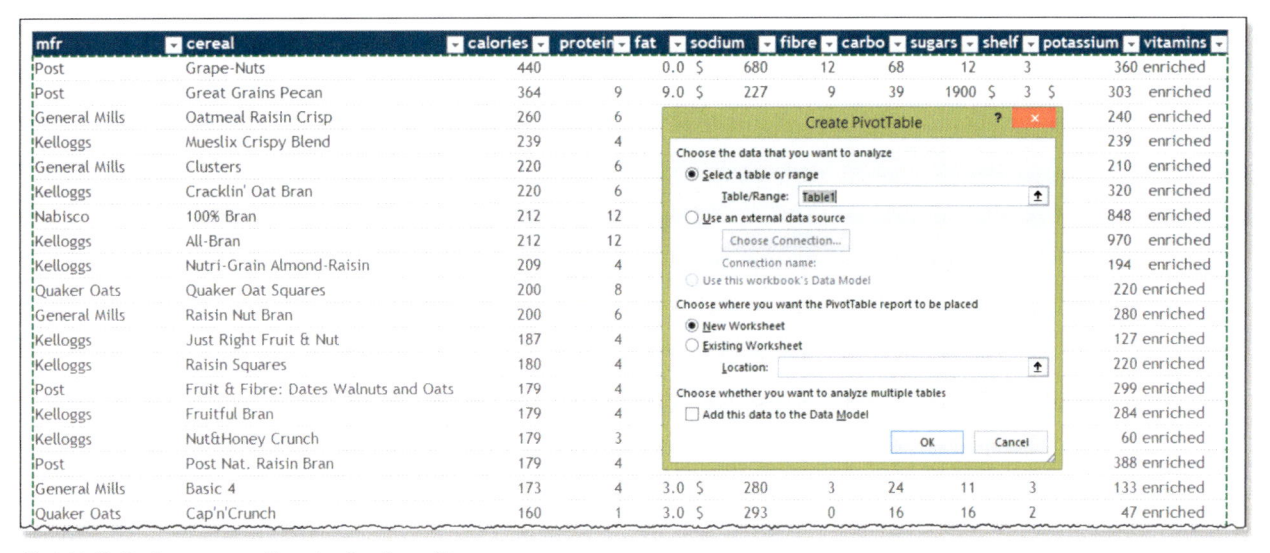

Fig 2.40: Verify data range and location for pivot table.

Step 1: Select any cell inside your dataset and navigate to **Insert ▸ Tables ▸ PivotTable**. In the **Create PivotTable** dialog, verify the range and choose where you want to drop your pivot table. I usually put it in its own worksheet **(Fig 2.40)**.

Step 2: Select any cell in your pivot table and navigate to **PivotTable Tools ▸ Design ▸ PivotTable Styles**. Click the down-facing arrow in the bottom-right corner of the **PivotTable Styles** menu to expand it. Right-click a style you want to modify and choose **Duplicate** from the contextual menu. Give your new pivot table style an intuitive name in the **Modify PivotTable Style** dialog **(Fig 2.41)**.

Tip: An easy to way to find elements that have been styled in some way is to look in the **Table Element** pane inside the **Modify PivotTable Style** dialog. If it's been styled, it will appear boldfaced. If you customize additional elements, they will also appear boldfaced.

Step 3: I want to give my header row a dark blue fill and white text. To start, select **Header Row** from the **Table Element** pane, then click the **Format** button. In the **Format Cells** dialog, click the **Fill** tab. Set **Background Color** to your preference. I used

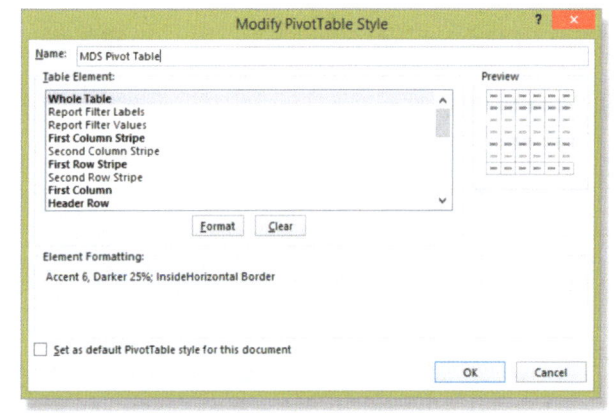

Fig 2.41: Give your new pivot table style an intuitive name.

blue. Open the **Border** tab and set **Presets** to preference or manually set your borders as we did in the Design a Table Style tutorial. I set **Presets** to **None**. Next, open the **Font** tab, then set **Color** to preference. I used white. Click **OK** to apply and close the **Format Cells** dialog. If you want to preview your new header, you'll need to click **OK** again to close the **Modify PivotTable Style** dialog. When you're finished previewing it, right-click your new custom style (which will now appear in the **Custom** section of the **PivotTable Style** drop-down menu) and choose **Modify** from the contextual menu to continue ~~obsessing~~ tweaking.

CHAPTER 2 • MASSAGE YOUR DATA

Step 4: Next, I want to add double lines to the top border of my total row and a thick line to the bottom border. To start, select **Grand Total Row** from the **Table Element** pane, then click the **Format** button. In the **Format Cells** dialog, open the **Border** tab. Under **Line**, set **Style** to the double line (bottom-right icon) and **Color** to preference, then click the top border in the **Border** preview window to apply it. I like the look of the thicker line when I have a total row applied.

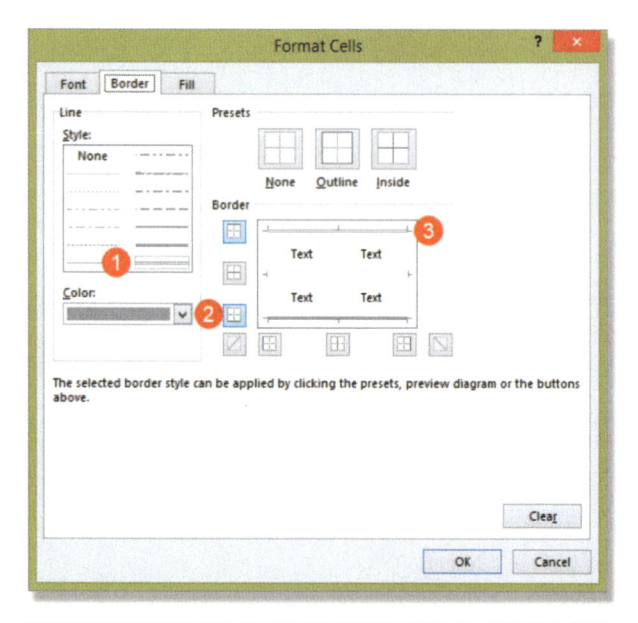

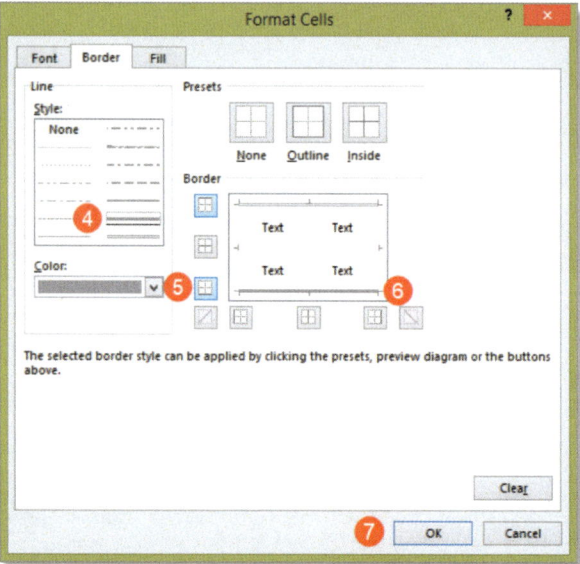

Fig 2.42: Customize Grand Total Row border.

If you want to add one to your pivot table style, while still in the **Border** tab, set **Style** to the thickest line (second from the bottom in the right column), set **Color** if you want a different color, then click the bottom border in the **Border** preview window. Click **OK** to apply **(Fig 2.42)**.

Step 5: Click **OK** again to close the **Modify PivotTable Style** dialog and preview your total row border format **(Fig 2.43)**. I like it, so I'll put a ring on it.

Learn more about how to customize your pivot table style from the **office.com** website: **bit.ly/mdsepc-pivot-table-format**.

Fig 2.43: Set the formatting options of your total row.

26

UNPIVOT DATA

There are times your data comes to you already pivoted—or partially pivoted—where you have values in your columns instead of the variable itself. If these columns are the only ones you'll be working with, no harm no foul—as we'll see with the Ivy League data in the Dumbbell Plot and Slope Chart tutorials. There are other times, however, that data needs to be unpivoted to make it play with the rest of the data in your dataset.

EXAMPLE FILE

Case in point: A dataset I downloaded from Open NYC Data was a hot mess **(Fig 2.44)**.

From an aesthetics perspective, there are quite a few issues with this dataset that trigger a data-induced panic attack for me. But those are fixed with relative ease. The issue that's not fixed quite so effortlessly is their analyst's unfortunate decision to spread a single variable (i.e., bridges crossing the East River) into four disparate columns (columns G:J in Fig 2.44).

Date	Day	High Temp (°F)	Low Temp (°F)	Precipitation	Brooklyn Bridge	Manhattan Bridge	Williamsburg Bridge	Queensboro Bridge	Total
5/1	Sunday	50.0	45	0.16	2,634	1,525	2,062	1,408	7,629
5/2	Monday	59.0	46	0.04	2,505	4,149	4,864	3,575	15,093
5/3	Tuesday	54.0	51.1	0.61	1,059	1,986	2,614	1,741	7,400
5/4	Wednesday	51.1	48.9	0.01	1,710	2,905	3,797	2,693	11,105
5/5	Thursday	57.0	46.9	0.00	2,582	4,196	4,971	3,619	15,368
5/6	Friday	53.1	48	0.54	652	1,453	2,122	1,612	5,839
5/7	Saturday	57.9	48.9	0.00	1,867	3,157	3,789	2,749	11,562
5/8	Sunday	64.9	50	0.16	1,916	2,927	3,948	2,463	11,254
5/9	Monday	72.0	55	0.00	3,650	6,591	7,386	4,951	22,578
5/10	Tuesday	62.1	51.1	0.00	3,384	6,283	7,335	4,796	21,798
5/11	Wednesday	73.9	53.1	0.00	3,883	7,216	8,231	5,327	24,657
5/12	Thursday	79.0	57.9	0.00	3,960	7,103	8,333	5,735	25,131
5/13	Friday	64.9	60.1	0.25	1,780	3,072	4,229	3,121	12,202
5/14	Saturday	71.6	60.8	T	2,556	4,042	5,479	3,951	16,028
5/15	Sunday	57.2	50	T	1,738	2,971	3,486	2,696	10,891
5/16	Monday	66.2	46.4	0.00	2,907	5,254	5,961	4,016	18,138
5/17	Tuesday	64.4	55.4	T	2,812	5,230	6,219	4,344	18,605
5/18	Wednesday	68.0	53.6	T	3,309	5,812	6,739	4,603	20,463
5/19	Thursday	71.1	57	0.00	3,474	6,214	7,485	4,763	21,936
5/20	Friday	75.0	55	0.00	3,976	6,994	7,776	5,221	23,967
5/21	Saturday	66.0	55	0.04	1,665	2,864	3,978	2,928	11,435
5/22	Sunday	69.1	53.1	0.09	2,117	3,729	4,633	2,956	13,435
5/23	Monday	77.0	57.9	0.02	3,928	6,843	7,769	5,135	23,675
5/24	Tuesday	73.0	59	0.18	1,482	3,019	4,260	2,392	11,153
5/25	Wednesday	87.1	64	0.00	4,244	7,869	8,959	5,906	26,978
5/26	Thursday	89.1	71.1	0.00	3,887	7,037	8,231	5,451	24,606
5/27	Friday	86.0	75	0.00	3,302	5,670	7,054	4,592	20,618
5/28	Saturday	90.0	72	0.00	2,664	3,998	5,327	3,895	15,884
5/29	Sunday	86.0	73	T	3,193	4,154	4,918	3,635	15,900
5/30	Monday	79.0	68	1.65	1,719	2,750	3,759	2,689	10,917
5/31	Tuesday	84.9	72	0.00	3,948	7,016	7,957	5,556	24,477

Cyclist Counts on East River Bridge Locations
May 2016 24-Hour Counts
Bicycle and Greenway Program

Fig 2.44: Columns G:J reveal a fatal flaw in how this data was structured.

CHAPTER 2 • MASSAGE YOUR DATA

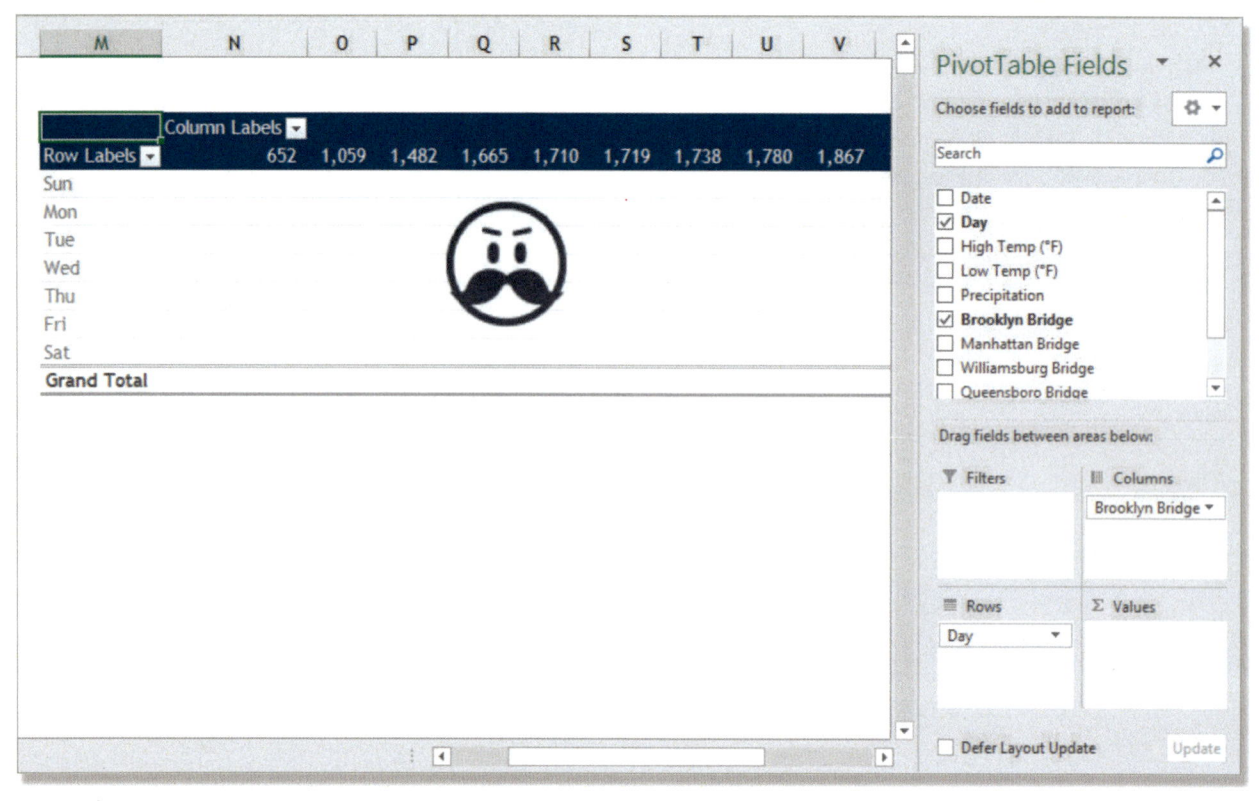

Fig 2.45: Separating a single variable into separate columns in your raw data wreaks havoc in a pivot table.

For maximum flexibility, this should be one column with a header of Bridge, with Brooklyn, Manhattan, Williamsburg, and Queensboro as its values.

If I wanted to create a line chart charting traffic by day, with each bridge as its own data series, I could select G6:J37, then create my line chart. But I want to answer the question, "Which days of the week received the most and least cyclist traffic for the month of May?" This dataset can't answer that question it its current form.

To aggregate by day of week, I need the dataset to be limber and in tiptop condition, which means it needs to be structured according to best practices. That requires giving each variable its own column. That way I could use a pivot table to easily aggregate my data by day of week. However, I couldn't bring this data into a pivot table because it's already partially pivoted. Ergo, instead of being able to pull Bridge into the **Columns** area of the **Fields List**—with each bridge populating to its own column—I could only pull individual bridges down. Disaster ensued **(Fig 2.45)**.

TUTORIAL

In the past you would have needed to use an R or Python script to fix an indiscretion like this—or if you're using Excel for Mac, at time of writing. But now you can use Excel to take a pooper scooper to your data using Power Query, which has been rebranded in Excel 2016 as Excel's Get & Transform tool. I'll walk you through the steps I took to unpivot this dataset.

28

Step 1: Remove all the funky formatting from the data. This is also a good time to apply number formatting. For example, I reformatted Day to ddd. This shortened Monday, for example, to Mon, which will fit more easily in a chart **(Fig 2.46)**.

LEARN MORE See the Customize Number Formatting section of the Design with Elegance chapter to learn how to finesse your number formatting.

Step 2: This next step is a hack. But changing the number formatting doesn't alter the underlying data. If you were to click on one of those cells, you would see a date in the Formula Bar. I wanted to be able to aggregate by day of week, to see which days of the week saw the highest bicycle traffic across all four bridges that cross the East River. Consequently, I selected the Day range, copied it and pasted it as values into...Word. You can paste as values by choosing **Keep Text Only** (last option) from the **Paste Options** pop-up that appears in the bottom-right corner of the range. Then I selected the range from Word by pressing Ctrl-A and pasted it back into my dataset, overwriting the Day column. Now, if you click one of the cells in this column and look in the **Formula Bar**, you'll see text **(Fig 2.47)**. In real life I would add it in as a new column, but I didn't want the following screenshots to get overly complicated. Life. It's all about compromises.

Date	Day	High Temp (°F)	Low Temp (°F)	Precipitation	Brooklyn Bridge	Manhattan Bridge	Williamsburg Bridge	Queensboro Bridge	Total
5/1/2016	Sun	50.0	45	0.16	2,634	1,525	2,062	1,408	7,629
5/2/2016	Mon	59.0	46	0.04	2,505	4,149	4,864	3,575	15,093
5/3/2016	Tue	54.0	51.1	0.61	1,059	1,986	2,614	1,741	7,400
5/4/2016	Wed	51.1	48.9	0.01	1,710	2,905	3,797	2,693	11,105
5/5/2016	Thu	57.0	46.9	0	2,582	4,196	4,971	3,619	15,368
5/6/2016	Fri	53.1	48	0.54	652	1,453	2,122	1,612	5,839
5/7/2016	Sat	57.9	48.9	0	1,867	3,157	3,789	2,749	11,562
5/8/2016	Sun	64.9	50	0.16	1,916	2,927	3,948	2,463	11,254
5/9/2016	Mon	72.0	55	0	3,650	6,591	7,386	4,951	22,578
5/10/2016	Tue	62.1	51.1	0	3,384	6,283	7,335	4,796	21,798
5/11/2016	Wed	73.9	53.1	0	3,883	7,216	8,231	5,327	24,657

Fig 2.46: Cleanup on aisle 2!

C6 | fx | Sun

	A	B	C	D	E	F
5		Date	Day	High Temp (°F)	Low Temp (°F)	Precipitation
6		5/1/2016	Sun	50.0	45	0.16
7		5/2/2016	Mon	59.0	46	0.04
8		5/3/2016	Tue	54.0	51.1	0.61
9		5/4/2016	Wed	51.1	48.9	0.01
10		5/5/2016	Thu	57.0	46.9	0

Fig 2.47: Sometimes the easiest path is the least elegant. #datahackjob #itsfiiine

CHAPTER 2 • MASSAGE YOUR DATA

Step 3: I only need Day and the four bridge columns, so I'll delete everything else **(Fig 2.48)**. Again, I don't usually take destructive steps like this. I'm only doing this because the following steps get pretty complicated, and I want to simplify the screenshots. So call off the data dogs, you purists out there!

Step 4: Okay, now we're ready to start unpivoting. The first thing we need to do is format the dataset as a table. To format as a table, select any cell in your dataset and navigate to **Home ▸ Format as Table** and choose your table formatting. I used my branded *MDS Table* preset.

Day	Brooklyn Bridge	Manhattan Bridge	Williamsburg Bridge	Queensboro Bridge
Sun	2,634	1,525	2,062	1,408
Mon	2,505	4,149	4,864	3,575
Tue	1,059	1,986	2,614	1,741
Wed	1,710	2,905	3,797	2,693
Thu	2,582	4,196	4,971	3,619
Fri	652	1,453	2,122	1,612
Sat	1,867	3,157	3,789	2,749
Sun	1,916	2,927	3,948	2,463
Mon	3,650	6,591	7,386	4,951
Tue	3,384	6,283	7,335	4,796
Wed	3,883	7,216	8,231	5,327
Thu	3,960	7,103	8,333	5,735
Fri	1,780	3,072	4,229	3,121
Sat	2,556	4,042	5,479	3,951
Sun	1,738	2,971	3,486	2,696
Mon	2,907	5,254	5,961	4,016
Tue	2,812	5,230	6,219	4,344
Wed	3,309	5,812	6,739	4,603

Fig 2.48: I pared the dataset down to the essentials.

💬 **LEARN MORE** See the Formatted Tables section of this chapter to learn how to customize Excel's tables to meet your needs and branding.

Step 5: With any cell in your dataset still selected, navigate to **Data ▸ Get & Transform Data ▸ From Table/Range (Fig 2.49)**.

Step 6: In the **Power Query Editor**, select the first column you want to combine, then Shift-click the last column to select the columns in between as well. Navigate to **Transform ▸ Any Column ▸ Unpivot Columns (Fig 2.50)**.

Note: Navigate to the **Transform** tab, not the **Transform** group under the **Home** tab.

Step 7: Excel will assign the new variable the header Attribute and newly aggregated metric column Value, which aren't particularly useful column labels, so let's rename them. You can do that by double-clicking the column header and typing in your new label **(Fig 2.51)**.

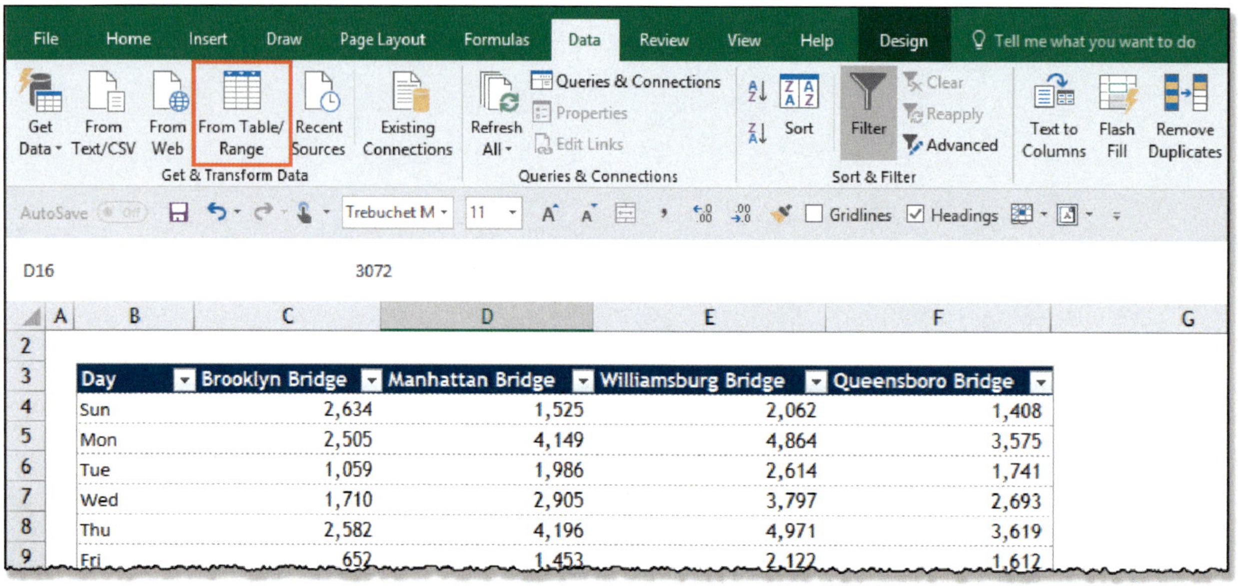

Fig 2.49: Start the transformation process with Excel's Get & Transform Data option.

CHAPTER 2 · MASSAGE YOUR DATA

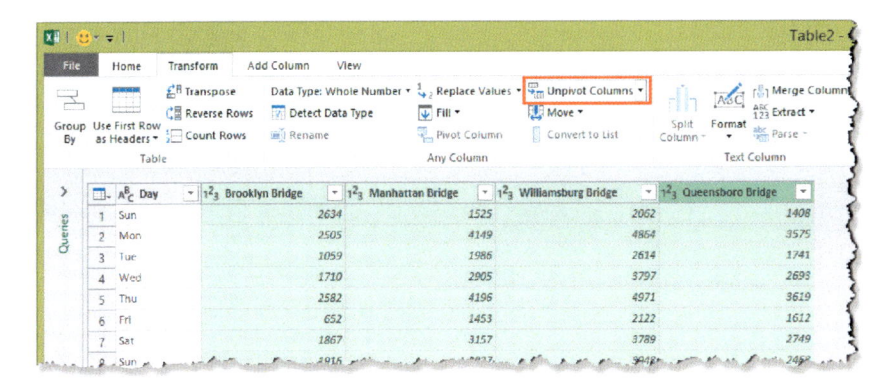

Fig 2.50: Unpivot columns using the Query Editor.

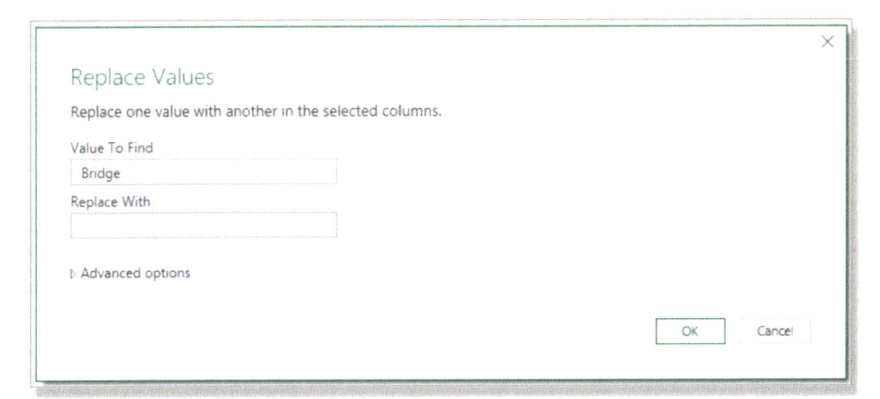

Fig 2.52: We get it. They're bridges.

Step 8: I also removed Bridge from each of the bridge values since it's unnecessary repetition, and anything unnecessary become chart junk downstream. You can perform a search and replace on your values by first selecting the column header, then choosing **Replace Values** from the same section. I replaced Bridge with nothing. Including the space before Bridge will prevent trailing spaces in your final dataset **(Fig 2.52)**.

Step 9: Okay, our data is now runway ready, so we can close out of the **Power Query Editor** by navigating to **Home** ▶ **Close** ▶ **Close & Load**. This will drop our squeaky-clean dataset into a new worksheet. I performed my usual worksheet cleanup: removed gridlines (yuk), added a margin column, added a worksheet title, and applied my *MDS Table* format I created for the book **(Fig 2.53)**.

Fig 2.51: Rename your column headers.

Day	Bridge	Cyclists
Sun	Brooklyn	2,634
Sun	Manhattan	1,525
Sun	Williamsburg	2,062
Sun	Queensboro	1,408
Mon	Brooklyn	2,505
Mon	Manhattan	4,149
Mon	Williamsburg	4,864
Mon	Queensboro	3,575
Tue	Brooklyn	1,059
Tue	Manhattan	1,986
Tue	Williamsburg	2,614
Tue	Queensboro	1,741
Wed	Brooklyn	1,710
Wed	Manhattan	2,905
Wed	Williamsburg	3,797
Wed	Queensboro	2,693
Thu	Brooklyn	2,582
Thu	Manhattan	4,196
Thu	Williamsburg	4,971
Thu	Queensboro	3,619
Fri	Brooklyn	652
Fri	Manhattan	1,453
Fri	Williamsburg	2,122
Fri	Queensboro	1,612
Sat	Brooklyn	1,867
Sat	Manhattan	3,157

Fig 2.53: Clean data at last!

CHAPTER 2 • MASSAGE YOUR DATA

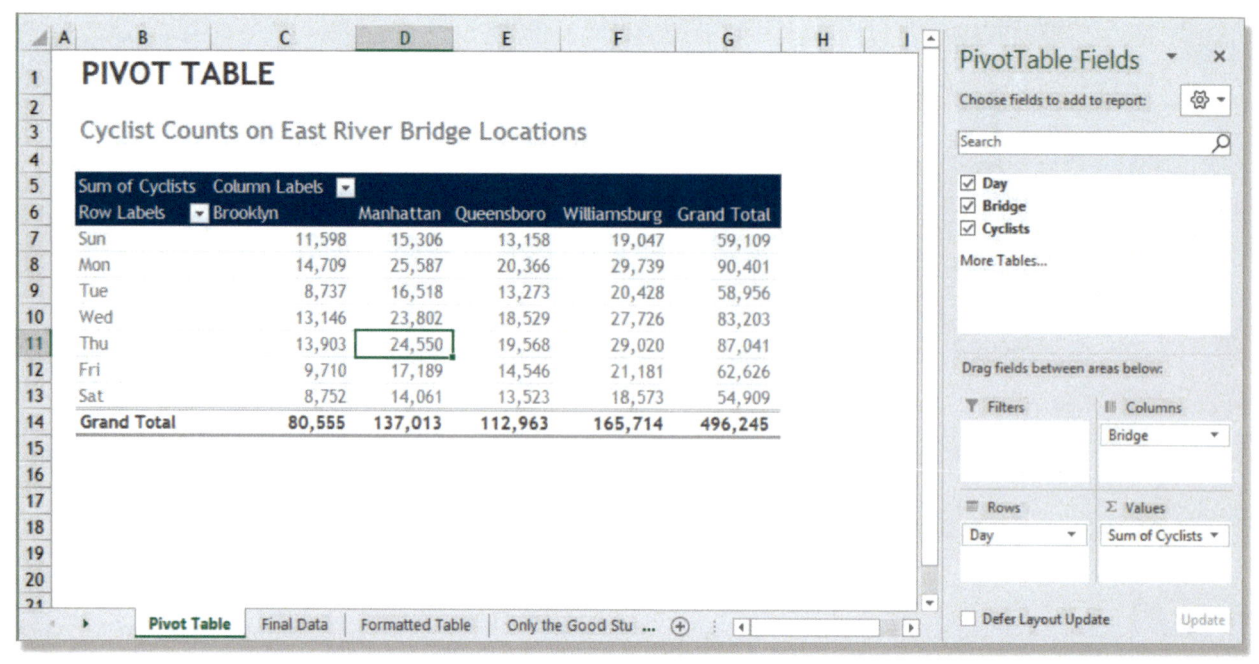

Fig 2.54: Now we can aggregate by day of week using a pivot table.

Step 10: Now I can create a pivot table from this dataset. I pulled Day into **Rows**, Bridge into **Columns**, and Cyclists into **Values (Fig 2.54)**.

> 💬 **LEARN MORE** See the Pivot Tables section of this chapter to learn how to rock pivot tables.

AVOID SECURITY WARNING

Excel creates a live connection between your new dataset and the original data you transformed, so you can always go back to do more cleanup and then refresh it. Also, Excel records each of the steps you took and allows you to modify them later. Learn more from Microsoft's *Getting Started with Get & Transform in Excel 2016* resource, which you can access on the **office.com** website: **bit.ly/mdsepc-transform-data**.

However, the dark side of that moon is you may receive a nasty warning when you open the file **(Fig 2.55)**.

When you click through you'll get more information about the warning…sort of **(Fig 2.56)**.

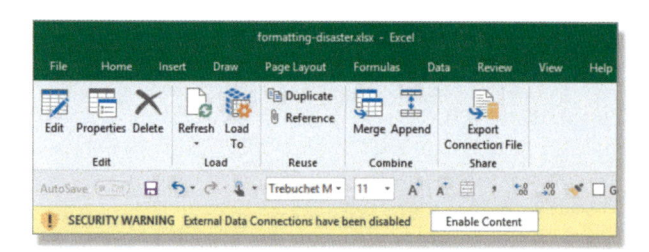

Fig 2.55: Be prepared for this annoying warning when your file includes a query.

Fig 2.56: Oh, that's very clear. Thanks, Microsoft!

CHAPTER 2 • MASSAGE YOUR DATA

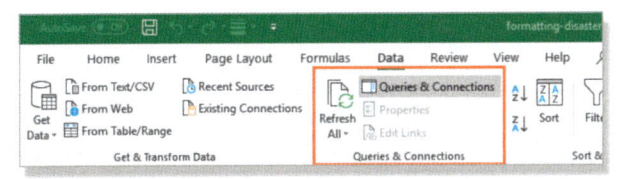

Fig 2.57: Labeling for this group has changed in Office 365.

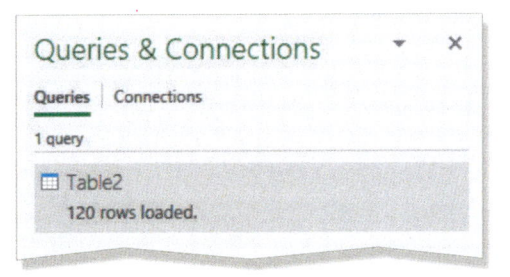

Fig 2.58: In Office 365, Microsoft reverted to a dialog. Because progress is overrated.

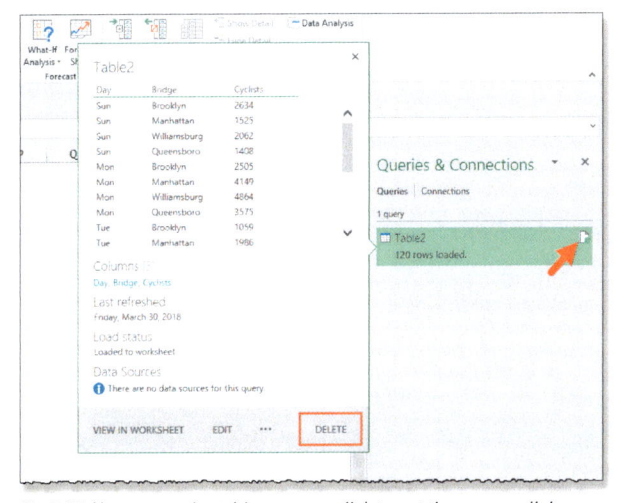

Fig 2.57: Hover over the table name or click to get the pop-up dialog.

To avoid these, you can delete the query. But only do this if you know you're finished tweaking it. I take this step if I'm sending the file to someone else because, to the uninitiated, these warnings can be disconcerting. To delete a query, navigate to **Data ▸ Queries & Connections ▸ Queries & Connections (Fig 2.57)**. This will open the **Queries & Connections** pane **(Fig 2.58)**. Hover over the table name that contains the query and click **Delete** at the bottom of the pop-up dialog, then confirm your deletion **(Fig 2.59)**.

Note: If you subscribe to Office 365, the Queries & Connections section has been renamed Connections, and it's been restructured a bit. But the tool still functions the same.

EXAMPLE VISUALIZATION
To see an example of how this dataset was used, check out the Stacked Column Chart tutorial.

DOWNLOADABLE FILES
If you'd like to see the final file without getting a nasty security warning—and have purchased the full workbooks—download the formatting-disaster-final.xlsx file. If you'd like to check out the query, download the formatting-disaster-query.xlsx file.

To view the query, enable the content by clicking the **Enable Content** button in the **Security Warning** pop-up. Then navigate to **Data ▸ Queries & Connections ▸ Queries & Connections**. Hover over the **Refresh** icon, then click **Edit** in the flyout menu. This will open the **Query Editor**, and you'll see the steps in the **Applied Steps** section of the **Query Settings** pane **(Fig 2.60)**.

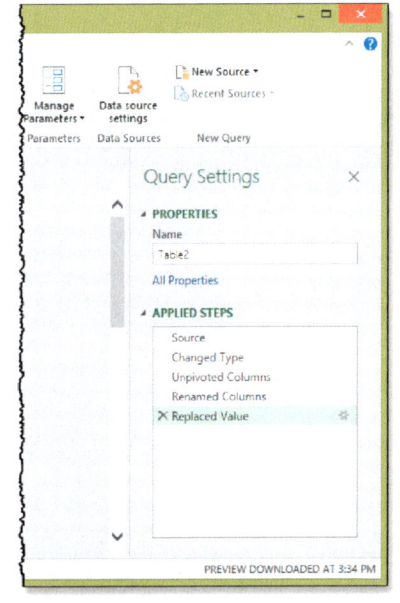

Fig 2.60: View the query steps in a workbook to review or modify.

33

FIX DATES

Sometimes data comes to you with dates that are formatted in a way that Excel doesn't recognize. The telltale sign Excel can't decipher your date as an actual date is if it's left aligned in the column. All number values, including dates, should be right aligned. Text values are left aligned.

There are a few common causes I see for this issue, and I'll give you some strategies for how to use the DATE, LEFT, and RIGHT functions to fix them. But first, let's introduce these functions.

FUNCTIONS YOU'LL NEED

DATE
The DATE function has three arguments: `year`, `month`, and `day`. Its syntax is expressed:

`DATE(year, month, day)`

LEFT / RIGHT
You will typically have each of these values trapped in text in a cell. That's where the LEFT and RIGHT functions come in. Their syntaxes are expressed:

`LEFT(text, [num_chars])`
`RIGHT(text, [num_chars])`

The `text` argument asks which cell contains the text we'll be extracting strings from. The `[num_chars]` argument is asking for the number of characters you need. You'll typically use 4 for year and 2 for month.

Ergo, the LEFT function will extract the number of characters you specify from the beginning of the string, and the RIGHT function will extract the number of characters you specify from the end of the string.

When we put it all together, if you have a month expressed as 201705—as you'll see in the Clustered Column Chart tutorial—and it's in cell A1, your function will look like:

`=DATE(LEFT(A1,4),RIGHT(A1,2),1)`

This function returns 5/1/2017.

The 1 could be any number; we just need something in there to satisfy the requirements of the DATE function. However, our dates in that chart won't include days, so we'll suppress those using custom number formatting.

You won't use the MID function in any of the tutorials in this book, but if you become a super analyst (or even a mediocre analyst), you will need to use it at some point. The only difference with the MID function is you need to feed it a starting point, which makes sense. Excel's just asking how many characters in do you want Excel to start counting?

Its syntax is expressed this way:

`MID(text, start_num, num_chars)`

> 💬 **LEARN MORE** See the Customize Number Formatting section of the Design with Elegance chapter to learn how to finesse your number formatting.

FORMATTING YOU'LL NEED

In the previous example, we transformed our text into a date format, but the formatting is wrong. Let's apply custom number formatting that will give us two numbers for month (so a leading 0 for months January through September), a forward slash, and two numbers for year:

`mm/yy`

That will format our date to look like this:

05/17

I'm partial to this formatting when I have room in my horizontal axis. If space is limited, I'll just use `m/yy` to remove the leading 0 from month or `m/d` to express dates in a single year along the x axis. Then it would appear as 5/1. If you're only reporting on data from a single year, the year can be specified elsewhere in the visualization, such as the title, subtitle, or annotation.

CHAPTER 3

Design with Elegance

There are a few techniques I will be using again and again throughout the tutorials in the Create Charts chapter. As I've said before, I'm a minimalist with design, but function wins out over form. There are times I will need to litter a chart with a text box that's larger than I feel comfortable with. But if there's an urgent issue that needs to be addressed, communicating that issue in a way that grabs management's attention outweighs my design preferences. However, in the absence of an exigent circumstance like that, I'm always looking for ways to weed out unnecessary pixels in my chart.

To that end, I'll give you some tools in this chapter that will help you format your charts in a way that will throw the spotlight where you want: on your data! Chart junk vies for that spotlight. By the time you get to the end of this chapter you'll have the skills you'll need to tamp it down.

You may choose to skip this chapter and just earmark it for when tutorials reference a section. But there's potential benefit to biting the bullet and reviewing the techniques listed here because you might find yourself in a situation where a recipient of one of your visualizations challenges you on a design decision, and you will need to be armed with alternatives to appease her concerns.

For example, let's say you remove gridlines from your chart, and your boss tells you she needs them to understand the chart. If you know the alternatives to just throwing gridlines into a chart, you could suggest some alternatives, like bringing the data into the chart with data labels. If she is recalcitrant, you can at least make those gridlines as unobtrusive as possible by setting them to a light color—like gray—or reducing their transparency.

TWO MAIN OBJECTIVES

When I'm assessing the effectiveness of a visualization's design, I'm mostly looking for three things:

- Does it align with the organization's brand?
- Is it free from clutter?
- Is it the best chart for what's being communicated?

The last item will be addressed ad infinitum in the Create Charts chapter. By the time you get to the end, you will have learned how to create 60+ visualizations in Excel. But I'll address the first two in this chapter.

ALIGNMENT WITH BRANDING

Many organizations have branding guides. And many more should have one but don't. Oftentimes brand managers in these organizations hawk over compliance with their branding guide but give nary a thought to the charts and other visualizations they publish and distribute to customers, clients, board members, and other stakeholders. This is an oversight—and one that I will address relentlessly throughout the book.

CHAPTER 3 • DESIGN WITH ELEGANCE

FREEDOM FROM CLUTTER

Microsoft doesn't do you any favors with their default visualizations when it comes to decluttering. They are rife with chartjunk, a term coined by Edward Tufte in his 1983 book T*he Visual Display of Quantitative Information.* Chartjunk—such as unnecessary gridlines (hate them!), axis labels that could be sent packing with data labels, 3-D effects, etc.—is distracting and can cause spikes in anxiety. It's like your viewers' eyes don't know where to land. They just dart back and forth, undirected, in search of relevance. I've seen some things. And some stuff….

Note: I'm not a fan of chart junk being one word, so I'll separate them in future references. It's not you, Edward; it's me. *laughs awkwardly*

This chapter will help you prepare your charts for presentation by handing you a tool belt you will use again and again in the Create Charts chapter.

A TALE OF TWO ORGANIZATIONS

DOJ'S DISASTROUS CHARTS

One of the saddest examples I've seen of the impact of this kind of oversight can be found on the U.S. Department of Justice (DOJ) website. The DOJ's branded palette contains an amalgam of powerful colors, as shown in the screenshot of their homepage below, taken in January 2018 **(Fig 3.1)**.

But then they use charts like the one in **Fig 3.2**, which was included in their *Asset Forfeiture Program* slideshow. You can view it here: **bit.ly/mdsepc-doj-chart-01**.

There are a number of issues with this visualization:

✦ It uses Excel's default theme.
✦ It uses Excel's default colors instead of its own.
✦ The markers serve no real purpose.
✦ The data table redefines unnecessary.
✦ The gridlines are unnecessarily thick.
✦ The axis label looks like it's seceding from the Union.
✦ The mean line is too thick and the wrong color.
✦ It's not clear what the mean line is an average of? Total deposits? Total expenses? Net?

Fig 3.1: The DOJ's branding is a treasure trove for a skilled data artist.

CHAPTER 3 · DESIGN WITH ELEGANCE

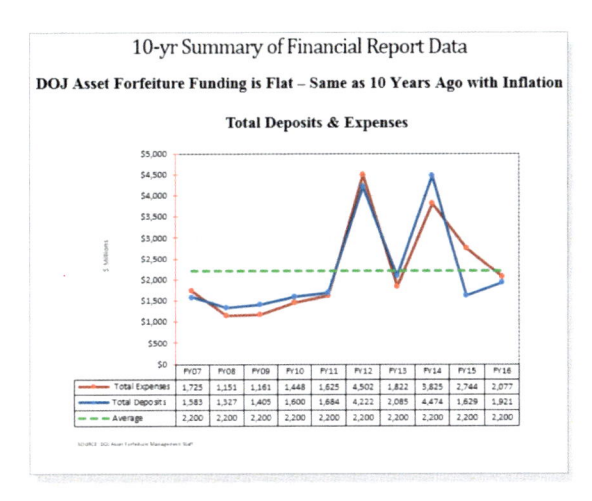

Fig 3.2: Not using their stunning color palette in their visualizations is a crime against data. Cuff 'em and stuff 'em!

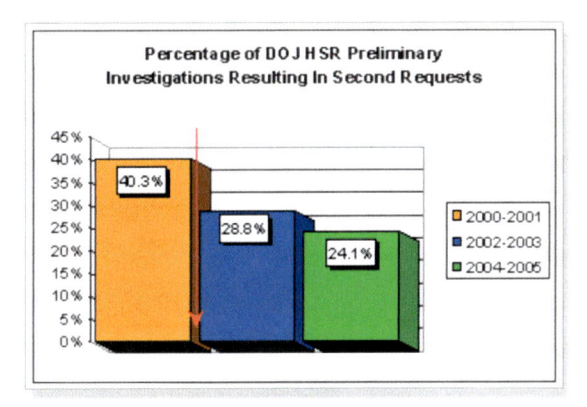

Fig 3.3: Just say no to 3-D charts—and neon! #chartmullet

They published an more egregious example of ugly data on their *Merger Review Process Initiative – Backgrounder* page **(Fig 3.3)**. You can view it on the DOJ website: **bit.ly/mdsepc-doj-chart-02**.

There are a number of issues with this visualization:

- It uses Excel's default theme.
- It doesn't use the DOJ's branded colors.
- It uses neon green. (Completely unrelated: my first car was neon green. And I had matching sunglasses.)
- It employs Excel's pretty horrible 3-D effect, which can be tougher to interpret and look more amateur than 2-D charts.

- There's no need for the axis labels because the values were brought into the columns using data labels.
- There's no need for gridlines for the same reason as above— especially not unnecessarily thick (and dark) lines.
- The border around the legend looks clunky.
- The red arrow is poorly formatted and confusing. It appears, from the preceding paragraph, it's supposed to denote when the Merger Review Process initiative was announced, but someone shouldn't need to read a paragraph of text to figure out why an annotation is on the chart.

I have to say, although I was a bit aghast at the lack of alignment with their powerful branding and basic chart design principles, I was encouraged by just how important these data visualization skills are. If the DOJ is struggling with making their data sexy, it should encourage you to see just how marketable these skills will make you!

THE MAKEOVER

We'll start with the easy fix. After accessing their branded colors and primary font using Chrome Developer Tools, I transformed their 3-D chart in just under three minutes **(Fig 3.4)**.

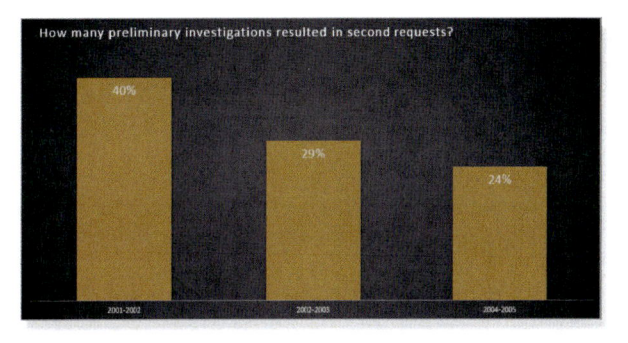

Fig 3.4: I transformed DOJ's 3-D chart using Excel's built-in styles and its colors and fonts.

💬 LEARN MORE See the Steps to Creating a Template section of the Set It.and Forget It chapter to find a brand's main fonts and colors.

39

CHAPTER 3 • DESIGN WITH ELEGANCE

Truth be told, I cheated on the background by using Excel's built-in chart styles (**Chart Tools ▸ Design ▸ Chart Styles ▸ Style 9**). Shh.

Tip: You can also access Excel's chart styles by selecting the paint brush icon that appears in the upper-right corner of a chart when you click anywhere inside a chart. The same options appear in a scrollable flyout menu.

I rarely use background colors for charts, especially ones with gradients, but it was a good opportunity to point out that you can use Excel's chart styles for inspirations, if nothing else. However, many of them are littered with chart junk, so proceed with caution. In reality though, the chart I would more typically create would be void of background color **(Fig 3.5)**.

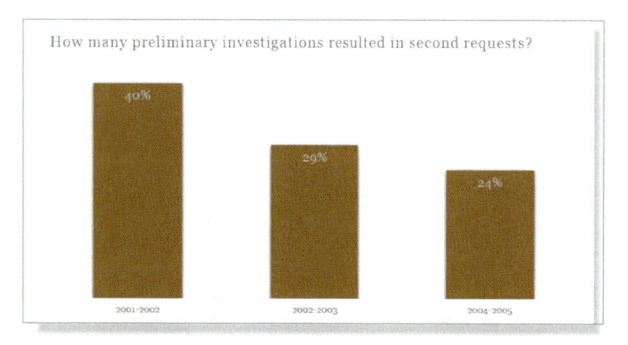

Fig 3.5: What the chart would look like with no background color.

I won't be using them in any of the visualizations in the Create Charts chapter. We'll build everything from scratch. Like how I used to make cookies until they came out with tubs of chocolate chip cookie dough.

Annnd I just lost my focus and am jonesing for cookies now....

For the second visualization (seriously, who uses data tables?), I built a few alternatives **(Fig 3.6)**.

The first one incorporates a mean line. Take note its subtlety compared to the one on the DOJ site. I also brought the data into the columns and bars to satisfy their lust for data tables.

The second chart took 12 minutes because it was a little more customized. But a diverging bar chart can be a great way to visualize metrics that are oppositional in nature. Sometimes recipients will spend more time digesting your data because they're mesmerized by the creativity of the presentation. And then they'll start wondering if they're paying you enough to prevent a competitor from poaching you.

But I digress.

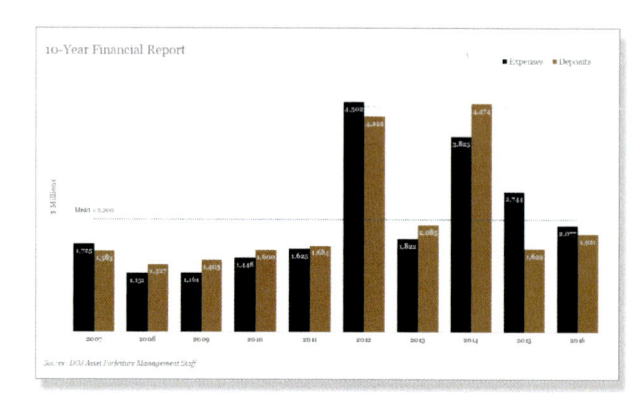

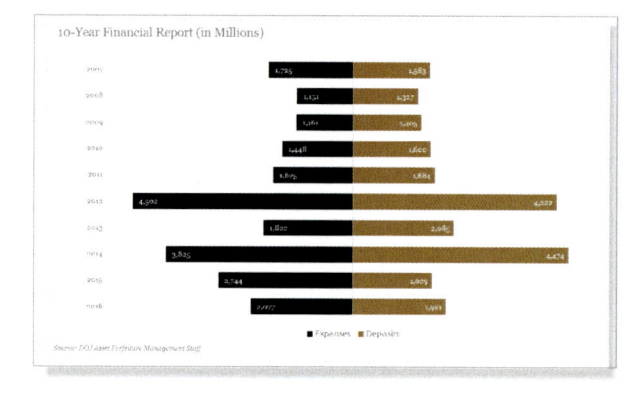

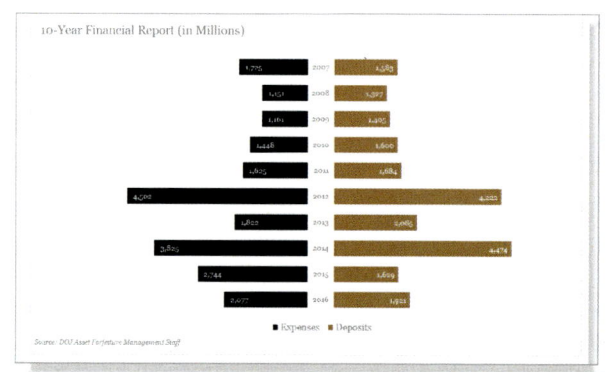

Fig 3.6: These are just a few streamlined alternatives to their line chart.

40

💬 **LEARN MORE** See the Diverging Bar Chart tutorial and the Diverging Stacked Bar Chart tutorial to learn how.

The last chart took me 40 minutes to figure out because it's super hacked. Excel doesn't give you the ability to position your data labels in between your diverging bar charts. And it really doesn't want to. But I walk you through how to accomplish this hack in the Diverging Bar Chart tutorial.

The moral of this story: Don't be a cautionary tale.

BE LIKE CHEVRON

Disclosure: I have no affiliation with Chevron, and they've never been a client. But I could hug the analyst responsible for branding their visualizations. Clearly a kindred spirit.

All the charts and visualizations I found on **chevron.com** reflected Chevron's branding. They were also elegant and relatively clutter free. When you're publishing your data publicly, you can't assume viewers have any kind of backstory. Consequently, annotations are key, and they use them effectively.

I'm not saying that they're perfect or that I endorse all their design choices. If we're being picky, there is a change I'd recommend for each of the charts:

✦ Sort the data before creating a donut chart from it. The colors also don't seem to follow a natural progression. **(Fig 3.7)**.

✦ Use a stacked bar chart over a stacked column chart because of the longer labels. I don't like making viewers tilt their heads to read labels **(Fig 3.8)**.

✦ You don't always need a vertical axis or gridlines

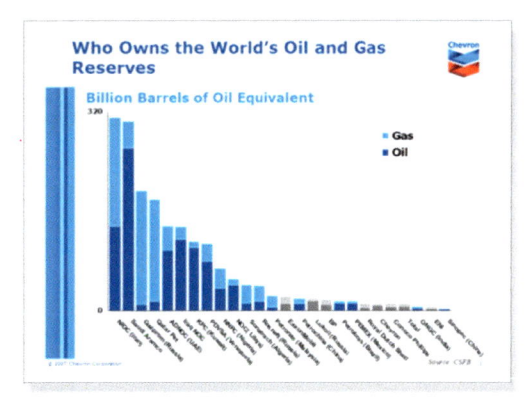

Fig 3.8: I'd use stacked bar charts.

with a line chart **(Fig 3.9)**. If the chart is simple, you can use markers with data labels embedded in them to pull the data into the line, as you'll see with Add Sizzle variation of the Line Chart tutorial **(Fig 3.10)**.

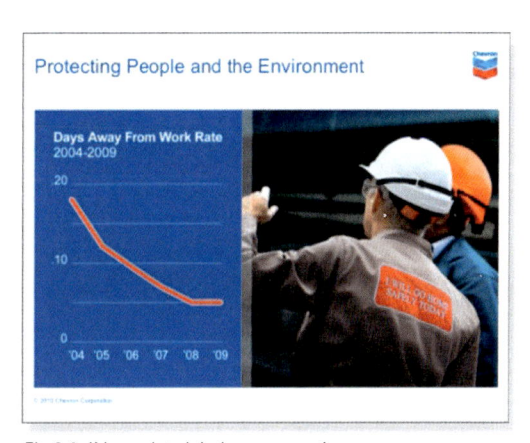

Fig 3.9: I'd use data labels over an axis.

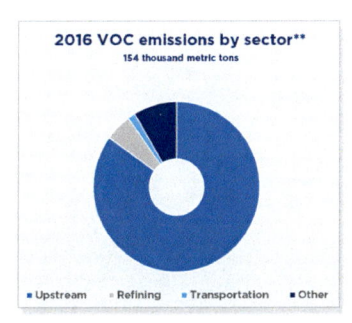

Fig 3.17 I'd sort the data first.

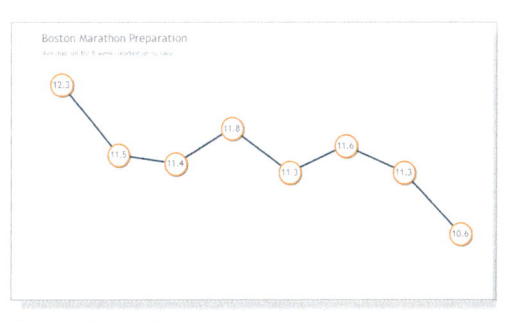

Fig 3.10: Or go full-bore minimalist.

CHAPTER 3 · DESIGN WITH ELEGANCE

- Use a thinner target line **(Fig 3.11)**.

That said, they are still a far cry from the DOJ's visualizations!

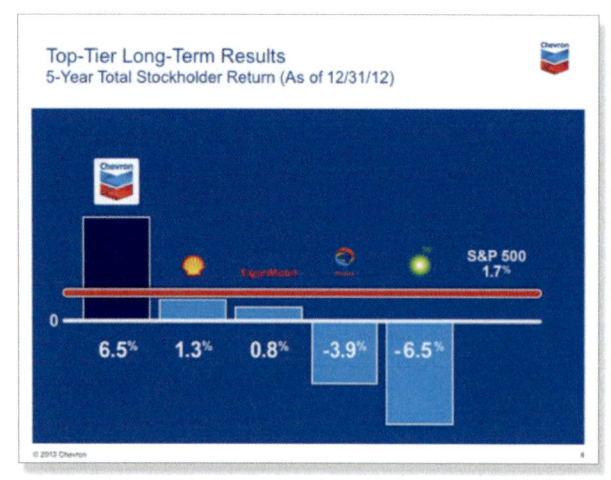

Fig 3.11: I'd user a thinner target line.

HOW TO ANALYZE AN ORGANIZATION'S BRANDING

When I'm analyzing a new client's branding I look a few places:

- *Their Homepage:* Most organizations put their best foot forward on their homepage of their website. I look at their accent colors and compare them to the logo. If your client is Google, finding their branded colors couldn't be easier. But you don't usually get quite that lucky.

- *Google Images:* I'll use Google Images specifically to search for images of charts and other visualizations. This query tells Google I want to search only the **chevron.com** domain, for example, for images about charts or data **(Fig 3.12)**.

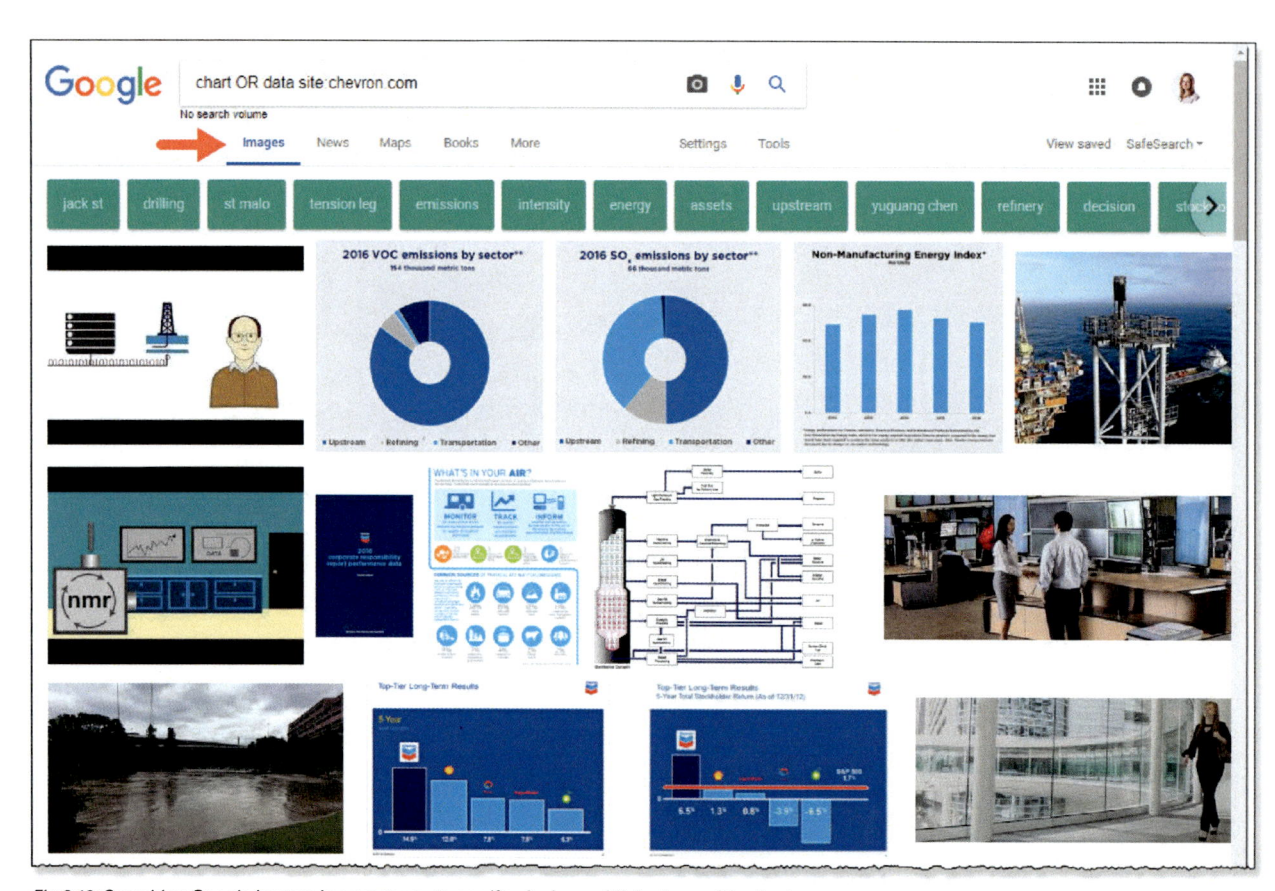

Fig 3.12: Searching Google Images is a great way to see if a site has published any of its charts.

42

- *Press Kit:* Most larger organizations have some kind of press kit they make available to journalists, and it's usually accessible from their About or Press page. I simply search Google and restrict my search to that domain, e.g., [press kit site:chevron.com]. (Remove the brackets.) If the site doesn't have a press kit per se, they usually have some kind of downloadable file. Organizations put a lot of effort into these kits, so you should find the best representation of their branding.
- *Investor Presentations:* Same as the press kit.
- *Downloadable PDFs:* Search Google, restricting your query to the organization's domain, but add [filetype:pdf] to your query. In this case, you don't need any keywords. You could search instead, for example, [filetype:pdf site:chevron.com]. Again, organizations usually brand PDFs they make available for download.

Okay, story time is over. Let's jump into some key techniques you'll come across in the Create Charts chapter.

HOW TO SUMMON FORMATTING OPTIONS

Since we'll be focusing on formatting options throughout this chapter, this is as good a time as any to share a formatting tip I use frequently. By frequently, I actually mean in every visualization I create, without exception.

It's Ctrl-1. It's the key to the formatting kingdom. Pressing Ctrl-1 will open the formatting options for whichever chart element you select. Microsoft has been moving away from pop-up dialogs in preference of a formatting pane that will open along the right edge of your workbook. You'll still come across some dialogs as you make your way through the tutorials in the book, but over time it will probably become like finding a Rhode Island license plate out in the wild.

Warning: Ctrl-1 doesn't work in some of the chart types that were introduced in Excel 2016. If it doesn't work, you will need to right-click and choose the format option from the contextual menu.

ALTERNATIVES TO KEYBOARD SHORTCUT

You have three alternatives to using the keyboard shortcut:

- *Right-click:* Right-click and choose **Format [element you right-clicked on]**. This is a reliable alternative, but sometimes I move the mouse when I right-click, causing me to misfire. Plus, I don't like wasting time looking for the format option in the contextual menu. But if you're a right-clicking maven, go for it!
- *Double-click:* Double-clicking on a chart element will oftentimes summon the corresponding formatting pane. However, at the time of writing, this technique doesn't work for data in cells or with objects, such as shapes or icons.
- *Ribbon:* When you select a chart element, you'll oftentimes see an additional menu appear in the ribbon (e.g., the **Chart Tools** or **Drawing Tools** menus). This new menu will serve up a **Format** tab that allows you to modify that element's formatting options.

ONE AND YOU'RE DONE

The first time you select a chart element to format it, you will need to open the formatting options using the keyboard shortcut or an alternative. Once the formatting pane (titled Format [Element You Selected]) is open, however, it will update whenever you select a new element in your chart.

Sometimes I forget about that workflow saver and press Ctrl-1 when I need to format a new element. It will update the format pane so no harm no foul. Over time you'll get more comfortable with just selecting a new element and looking eastward.

NOT JUST FOR CHARTS

Ctrl-1 doesn't just work in charts; it works with whatever you have selected. For example, if you select a cell or range of cells, you'll trigger the Format Cells dialog (no formatting pane for you, cells); if you're in a pivot table, you'll trigger pivot table formatting options; if you've selected a shape, you'll trigger the Format

CHAPTER 3 • DESIGN WITH ELEGANCE

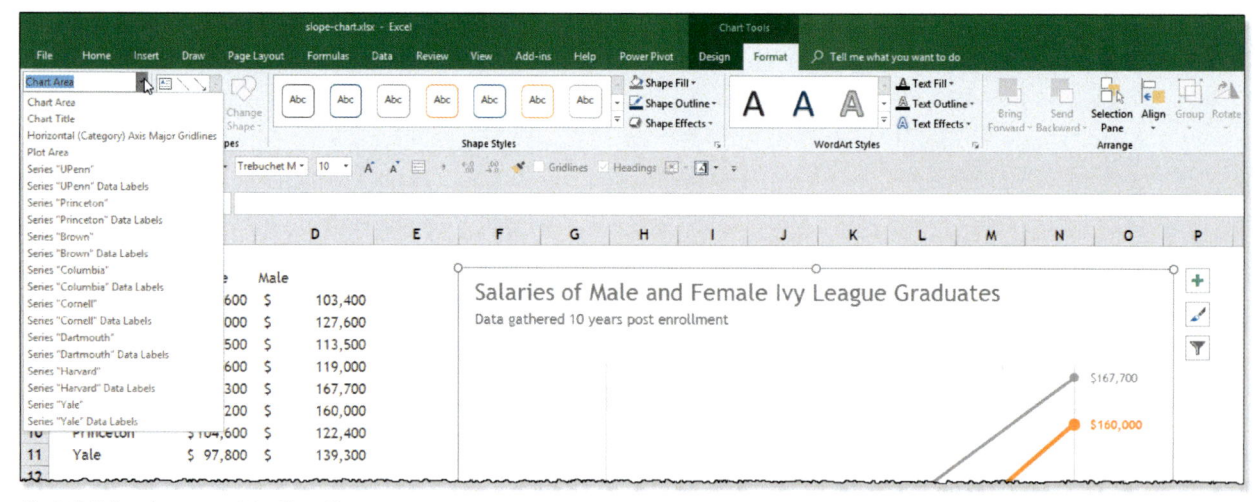

Fig 3.13: Take advantage of the Chart Elements menu to select hard-to-grab elements.

Shape pane. You get the idea. You can also get to formatting options through the contextual menu (by right-clicking). I just have a strong preference for Ctrl-1 because of its efficiency, so that's what I'll use throughout the tutorials in this book.

TIP FOR HARD-TO-REACH CHART ELEMENTS

There are times you may have a difficult time selecting chart elements. This might be because you've added a data series with a percentage (decimal) vis-à-vis a data series with whole numbers (e.g., website revenue vs. conversion rate or home sales vs. interest rate). When your data series is hugging the bottom of the ocean floor, it can difficult to select.

If this happens to you, click anywhere inside your chart. It can even be in the **Chart Area**. This will trigger the **Chart Tools** auxiliary menu. Open the **Format** tab and navigate to the drop-down menu in the far-left edge of the menu (under the **Current Selection** group). The visible item will be the chart element you clicked. Click the down-facing arrow to see all your chart elements and click to select one **(Fig 3.13)**.

I wish I had known about this tip earlier in my career as an analyst. It would have saved me some time and frustration zooming in on my data series until I could finally grab the elusive little sucker.

KEY FORMATTING TECHNIQUES

OVERALL AESTHETICS
Stick to Branded Colors

I can't underscore this point enough. Every time I see a website with a gorgeous color scheme publish visualizations with Excel's [hideous] default colors, a piece of me dies inside. It's exhausting.

I can hear your protests now: "Annie, my site only has two branded colors, but my visualization has seven categories…It takes a lot of time to manually change the colors in each of my charts…I don't know the RGB values for my branded colors…I don't even know what an RGB value is…."

Well, for one, if your site's palette doesn't include gray, you can throw that in. You'll need gray if you're building elegant visualizations, in my opinion. If you still need more, I highly recommend using tints of your branded

CHAPTER 3 • DESIGN WITH ELEGANCE

colors over foreign colors. Tints are the colors that result from adding white.

Aside: Shades are the colors that result from adding black, but I work more with tints than shades in building visualizations. As always, you do you!

Create Tints and Shades

Excel gives you the ability to create tints and shades with a slider in its **Colors** dialog that you can get to in a myriad of ways in Excel. Drag that slider up, and you create a tint; drag it down and you create a shade **(Fig 3.14)**.

I rarely use this option because I'm a stickler when it comes to precision and consistency, and I don't like that you can't titrate white or black into your color with greater specificity. I wish you could enter percentages. But since you can't, I'll outline a few options you can choose from.

First of all, to get to the **Colors** dialog, select **More Colors** from whatever pane or dialog you're in **(Fig 3.15)**. Excel gives you two tabs of color options to choose from **(Fig 3.16)**.

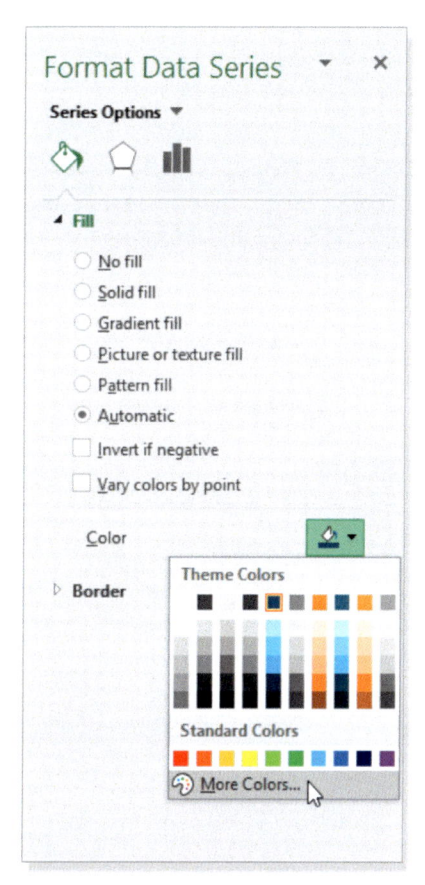

Fig 3.15: Select More Colors from any color picker menu.

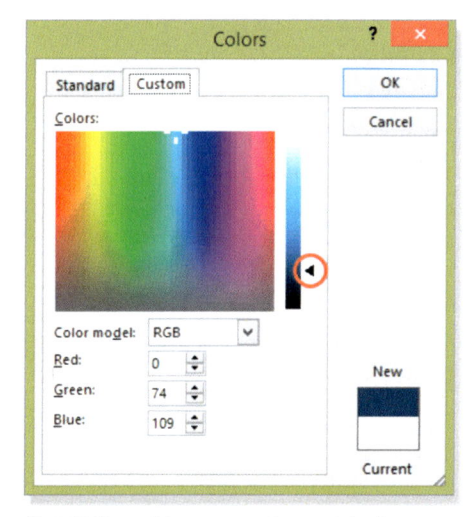

Fig 3.14: Drag slider to create tints and shades.

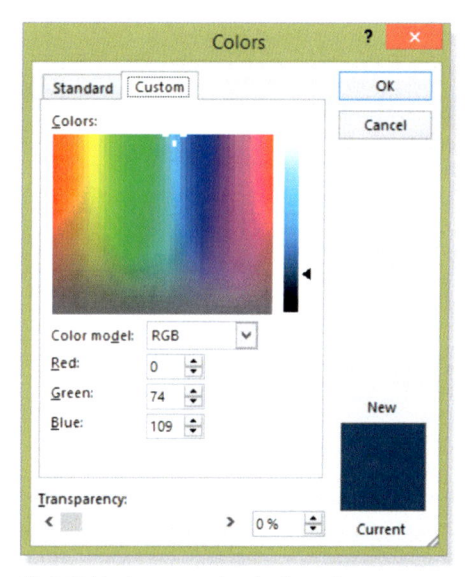

Fig 3.16: You have two tabs of color options to choose from.

45

CHAPTER 3 • DESIGN WITH ELEGANCE

Three Approaches

THE RESPONSIBLE (ADULT) OPTION

One great tool that gives you the tint values of your branded colors is the Hex Color Tool at **hexcolortool.com**. To get started, either enter the hexadecimal code associated with your color in the first field under **Enter a color** or enter the RGB values into the next field over **(Fig 3.17)**.

Fig 3.17: Find tint values for your theme colors using hexcolortool.com.

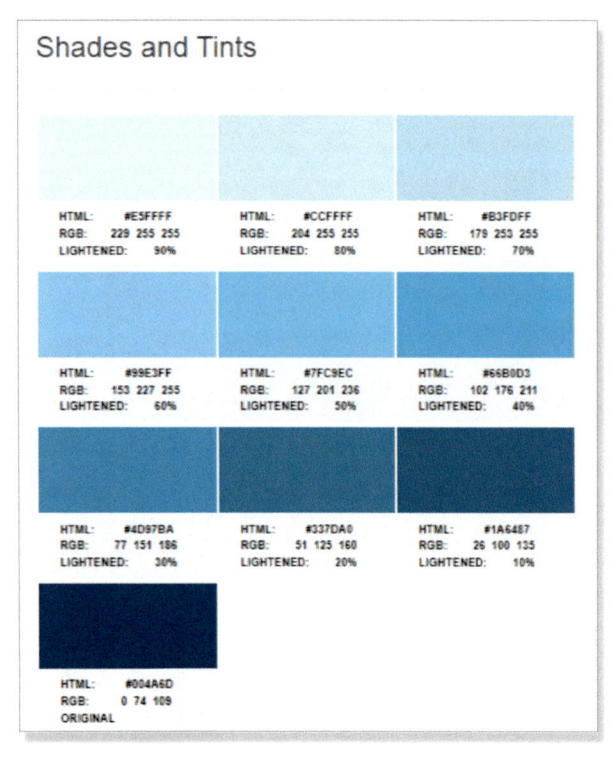

Fig 3.18: This is my favorite way to find tint colors.

If you only have the RGB values for your color, replace the R,G, and B values in the RGBA field. It's unlabeled, but the persistent tooltip inside the field reads 'rgba(x,y, z, 1)'. (Artistic Impression: 2.7) Replace x with your R value, y with your G value, and z with your B value. Leave the last parameter set to 1.

Tip: Only use the text boxes to the right of the sliders if you have HSL values. I kept trying to enter my RGB values and banged my head against that wall for a while before I figured that out.

Set what percent you want to lighten your branded color by using the drop-down menu, then click the **Lighten** button. You'll see your tints to the right. You can see examples from the dark blue color used in this book **(Fig 3.18)**.

I really like how the tool gives you both the hexadecimal and RGB values for each of the colors. Rinse and repeat for each of your branded colors. Store these somewhere because you'll need them for quick reference.

Note: Excel for Windows doesn't allow you to use hexadecimal codes to set colors. (It's the only advantage of Excel for Mac over the Windows version…Well, that and Mac users can choose colors using crayons with labels like *Bubblegum*, which is a big hit among Excel users under the age of 10, I'm sure.) Therefore, you'll want to record your RGB values. Also, dropping a hexadecimal value into Google (don't forget the #) will usually return the RGB values for that color. What a time to be alive!

MANUALLY ADJUST RGB VALUES

When finding incrementally lighter (or darker) colors to complement your branded palette, you can increment (or decrement) each of your RGB values by the same amount. For example, the orange I use for the visualizations in this book has the RGB value of 238, 132, 52. If I wanted to make it lighter, I could add 10 to each of those values, giving me an RGB value of 248, 142, 62.

CHAPTER 3 · DESIGN WITH ELEGANCE

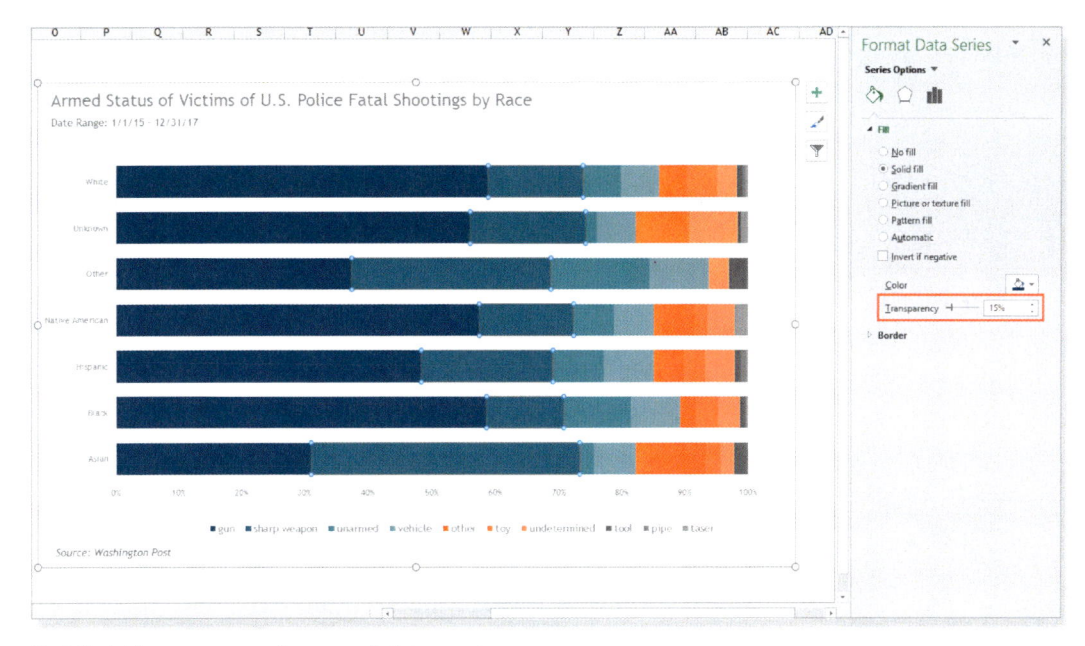

Fig 3.19: Adding transparency is a pauper's tint generator.

And if I wanted to make it darker, I could subtract 10 from each of those values, giving me an RGB value of 228, 122, 42.

THE KINDA NINJA (LAZY) OPTION
I probably shouldn't open myself up to scrutiny over this hack. Discretion is the better part of valor and all that. But sometimes it's not worth creating a full palette of colors to choose from, especially if you just need to create a tint or shade for an ad hoc chart (i.e., one you'll only use once and won't be templatized). In those cases, I create a poor man's palette on the fly by applying transparency to my branded colors using the same amount each time. I have an affinity for adding transparency in 8–15% intervals. Ymmv.

Whatever you choose, you should keep it the same for each color. Any time you have the option to add a fill color—whether to a chart element, shape, symbol, etc.—you can add transparency **(Fig 3.19)**. If your object is against a white background (which they are by default), you're adding white to your color.

I know, mind blown, right?

Warning: If anyone ever decides to throw your chart against a formatted background, your shenanigans will come to light. If that's a risk you can live with, go for it! With the skills you'll learn in this book, you should have no problem finding another job!

Design for Colorblind Users
According to the National Eye Institute, the most common type of hereditary color blindness is red-green. Also, men are more likely to be colorblind than women because the genes responsible for the most common, inherited color blindness are on the X chromosome. And since we chicks have twice as many X chromosomes (HOLLAH!), we're infinitely more qualified to decide if your tie clashes with your socks. (Okay, I might be taking liberties.) Learn more about what researchers actually said about color blindness: **bit.ly/mdsepc-color-blindness**.

The problem is many cultures, including the U.S., use greens and reds to express positive and negative metrics, respectively. You will need to decide, as an organization, how your chart designs will comply with usability best practices.

CHAPTER 3 • DESIGN WITH ELEGANCE

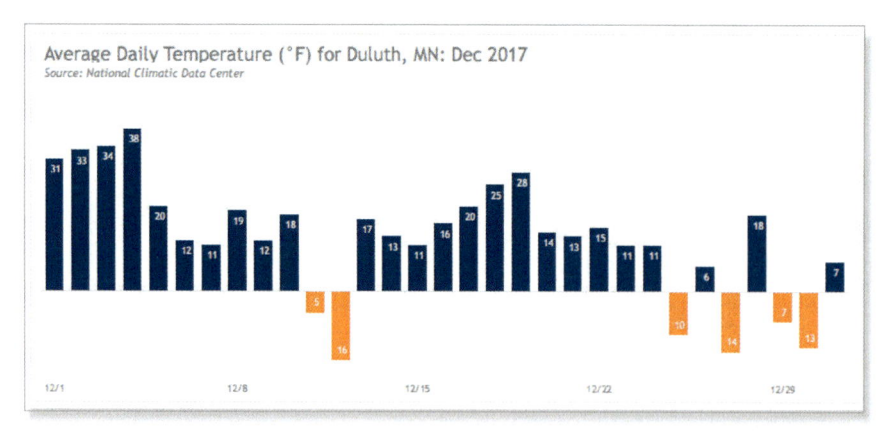

Fig 3.20: It's clear that the orange bars indicate negative metrics by their position and the data labels.

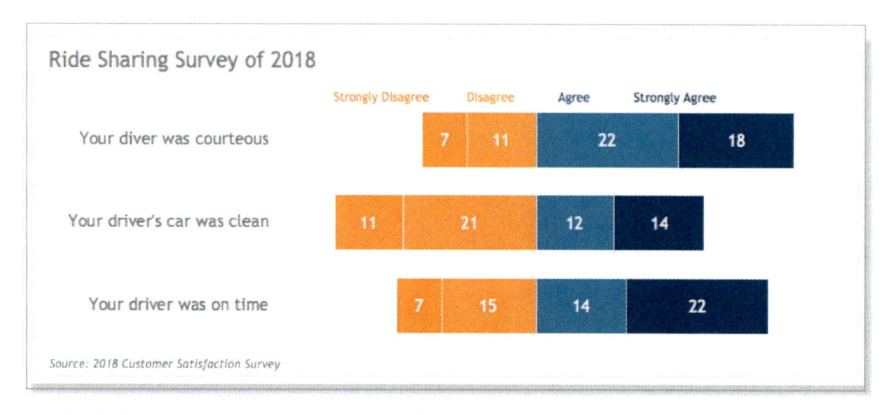

Fig 3.21: I didn't need to use red and green to indicate positive and negative survey responses.

I'm not a fan of using red and green in charts (unless they're your branded colors) and don't use them in any of the visualizations in this book. I restrict all my colors to the three branded colors in my palette. Instead, I rely on positioning, data labels, annotations, and chart titles to express negative metrics. For example, in the Baseline Delta Chart tutorial, you'll see that you can use positioning of your negative columns and data labels to clearly communicate negative values. I even customized the number formatting of the data labels to remove the negative signs. If the bars dip below 0 and are a different color, you don't need to also have a negative sign **(Fig 3.20)**.

💬 **LEARN MORE** See the Customize Number Formatting section of the Design with Elegance chapter to learn how to finesse your number formatting.

Similarly, I use position and tints to indicate different levels of positive and negative survey responses **(Fig 3.21)**.

To see how people with different types of color blindness would see your visualization, you can use a tool like the Coblis Colorblindness Simulator (available at **bit.ly/mdsepc-cb-test**). Just take a screenshot of your chart and upload it. You might be surprised by what you see!

Point Out Actionable Insights

Storytelling with data is all the rage. When you're using data to tell a story, your chart title serves as a headline of sorts. The rule I follow, in general, is if I'm doing ad hoc analysis, I gravitate more toward headline-esque titles, e.g., "Social Traffic is En Fuego" or "Sales in the Southeastern Region Have Stalled." Then I'll use a subtitle to indicate what I'm charting.

However, if I'm building out a dashboard—especially one that updates automatically using an API connection—I tend to flip that approach on its axis (data analogies for days!) and use the title to indicate what's being charted and a subtitle or text box to tell the story. This is just my approach. Over time, you will need to figure out what works for you.

The important takeaway is that you always point out what you want your viewers to take away from the visualization. You might think it's glaringly obvious, but you've been all up in the data. Your viewers haven't. Therefore, be cognizant of that and err on the side of over-communication, even if these insights

CHAPTER 3 • DESIGN WITH ELEGANCE

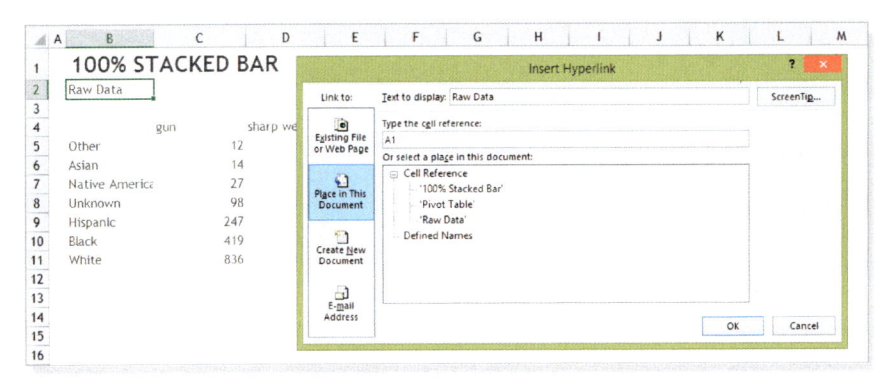

Fig 3.22: Link to other cells/worksheets to make your document easy to navigate.

are extracted and published in a separate worksheet. I recommend linking to the visualization you reference if you tease out findings in a separate worksheet. To link to another worksheet in your workbook, take the following steps: press Ctrl-K ▶ select **Place in this Document** in the **Insert Hyperlink** dialog ▶ navigate to your cell (or select your worksheet in the preview pane if you just want to link to cell A1 in that sheet) ▶ click **OK (Fig 3.22)**.

Customize Number Formatting

There are times in this book you will need to use custom number formatting. You can access number formatting options by selecting a value in a cell(s) or chart element

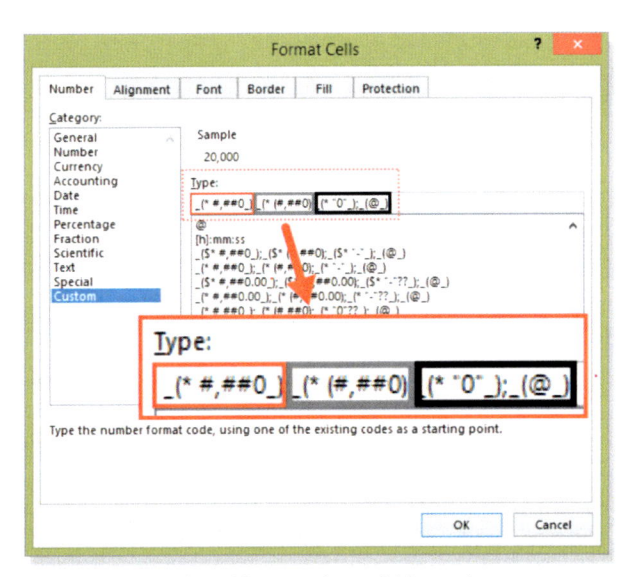

Fig 3.23: From L to R: positive, negative, and 0 formatting.

(e.g., data label or axis) and pressing Ctrl-1. Excel's custom number formatting options are robust, and we'll barely scratch the surface in this book.

The main thing to keep in mind is that Excel separates each of the codes with semicolons **(Fig 3.23)**. The first set of code determines the formatting for positive numbers (red outline), the second set determines the formatting for negative numbers (gray outline), and the third set determines the formatting for 0 (black outline).

The simplest way to customize number formatting is to tweak the code Excel generates, as we'll see in examples later in this section.

Use Cases
Working with Accounting Number Format

I use Excel's accounting number format quite a bit because revenue is oftentimes the bottom-line metric organizations care most about. But there are a couple customizations I almost always opt for:

- *Remove decimals:* In my line of work, I'm usually working with revenue data that's aggregated. So, when dealing with large revenue amounts, the decimals are unnecessary and hence become noise. I remove them using the **Decrease Decimal** button under **Home ▶ Number** (last one on the right).

- *Replace hyphen with 0:* Excel's default accounting number format uses hyphens for the value 0. So, instead of a 0 in the bottom-left corner of a chart, you'll see a hyphen, which I *really* don't like. It reminds me of the "Which one of these is not like the others" game on *Sesame Street*. I *always* pull up the formatting options and change the hyphen to a 0. The easiest way to switch them out is to apply the number format you want from the

49

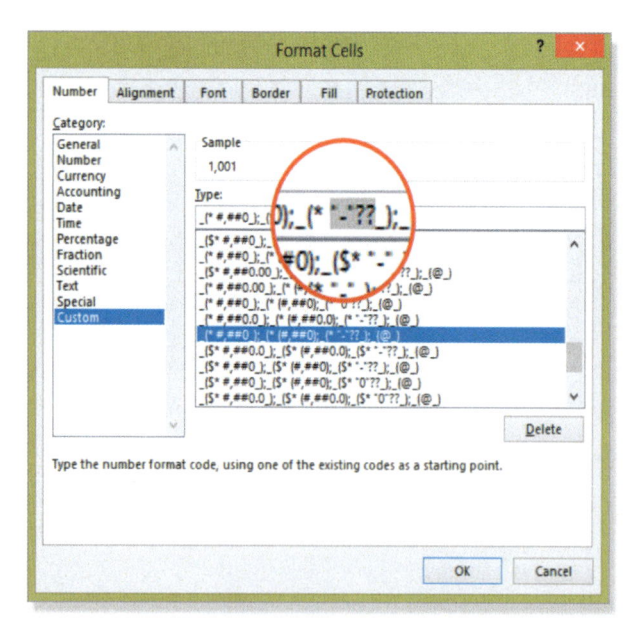

Fig 3.24: Evict those unsightly hyphens without notice, if you're charting accounting data.

number formatting buttons under the **Home** tab, then press Ctrl-1 to open the **Format Cells** dialog. Under the **Number** tab, set **Category** to **Number** and carefully select "–"?? in the 0 formatting code (i.e., after the second semicolon) and replace it with a 0 **(Fig 3.24)**. Be careful not to remove the space before the first quotation mark, or your 0 won't align with your positive and negative values.

* *Remove negative formatting:* You'll see an example of where I do this in the Baseline Delta Chart tutorial.

* *Assign special formatting:* Although we don't use this strategy in the book because it falls more under data entry, you could format numbers like social security numbers (in the U.S.) or product IDs. For social security numbers, you would set it to ###-##-####. This way, if a user needs to enter a column of social security numbers, he won't need to include the hyphens.

Adding a Thousands Separator

I use thousands separators (essentially, commas for numbers) for any numbers > 999. But I also tuck in most blouses, blow dry cold necklaces, and fluff my pillows before I can use them. So there's that.

Some [wrong] people eschew thousands separators until they hit five digits. But whatever you decide for yourself, when you need them, you can access the **Comma Style** button by navigating to **Home ▸ Number**. This style adds sophisticated custom number formatting code that helps your numbers align. If I'm not working with currency (which uses the same highly customized formatting, only with currency symbols), I'll frequently apply this formatting just to trigger these finessed formatting options, then go in and customize it to take it the rest of the way.

Warning: Applying the thousands separator also switches out 0s with hyphens, so if you prefer a 0 over a hyphen, you'll need to customize the format, as I demonstrated in the previous section.

Employ the TEXT function

We will also use custom number formatting code inside TEXT functions to trick out our visualizations with chart elements, such as chart titles and annotations that update dynamically—as you'll see in the Bar Chart tutorial—or reference a cell containing text and a calculated value—as you'll see in the Scatter Plot tutorial, where we pull the correlation coefficient into our chart.

Simplify Large Numbers

Having very large numbers in a chart axis can make the axis cumbersome, so I'll oftentimes express values in thousands or millions. To do this, first apply the thousands separator to format your data, then take the steps below, depending on your values.

Alternatively, you can use the **Display options** feature in the **Format Axis** pane to indicate that you've simplified larger numbers in your axis **(Fig 3.25)**.

Thousands

If you want to simplify thousands, you would carefully remove the highlighted sections from the **Format Code** field **(Fig 3.26)**.

Take note that I also customized the code for 0 in the screenshot above by replacing "-"?? with 0. If you're working with currency, you would first apply accounting number formatting to your data, then take the same steps **(Fig 3.27)**. The accompanying screenshot shows what it would look like when you're finished.

And if you want a decimal place, don't remove the decimals before customizing your number formatting.

You can see an example of simplifying thousands values in the Line Chart tutorial **(Fig 3.28)**.

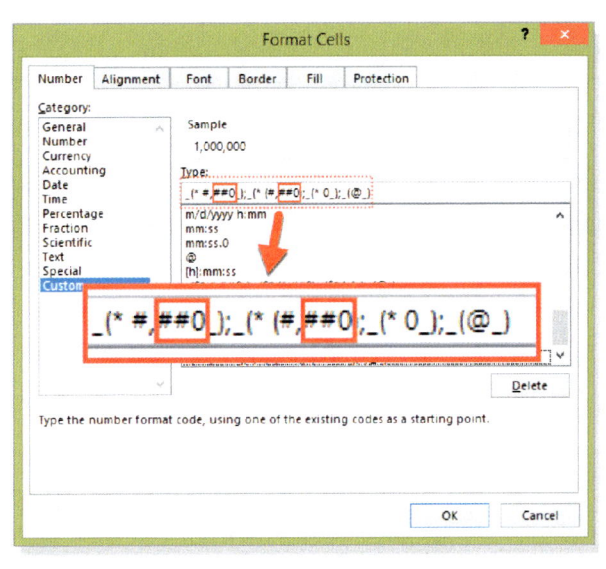

Fig 3.26: Simplify large numbers using the format code.

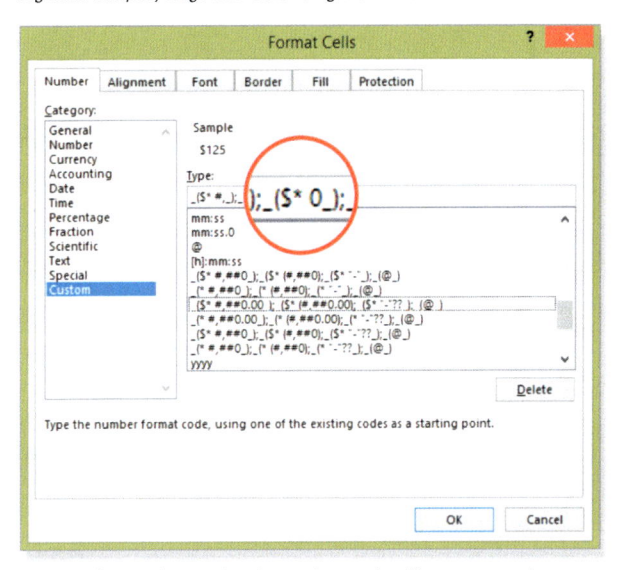

Fig 3.27: Customize number formatting to simplify currency values.

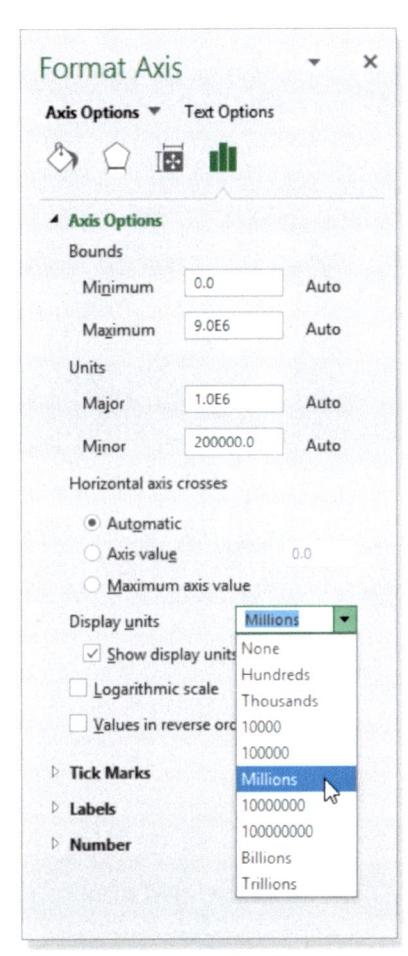

Fig 3.25: You can also use the Display units feature to simplify larger numbers.

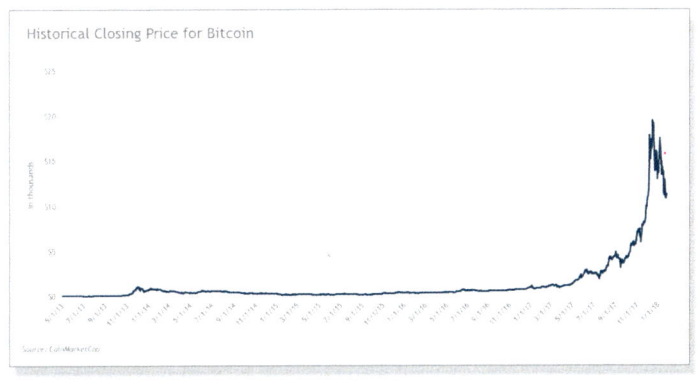

Fig 3.28: Axis labels are shown in thousands with an axis title indicating the truncation.

CHAPTER 3 · DESIGN WITH ELEGANCE

Millions

If you want to express numbers in millions, simply add another comma **(Fig 3.29)**. This will make your first comma sliiide to the left.

If you simplify large numbers in your axis, be sure to communicate that somewhere in the chart. A few options you have are in the chart title (if the title isn't too long), a subtitle, a text box, or an axis label.

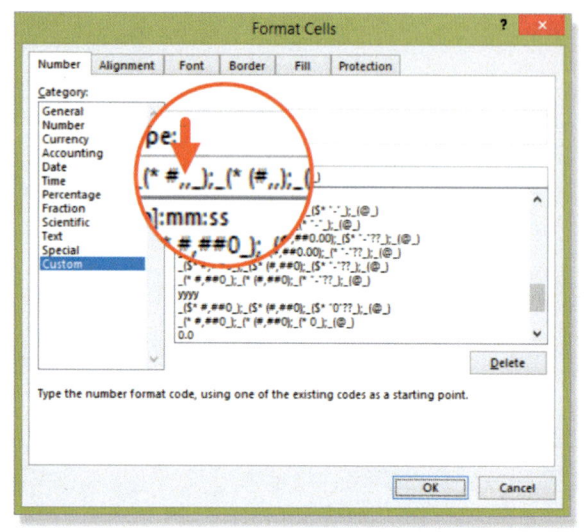

Fig 3.29: Throw in an extra comma to simplify by millions.

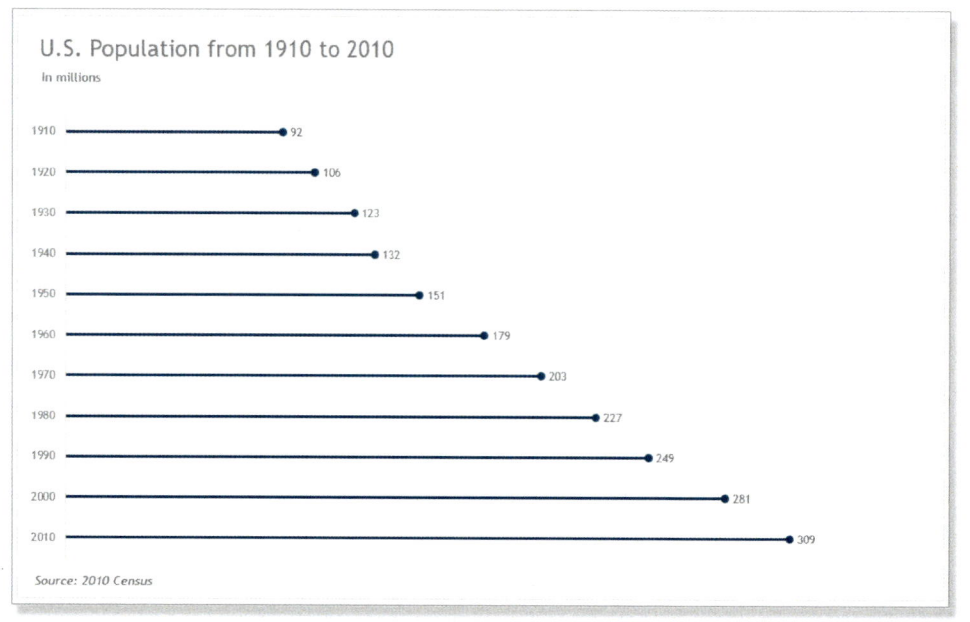

Fig 3.30: Values in data labels are simplified to millions.

You can see an example of simplifying millions values in the Lollipop Graph tutorial **(Fig 3.30)**.

Reformat Negative Numbers

There are times you'll use negative values to position your data to the left of the vertical axis, as you'll see in the Diverging Bar Chart tutorial. To do this, you'll just make sure the code for negative numbers (after the first semicolon) matches the code for positive numbers. This usually means removing parentheses or the negative sign.

Fix Alignment

I don't know about you, but when I was a kid we would randomly pretend that certain areas of the playground—usually under the jungle gym—were filled with lava. Unless you wanted to be screamed at by a mob of kids doped up on sugar-laden school lunches, it was in your best interest to avoid these pop-up lava pits.

Where am I going with this?

The first time I added data labels to markers in a chart, I was perplexed by the alignment and would wonder to myself, *Why does it look like the left side of the marker is a lava pit from years gone by?* It was because the number formatting was programmed to add a space in front of positive numbers **(Fig 3.31)**.

The "_(" at the very front of the number code is the drama queen. The underscore tells

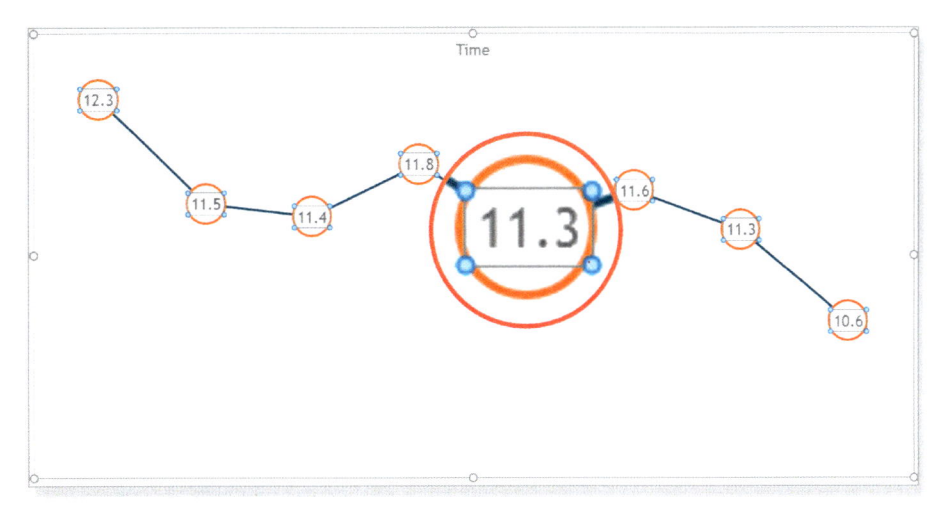

CHAPTER 3 • DESIGN WITH ELEGANCE

Fig 3.31: Your number formatting may push the numbers in the data labels a hair to the right of center.

💬 **LEARN MORE** See the Customize Number Formatting section of this chapter to learn how to customize your number formats.

Excel to leave a space the width of whatever character comes after it—in our case, a parenthesis. This allows positive numbers to line up with negative numbers that are formatted with parentheses. For charts with both positive and negative values in your data labels (which you'll see in the Baseline Delta Chart tutorial), you'll need to remove this dynamic duo for both positive and negative values.

To resolve this issue, select your data labels by clicking one of them, then press Ctrl-1 to open the **Format Data Labels** pane. Next, open the **Label Options** tab. Under **Number**, deselect **Linked to source**, which will reveal the number formatting code in the **Format Code** field.

You can delete the offending underscore-parenthesis teams for the positive and negative number codes, and add your custom code into the **Type** field by clicking the **Add** button **(Fig 3.32)**.

Alternatively, you could simplify your code to what you know you'll need, which is what I did to clean up the data labels in my tricked-out line chart **(Fig 3.33)**. I used:

0.0;-0.0;0

Just keep in mind this simplified format won't be as robust as what we've been using.

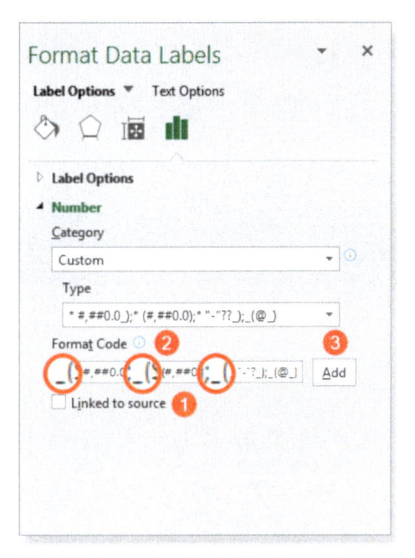

Fig 3.32: Removing the "_(" in the number code may fix the funky alignment.

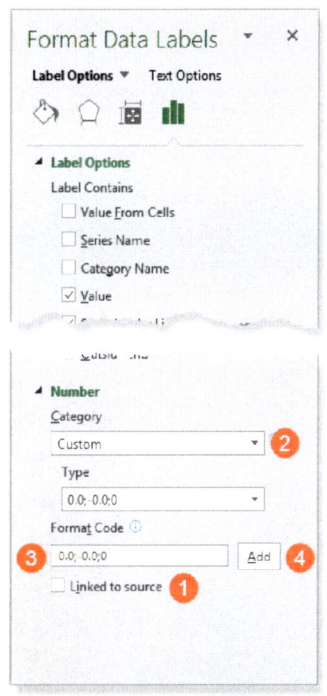

Fig 3.33: You can also simplify your number formatting to realign text.

53

CHAPTER 3 · DESIGN WITH ELEGANCE

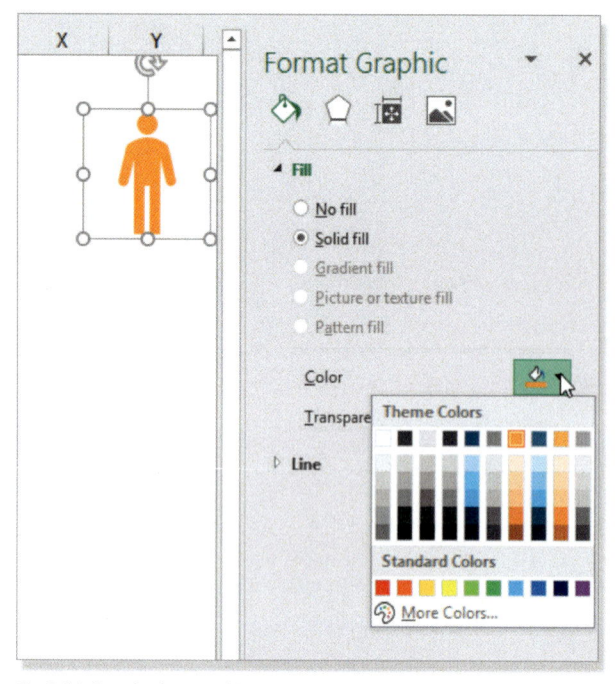

Fig 3.34: Recolor icons using your branded colors.

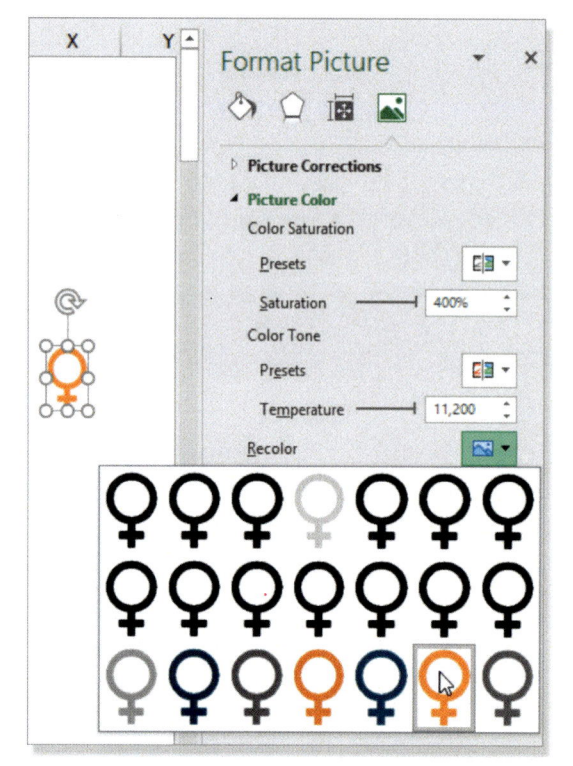

Fig 3.35: Recolor images to augment your visualizations to align with your branding.

Colorize Icons

If you use an icon from Excel (as you'll see in the Pictograph tutorial), you can easily change its color. First, select it, then press Ctrl-1 to open the **Format Graphic** pane.

Open the **Fill & Line** tab and set **Fill** to **Solid fill**, then set **Fill Color** to your preferred color **(Fig 3.34)**.

> **LEARN MORE** See the Stick to Branded Colors section of this chapter to learn about the importance of restricting your design choices to your branded colors.

If you're using an image from elsewhere (as you'll see in the Dumbbell Plot tutorial), you can change its color to align with your branding. First, select the image, then press Ctrl-1 to open the **Format Picture** pane, then open the **Picture** tab. Click the **Recolor** drop-down menu and choose from your branded colors **(Fig 3.35)**.

If you're not using a monochromatic image (your lust for color may be your undoing!), you may need to tinker with more advanced settings, which you can get to from the **Picture Color** section of the **Picture** tab in the **Format Picture** pane **(Fig 3.36)**.

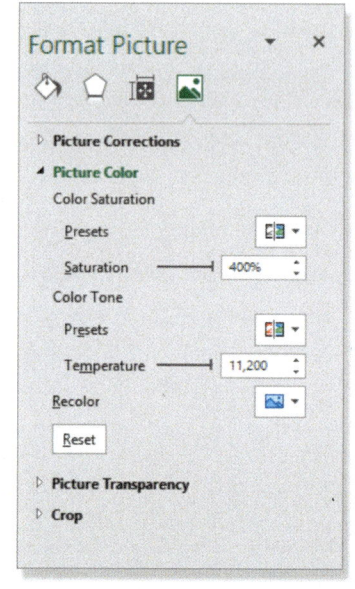

Fig 3.36: Additional settings are available in the Format Picture pane.

54

Paint Formats

Sometimes you go to great lengths to format your data, just to do it all over again for another dataset. A great tool I use all the time is Excel's **Format Painter**. You can access it by navigating to **Home ▶ Clipboard ▶ Format Painter**.

If you select one cell, Excel will use that cell's formatting to paint another cell or range of cells. If you select several columns of data with a variety of formats (number, fill, border, conditional formatting, etc.) and paint over a range of cells with the same number of columns (or rows), Excel will apply the formatting from the first dataset's columns to the second dataset.

You can also use the Format Painter to copy column and row widths. To do this, select the column or row heading, click the **Format Painter**, and apply to another column(s) or row(s).

Warning: If you use the Format Painter to set column width or row height, Excel will apply any other formatting that was applied to the column/row you sampled. For this reason, I'll usually only sample blank columns/rows. If you need to sample a column that is formatted, you'll want to use the Paste Special option.

LEARN MORE See the Copy Column Widths section of this chapter to learn how to replicate column widths.

Finally, double-clicking the **Format Painter** will allow you to use it more than once. When you're finished, click the **Format Painter** again or press the Esc key to send it back into retirement.

Where this can be especially handy is with visualizations that require advanced formatting, such as the heatmap, which you'll see in the Heatmap tutorial. Because a heatmap is a table and not a chart, you can't save it as a chart template. However, you can save it in your workbook of visualizations you can't save as templates, then apply its formatting to a new dataset with similar data, saving you a significant amount of time. And the Format Painter works across workbooks, meaning you can load it up with formats from one workbook, switch to another workbook, and apply your formatting.

CHART AESTHETICS
Add Data Labels

In my quest to reduce clutter, I'm always looking for the opportunity to get rid of as many elements in my charts as possible. One of the top elements on my radar is the value axis. The best way to eliminate the need for the value axis is to bring your values into the chart via data labels.

How to Add

To add data labels, click anywhere in your chart and click the plus sign that appears in the upper-right corner of your chart to open the **Chart Elements** flyout menu. Click the right-facing arrow next to **Data Labels** and choose **More Options** to open the **Format Data Labels** pane. Under the **Label Options** tab you'll be able to set what you want to appear in your data labels under **Labels Contain** and your data labels' position under **Label Position**.

If you don't want your data labels to appear outside the bars—like they locked their keys in the car and are stranded until AAA shows up—click the right-facing arrow next to **Data Labels** and choose **Inside End**. That's my go-to setting for most charts, unless a bar or column is too short.

If I'm adding them to markers, I'll typically blow up the marker and position them in the center. You can see examples of these in the Line Chart tutorial and the Dumbbell Plot tutorial.

Making Room

If you're working with columns or bars and they're not wide enough to accommodate your labels, you can fatten them up. You'll learn how in the Adjust Chart Series Spacing section in this chapter.

CHAPTER 3 • DESIGN WITH ELEGANCE

You also have the option to customize the number formatting of your data labels in the **Number** section under the **Label Options** tab of the **Format Data Labels** pane. To customize, enter your code in the **Format Code** field and click the **Add** button to the right.

> 💬 **LEARN MORE** See the Customize Number Formatting section of this chapter to learn how to finesse your number formatting.

After you set the formatting for one series, you don't need to continue doing this. When you apply it to another data series, you can set **Category** to **Custom**. Then you will see your custom code in the **Type** dropdown. Simply select it to apply it. You can see an example of using custom number formatting in the Dumbbell Plot tutorial **(Fig 3.37)**.

Format Painter Trick
If you need to select each of your data series individually to format your data series (as we do in the Dumbbell Plot tutorial), this could get repetitive. In those cases, I format the data labels in the first series, then double-click on the **Format Painter** (under **Home** ▸ **Clipboard**) and click a data label from each of the other data series to paint them. Double-clicking the **Format Painter** will allow you to use it more than once.

> 💬 **LEARN MORE** See the Paint Formats section of this chapter to learn how to format using the Format Painter.

Note: Unfortunately, you can't use the Format Painter to apply custom number formatting, only simple elements like font type, color, and size.

Reassign Values
Sometimes you might need to point a data label to a different cell in your data. Excel for PC has a great option in the **Format Data Labels** pane, under the **Label Options** tab (under **Label Options**) called **Value from Cells (Fig 3.38)**. Simply click the **Select Range** button and click-and-drag over the data you want to use in your data labels **(Fig 3.39)**.

At the time of writing, Excel for Mac doesn't offer this feature, so if you use both and don't see it, you're not crazy. At least not in this regard!

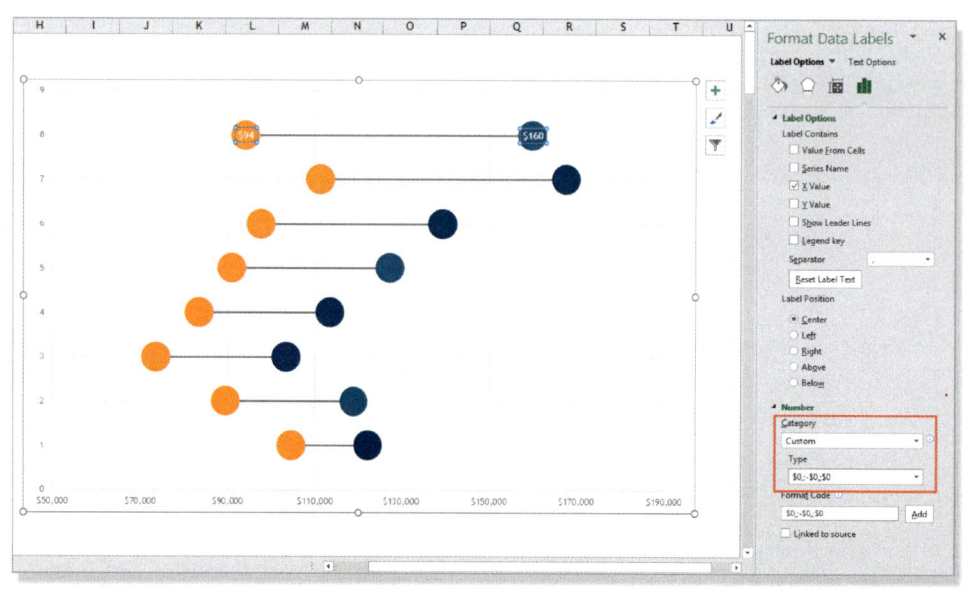

Fig 3.37: Choose your custom number formatting from the Type drop-down menu.

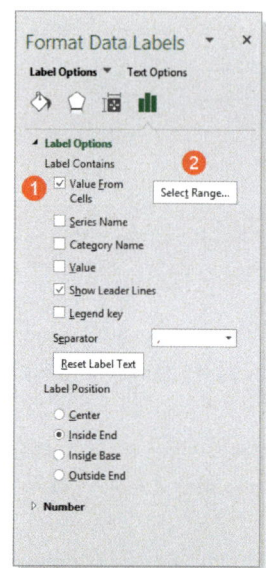

Fig 3.38: You can reassign data label values with the Values from Cells option.

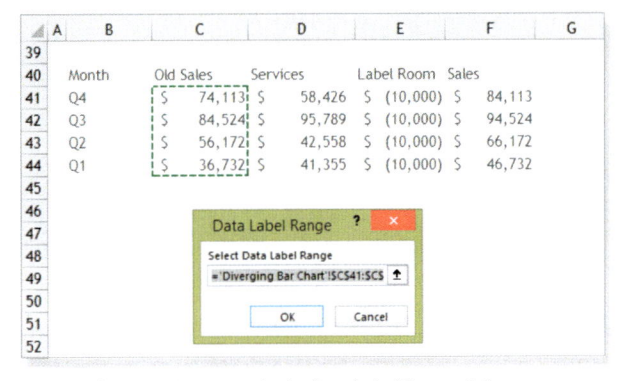

Fig 3.39: Set new data range in the Data Label Range dialog.

Hopefully the Excel for Mac team will bring this feature to the Mac version one day. But until then, you'll need to select each data label individually and point it to the new cell with a formula.

Ditch Gridlines

You don't need gridlines in your worksheets, especially if you format your data as tables. To rid your worksheets of them, select any cell in your worksheet, then navigate to **View ▸ Show** and deselect **Gridlines**.

You can rid your charts of them by selecting one and pressing Delete or lighten them by pressing Ctrl-1 and either increasing **Transparency** or setting **Color** to a lighter color in the **Format Major Gridlines** pane **(Fig 3.40)**.

💬 **LEARN MORE** See the Formatted Tables section of the Massage Your Data chapter to learn how to customize Excel's tables to meet your needs and branding.

Modify Chart Title

Excel will pull the metric you're charting into your chart title, but a sole metric is rarely adequate. You want your chart title to be descriptive. You may also want to add a subtitle or annotation with more info about the data source, methodology, etc.

You can also create a title that dynamically updates with new info, such as the current month. To do this, you need to create a formula in a cell that concatenates (i.e., glues together) the different pieces that will become your title. These formulas can get complicated fast, but you can use helper cells to build your title piece by piece and then stitch them together in a cell with your final formula. Learn more about using helper columns in my post on the Search Engine Land website: **bit.ly/mdsepc-helper-columns**.

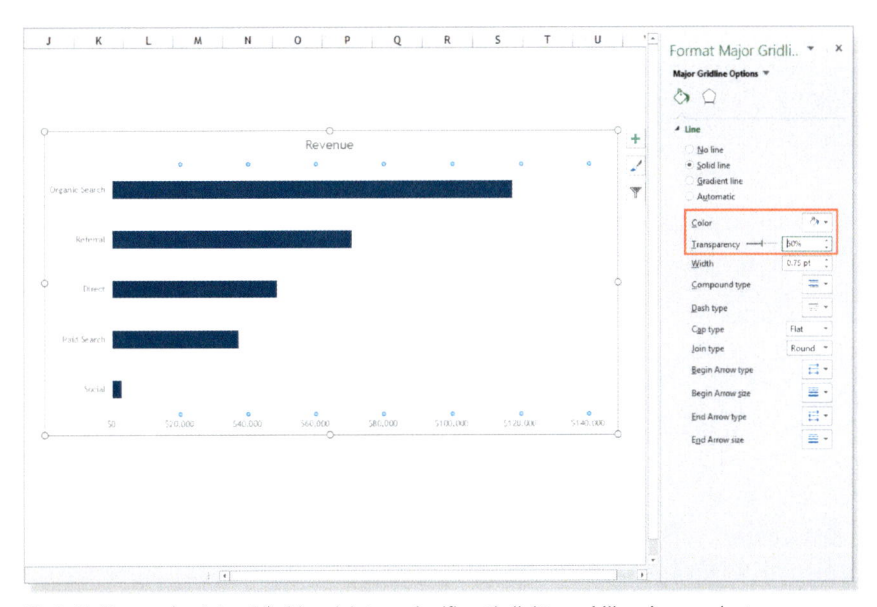

Fig 3.40: You can (and should) either delete or significantly lighten gridlines in your charts.

CHAPTER 3 · DESIGN WITH ELEGANCE

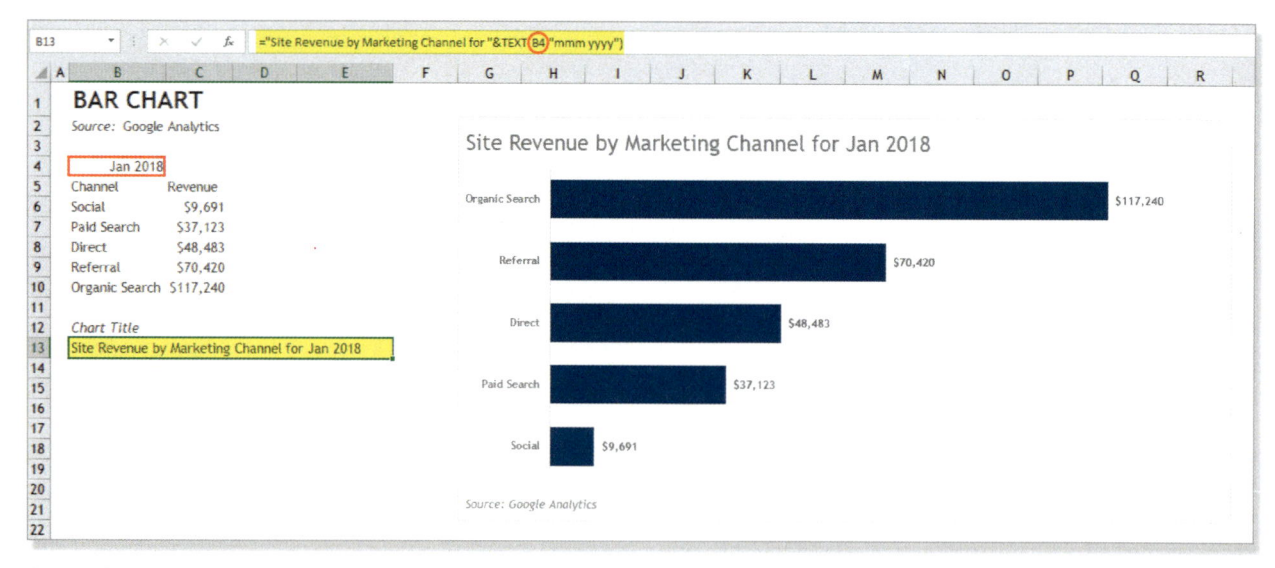

Fig 3.41: Create dynamic titles with the TEXT function.

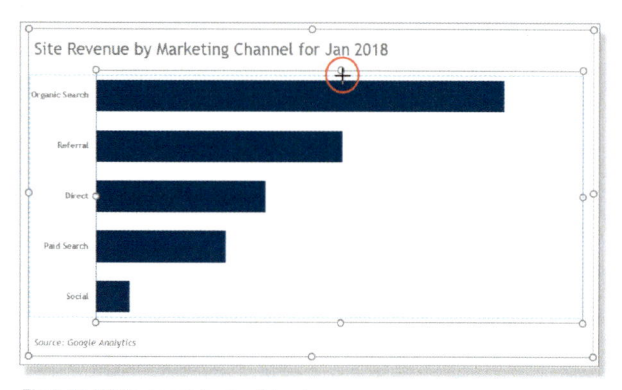

Fig 3.42: White space is your friend.

When combining text and a value that updates in a cell, you'll usually need to wrap the value reference in a TEXT function. The TEXT function lets you change the way a number appears by applying formatting to it with number format codes **(Fig 3.41)**.

When you're finished formulating your chart title in your helper cell, select your chart title chart element, then enter an equal sign in the **Formula Bar** and select the helper cell to link them.

Give Chart Title Space

I personally don't like how Excel's default format is to have the chart title hug the top boundary of the **Plot Area** in a chart. I like to add a little white space. If you want to add white space, click inside the **Plot Area** and drag the sizing handle positioned at the top of the **Plot Area** down a bit **(Fig 3.42)**.

Add Annotations

You can add additional information, such as source for your data, methodology, observations, and even additional data by inserting a text box.

How to Add

To add a text box to your chart, click anywhere inside the **Chart Area** and navigate to **Insert ▸ Text ▸ Text Box**. Drag toward the bottom-right corner of your worksheet until it's about the size you need. You may need to make room for your text box it by selecting the **Chart Area** and/or **Plot Area** and dragging the sizing handles to taste.

Formatting Text Boxes

The default formatting of text boxes is pretty clunky. There are a few steps I recommend before taking them out in public. To open the formatting options for a text box, select it, then press Ctrl-1 to open the **Format Shape** pane.

58

CHAPTER 3 • DESIGN WITH ELEGANCE

Fix Funky Margins

To adjust the settings, select the outside edge of your text box and select the **Size & Properties** tab of the **Format Shape** pane. In the **Text Box** section, select **Resize shape to fit text**. This will ensure your text doesn't break the boundaries of your text box (and should be the default setting, in my opinion). Next, set **Top margin** and **Bottom margin** to .1 to match the **Left margin** and **Right margin (Fig 3.43)**. These options, when used in concert, ensure that your text will be centered in the text box.

Format Border

The default text box border is unremarkable. Experiment with different settings in the **Format Shape** pane **(Fig 3.44)**. I formatted the text boxes in the book using the following settings under **Fill & Line**:

- **Line**: Solid line
- **Line Color**: blue
- **Width**: 2.25 pt
- **Dash type**: Round Dot

I also added a drop shadow to give my text box a little depth **(Fig 3.45)**. Under the **Effects** tab, I set **Shadow Presets** to **Outer** ▶ **Offset: Bottom Right** (the first option).

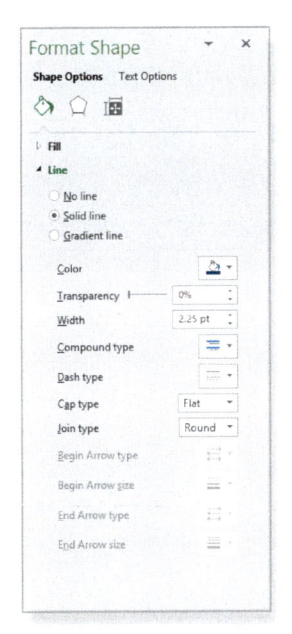

Fig 3.44: Soup up your text box with some Boujee borders.

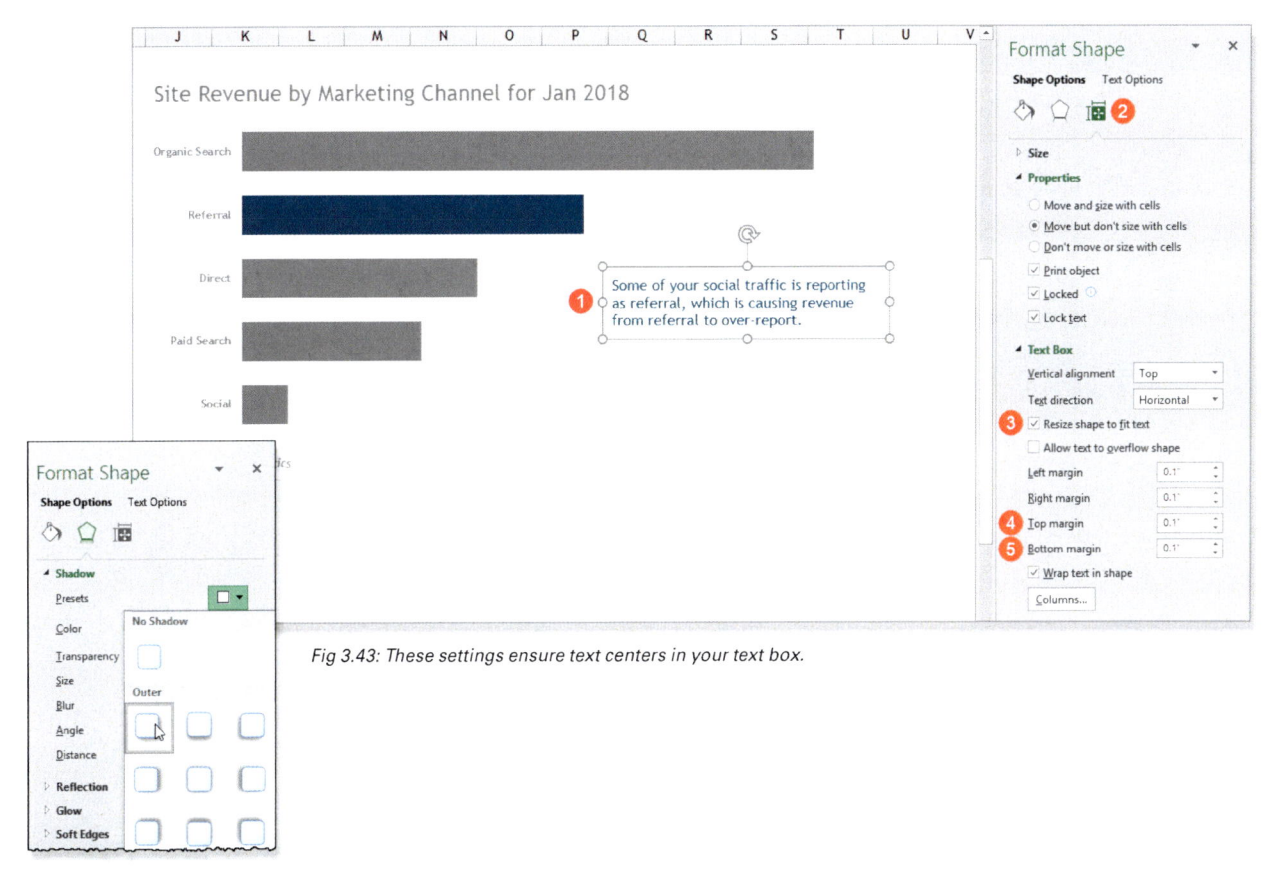

Fig 3.43: These settings ensure text centers in your text box.

Fig 3.45: Add a subtle drop shadow to your text box to create a sense of depth.

59

CHAPTER 3 • DESIGN WITH ELEGANCE

Set Default Text Box Formatting

If you want to reuse your text box, you can set it to be your default text box by right-clicking on it and choosing **Set as Default Text Box** from the contextual menu **(Fig 3.46)**. The super bummer is that this will only set the text box as the default for that file. Grr. But we will fix that with the mother of all Excel hacks.

> 💬 **LEARN MORE** See the Templatize Everything section of the Set It and Forget It chapter to learn how to save time and effort with templates.

Pull in Values

Another cool thing you can do with text boxes is dynamically pull in text or a value that resides in a cell. To do this, select the text box, then enter an equal sign in the **Formula Bar**, select the cell with the text or value you want to pull in, then press Enter to activate the formula **(Fig 3.47)**.

If you want to combine values and text, you'll need to stitch everything together in a cell in your worksheet. First, enclose text in quotation marks. Then enter an ampersand (&). Finally, wrap the cell that contains the

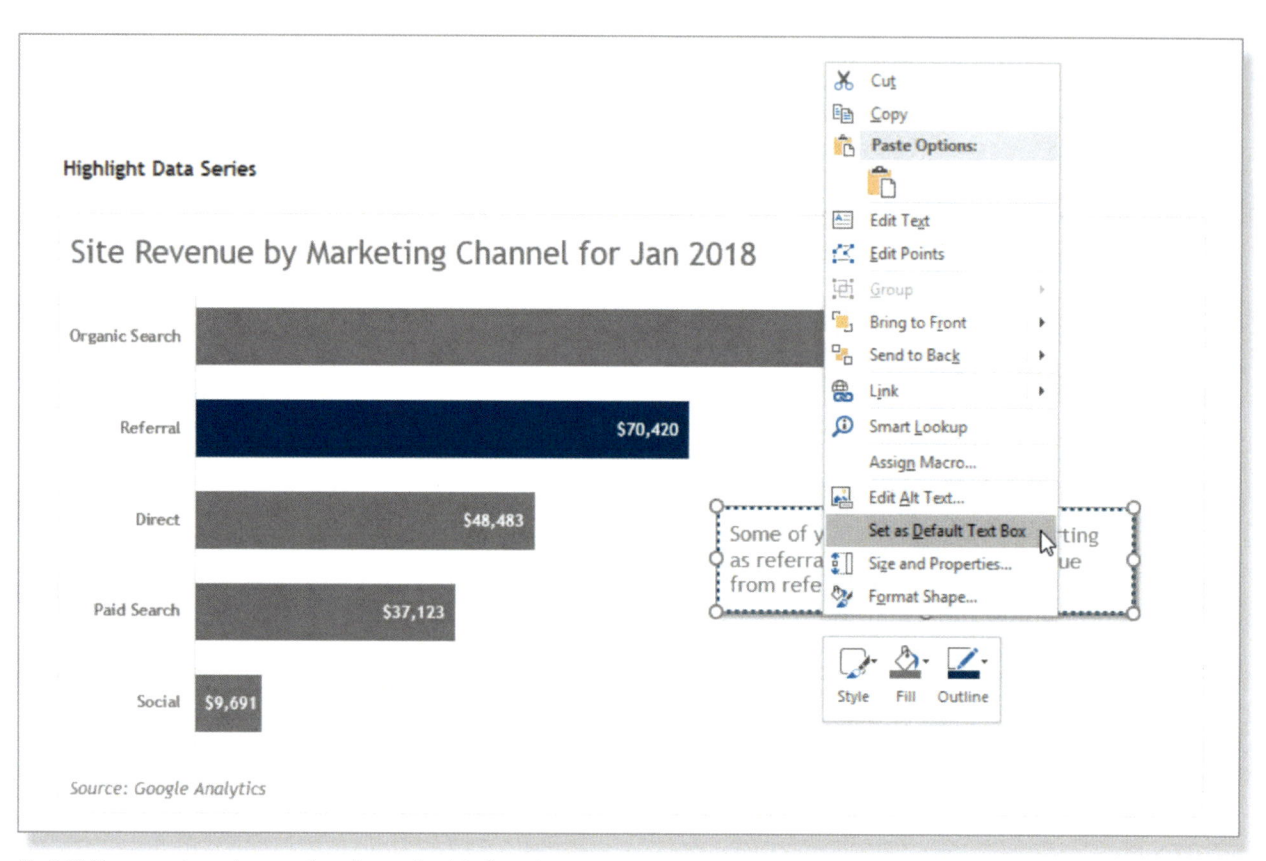

Fig 3.46: You can set your formatted text box as the default text box.

60

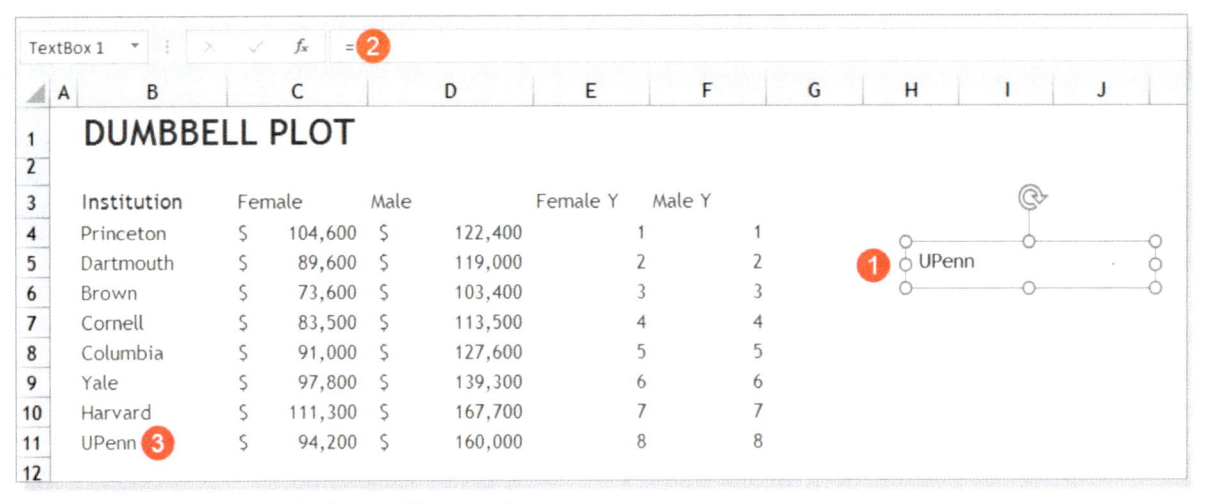

Fig 3.47: Use a formula to pull a value from a cell into a text box.

value you want inside a TEXT function. You will need to include number formatting in the function.

For example, in the Scatter Plot tutorial, I dropped the following formula into cell E5:

`="r = "&TEXT(E4,"0.00")`

A few notes about this technique:

- If you specify only one section of format code—as I do in the above example—the code is used for all numbers. Negative numbers, in this case, will use the negative sign (as you'll see in the Scatter Plot tutorial).
- If you specify two sections of format code, the first section of code will be used for positive numbers and zeros, and the second for negative numbers.
- If you skip code sections in your number format, you must include a semicolon for each of the missing sections of code.
- If you forget which function you need in order to pull both text and a calculated value into a text box, just think TEXT function, *text* box…You won't be judging my silly mnemonic devices when you realize the Google searches I just saved you the first three or so times you try to do this!

- Any text needs to be wrapped in quotation marks. Remember to add in spaces so the value generated from your TEXT function doesn't get too awkward with your text. This makes perfect sense until you go to concatenate.
- You use an ampersand (&) to join text with calculated values.
- In cell E4, I've used the CORR function to calculate the correlation coefficient for my dataset:

  `=CORREL(B4:B253,C4:C253)`

- Even if I restrict my value to two decimal places in E4, unless I use the TEXT function, Excel will pull in the full value **(Fig 3.48)**.

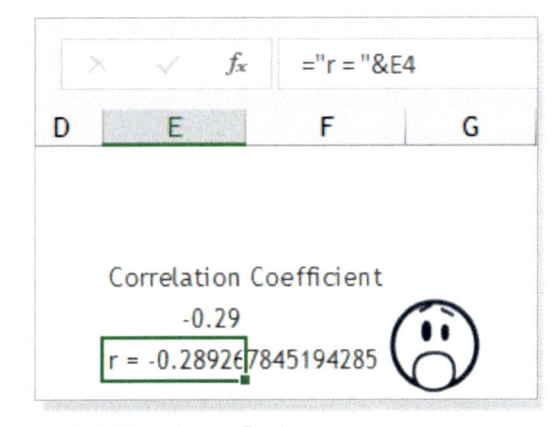

Fig 3.48: Whoa. At ease, Excel.

CHAPTER 3 · DESIGN WITH ELEGANCE

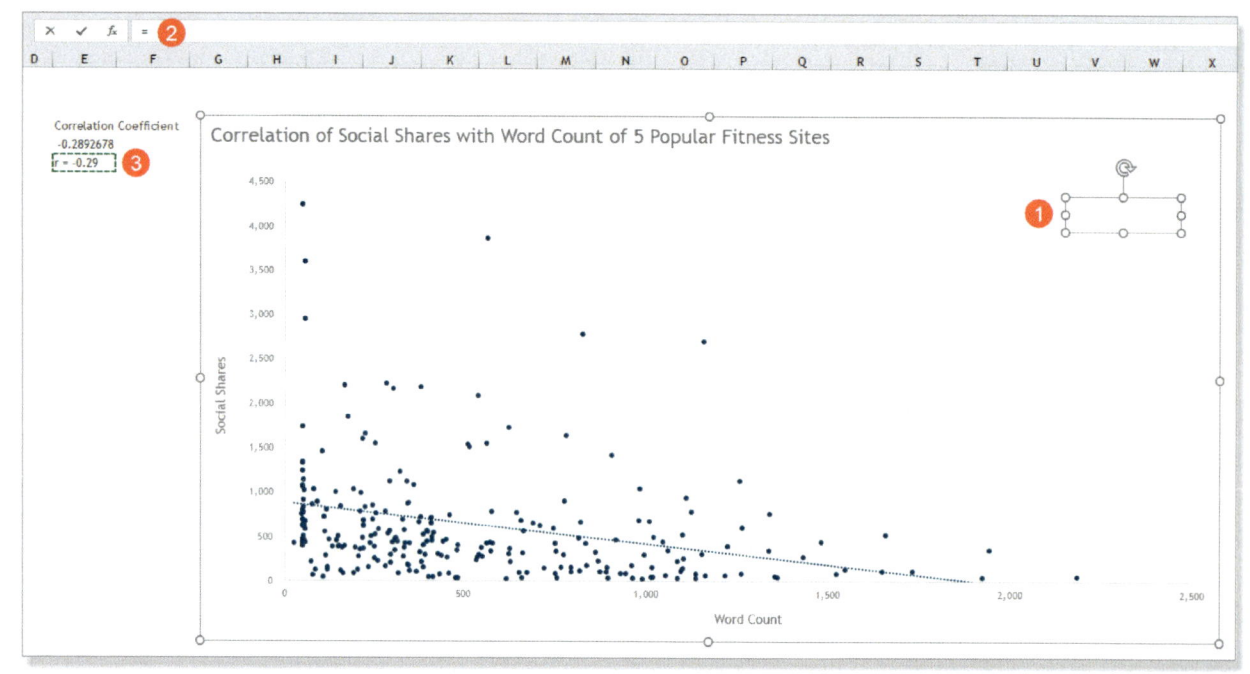

Fig 3.49: Use a function to dynamically pull content into a text box.

- Remember to wrap your number formatting in quotation marks. I forget to do this more than I remember.

- Remember to add any symbols you need inside your number formatting, e.g., currency, thousands separator, percentage sign, etc.

- To embed your text box in your chart so that it moves and resizes with it, you need to select your chart before drawing it. If you just draw it on top of your chart without selecting it or create it outside your chart and drag it in, your text box will not be embedded. It will sit in a layer on top of your chart. This isn't ideal. If you created your text box outside your chart, you'll need to cut it, select your chart, and paste it in. Excel will drop it in the upper-left corner of the **Chart Area**. Just drag it where you want it.

- To pull your final value into your chart, first draw your text box inside your chart. Next, select it by clicking its border, then enter an = sign into the **Formula Bar** and select the cell that contains your TEXT function your TEXT function **(Fig 3.49)**. When you press Enter, it will appear in your text box. Then you can format it to taste.

> 💬 **LEARN MORE** See the Customize Number Formatting section of this chapter to learn how to finesse your number formatting.

Add Border to Touching Data Series

With some charts, such as any of the stacked charts or the pie chart, your series may touch. Sometimes Excel is cool and automatically adds a border between series. You'll see that in the Pie Chart, Donut Chart, Histogram, and Pareto Chart tutorials. But other times, Excel is asleep and lets the series touch. And sometimes when they touch, the honesty's too much. Lulz. #productofthe70s

I like the extra contrast the border adds, especially with charts with many series, where you may be working with tints of your branded colors, as you'll see in the 100% Stacked Area Chart tutorial **(Fig 3.50)**.

62

CHAPTER 3 · DESIGN WITH ELEGANCE

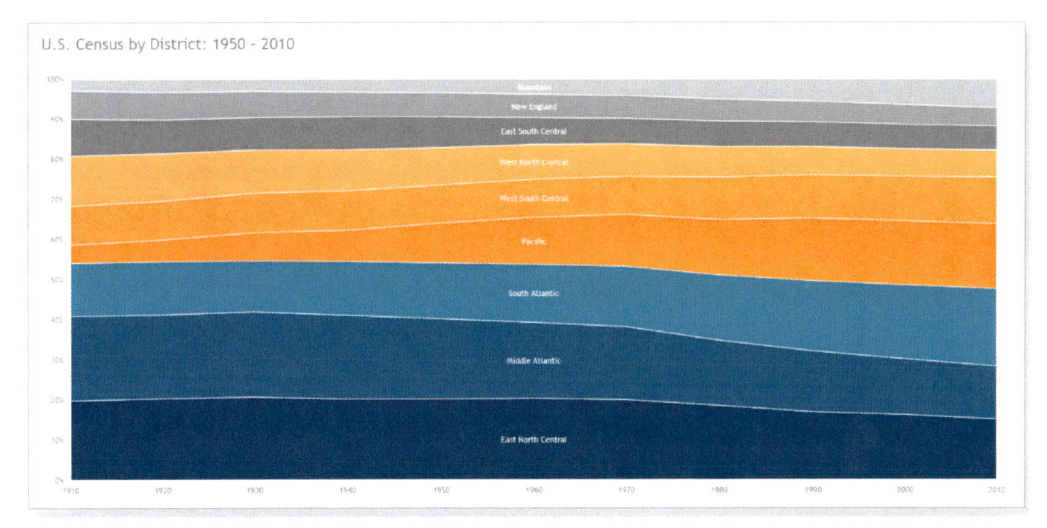

Fig 3.50: Series borders add a touch of class to your charts.

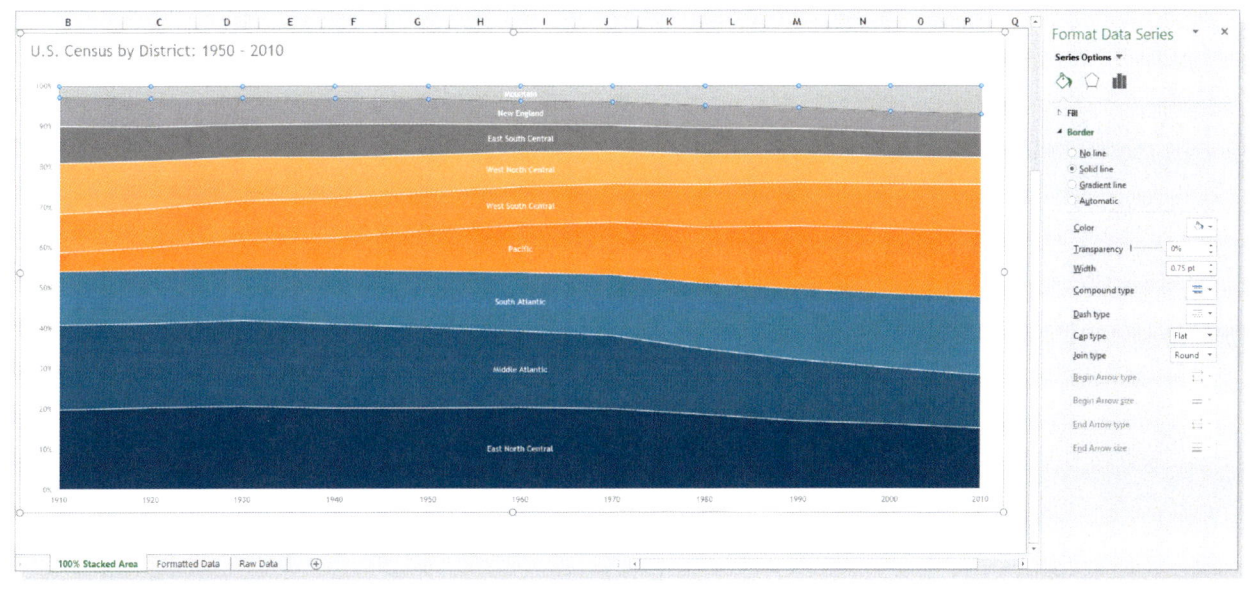

Fig 3.51: Use half the border line width you need because borders are additive.

To add a border to a data series, select the series and press Ctrl-1 to open the **Format Data Series** pane. In the **Fill & Line** tab, set your border parameters **(Fig 3.51)**. I typically use the settings below:

+ **Border**: **Solid line**
+ **Outline color**: white
+ **Width**: .75 pt

Tip: When you apply the border to your first series, it will look pretty lame. But when you apply the same border to your next data series, you'll notice that the width of the border will be doubled because the borders are adjacent, not overlapping. Then your border width should work—although feel free to tweak and test!

63

CHAPTER 3 · DESIGN WITH ELEGANCE

Align Strategically

Microsoft centers most everything by default, but this isn't necessarily the optimal choice. For example, if you pull together a number of charts in a dashboard, it can be more visually distracting to have chart titles floating all over the place, which is how they look when they're centered in each of your chart widgets.

Anything can be moved, so don't be afraid to move things around. To move the title, drag it to where you want it. Hold down the Shift key to constrain movement to a horizontal or vertical plane. Sometimes Shift-dragging will allow you to move an object at 45° angles.

Align Options

We will use the Align menu quite a bit in tutorials **(Fig 3.52)**. You can access it from the Format tab of whatever you're working with (e.g., Drawing Tools, Chart Tools). You can also usually access it through the contextual menu, which you summon by right-clicking on the object you want to edit.

The Align menu is organized in a helpful way, but its options can be confusing so I'll break them down:

- *Horizontal:* This is the first set of options. If you have objects that are distributed vertically, the options in this section are what you need—even though it's categorized as Horizontal. (See the pinch?) If you're lining up text boxes, left-aligning is the most

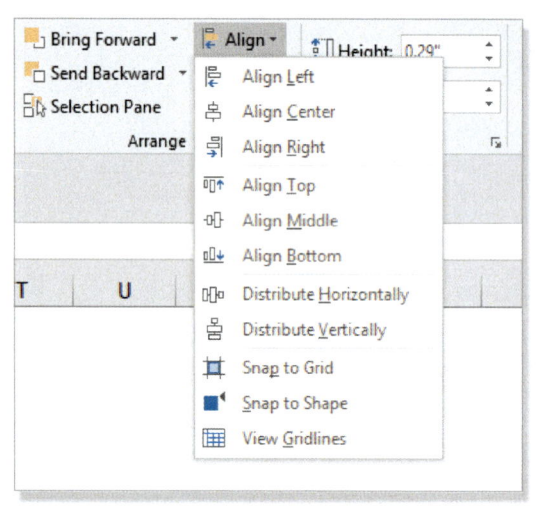

Fig 3.52: The Align menu is helpful in adding finesse to your visualizations.

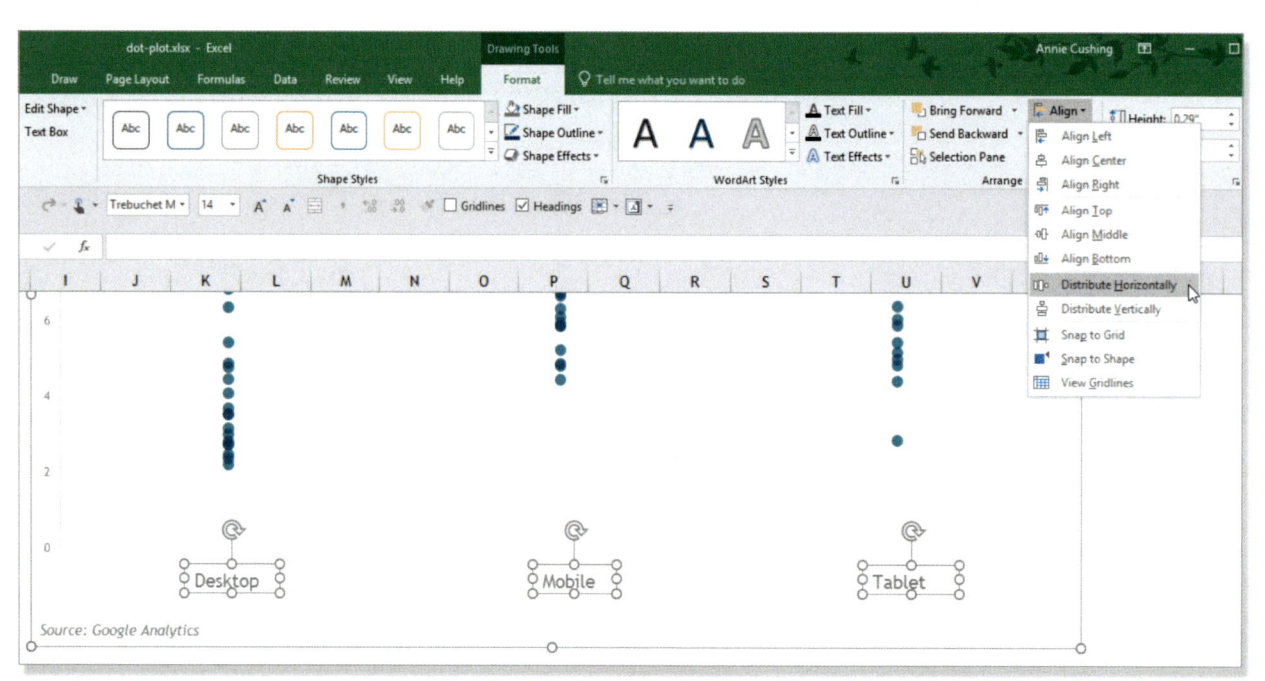

Fig 3.53: Use the Distribute Horizontally option to ensure your faux axis is runway ready.

reliable. If you want to use one of the other options, you need to ensure you don't have excess space between your text and the right edge of the text box because it will be factored in and will throw off the alignment.

* *Vertical:* This is the second set of options. Same as above, but they help you align objects that are positioned horizontally.

* *Distribute:* These options allow you distribute objects equidistantly—either horizontally or vertically. You'll see an example of how we use the Distribute Horizontally option to mimic horizontal axis labels in the Dot Plot tutorial **(Fig 3.53)**.

Snap to Options
Snap to Grid

If you want to snap to the grid when you're moving objects (charts, text boxes, shapes, etc.), you can enable this option from the **Align** menu by selecting **Snap to Grid**. Then items will automatically try to snap to the grid as you drag them.

Alternatively, if you hold the Alt key down while you drag objects, Excel will snap your object to the grid. I actually prefer this method.

Alt-dragging also works when you're dragging sizing handles, e.g., to line your chart's edges up with Excel's grid and other charts. You don't need to turn on gridlines for this to work. I don't even think about it. I just Alt-drag like a fool. But if you want to turn on gridlines, you can do that by navigating to **View ▸ Show ▸ Gridlines**.

Warning: It's one or the other. Alt-dragging won't work if you've enabled Snap to Grid.

My typical workflow is to snap the upper-left corner of the chart to the grid, then Alt-drag the bottom-right corner of the chart to snap the right and bottom borders to the grid.

Tip: If you want to make charts the same size, you can drag over a range next to the chart to see how many cells wide and/or long it is. If there's data in those cells, you can look at **Count** in the **Status Bar (Fig 3.54)**.

If there's no data in the cells, it's a little trickier but still doable. Just click-and-drag over the range, but before you let go of the mouse, take a gander at the **Name Box (Fig 3.55)**. It will tell you the number of columns and rows in the active selection. It will take the form of rows x columns.

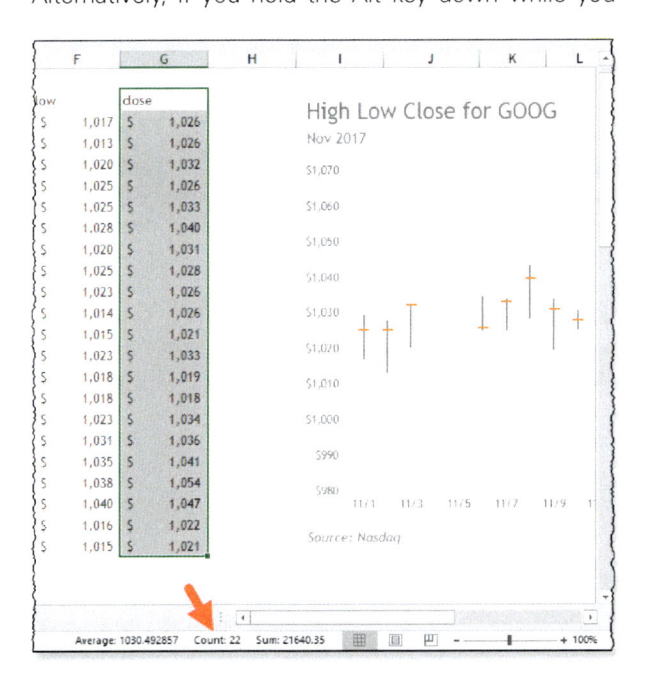

Fig 3.54: Use Count to calculate the length of a range.

Fig 3.55: Employ the Name Box to count empty cells.

CHAPTER 3 • DESIGN WITH ELEGANCE

Snap to Shape

There are times you can't just select the items you want to align and use an option from the Align menu. In those instances I'll drag out a shape—usually a very thin line—and use Excel's Snap to Shape feature to align my chart elements. This is a helpful trick when aligning your chart title with other text boxes, like a subtitle and/or text box (as you'll see in most of the charts in this book). Aligning data labels can be also problematic, as you'll see in the Add a Target Line variation of the Line Chart tutorial.

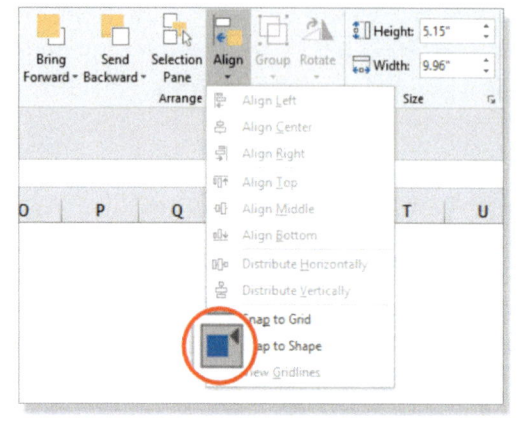

Fig 3.56: Turn on Snap to Shape if you need to align objects to a common element.

First, Shift-drag out a line (**Insert ▶ Illustrations ▶ Shapes ▶ Lines ▶ Line**). Holding down the Shift key constrains the angle of the line you're drawing to 45° angles, giving you perfectly horizontal, perpendicular, or diagonal lines. Then select the object you want to line up with your line and turn on **Snap to Shape** by navigating to **Chart Tools ▶ Format ▶ Arrange ▶ Align ▶ Snap to Shape (Fig 3.56)**. There's a very subtle highlight that's applied to the icon when you enable it.

Now drag your object toward the line until it snaps into position. Rinse and repeat for each object. Once everything is lined up, delete the line.

Warning: If you turn on Snap to Shape before you drag out your line you might have a difficult time positioning it where you want. Also, make sure Snap to Grid is turned off.

Group to Preserve Your Work

Once you get everything aligned, I highly recommend grouping your objects before you deselect them—also available from the **Arrange** group under the **Format** tab or the contextual menu.

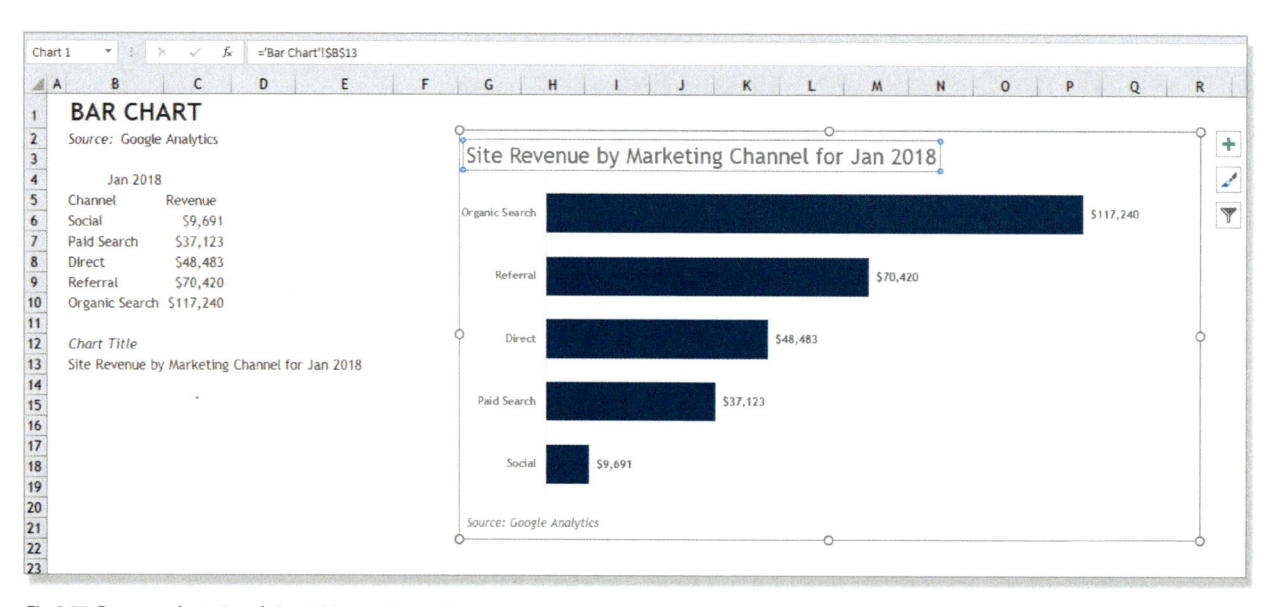

Fig 3.57: Bump up font size of chart titles and legends.

66

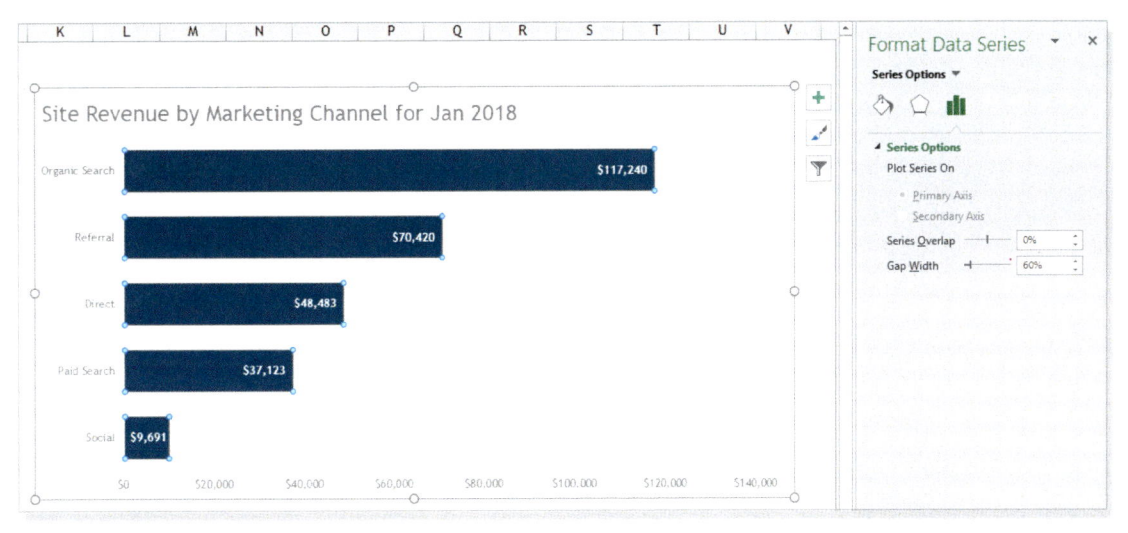

Fig 3.58: Fatten up your bars by reducing Gap Width.

At the time of writing, with the [half-baked] charts that are new to 2016, you can't add text boxes inside a chart (meaning they won't move with your chart), so you need to group them with your chart. Grouping objects with your chart is kind of like adoption for data. Aww!

Resize Fonts

I usually like to bump up the font size on chart titles to make them stand out more **(Fig 3.57)**. I also sometimes increase the font size in legends. To adjust the font size, select the item and navigate to **Home ▶ Font** and choose the size from the drop-down menu. Or use the **Increase Font Size** or **Decrease Font Size** buttons to the right for incremental changes.

Adjust Chart Series Spacing

Bar and column charts sometimes look a little better if the bars are wider. To adjust them, select a bar to select the entire data series and press Ctrl-1 to open the **Format Data Series** pane. Next, open the **Series Options** tab and reduce **Gap Width** using the slider or entering the percentage **(Fig 3.58)**. I'll usually fatten up my bars to make room for data labels. With the extra space, you may also want to bump up the font size of your labels and/or make the font bold.

Thin Out Axes

There are two issues I commonly see with axis labels:

- *Labels are turned sideways:* This is one of my greatest pet peeves with chart design. Sometimes this happens because the chart designer used a [vertically oriented] column chart when she should have used a [horizontally oriented] bar chart. This is addressed in the Bar Chart and Column Chart tutorials. But, as a general rule, if your x axis isn't charting a time element, use bar charts with datasets that have longer labels.

- *Axes are unnecessarily dense:* Sometimes your axis labels are unnecessarily detailed, exceeding that optimal noise/signal ratio we're aiming for. Excel allows you to set the interval for your axis to give you some breathing room. To access these options, select your axis and press Ctrl-1 to open the **Format Axis** pane.

We push Excel a little too far in the Cycle Plot tutorial—causing it to crumble into a ball on the floor and scream uncontrollably like a toddler who missed naptime—but most times this technique is fairly reliable. That said, there are some nuances, depending on the type of data that's referenced in your axis, so we'll address the strategies for each data type separately.

CHAPTER 3 • DESIGN WITH ELEGANCE

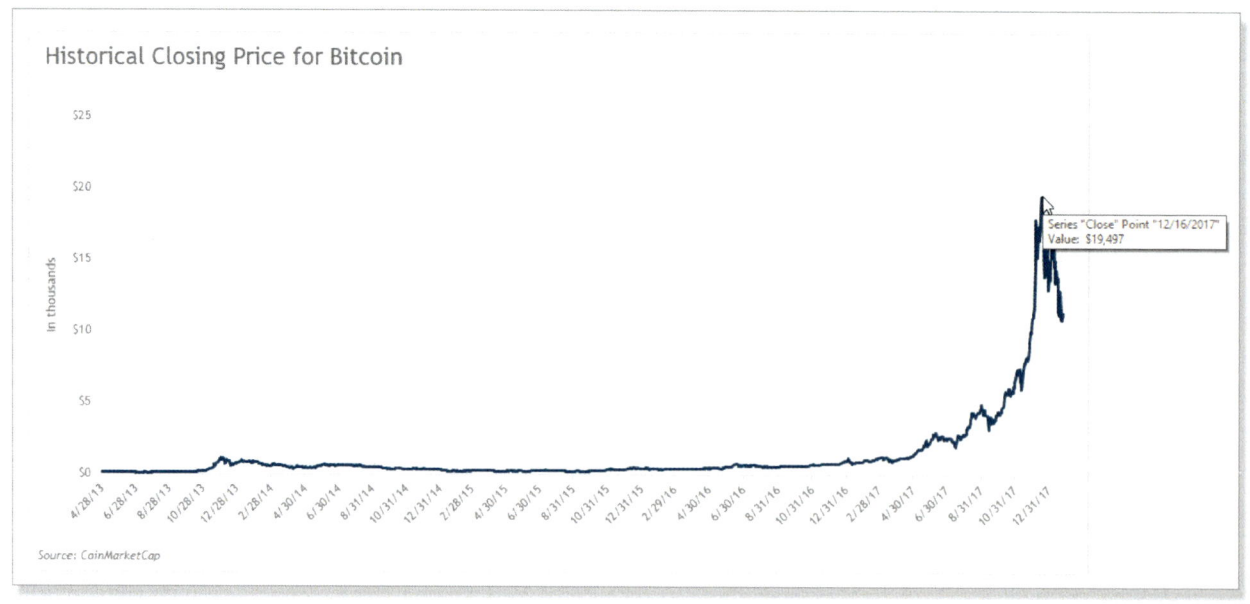

Fig 3.59: A viewer can always get the exact date by hovering over a data point.

Date Axes

I often see charts with dates in the horizontal axis, but the chart designer will do silly things like have the full date representing each day of the month, for example, and then turn the labels on their side because there's no room in the axis. I call it silly because indulge me for a second: You're reporting on the performance of something from last month, and you're telling me the month started with 6/1/2018, followed by 6/2/2018, followed by 6/3/2018…See where I'm going with this?

You can put the year (and even month) elsewhere in your worksheet: the chart title, the worksheet title, an annotation…*anywhere* but your axis labels, where real estate is precious.

With year taken care of, we can still shave our axis labels down more. You could easily skip every other day, and no one will panic and ask, "Wait, what about even days? What kind of dystopian society did I wake up in today?" I've seen charts skip as much as six days at a time, and I've still been able to interpret the chart with no issue. If a viewer wants to see the exact date Bitcoin experienced that spike (or precipitous drop), he can hover over a data point **(Fig 3.59)**.

To this end, Excel gives you some extra options with a date axis. The one I use most is the **Major Units** setting in the **Format Axis** pane. Regardless of what format my dates take (e.g., month, days of week, date), I usually bump **Major Units** up from 1 to 2 **(Fig 3.60)**. If you set your number formatting to mmm for months only, ddd for days of the week, mm/dd for month and day (with leading 0s), or m/d, skipping one unit will result in a lighter axis without sacrificing clarity. You'll put this technique to good use in the Line Chart tutorial.

Fig 3.60: Thin out a date axis by bumping up the Major Units setting.

CHAPTER 3 · DESIGN WITH ELEGANCE

💬 **LEARN MORE** See the Customize Number Formatting section of this chapter to learn how to finesse your number formatting.

Value Axes

For all other axis-thinning options, you'll just adjust the **Major Units** under **Axis Options**. I often find doubling the default value gives me the light and airy axis my minimalist soul craves. In the accompanying screenshot, the original value was 5000. I bumped it up to 10000 **(Fig 3.61)**. (Thousands separators need not apply in these settings.)

Warning: If you plan to create a chart template from your chart, you should first create the template (right-click ▶ **Save as Template**) before customizing your axis. If you customize your value axis (usually the y axis) *before* you create the template, you will hardcode the major units for all future charts.

💬 **LEARN MORE** See the Create a Store of Chart Templates section of the Set It and Forget It chapter to learn how to work with templates.

Text Axes

You may be wondering why you would want to thin out a text axis. In most cases a text axis is the default for categorical data: region, marketing channel, country, employee, gender, etc. You obviously wouldn't want to remove every other region or employee in your department. However, sometimes you will have dates show up as text. For example, if the dataset you're handed has hardcoded months as abbreviations (Jan, Feb, Mar), Excel will treat them as text. Same with days of the week. (I've seen it all and have suffered so you don't have to! #messiahcomplex)

You can see an example of a date column that is treated as text in the Lollipop Graph tutorial, as well as the Add a Target Line variation of the Line Chart tutorial. In this case, you can thin them out in the **Labels** section of the **Axis Options** tab of the **Format Axis** pane.

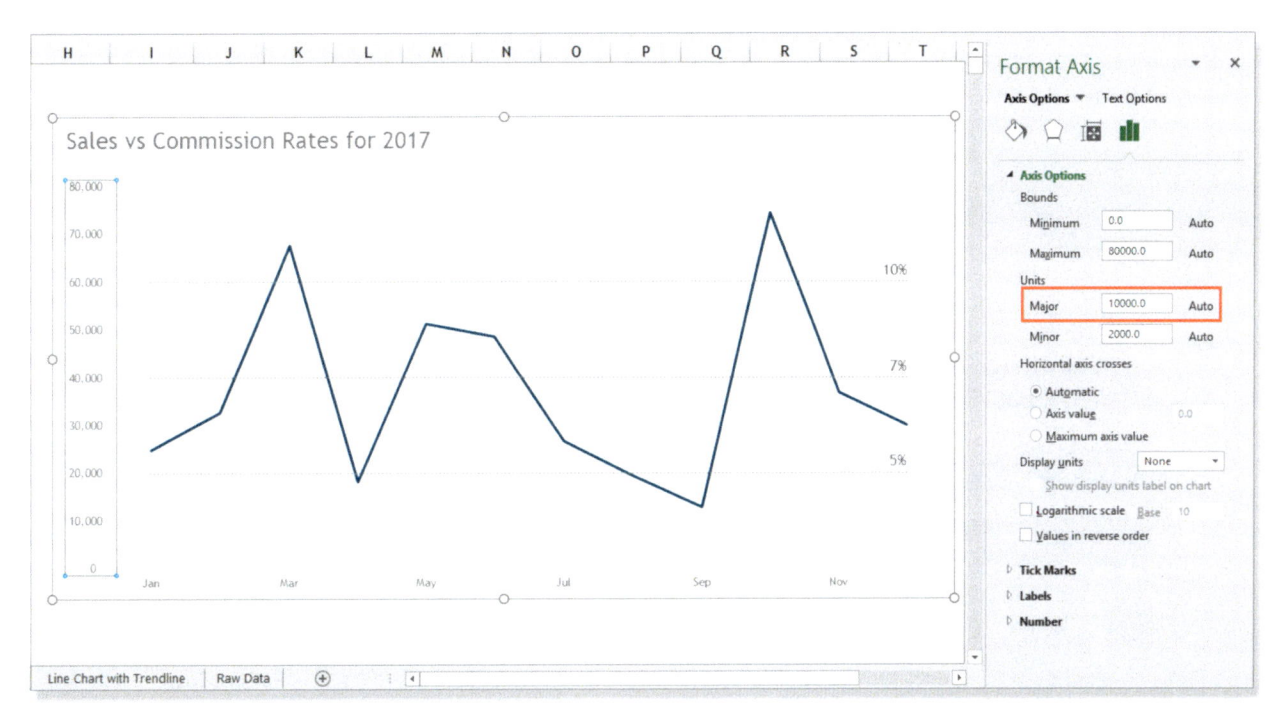

Fig 3.61: Hardcode Major Units setting to thin out a value axis.

CHAPTER 3 • DESIGN WITH ELEGANCE

Set **Specify interval unit** to your desired amount **(Fig 3.62)**. I usually set mine to 2, but if you're working with months you could just as easily show only the first month of each quarter without loss of understanding.

Remove Unnecessary Axes
If you use data labels in your charts, you may no longer need your axis **(Fig 3.63)**. To remove it, select it and press Delete on your keyboard.

Add an Axis Title
Sometimes it's important to add an axis title to your chart. One reason might be that you have two axes, as you do with a combination chart.

Warning: If you're creating a scatter plot you'll need to manually add axis titles. I'm surprised Excel doesn't add them by default because there's no way to know which value is plotted on which axis without them.

To add one, select the chart and open the **Chart Elements** flyout menu in the upper-right corner. Click the right-facing arrow next to **Axis Titles** and select the axis you want to add a title to. Excel nudges the title a little too close to the axis for my taste—at least for vertical axes—so I drag it away while holding down the Shift key to keep it vertically centered. To enter your text, either double-click inside the title box and start typing or select the title box and type your title in the **Formula Bar**.

Add Markers
Some chart types consist primarily of markers, as you'll see in the Dot Plot, Dumbbell Plot, and Scatter Plot tutorials. In other charts markers are merely an augmentation, as you'll see in the Slope Chart, Sparklines, Line Chart, and Stock Chart tutorials.

If your chart has a marker option, you'll be able to add them by selecting your data series and pressing Ctrl-1 to open the **Format Data Series** pane. The **Marker** options are housed under the **Fill & Line** tab.

Excel gives you a few formatting options for your markers:

+ *Type:* You have a number of shapes to choose from. This book will only use the circle option. I personally

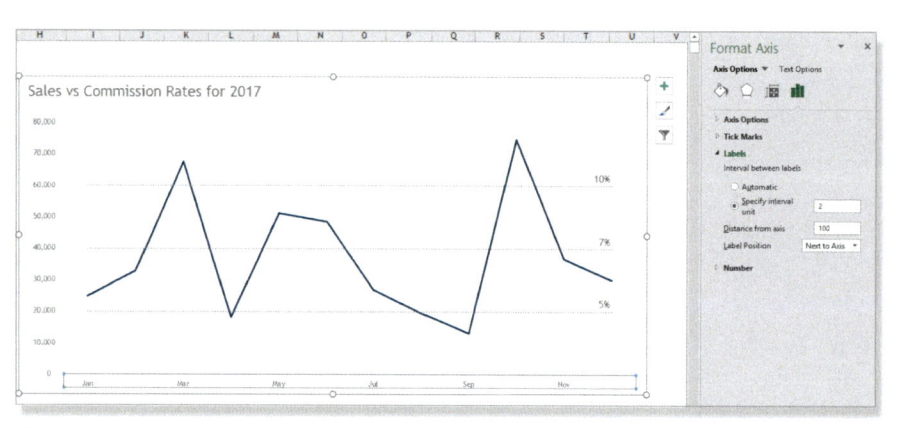

Fig 3.62: Thin out text axes by setting the interval unit.

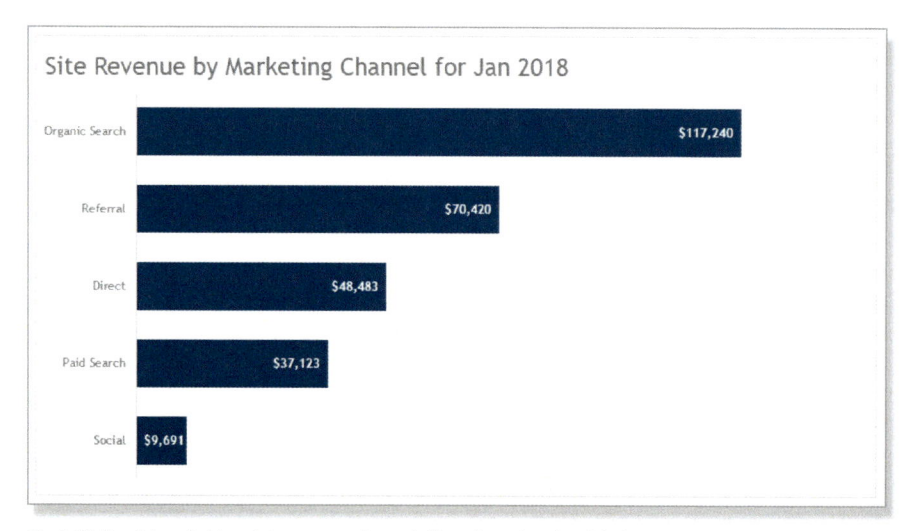

Fig 3.63: Don't be afraid to delete your value axis if you're using data labels.

find charts that mix types busy and amateur. Also, one marker type might look bigger or smaller than another marker type of the same size. Just because you *can* do something doesn't mean you *should*.

* *Size:* Don't be afraid to play with this setting. Using larger markers can give your chart a trendy feel. Because most Excel users don't play with size, you'll automatically show yourself to be a cut above.

* *Fill:* You can play with both the color and transparency of your fill color. I will sometimes add transparency to my marker fill to make my border pop or reduce the visual impact of crowding in charts, as you'll see in the Dot Plot and Bubble Chart tutorials.

* *Border:* I use borders in several scenarios: if I want a white fill (as you'll see in the Add Sizzle variation of the Line Chart tutorial), if I'm increasing the transparency of my markers (as you'll see in the Bubble Chart tutorial), or if I'm increasing the transparency of a data series (as you'll see in the Use Area Chart variation of the Combination Chart tutorial).

Hack Excel's Charting Engine

As you'll see in the Slope Chart tutorial, (which spotlights one of the more frustrating datasets in the book), there are times you'll need to modify a data series (i.e., a column of data in a dataset). And, as you'll see in the Dumbbell Plot tutorial, Add Categories variation of the Scatter Plot tutorial, and Add a Mean Line variation of the Bar Chart tutorial, there are other times you'll need to create a data series from scratch. To do these things, you'll need to summon the Select Data Source dialog. I'll dedicate this section to demystifying what I believe to be Excel's most confusingly labeled dialog.

Scariest Dialog in Excel?

This dialog was one of the more intimidating menus in Excel to me for years. I only went into it and modified it if a tutorial walked me through the steps. And even then, I was convinced when I clicked **OK** to close it, my chart would vomit all over itself like a college kid after playing beer pong for the first time.

But the **Select Data Source** dialog—which you get to by right-clicking anywhere inside a chart and selecting **Select Data** from the contextual menu—is to charts what creepers are to auto mechanics. (Not this kind of creeper: **bit.ly/mdsepc-creep**.) It's how you access the underbelly of your chart to restructure it **(Fig 3.64)**.

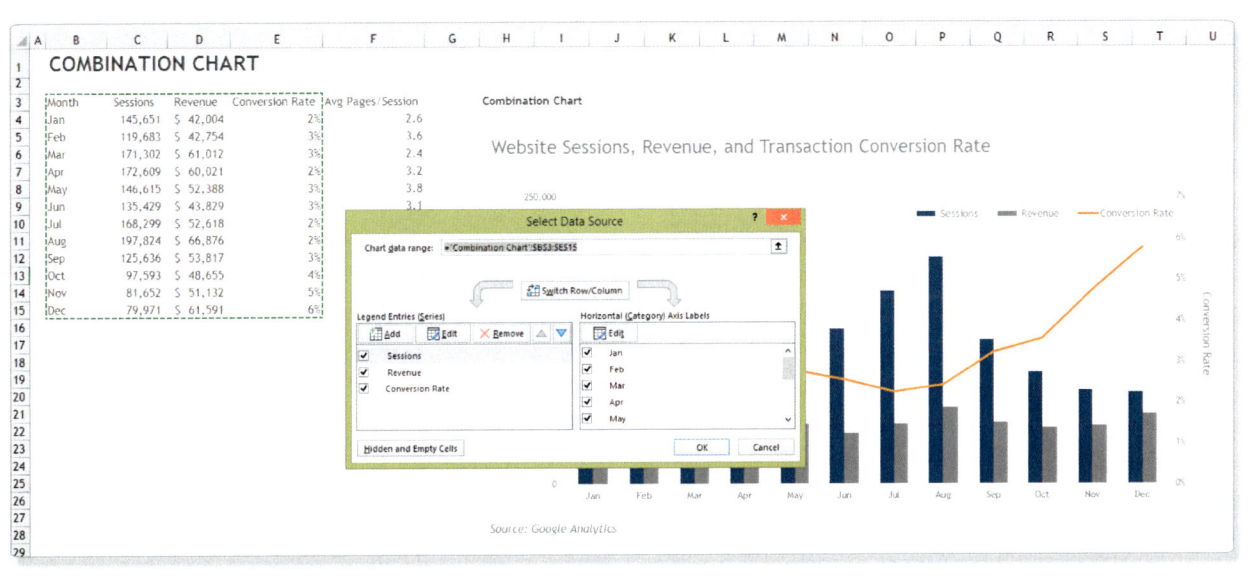

Fig 3.64: The Select Data Source dialog will become your BFF!

Anatomy of the Select Data Source Dialog

Let's break down the Select Data Source dialog a bit. I think one of the problems with this dialog is Microsoft engineers got a little…enthusiastic…with the labels. But they actually make sense once you learn to interpret them. I kind of picture the engineers calling for a brainstorm meeting to determine how to label the features in this dialog and then shoehorning all the terms on the whiteboard into the dialog.

Chart Data Range

Ah. The most intuitive label in the dialog, in my opinion. No complaints on this one. As the label suggests, this field references the range of cells in your dataset that are included in the chart. If you click inside this field, your range will suddenly find itself surrounded by an army of marching ants.

These columns are oftentimes contiguous (touching) but don't have to be. You can click-and-drag over a column of data, hold down the Ctrl key and click-and-drag over another column. Technically, this can be a row of data, if your data is structured using rows instead of columns, but all the datasets used in this book are structured using columns, which is more the norm. If your columns aren't contiguous, you'll see them separated with a comma in the Chart data range field **(Fig 3.65)**.

You'll notice a little icon next to this field with an up-facing arrow. You'll see these all through the Select Data Source dialog. If you click it, the Select Data Source dialog will be reduced to just that field, which is intended to make it easier for you to click-and-drag over the data range you want represented in the chart **(Fig 3.66)**.

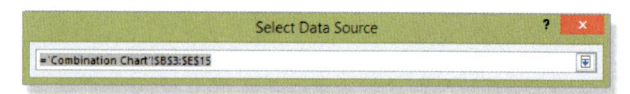

Fig 3.66: Excel allows you to remove all distraction when selecting a new data range.

You don't need to click this button to select a range. You can, instead, click inside the field, delete what's in there, if anything, and click-and-drag over the range of columns in your dataset you want included in the chart.

Legend Entries (Series)
THINK METRICS PLAYGROUND

This area of the Select Data Source dialog was the most confusing to me. This is just where your data series reside. And because data series (e.g., individual lines in a line chart or columns of the same color in a clustered column chart, etc.) are differentiated in the legend, Microsoft engineers decided to lead with that. But the legend is more secondary. You don't open Excel, crack your knuckles, and tell yourself, "Today's the day I put the legend in legendary!"

I suggest that you just think of this section as simply where your metrics go: runs batted in, miles, interest rate, website visits, revenue, etc. If you can measure it with a number, it most likely goes in this section. Therefore, whenever you go into the Select Data Source dialog (which you will do a lot throughout the course of this book), maybe switch out that label in your mind with something like *Metrics*. It might help.

ADDING AND EDITING DATA SERIES

You can add a new data series by clicking the **Add** button and edit a data series by clicking the **Edit** button. You won't usually need to do that in Excel. In most cases, Excel will use your second through your last column from your chart data range to populate this area. But we'll click the **Add** button to see what

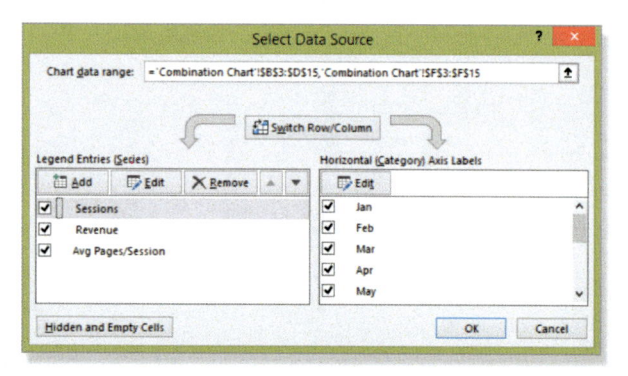

Fig 3.65: Non-contiguous ranges are separated by a comma.

CHAPTER 3 · DESIGN WITH ELEGANCE

Fig 3.67: Click the Add button to manually add a data series.

wonders await us in the **Edit Data Series** dialog **(Fig 3.67)**.

To set **Series name**, you can either type your series name in (not recommended) or click inside the field, then select the column header of the column you're adding. To set **Series values**, delete the ={1} that auto-populates this field, and click-and-drag over just the values of your new column of data. To speed this up, select the first cell, then hold down the Ctrl and Shift keys while pressing the down arrow key on your keyboard to select all contiguous cells in your column. If you have blanks in your data (tsk tsk), it might be faster to scroll to the bottom of the column and Shift-click the last cell. (If it's a really large dataset, I'll scroll way down, to where I know there are blank cells, then Ctrl-click to pop up to the closest non-blank cell.) Then click **OK** to close the **Edit Series** dialog.

WHEN YOUR DATA SERIES HAS X AND Y VALUES
If you're working with a chart type that uses x and y values (as you'll see in the Dot Plot, Scatter Plot, and Bubble Chart tutorials), the **Edit Series** dialog will split the series values into x values and y values **(Fig 3.68)**. When you click inside the **Series X values** field, the marching ants will surround the values you're plotting along the x (horizontal) axis. This may cause some flashbacks to middle school algebra. If so, I apologize. Nobody deserves that.

With a scatter plot, for example, individual data points are determined by both their x value and y value. But

Fig 3.68: If you have x and y values, your Edit Series dialog will add an extra field.

we'll also use this ability to set x and y values to create charts Microsoft engineers didn't anticipate us greedy little Excel users to try to create in Excel, as you'll see in the Dot Plot tutorial. Muahaha!

Horizontal (Category) Axis Labels
THINK CATEGORIES PLAYGROUND
This label reminds me of bangs that aren't texturized. #thicc This area of the dialog is very simply where your categories go: department, region, team, school, smart phone brand. You get the idea. And don't let the word *horizontal* throw you. Axis labels aren't always in the horizontal axis **(Fig 3.69)**.

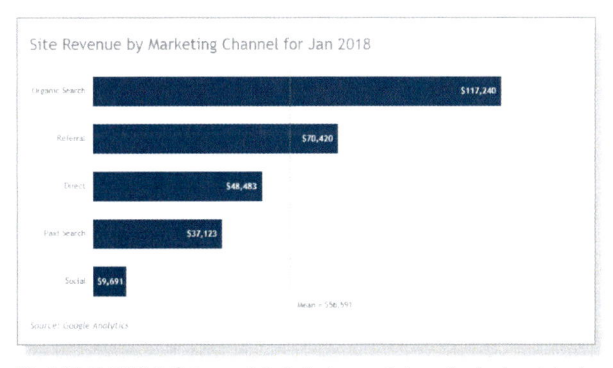

Fig 3.69: Exhibit A: Category labels that are not along the horizontal axis.

73

CHAPTER 3 · DESIGN WITH ELEGANCE

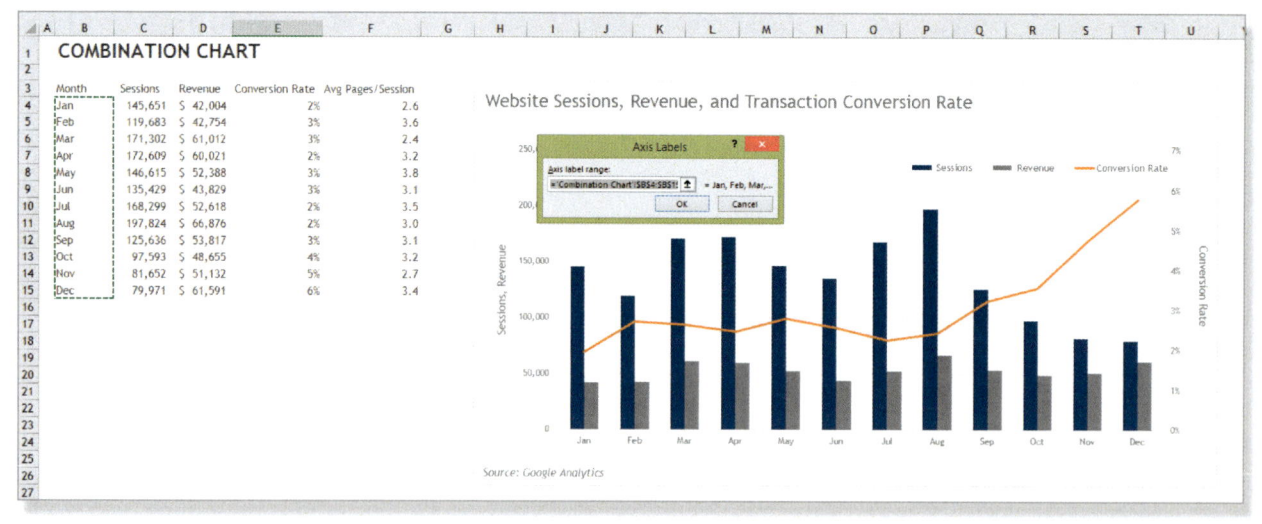

Fig 3.70: Excel only needs the categories; the header need not apply.

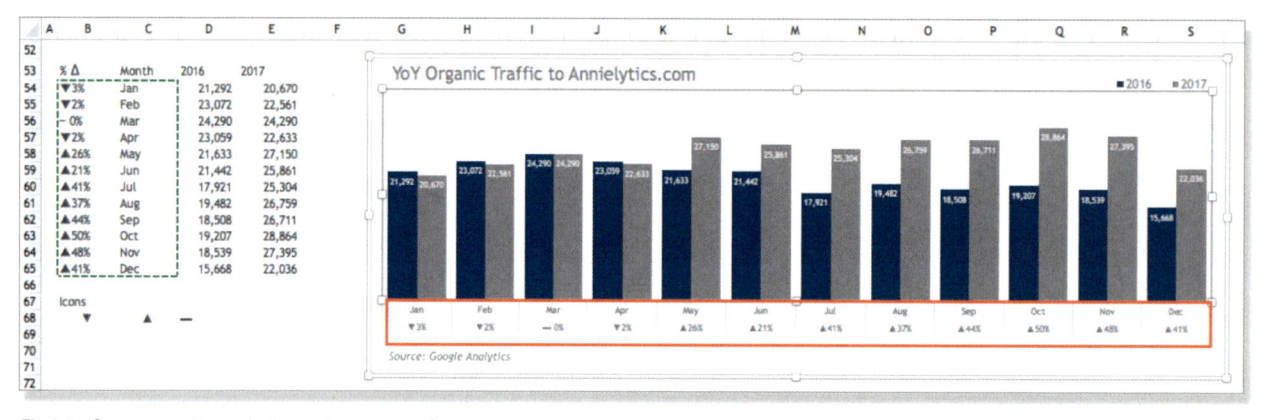

Fig 3.71: Can you say bogo deal on axis categories?

By default, Excel uses the first column of your dataset to fill this area of the **Select Data Source** dialog. If you click the **Edit** button, you'll see marching ants around the categories you're charting. Excel doesn't care about the column header, only the categories. So, if you must manually select your categories, don't include the header in the selection **(Fig 3.70)**.

CATEGORIES GONE WILD

With some of the non-standard charts we'll build, you'll see the flexibility of the **Horizontal (Category) Axis Labels** field. For example, you'll see in the Show % Change variation of the Clustered Column tutorial that we will pass two columns into our categories **(Fig 3.71)**.

Loosely related: You will adore pivot tables [almost] as much as I do when you see how easy it is to calculate year-over-year % difference! Go ahead and create this chart for your boss and see what happens. I'm telling you...Nothing says job security quite like a YoY % delta chart!

74

CHAPTER 3 • DESIGN WITH ELEGANCE

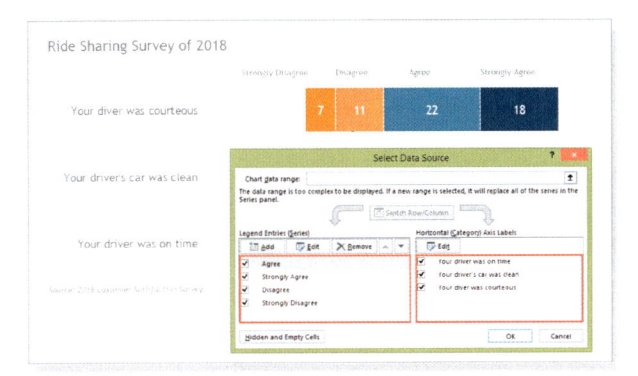

Fig 3.72: Two! TWO categories! Ah ah ahhh!

Fig 3.73: The diverging stacked bar chart is a great way to visualize survey data.

And in the Diverging Stacked Bar Chart tutorial we'll use multiple category variables: survey questions and survey responses **(Fig 3.72)**. This is a hacked chart type but oh so sexy—and a great way to visualize survey data **(Fig 3.73)**!

Switch Row/Column

As you might imagine—because, you know, it's an intuitive label—this button allows you to switch your rows and columns **(Fig 3.74)**.

We'll need to use it in our 100% Stacked Bar Chart tutorial (if you're following along with the sample dataset) because sometimes Excel will ignore the way your data is structured and do what it thinks is best. It'll make sense when you read through the tutorial.

We'll also use it create some sophisticated, non-standard charts, as you'll see in the Slope Chart tutorial.

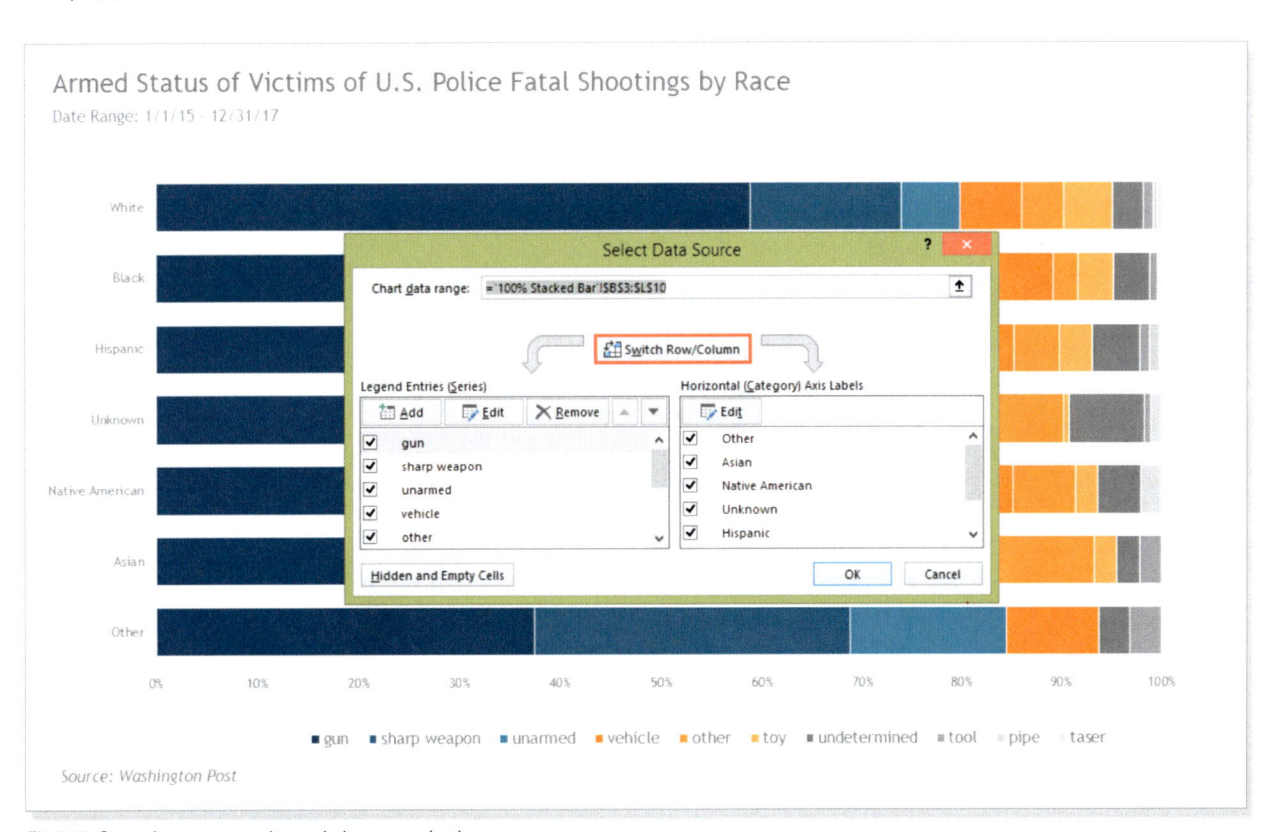

Fig 3.74: Sometimes you need to switch rows and columns.

75

CHAPTER 3 • DESIGN WITH ELEGANCE

Add Data Series to Chart

You will occasionally need to add new data series to charts. You have three ways to add them:

- *Drag sizing handles:* When you click anywhere in your **Plot Area**, you'll see purple, red, and blue highlighting in your dataset. These indicate the categories, series name, and series values, respectively. Click-and-drag the handle in the bottom-right corner of the blue outlined area to add contiguous (touching) data to the chart.

- *Copy-and-paste:* Click-and-drag over your new column (including the column header) to select it. Next, copy it, select your chart's **Plot Area**, and paste it into your chart.

- *Select data:* Right-click anywhere in the **Plot Area**, choose **Select Data** from the contextual menu, and add your new data series by clicking the **Add** button in the **Legend Entries (Series)** section. See the previous Anatomy of the Select Data Source Dialog section to learn how to use this dialog. I break it down in the simplest terms.

COPY-AND-PASTE TIPS

You will employ a variety of paste options throughout the tutorials in this book. For starters, it's important to know that the keyboard shortcut to copy is Ctrl-C and paste is Ctrl-V. But there are several advanced paste features you may use in creating visualizations.

For consistency's sake, we'll access all the paste options by right-clicking in a cell and choosing **Paste Special ▶ Paste Special** to open the (wait for it....) **Paste Special** dialog.

Paste as Values

We will use pivot tables to get our data in the structure we need for many tutorials. But if you create a chart from a pivot table and there's a pivot chart version of your chart option, Excel will use a pivot chart. The problem with this is pivot charts are difficult to finesse, in my opinion. I've never preferred them because of their design limitations. The way we'll get around this is to copy only what we need from the pivot tables we create and paste as values into the worksheet we'll be building the chart in. In real life I never put raw data

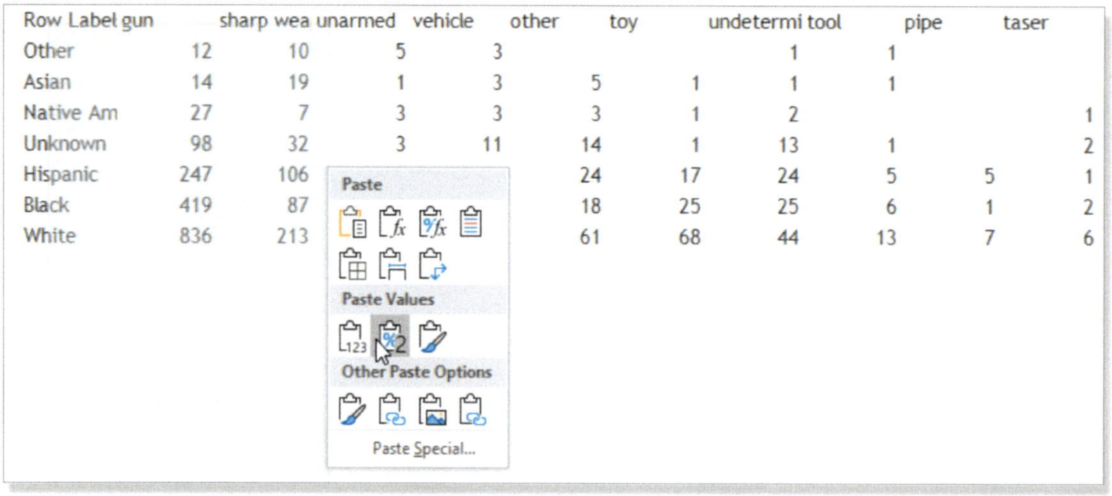

Row Label	gun	sharp wea	unarmed	vehicle	other	toy	undetermi	tool	pipe	taser
Other	12	10	5	3			1	1		
Asian	14	19	1	3	5	1	1	1		
Native Am	27	7	3	3	3	1	2			1
Unknown	98	32	3	11	14	1	13	1		2
Hispanic	247	106			24	17	24	5	5	1
Black	419	87			18	25	25	6	1	2
White	836	213			61	68	44	13	7	6

Fig 3.75: The Paste Options button is my favorite way to paste as values.

in the same worksheet as my visualizations. I only do it in the charts for this book so that it will be easy for neophytes to reference the data that's being charted.

You can copy/paste as values using the **Paste Special** dialog via the contextual menu. Choose **Values & Number Formatting** to maintain your number formatting, if you customize it before pasting. If you haven't customized it, you can just choose **Values**. With experience, decisions like this will become second nature.

Alternatively, you can paste your range, then choose **Values** or **Values & Number Formatting** from the **Paste Options** button that appears in the bottom-right corner of your pasted range **(Fig 3.75)**.

Calculate on the Fly

Excel allows you to do quick math by copying a value from a cell, selecting a range you want to apply the calculation to, and choosing a calculation option in the **Paste Special** dialog. You can add, subtract, multiply, and divide. You'll see how we use this option to make values negative to move bars to the left of our y axis in the Diverging Bar Chart tutorial.

Copy Column Widths

Sometimes, when you're cleaning up your workspace, you might find that you have columns that have variable widths because of data that once resided there. And when you're creating dashboards, you'll sometimes want to standardize your column widths across some of your visualizations. You can access this option in the **Paste Special** dialog, under **Paste ▶ Column widths**.

Transpose

As you saw in the Hack Excel's Charting Engine section, if your data charts with the wrong configuration, you can use the **Switch Row/Column** button. But you can switch your rows and columns before you create your chart with the **Transpose** option. Access this option at the southern tip of the **Paste Special** dialog.

Paste Visible Cells Only

If you copy-and-paste a range with hidden rows or columns, they become unhidden. We don't have the option to paste only the visible cells in the **Paste Special** dialog, but we do in the **Go To Special** dialog, which you can get to by pressing Ctrl-G and clicking the **Special** button or by navigating to **Home ▶ Editing ▶ Find & Select ▶ Go To Special (Fig 3.76)**.

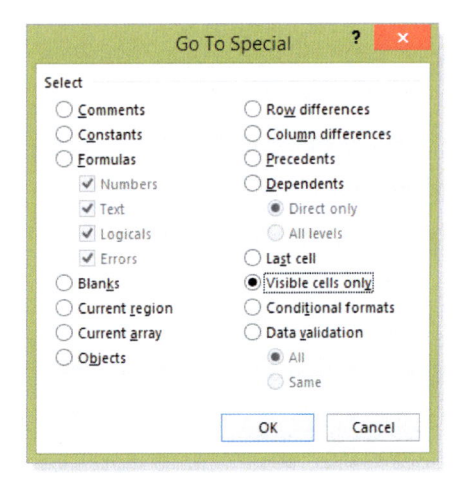

Fig 3.76: Select only visible cells from the Go To Special dialog to prevent pasting hidden cells.

Move without Pasting

When you select a range, you can drag it without copying-and-pasting, but it's a little tricky. And I can't grab a screenshot because of conflicting keyboard shortcuts, so I'll have to walk you through it sans screenshot.

First, select a data range you want to move, then hover over any edge of the selection. Your cursor will become a hand. Now Shift-drag the selection where you want it. You'll see guides letting you know where you're about to drop the range.

Paste Chart Formats

Let's say you format a chart exactly the way you want it (the chart on the left). Then you want to create another chart just like it, so you create the second chart…the kinda derpy chart on the right **(Fig 3.77)**.

You can simply copy its format and paste it onto the new chart. To start, select your initial chart, then follow these steps: Copy it ▶ select new chart ▶ navigate to **Home ▶ Clipboard ▶ Paste ▶ Paste Special**. This will open a tiny version of the **Paste Special** dialog **(Fig 3.78)**. Choose **Formats** and click **OK** to close the dialog **(Fig 3.79)**.

The only thing that didn't copy over was my Chart Area border, but that's easily fixed.

As you'll see in the Set It and Forget It chapter, if you plan to use a chart type more than once, you should create a chart template from it. Learn more in the Create a Store of Chart Templates section.

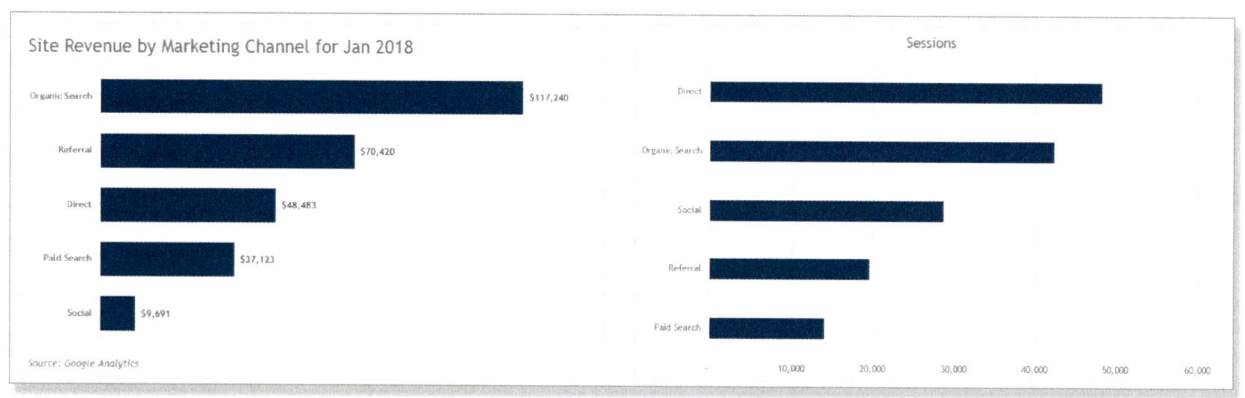

Fig 3.77: We'll copy the format from the chart on the left onto the chart on the right.

Fig 3.78: Aww, a baby Paste Special dialog!

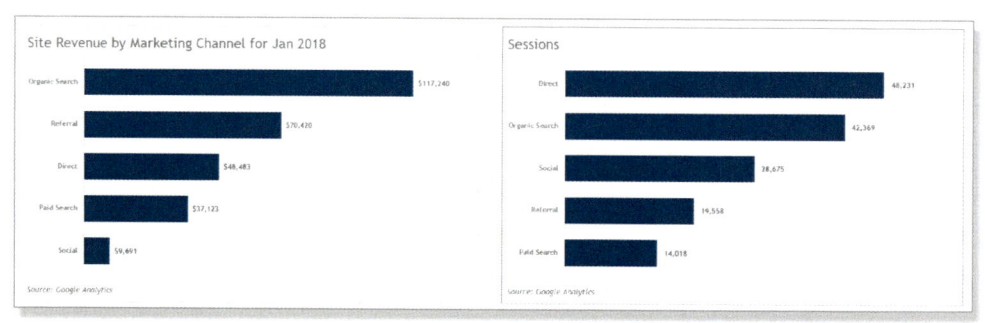

Fig 3.79: Copying chart formatting isn't a perfect solution, but it's a great option for ad hoc analysis.

CHAPTER 3 · DESIGN WITH ELEGANCE

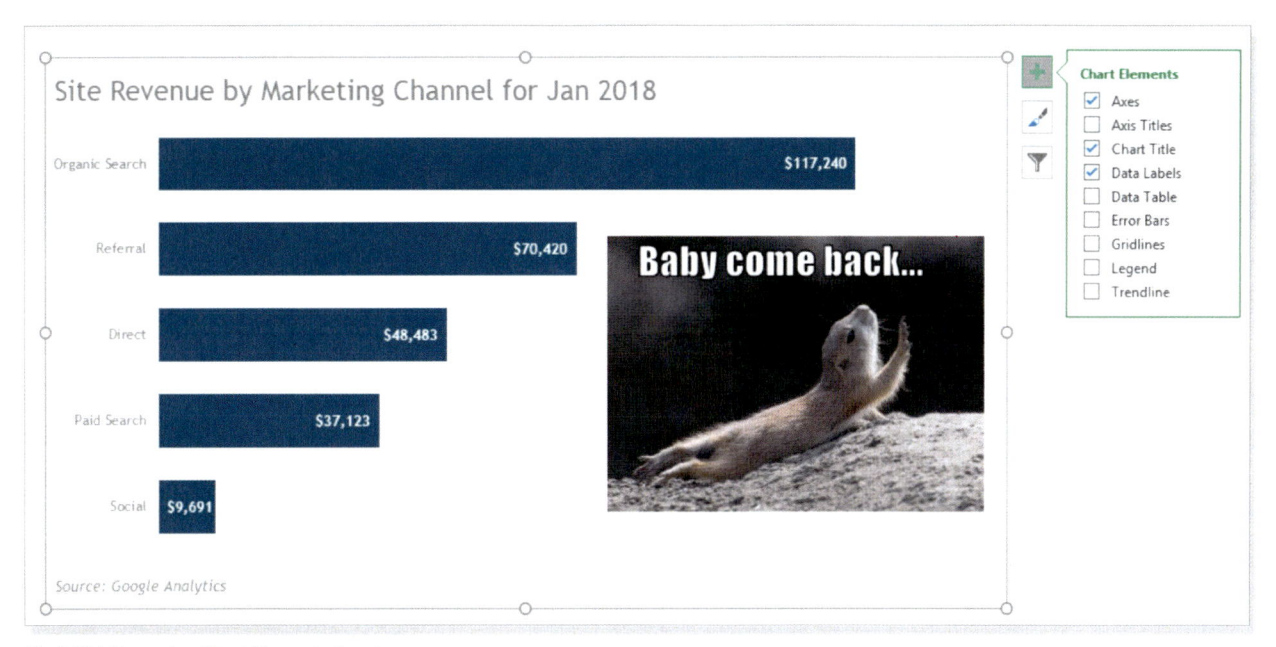

Fig 3.80: Life needs a Chart Elements flyout menu.

BRING SEXY BACK

If at any point you regret your minimalist ways and want to add elements back in, you can do that by clicking anywhere in your chart and selecting the chart element(s) you want to bring back into the fold from the **Chart Elements** flyout menu **(Fig 3.80)**.

Alternatively, you can add chart elements back in by navigating to **Chart Tools ▶ Design ▶ Chart Layouts ▶ Add Chart Element**.

79

CHAPTER 4

Set It and Forget It

GOOD DATA GONE BAD

HIDDEN SCHMIDDEN

In 2008, Lehman Brothers was facing imminent bankruptcy and chose to sell some of its contracts to Barclays. A second-year associate at the law firm Lehman Brothers contracted to handle the minutia assigned a first-year associate the task of reformatting a list of contracts from an Excel spreadsheet to a PDF.

One problem: The Excel file contained a hidden column with a list of 179 contracts Lehman Brothers wasn't including in the deal. This golden column became unhidden in the conversion, creating a massive headache and embarrassment for both Lehman Brothers and the law firm. In the end, Barclays had to swallow the losses on the 179 contracts.

LEARN MORE bit.ly/mdsepc-lehman.

$125 MILLION UP IN SMOKE

In 1999, Lockheed Martin helped build and operate NASA's Mars Climate Orbiter, worth $125 million. Its engineers provided acceleration data in the English system of measurement (e.g., inches, feet, and pounds) but the navigation team at the Jet Propulsion Laboratory (JPL) used the metric system in its calculations. No one caught the mismatch, and JPL engineers mistook acceleration data measured in pound-seconds for a metric measure of force, newton-seconds. As a result, the orbiter wandered aimlessly through space and eventually burned up in the Martian atmosphere.

LEARN MORE bit.ly/mdsepc-mars.

WHO NEEDS AUTOMATION ANYWAY?

In 2003, JP Morgan needed a new value-at-risk (VaR) model and hired a mathematician and model developer to create it. He used Excel, but instead of automating the process, users had to copy-and-paste values from one worksheet to the next. Canadian power generator TransAlta used this model, and a pasting error caused a shift in the data, which led to the model dividing by a sum instead of an average. As a result, TransAlta underestimated its risk and bought more U.S. power transmission hedging contracts at higher prices than it should have, costing them $24 million USD.

LEARN MORE bit.ly/mdsepc-transalta.

NEED FOR CODIFIED PROCESS

As with most things in the 21st century workplace, repetitive tasks chip away at an organization's efficiency. They also introduce risk that needs to be mitigated, as we saw in the data horror stories ripped from the headlines detailed here.

But the scenarios below could happen to any organization:

- An intern visualizes a dataset with 23 data points with a pie chart, obfuscating an important finding.
- An actuary calculates the median on a column of data, forgetting that there are hidden rows quietly weighing into the calculation.
- An analyst for a real estate agency calculates the average value of homes for sale in a neighborhood

without looking at a distribution of the data, and the price of a celebrity's mansion skews the data and misinforms analysis downstream.

- Data from a study is massaged into conclusions that are published in a top industry journal and reported on by news stations and websites, just to have a competitor point out that the data should have been weighted.

You can't avoid every data disaster, but scripting processes for the collection, analysis, and visualization of data empowers employees to more easily maintain both data and branding integrity. Codifying these processes go a long way toward avoiding the more devastating disasters.

Beyond the grandiose meltdowns that tank profits and careers, there are the everyday losses of productivity because each person in your organization creating visualizations needs to learn (and relearn at times) how to create said visualizations. I've seen organizations spend in the five digits per month on marketing campaigns just to get text-based reports and/or reports with rudimentary visualizations. No branding, no finesse, and chart junk for days.

I've also seen important findings miscommunicated because there was a fundamental lack of understanding about how to interpret the data. Over-reporting something like year-over-year gains because you didn't understand the formula can subject an individual or company (depending on the recipients) to the risk of censure and even termination.

Enough said. Let's jump into the gritty details of how to make your processes and output repeatable.

TEMPLATIZE EVERYTHING

You've learned how to create formatted tables, pivot tables, and text boxes that will support the visualizations you'll learn how to build in the next chapter. I've also stressed the importance of designing with your brand in mind—and even data shamed the DOJ. (Sorry, guys! It wasn't personal!) Exceptions to this rule should be extremely rare. If your manager or client specifically requests a visualization that's not on brand and is recalcitrant upon appeal, there's not much you can do. In situations like that, I will accommodate these requests—but under duress. But those instances should be the exception, not the rule.

The good news is your sexy charts can be saved as templates. As you'll see in this chapter, creating a chart template makes it accessible to you any time you use Excel on that machine—and other machines if you move these template files. But you may find yourself using other objects in Excel—such as formatted tables, pivot tables, text boxes, shapes, and icons—to augment your workbooks. Even though Excel falls short in making these objects readily available from workbook to workbook (*sigh*), I'll show you some hacks and workarounds that will allow you to access these objects from a global template.

Although the next chapter is arguably the most exciting, because of all the cool visualizations you'll learn how to rock in Excel, I can say without hyperbole this chapter is the most important. I would even go so far as to say that templates are Excel's best-kept secret. Even if you were just comfortable with a handful of visualizations, this chapter could save you hours over the course of a year, depending on how much your job requires you to work in Excel.

CREATE A STORE OF CHART TEMPLATES

You may find, in your sojourn, that there are some charts you develop well-worn paths to and may even want others in your organization to be able to use, for the sake of consistency. It takes time to get your charts exactly the way you want them. And any time spent on repetitive tasks is wasted time.

CHAPTER 4 • SET IT AND FORGET IT

To save you and your team this repetition of effort, you can save your favorite charts as templates. Chart templates save time and aid you in maintaining a consistent look and feel to your organization's visualizations.

How to Create a Chart Template

Once you get your chart the way you want it, just right-click on it and select **Save as Template (Fig 4.1).**

I know. It's almost anticlimactic to give this its own section. But it's just this easy.

There are a few caveats to be aware of though:

+ At the time of writing, none of the charts new to 2016—namely the Histogram, Waterfall, Pareto, Box & Whisker, Treemap, and Sunburst charts—can be saved as chart templates. I'm hoping this is an oversight and will be fixed in a future update.

+ Not all customizations will be saved in a template. For example, the Clustered Column with Mean Line Chart doesn't save as a template. Only the bar chart survives. Consequently, you'll want to test the templates from the book, if you purchase them, as well as any you create for yourself or your organization, and fix any issues before approving them for use by others in your organization.

+ Be careful not to tweak individual settings before saving as a template. We've all been there: You have a design element that won't cooperate—such as a data label that doesn't fit inside one bar too well—so you grab it and nudge it over a bit. (Based on a true story.) No one will know, right? Well, if you save that chart as a template, someone will know because all future charts you create based on that template will have that last data label nudged a little to the left. D'oh!

+ Don't modify any colors in your chart before spawning off a template; they will be hardcoded into the template. If you're creating charts for your organization, you should be working from a customized theme. If you are, Excel will use your theme colors.

💬 **LEARN MORE** See the Stick to Branded Colors section of the Design with Elegance chapter to learn about the importance of restricting your design choices to your branded colors.

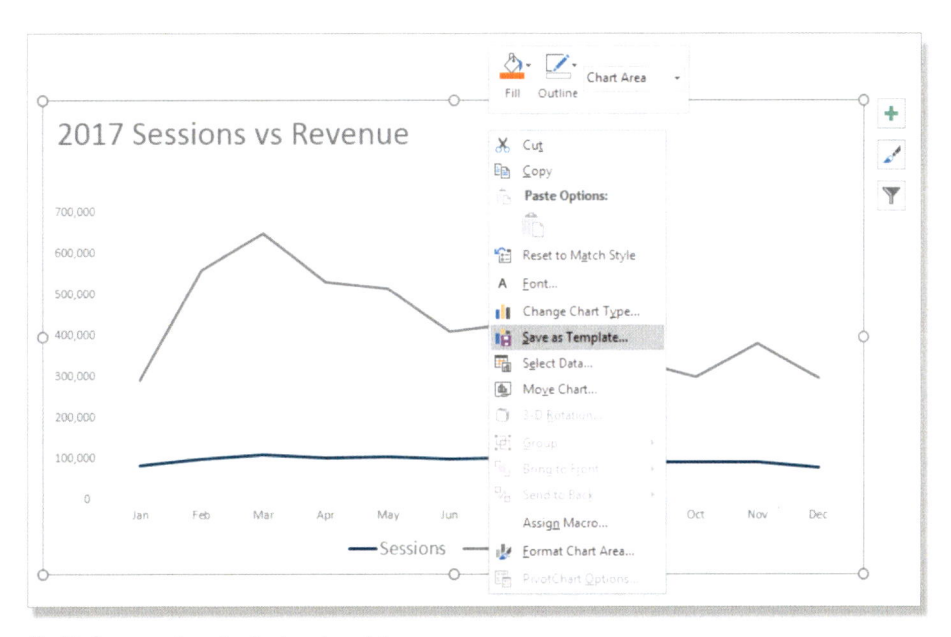

Fig 4.1: Save your favorite charts as templates.

83

- If you go crazy with customizations, don't wait until you get to the end of your souped-up chart to create a template. It can be frustrating to have to walk back your changes to figure out where your template broke.

- If customizations are lost in your chart template, document the changes that will need to be added in for the benefit of others using your chart template downstream. Check out the Clustered Column Chart tutorial to see an example of how I've done this. Each tutorial has a Template(s) section.

How to Apply a Template

To apply a chart template to a new dataset, you won't be able to use the ribbon. You'll need to hit up the menu. Navigate to **Insert ▸ Charts ▸ See All Charts** (the arrow in the bottom-right corner of the **Charts** group). In the **Insert Chart** dialog, open the **All Charts** tab and select **Templates** from the left sidebar. When you hover over a template, you'll see a preview, which is helpful, but not always an accurate picture of what your chart will ultimately look like.

Tips

Here are some tips to help ease the learning curve with chart templates:

- Chart templates use the file extension .crtx and are saved to the following directory: `C:\Users\[name]\AppData\Roaming\Microsoft\Templates\Charts`.

- Give your chart templates intuitive names. For example, I recommend including the chart type (bar, radar, scatter, etc.) and any embellishments (e.g., a mean line). For example, if you create a column chart with a mean line, you might want to save it as column-mean.crtx.

- Prepend chart template names with categories to make them easier to sift through. For example, you may want to have categories for different departments in your organization (e.g., finance-column.crtx) or clients (e.g., clientx-column.crtx).

- Chart titles won't carry over to your new charts. However, if your chart contains a text box (like with your data source or observation), all the text in it will be included in every chart you create from that template. For that reason, you'll want to remove the actual data source or observation so that it doesn't appear in every chart you create. For example, I frequently include the data source in my charts, so I leave a text box with Source: in my chart (usually the bottom-left corner) to prompt me to add my data source, but I remove the actual source before saving off the template. If you think you might want to use a version of that chart without a source text box, you might want to save a version of that template off that doesn't include the text box at all.

- Anything you've removed from your chart before you save it as a template can be easily added back by selecting the chart and reselecting that item from the **Chart Elements** flyout menu that appears as a plus sign in the upper-right corner of your **Chart Area (Fig 4.2)**.

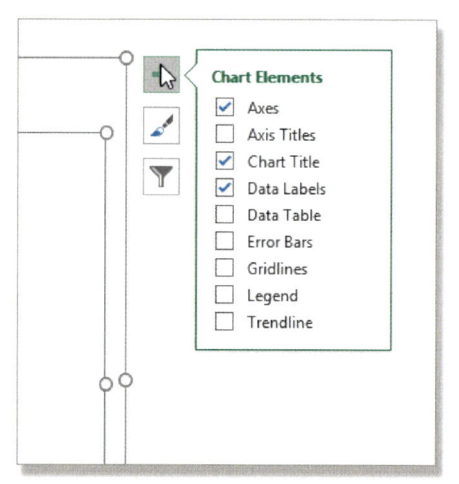

Fig 4.2: Add chart elements back in from the Chart Elements flyout menu.

- If a chart won't save as a template, I recommend saving it in a file where you store all your fancy charts that won't save as templates. You could maybe call this file the Island of Misfit Data. (I mean, that's what I would do, but you might have less job security—or greater maturity—than I do.) If you go to all this trouble, I'd recommend making this file available to others, so other people in your organization can benefit from your hard work and expertise.

CREATE A BRANDED TEMPLATE

The templates we'll cover in this section have nothing to do with chart templates; they're completely separate beasts. And there's no overlap between them, so don't get confused with the shared terminology. I'll sometimes refer to them as global templates, but I'm borrowing from a term used more in Word than Excel. However, I'll use it if I think there's any potential for confusion with chart templates.

Broken but Hackable

Global templates in Excel are under-baked in their default functionality. It's great that we can save off charts as chart templates with relative ease. You can also save macros, toolbars, custom menu settings, and shortcut keys in Excel's global templates. You cannot, however, save formatted tables, pivot tables, text boxes, icons, or shapes as global defaults; you can only save them in the file you're working in.

However, Debra Dalgleish—who runs one of my favorite Excel blogs, **contextures.com**—published a pivot table hack that made me wonder if I could do the same thing with other objects in Excel that fall outside the reach of templates. (You can read Debra's post here: **bit.ly/mdsepc-debra-hack**.) I felt like a mad data scientist, testing Excel's limits. But I eventually figured out how to cram all my styles into a global template.

Steps to Creating a Template

I'll break the process of creating and distributing your branding strategy into four steps.

Step 1: Collect your brand colors and fonts
Hunt Down the Branding Guide

Most organizations have their brand colors and fonts stored in a branding guide somewhere. This guide usually lists the organization's colors, fonts, and even approved symbols. For example, I would imagine Accenture has various versions of its logo (which looks like a greater than sign) that it makes available to its designers and content creators. Use this document to generate your template.

Use Chrome Developer Tools

In the rare case you don't have access to this document (or it doesn't exist), you can use Chrome Developer Tools (CDT) to find out the colors and fonts a website uses—or a tool like it.

I assumed that the DOJ would be unwilling to share their branding guide with me so I could recreate its charts using their branded colors and fonts. No one sets out to be a cautionary tale. So I used CDT instead to get everything I needed. Discretion is the better part of valor.

To find a color or font, right-click on an element on the web page and look in the **Styles** tab, which hugs the right edge of the browser. This is where you can access the website's custom style sheets (CSS), which is where most websites store the colors used on their website.

Colors are especially easy to find. Just scan for a color bar in the **Styles** tab. The hexadecimal value is to the right. Excel can't accept hexadecimal values; it uses RGB (short for red, green, blue) values. A quick trick to acquiring the hex value is to drop it into Google with the hash (e.g., #004A6D). Most times the RGB value comes up in the search results page.

CHAPTER 4 • SET IT AND FORGET IT

Fig 4.3: Find a website's branded colors and font(s) by examining its code.

To find the font, I search for the tag font-family. If there's more than one reference to font-family, look for the one that's not crossed out **(Fig 4.3)**. Rinse and repeat for each of the elements that contain the color/font you want to obtain. You don't even need a search warrant. #dojhumor

Step 2: Create a branded theme
One of the worst mistakes you can make in Excel, from a design standpoint, is to use Excel's default settings. That includes its default colors and fonts. With that in mind, let's cover a few steps to avoid becoming a data viz casualty.

Customize Colors
SET EXCEL'S COLOR THEME
Once you have your branded colors from the keeper of your company's branding, you'll want to bring them into Excel by creating a custom color palette. To do this, navigate to the **Page Layout ▸ Themes ▸ Colors ▸ Customize Colors**.

This will open the **Create New Theme Colors** dialog. Here you'll be able to drop your branded colors. There's a preview in the top-right corner of the menu, but it can be difficult to interpret.

For this book, I'll be using a palette that a designer friend, Kristina Sambucci, created for me **(Fig 4.4)**.

Fig 4.4: Use your custom color palette to build all your visualizations.

Note: I actually fat fingered the gray color in converting it from hexadecimal values to RGB (something you need to do on a PC) but was already finished creating all the screenshots when I discovered it. This book contains just under 500 screenshots. I wasn't about to redo them for a tweak that's barely discernible to the

86

CHAPTER 4 • SET IT AND FORGET IT

human eye. If we're being honest, I wouldn't redo them even if was glaring.

All visualizations will use this palette to demonstrate how I align every visualization in this book with a singular (albeit completely fictional) brand. Maintaining this kind of alliance requires setting policy within your organization that visualizations use only your branded colors.

Pictured here is the custom theme I created for this book **(Fig 4.5)**.

Tip: If you drop one of your custom colors into the first drop-down (**Text/Background–Dark 1**), you could have funky monkey colors in your chart titles, gridlines, chart border, etc. **(Fig 4.6)**. So, if you want more standard grayscale colors for these chart embellishments, you'll want to stick to black or a dark gray.

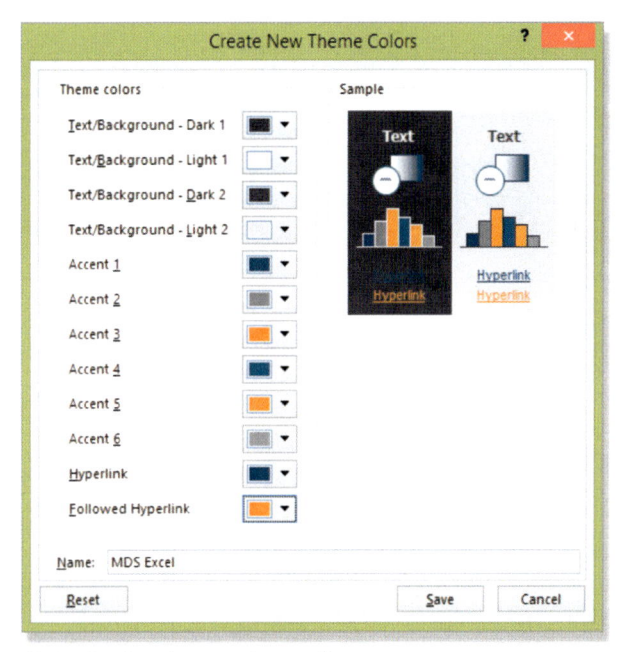

Fig 4.5: Set the colors your theme will use.

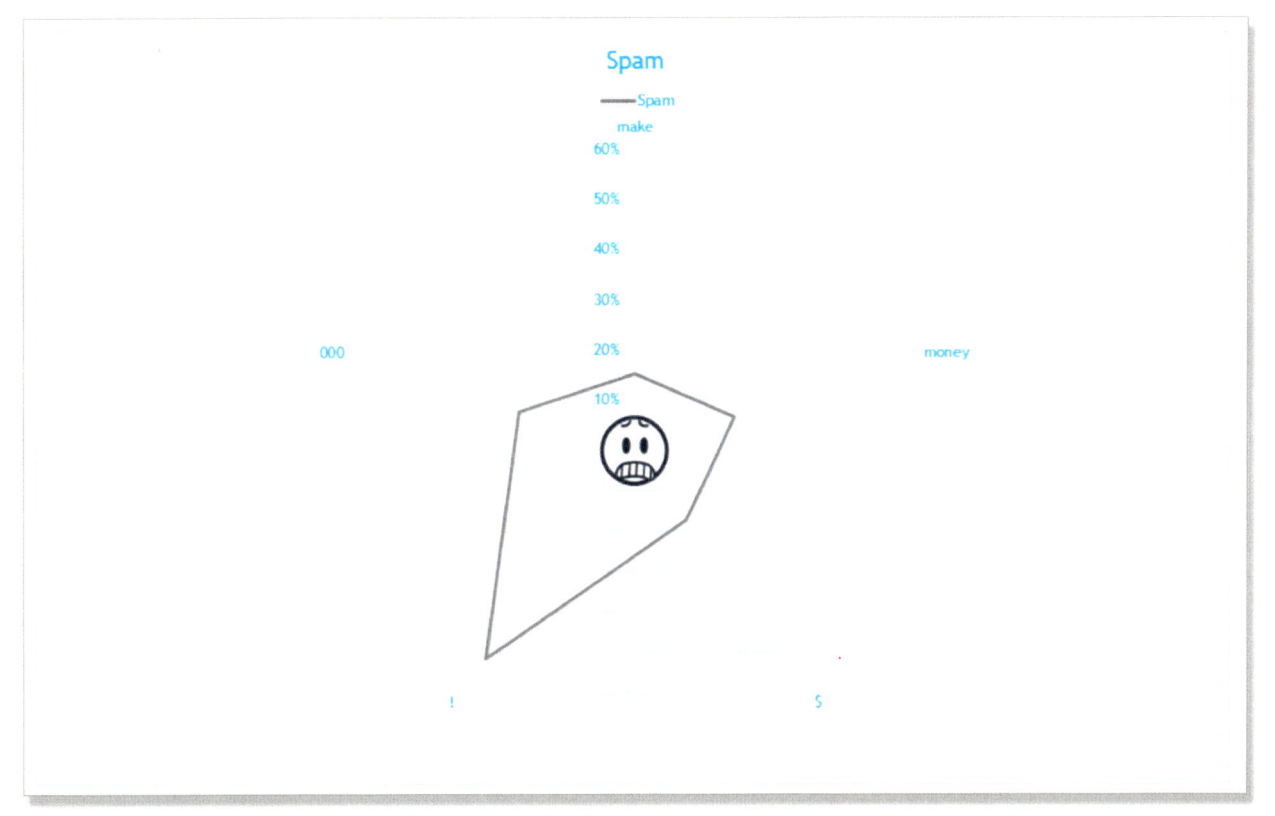

Fig 4.6: Customizing your theme colors can produce unexpected results.

87

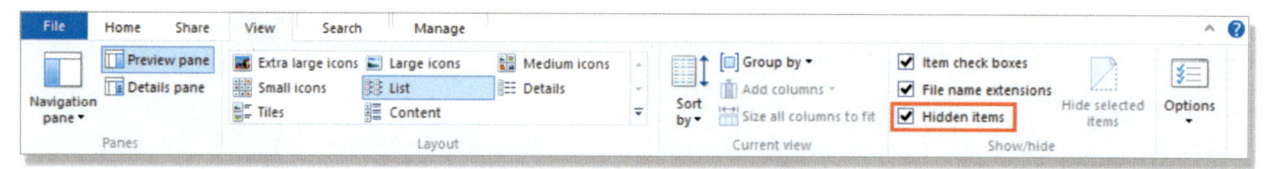

Fig 4.7: You'll need to enable hidden items in Windows Explorer to access the AppData directory.

MAKE THEME COLORS GLOBALLY AVAILABLE

To make your custom theme available to all in your organization, you just need to drop it into Microsoft's Theme Colors directory, which you can get to by navigating to:

```
C:\Users\[Your User Name]\AppData\
Roaming\Microsoft\Templates\
Document Themes\Theme Colors
```

If you don't see the AppData directory, you'll need to opt into showing hidden items by navigating to **View ▸ Show/hide** and selecting **Hidden items** in Windows Explorer **(Fig 4.7)**.

When you create a custom color theme in one Office program, it will be available in others because they all reference this Theme Colors directory. In Word it's available under **Design ▸ Colors**. In PowerPoint, it's available under **Design ▸ Variants ▸ Colors**.

You can learn more about Excel themes on the **office.com** website: **bit.ly/mdsepc-ms-themes**.

Customize Fonts

While you're at it, you might as well also use your company's font, with a couple of caveats:

- Office doesn't support third-party fonts.
- To ensure your fonts work across platforms, you may want to limit your font choice(s) to those supported by both Mac and Windows computers. You can find a list of these fonts on the **coderwall.com** website: **bit.ly/mdsepc-common-fonts**.

To set your fonts, navigate to **Page Layout ▸ Themes ▸ Fonts ▸ Customize Fonts**. In the **Create New Theme** **Fonts** dialog, set **Heading font (Latin)** and **Body font (Latin)** to fonts of your choosing, then name your theme fonts in the **Name** field and click **Save** to close the dialog **(Fig 4.8)**. The visualizations in this book use Palatino Linotype for headings and Trebuchet MS for body.

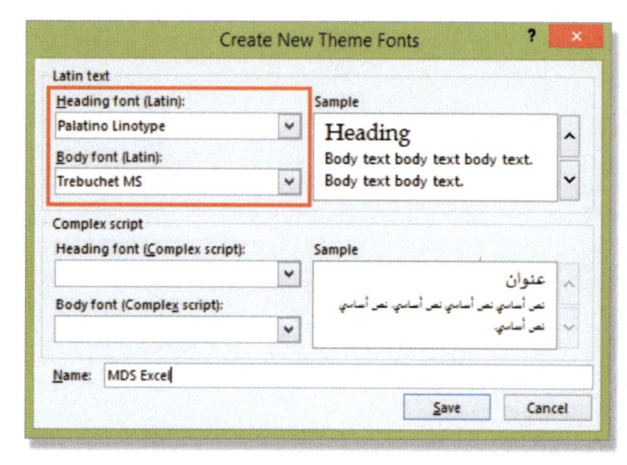

Fig 4.8: Customize your theme fonts.

Customize Effects

Excel comes with a set of 15 preset effects. You get to these by navigating to **Page Layout ▸ Themes ▸ Effects**. For each preset, you'll see a circle, arrow, and rectangle. These icons give you a teeny tiny preview of Excel's subtle, moderate, and intense themes, respectively.

A BETTER WAY TO TEST

I recommend testing them in a file on actual shapes where you can see the different effects each preset uses.

I'll break down the steps I took.

Step 1: Create three shapes in a worksheet by navigating to **Insert ▸ Illustrations ▸ Shapes**. I used the **Rectangle Rounded Corners** shape **(Fig 4.9)**.

Fig 4.9: Start by dragging out three shapes.

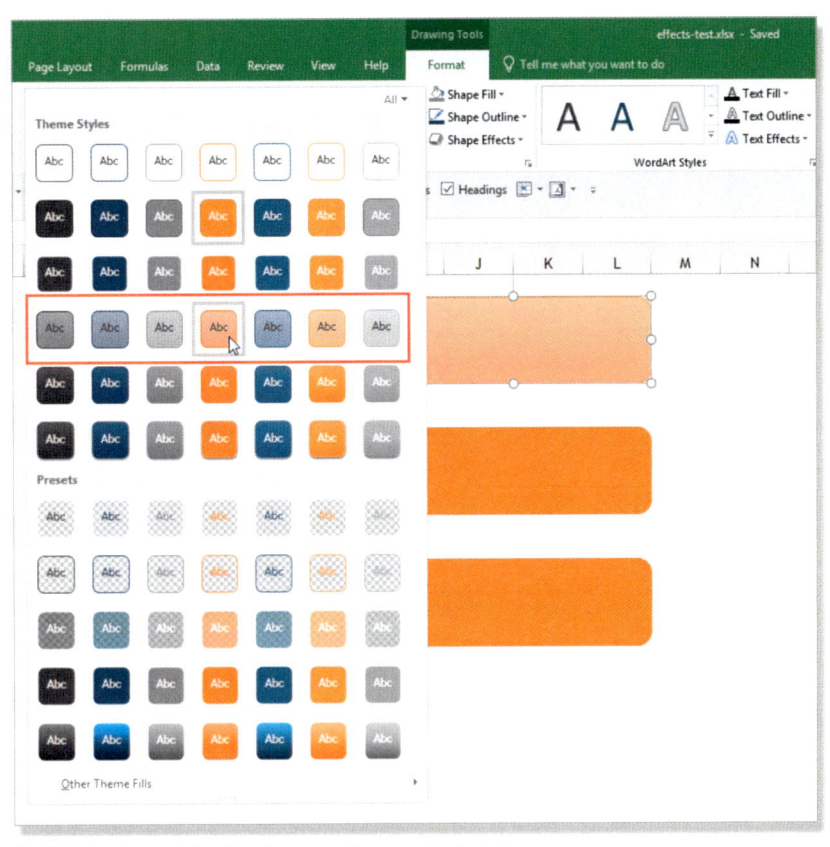

Fig 4.10: Apply a subtle effect from your theme to the first shape.

Step 2: Now we'll assign them an effect from each of these categories. To assign a subtle effect to the first shape, select it and navigate to **Drawing Tools ▸ Format ▸ Shape Styles ▸ Format Shape** (the arrow in the bottom-right corner of the preset window) ▸ any preset in the fourth row down **(Fig 4.10)**.

Warning: It's not terribly obvious that Excel puts these different categories of effects into separate rows. Usually Excel does a better job with labeling these categories.

Step 3: Rinse and repeat with the moderate and intense effects, using one of the presets from the fifth and sixth rows, respectively **(Fig 4.11)**.

Fig 4.11: Each shape represents an effect level.

89

CHAPTER 4 • SET IT AND FORGET IT

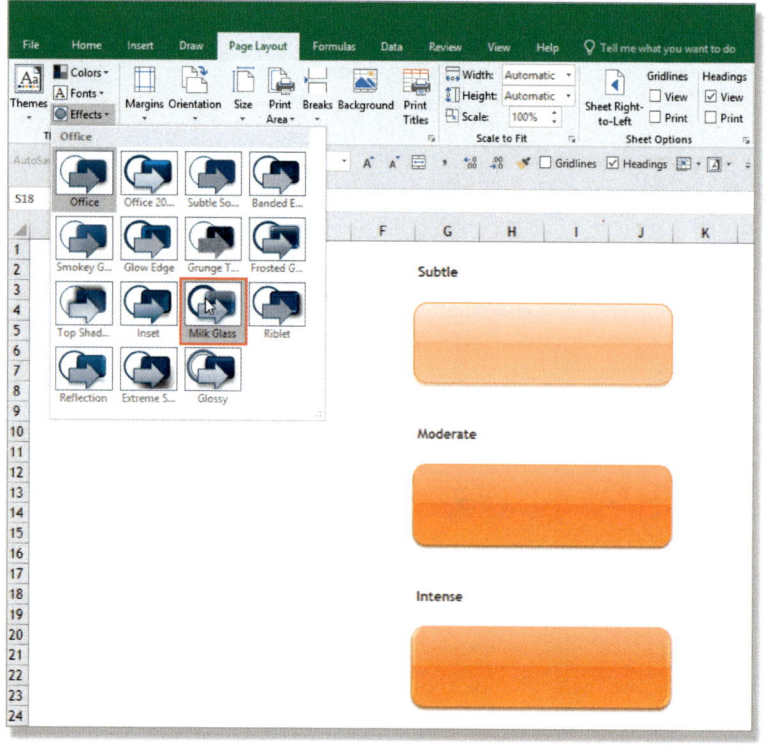

Fig 4.12: I see you, Milk Glass effects!

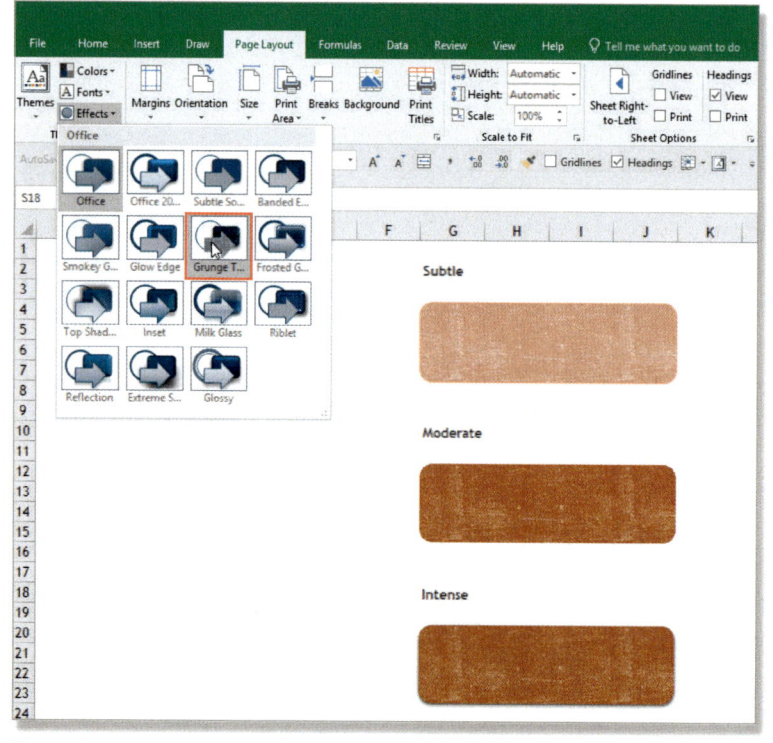

Fig 4.13: Want an effect to match your poorly groomed hipster beard? Excel's gotchu!

Step 4: Now, to see a preview of each preset's effects, hover over each of them and watch your shapes update. For example, here's a preview of the Milk Glass preset **(Fig 4.12)**.

And, if you run a super hipster startup and want your visualizations to have an edgy, grungy texture to them, Excel will let you do that **(Fig 4.13)**. Excel, you off the chain!

I like the default effects, so I used the Office effect in my theme. I wanted to like one of the other ones more, so I had an excuse to customize. I used to think I was cool and edgy, but Excel proved me wrong.

Credit: Shout out to Bill Jelen ("Mr Excel") for the inspiration to test effects on shapes. You can see his video here: **bit.ly/mdsepc-theme-effects**.

Save Your Theme

Once you've set your color, font, and effects, you'll want to save your theme by navigating to **Page Layout ▸ Themes ▸ Themes ▸ Save Current Theme**. This will save your theme as a .thmx file and drop it in the following folder:

`C:\Users\[Your User Name]\AppData\Roaming\Microsoft\Templates\Document Themes`

You'll need your network administrator to make this theme file universally accessible.

90

CHAPTER 4 · SET IT AND FORGET IT

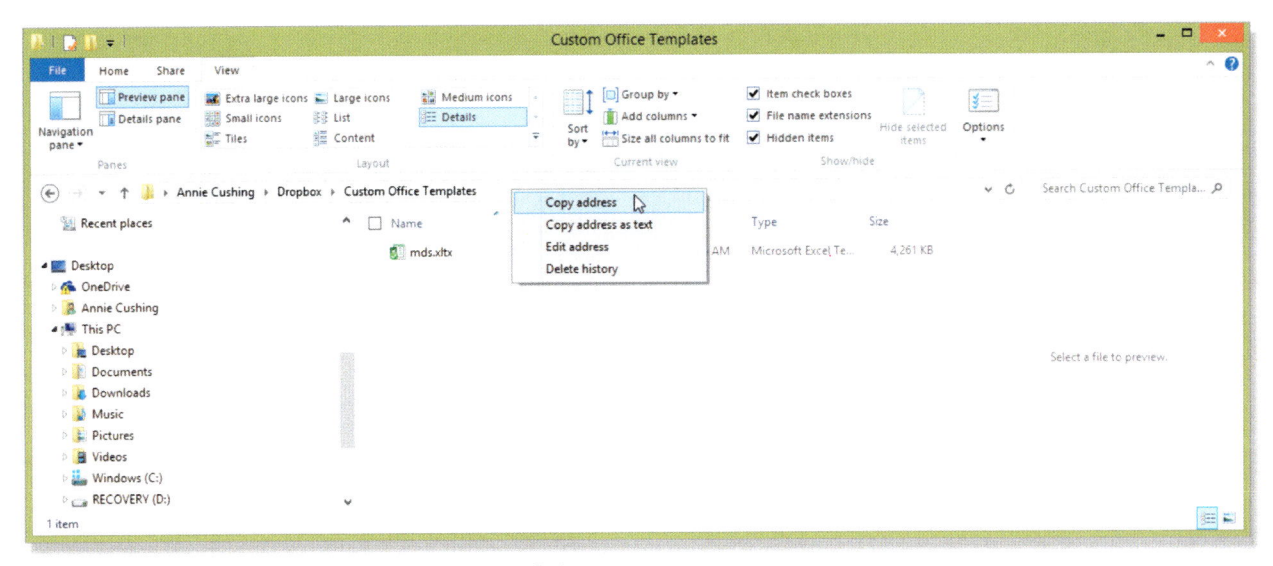

Fig 4.14: Grab the file path to your default personal templates folder.

💬 LEARN MORE See the Distribute Themes and Templates section of this chapter to learn more about making your themes and templates available to your organization.

Step 3: Create a global template
Okay, all aboard the Hack Train to Hackville!

The best way to explain what we'll be doing in this step is with the kids' game, Tag. Bear with me a minute. I'm too visual for my own good and think about everything in analogies. But when you're 7 and playing Tag, someone only needs to touch you for a split second for you to become It.

Saving items to a global template is a lot like Tag. All your formatted item needs to do is touch this magical file, and it will be available in your template. And if you set that item (table, pivot table, text box, etc.) to be the default style, it will be the default style for any workbook created from that particular template.

This analogy breaks down a bit in that you don't delete children who are tagged. But you can delete your objects once you've set them as your default (available from the contextual menu).

Let's break the process down a bit. I will also share some of the pitfalls I tumbled into unceremoniously as I was testing.

Step 1: If you've never saved a workbook as a template, you'll need to first set your default personal templates location. First, create the folder you want to use somewhere that's both easily accessible and backed up to the cloud. (You don't want to lose this folder in a crash!) I created a folder I called Custom Office Templates in Dropbox, so I'd be able to access it from any of my computers. If you're working on a network, I recommend setting your folder somewhere accessible to all who will need it.

Step 2: You'll need the file path to feed Excel. To copy it to your clipboard, right-click the **Navigation Bar** in Windows Explorer and choose **Copy address** or **Copy address as text (Fig 4.14)**.

91

CHAPTER 4 · SET IT AND FORGET IT

Fig 4.15: Let Excel know where your template file(s) is stored.

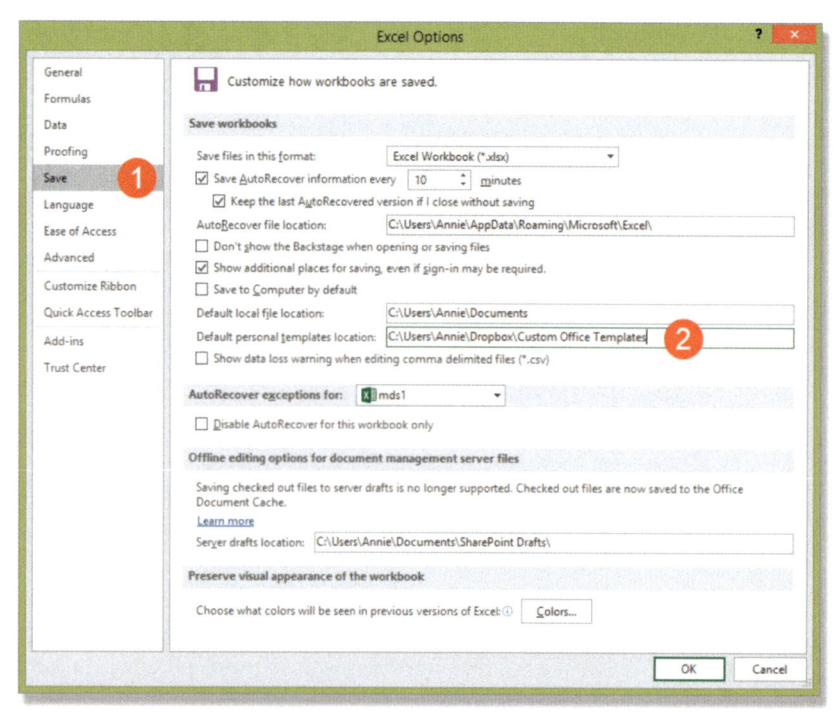

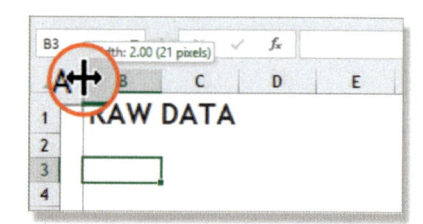

Fig 4.16: I give every worksheet I work with a 2 pt margin, so I added it to my template.

Step 3: Back in Excel, navigate to **File ▸ Options**, which will open the **Excel Options** dialog. Select **Save** in the left navigation menu, then set **Default personal templates location** to your folder by pasting the file path address you copied in the previous step **(Fig 4.15)**. Then click **OK** to close the **Excel Options** dialog.

Note: Excel will add a trailing backslash to this file address after you save it.

Step 4: Open a new file (Ctrl-N). Format your worksheet(s) to taste. There are a few things I always do to a file when I start working on it, so I'll take these steps with my template file:

✦ Apply my custom theme, which I select by navigating to **Page Layout ▸ Themes ▸ Themes ▸ Custom**. If you apply your theme to the global template, you won't need to do that for each new file. If you haven't created a theme yet (tsk tsk), make sure you come back and apply it to your global template at some point.

✦ Remove gridlines (**View ▸ Show ▸ Gridlines**). Aside from using them to line up my visualizations (although you can Alt-drag them to snap to the grid without enabling gridlines), I see no good use for gridlines. If you use formatted tables, your data will have lines without assaulting viewers with an entire worksheet of gridlines.

💬 **LEARN MORE** See the Align Strategically section of the Design with Elegance chapter to learn how to align objects in your workbook.

✦ Title each worksheet using bold font and set it to 20 pts. Because I always put my raw, unformatted data in a worksheet I call Raw Data, I'll include it in my template and label the worksheet tab. Now, when I add new worksheets to any workbook created from my template, I can copy this worksheet (right-click on tab ▸ choose **Move or Copy** ▸ select **Create a copy** ▸ click **OK**).

✦ Create a left margin for my worksheets by dragging the boundary of column A **(Fig 4.16)**. Do you read books or websites where the text comes right up to the edge? No? Then maybe don't do it in your spreadsheets either?

CHAPTER 4 • SET IT AND FORGET IT

Step 5: If you've created any formatted objects, such as a text box, shape, or icon, copy them from wherever you created them and paste them into your template file. Set them as defaults by right-clicking on the object and choosing **Set as Default [Object]** from the contextual menu **(Fig 4.17)**. Then delete them. Anything you leave undeleted will appear in every file you create from that template.

Warning: With text boxes, you need to right-click on the border. If you right-click inside the text box, you'll get options for your text.

Step 6: If you've created a custom formatted table you might want to access from any file you create from this template, paste it into your file, then delete it. Now navigate to **Home ▸ Styles ▸ Format as Table** and open the drop-down menu. Like magic, your custom table formatting is there! If you use the keyboard shortcut Ctrl-T to create formatted tables, set it as your default. To do this, right-click on your custom style under **Home ▸ Styles ▸ Format as Table** (or **Table Tools ▸ Design ▸ Table Styles** if you've selected a cell inside your formatted table), then choose **Set As Default** from the contextual menu **(Fig 4.18)**.

Step 7: If you've created a custom pivot table style, you can't just copy-and-paste it into your template file. Instead, you need to copy the worksheet containing your custom pivot table into your template file. To do this, right-click on the tab of the worksheet and choose **Move or Copy** from the contextual

Fig 4.17: Paste object in, set it as default, then make it walk the plank.

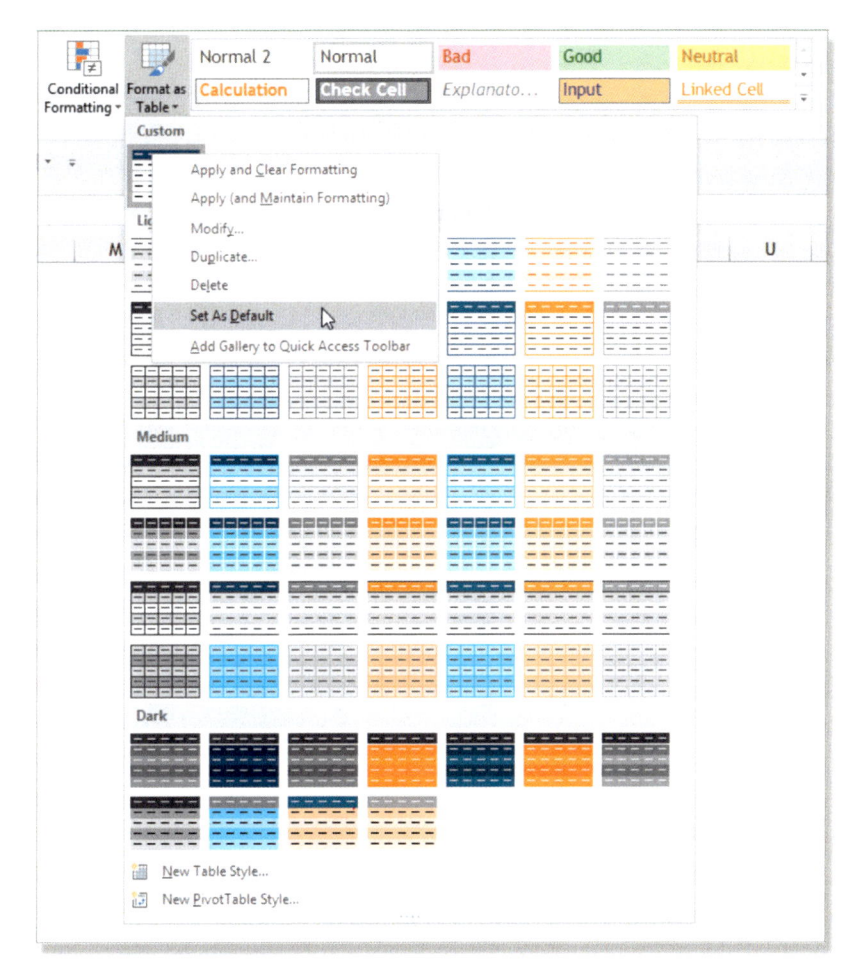

Fig 4.18: Set your formatted table as the default table style.

93

CHAPTER 4 · SET IT AND FORGET IT

menu. In the **Move or Copy** dialog that pops up, choose your template workbook from the drop-down menu and select **Create a copy (Fig 4.19)**. If you don't select this checkbox, Excel will move your pivot table worksheet from your original file, prompting a custody battle between your workbooks. When you click any cell inside your pivot table, you should see your custom pivot table style in the **Design** tab of the **PivotTable**

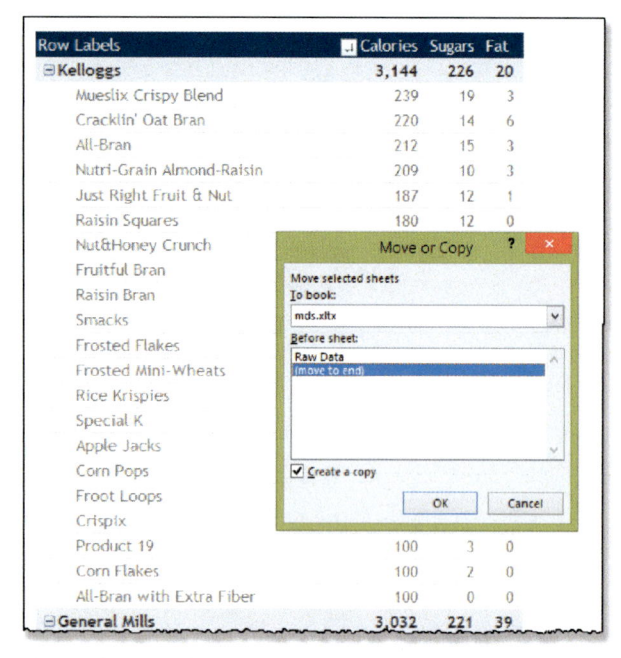

Fig 4.19: Copy your formatted pivot table and select your template file

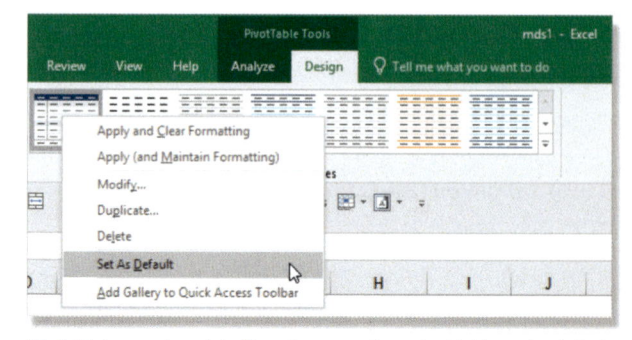

Fig 4.20: In your template file set your custom pivot table as the default.

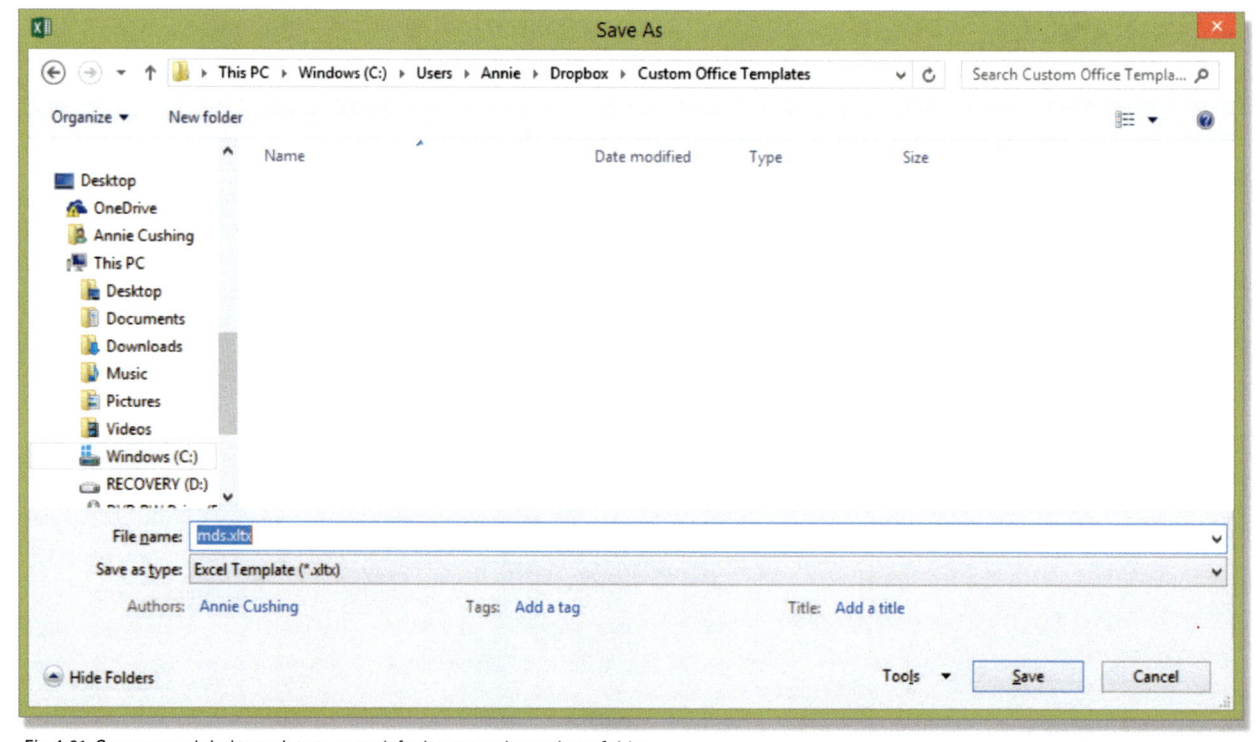

Fig 4.21: Save your global template to your default personal templates folder.

94

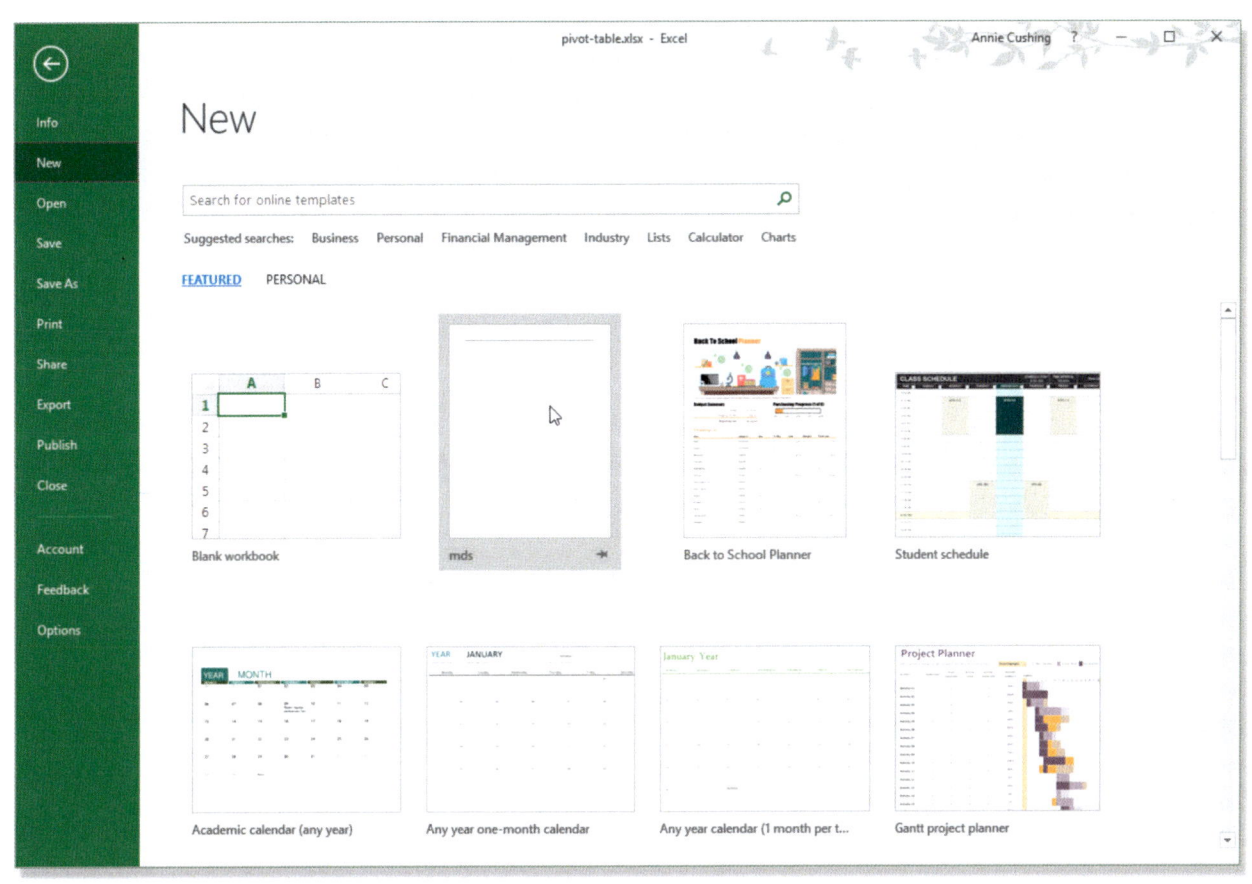

Fig 4.22: Create a file from your template.

Tools menu. Right-click it and choose **Set As Default** **(Fig 4.20)**. Then delete that worksheet.

💬 **LEARN MORE** See the Design a Pivot Table Style section of the Massage Your Data chapter to learn how to align your pivot tables with your branding.

Step 8: Think through any other repetitive tasks you always do to a new workbook, as well as objects you frequently work with: arrows, boxes, colorized icons, etc. This is your chance to stockpile them into your template. Paste each of them into your template file, set them as the default, then delete them.

Step 9: When you're finished modifying your template, navigate to **File ▸ Save As ▸ Browse**, which will open the **Save As** dialog. Navigate to your default personal templates folder, then set **Filename** to an intuitive template name and choose **Excel Template (*.xltx)** from the **Save as type** drop-down **(Fig 4.21)**.

Step 10: To create a file from your template, instead of using Ctrl-N, you'll need to navigate to **File ▸ New** and select your template **(Fig 4.22)**. You'll also see a slew of templates Excel makes available, which can be a source of inspiration for design elements.

Warnings from My Testing

✦ You can modify your template file by opening it and making your changes, but Excel imposes an odd restriction: If you open your template by double-clicking on it in Windows Explorer, Excel

CHAPTER 4 • SET IT AND FORGET IT

quietly creates a copy of your template file **(Fig 4.23)**. There's no indication that Excel has spawned off a copy of your template other than Excel quietly appending a 1 to the end of the filename. It's Excel's way of saying, "I'll protect you from yourself. You're welcome." So, if you want to modify your template, you'll need to either open it by navigating to **File ▶ Open** or right-clicking the filename in Windows Explorer and choosing **Open** from the contextual menu. If you forget and discover you've modified a copy of your template file, just overwrite your original file.

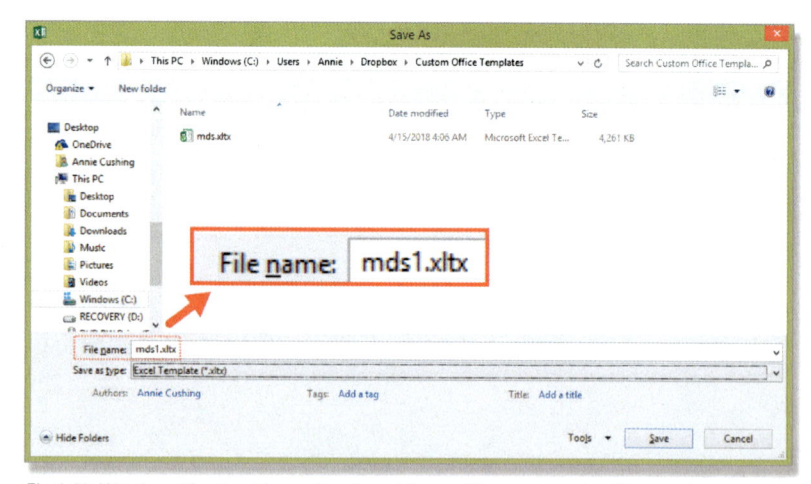

Fig 4.23: Watch out for Excel's sneaky attempt to modify your template filename!

Credit: Shout out to Debra Dalgleish from **contextures.com** for providing the workaround to avoid creating a copy of your template file.

- You can't create a file from your template if the template file is open. You'll get a nasty warning **(Fig 4.24)**.

- During a couple of my tests, I couldn't copy a pivot table worksheet into a template file **(Fig 4.25)**. I would get an error message that said, "Permissions to this workbook are restricted." But other times I was able to add a worksheet containing a pivot table without issue. If you get this error, I discovered a workaround: Create a standard workbook that's based on your template following step 10 in the global template tutorial. Then copy your pivot table worksheet into that workbook. Finally, delete that worksheet, and save your workbook as a template in your template folder. This will overwrite your template file.

- After modifying a template file, test all your formatted objects to make sure they still work. In two of my tests, my formatted text box went MIA. #irishgoodbye

- If you ever see a regular Excel file (*.xlsx) in your template folder, you most likely forgot to update the filetype when you saved it. Open it and save as a template file. I may or may not know this from personal experience.

Fig 4.24: Close your template file before creating a file from it.

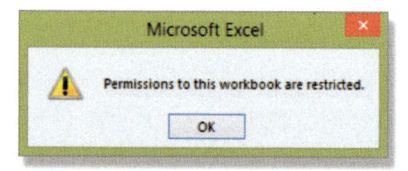

Fig 4.25: Copying a pivot table worksheet into your template can cause data drama.

DISTRIBUTE THEMES AND TEMPLATES

This critical part of the process will fall squarely under the role of your organization's network administrator. But the organization doesn't benefit from the global templates, chart templates, or themes unless they're available to employees and possibly even contractors, if they're responsible for reports that include branded visualizations. Ergo, making these resources available to employees will need to be baked into the organization's visualization workflow.

96

I reached out to Microsoft engineers on Twitter to see if there's an easy way to update these files without sending them out to everyone in your organization who needs access to them. They didn't know.

So I asked my network on Facebook if there was a more elegant solution. (I want to make sure I've turned over every stone for my readers!) You can follow that thread here: bit.ly/mdsepc-template-post.

Paul Wilder, who's a network administrator, answered my question this way:

> I just looked at Excel chart templates briefly and it seems that Microsoft has locked down custom user templates to only be selected from C:\Users\<user_name>\AppData\Roaming\Microsoft\Templates\Charts; so custom chart templates saved on the computer under one user account wouldn't even show up for another user on the computer unless the user manually copied the file to their own …\Templates\Charts folder.
>
> I attempted to create both a sub-folder shortcut and a specific template file shortcut to a network share without success. I even tried replacing the \Charts folder for my user profile with a network shortcut without success.
>
> It also seems there is no way to change this default folder location. Unfortunately, it would appear there isn't a way to make it work, short of convincing Microsoft to allow the selection of an alternate folder in a future product revision.

He added this in a message:

> If someone was really determined to make it work, something could likely be scripted to push content to domain-connected computers from a file server location. However, that would be a file-based solution, not actually anything you could do via Excel.

If you're a system admin and come up with a workable solution that I can link to in future revisions of the book, please jump on that thread. I'll be happy to credit you.

I also highly recommend creating a file that contains unhidden copies of all the elements in your global template. This way, you have a record of your customized objects and something to reference when you want to test all your objects. I would also include a worksheet with a copy of all the charts you used to create your chart templates, for the purpose of documentation.

After you're finished pulling all your template items through the keyhole backwards, make this file available to everyone in your organization, so they know the branded elements they can choose from. The more instruction you can offer, the better your chances are that employees will maintain the organization's branding in all the visualizations they create—and even contribute elements and charts to the template file!

ASSIGN SOMEONE TO MANAGE PROCESS

If not managed properly, even the best systems will succumb to chaos and fall into disrepair. For this reason, it's essential to assign a gatekeeper to facilitate and manage the creation of your organization's visualizations, including formatted tables, pivot tables, charts, text boxes, shapes, symbols, smart art, and other objects your organization frequently uses in visualizations. This person should be savvy in creating elegant visualizations. Learn how to use the techniques detailed in this book, and that person might just look a whole lot like YOU!

Imagine if the DOJ had someone who was the keeper of their data visualizations—someone who inspected their visualizations for accuracy, branding, and elegance before they were published. Imagine if said gatekeeper created chart and global templates, as well as theme files, and made them accessible to everyone else in the organization. I would have had to keep looking for a cautionary tale to use in the book!

CHAPTER 5

Create Charts

Roll up your sleeves because in this chapter we'll jump into learning how to create 60+ visualizations in Excel! Some are charts that are built into Excel, whereas some are the result of clever Excel users figuring out how to hack Excel to get the chart types they need to make their data tell a compelling story.

There are several approaches I could have taken in organizing these charts. The task was not an insignificant endeavor. Obviously, many of the charts could easily fit into more than one category. But after attending an Overthinkers Anonymous meeting, I moved on from my internal torment and let the chips fall where they may.

Also, I formatted each of these sections strategically. When I'm trying a new recipe, I'm not interested in reading an intro paragraph about how I could regale my guests with Parmesan-crusted kale chips I'm about to attempt to make; I want to jump right into the recipe. The only difference here is, instead of jumping right into the tutorial(s), I provide some basic information about each chart type.

However, notes about the charts are distilled down to bullet points to make them more scannable. My goal is to cut out the fluff (within reason) and provide a desktop resource you can refer to as needed.

COMPARISON

BAR CHART

Uses

Bar charts are best used to:

- compare data across discrete groups
- plot datasets with categories that contain long labels because of its horizontal orientation
- visualize data with no time component

Fig 5.1: Our final chart.

CHAPTER 5 • CREATE CHARTS

Tips
- Consider sorting your data to make it easier to read because it saves viewers' eyes from darting back and forth to figure out the order of your data.
- I tend to sort my bars so that the longest ones are at the top of the stack in the chart to emphasize the most significant finding. But I've seen others effectively argue their case for building up to their largest value.
- Look for opportunities to simplify your bar charts by removing gridlines, simplifying your axes, and bringing data into the bars.
- Consider a chart type with less ink if your data has a high number of categories. Too much ink can be visually overwhelming and can even contribute to the Moiré effect, where your bars become almost indistinguishable. Using a lollipop graph can be a great way to reduce pixels.

💬 **LEARN MORE** See the Lollipop Graph section of this chapter to learn how to create one.

- Remove the legend for datasets with one series. If only one person shows up for your birthday party, he doesn't need a name tag.
- Sort your data in ascending order to get your bars in descending order and vice versa.
- Make one of your bars a different color if you want to highlight an observation. Even better: Couple it with an annotation in a text box.

Category / Metric
1 category / 1 metric

Tutorial
Step 1: Select any cell inside your dataset and navigate to **Insert** ▸ **Charts** ▸ **Insert Column or Bar Chart** ▸ **2-D Bar** ▸ **Clustered Bar (Fig 5.2)**.

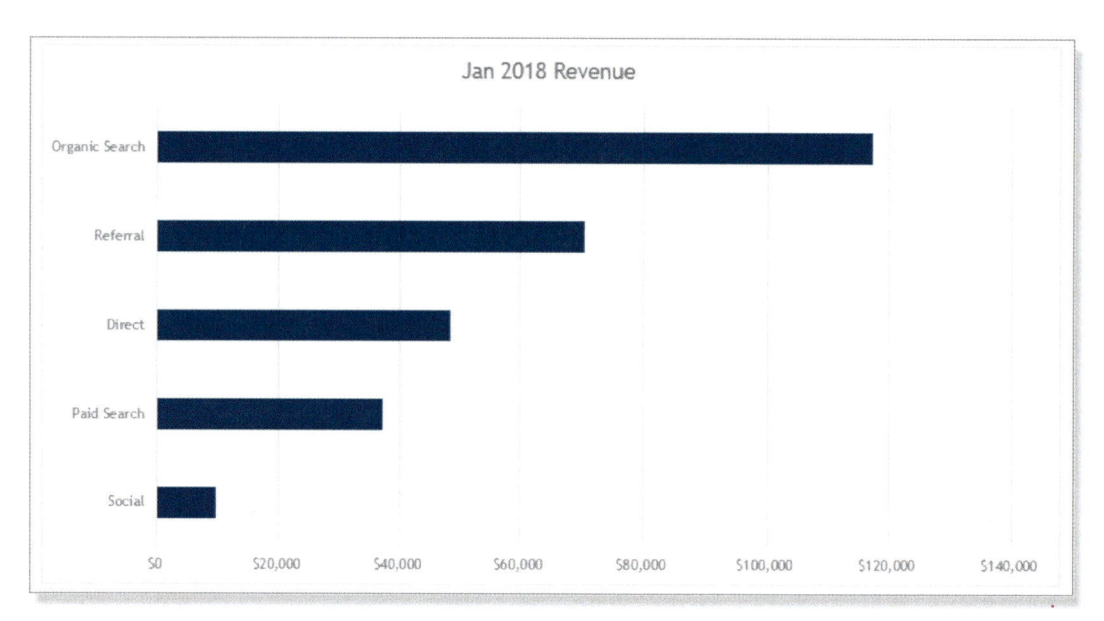

Fig 5.2: Our initial chart.

CHAPTER 5 • CREATE CHARTS

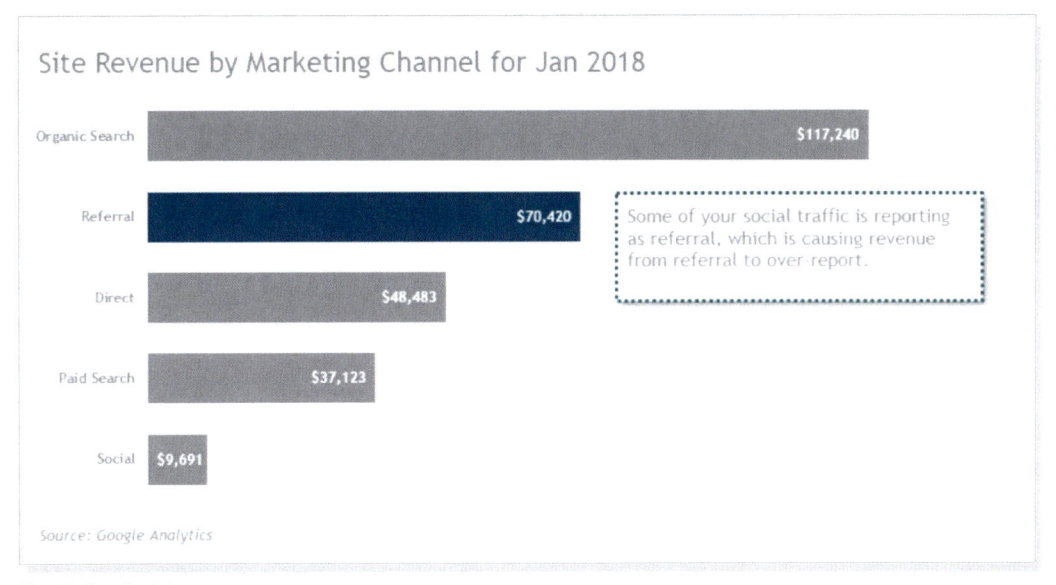

Fig 5.3: Our final chart.

Step 2: Clean up your chart to taste. I cleaned mine by taking the following steps:

+ made title more descriptive
+ left aligned chart title
+ bumped up title font size
+ made chart title dynamic
+ added annotation for data source
+ removed gridlines
+ made bars wider
+ added data labels
+ removed horizontal axis

LEARN MORE See the Design with Elegance chapter to learn how to perform these refinements.

Variations
Highlight with Color
You can use color to highlight a particular data point (a single point in a data series) by graying out the other data series and adding one of your darker colors to the data series you want to spotlight. We'll pick up where the main tutorial left off since this is an embellishment to it.

Step 1: Select one bar to select the entire data series, then press Ctrl-1 to open the **Format Data Series** pane. To change the fill color of your data series to gray, open the **Fill & Line** tab and set **Fill Color** to your theme's gray color **(Fig 5.4)**. Your theme colors appear at the top of the **Fill Color** drop-down.

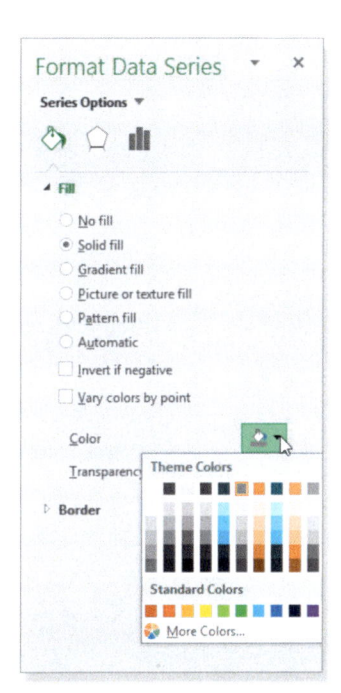

Fig 5.4: Set fill color of data series to gray.

101

CHAPTER 5 • CREATE CHARTS

Step 2: With your data series still selected, click the bar you want to highlight again to isolate it, and set its fill color to one of your theme colors **(Fig 5.5)**.

Step 3: When you highlight a particular data point, it's a good idea to accompany it with a text box. I added my branded text box **(Fig 5.6)**.

> 💬 **LEARN MORE** See the Add Annotations section of the Design with Elegance chapter to learn how to create text boxes.

Add a Mean Line
You can add a mean or target line to your bar chart to help readers see how your data measures against a goal or a measure of central tendency, such as the mean (aka average). You could also use median if your dataset has extreme outliers that can skew the mean—or even mode if your audience is comfortable with statistics.

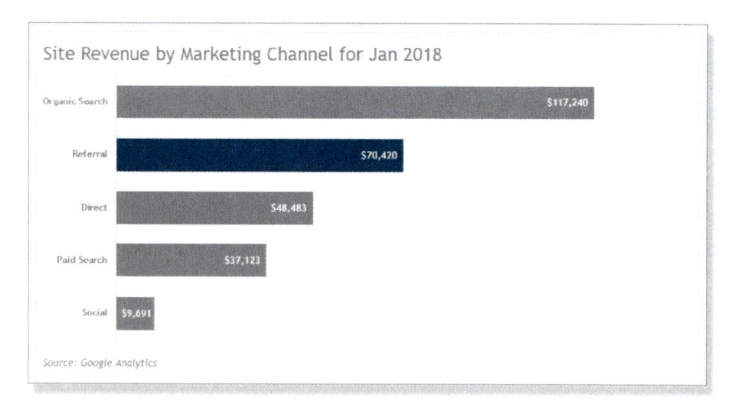

Fig 5.5: Highlight a data point by changing its fill color.

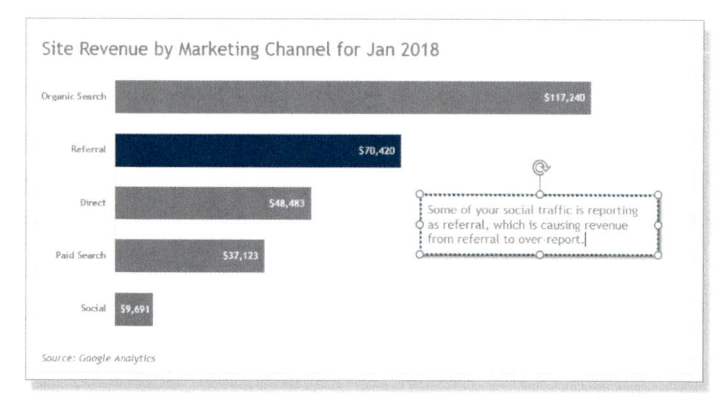

Fig 5.6: Consider adding an annotation to explain why you're highlighting a data point.

Fig 5.7: Our final chart.

CHAPTER 5 • CREATE CHARTS

Hundred-Foot View

We'll create an auxiliary dataset that we'll use to add a scatter plot with lines to our bar chart. Our x values will be the mean, and our y values will be 0 and 1. Given the size of our y values, we'll need to put this data series on a secondary axis and set the maximum value of that axis to 1. This data series will give us a vertical line that uses the mean to align itself along the x axis and extends the full height of the **Plot Area**. Then we'll delete the secondary axis to hide it. Finally, we'll format it and add a data label that points to a cell that combines text and our mean value.

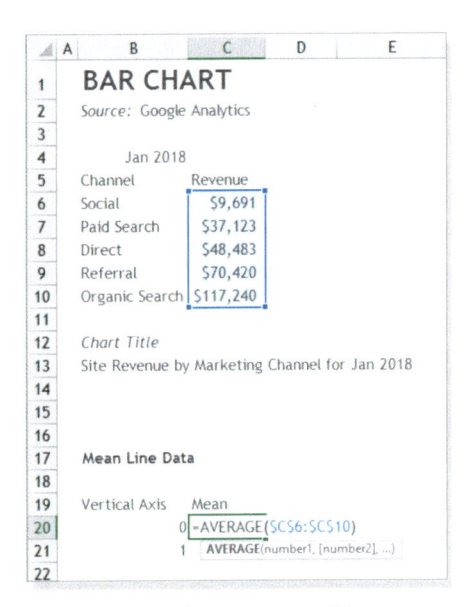

Fig 5.8: It's weird, I know. Just stay with me.

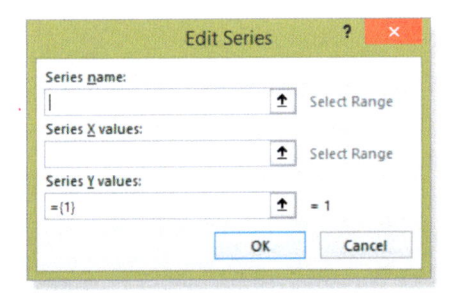

Fig 5.9: Add new data series for mean line.

Step 1: Create a table with two columns and three rows. In your first column (what I labeled Vertical Axis), enter 0 in the first row and 1 in the second row. This sets the span of our vertical mean line in the chart. In the second column (what I labeled Mean) calculate the mean, or average, for the metric you're charting **(Fig 5.8)**.

Step 2: Right-click anywhere inside the chart's **Plot Area** and choose **Select Data**. In the **Select Data Source** dialog, click the **Add** button under **Legend Entries (Series),** which will open the **Edit Series** dialog **(Fig 5.9)**.

Step 3: Set **Series name** by first clicking inside the field, then selecting the column header for your mean column. Set **Series X values** by clicking-and-dragging over the column on the right with your mean or target data. Set **Series Y values** by dragging over the values in your Vertical Axis column—or whatever you called it **(Fig 5.10)**. Click **OK** to close the **Edit Series** dialog and **OK** again to the **Select Data Source** dialog.

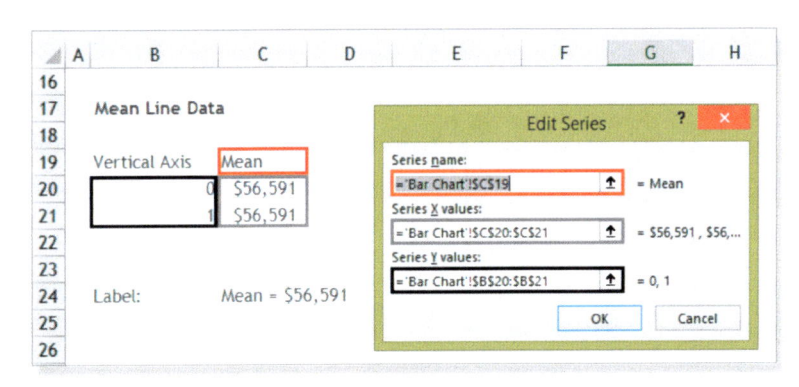

Fig 5.10: Create the data series for your mean line.

103

Step 4: Your mean (or target) data series will show up as two bars at the bottom of your chart. Right-click on either bar and select **Change Series Chart Type** from the contextual menu **(Fig 5.11)**. If you can't grab it, with your chart selected, navigate to **Chart Tools ▸ Format ▸ Current Selection** and select it from the drop-down menu.

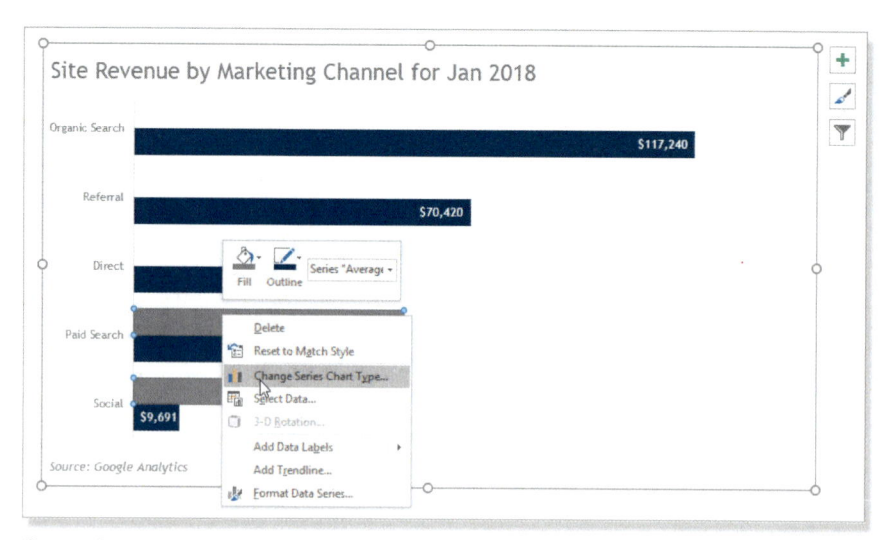

Fig 5.11: Change the chart type.

Step 5: In the **Change Chart Type** dialog, set Mean (or Target) to **Scatter with Straight Lines** by selecting it from the drop-down menu, and select **Secondary Axis (Fig 5.12)**. Click **OK** to close the **Change Chart Type** dialog.

Step 6: Now let's set the maximum value of the secondary axis (the axis to the right of our chart) to 1 so that our vertical line stretches to the top of the **Plot Area**. To do that, select the secondary axis and press Ctrl-1 to open the **Format Axis** pane. In the **Axis Options** tab, under **Bounds**, set **Maximum** to 1, then delete the secondary axis **(Fig 5.13)**. It's served its purpose.

Step 7: Now let's add some finesse to that mean line. It's kind of thick and chunky. Select the line to pull up the **Format Data Series** pane. Select the **Fill & Line** tab, then reduce **Width**. I set mine to 1.5 pt. I then set **Dash type** to **Round Dot (Fig 5.14)**.

Step 8: If you want to get really crazy, you can add a data label that identifies the metric and value. We'll create a data label that says, "Mean = [Avg value]." First, let's program the text we want in the data label in a cell. I used this formula:

`="Mean = "&TEXT(G5,"$#,##0")`

The number code in the formula sets how your text will be formatted. You always need to do this if you dynamically reference data from a cell in any kind of text box (title, data label, text box, etc.).

Tip: To check my code, I just formatted the data in my table the way I wanted it, then selected the cell and pressed Ctrl-1 to pull up the **Format Cells** dialog. If you aren't working with currency, remove the dollar sign, and if you want two decimals, add ".00" to the end:

`="Mean = "&TEXT(G5,"#,##0.00")`

> 💬 **LEARN MORE** See the Pull in Values section of the Design with Elegance chapter to learn how to use the TEXT function to dynamically update chart elements.

Step 9: Now that we've stitched together the text that will go into our data label, let's add our data label. First, select the mean line, then select **Data Labels** from the **Chart Elements** flyout menu. Doing this will add a 0 to the bottom of the line and a 1 to the top. This will never do. To resolve this issue, select one of the data labels to select both and press Ctrl-1 to open the **Format Data Labels** pane, if it doesn't open automatically. In the **Label Options** tab, under **Label**

CHAPTER 5 • CREATE CHARTS

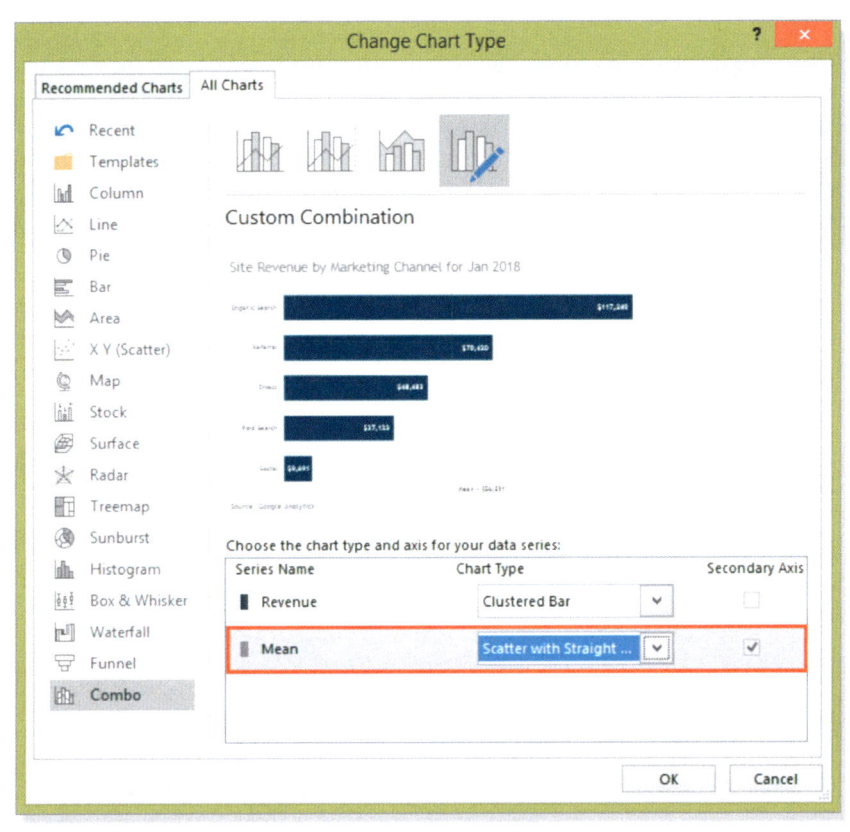

Fig 5.12: Create a scatter plot that connects the two dots with a line.

Fig 5.13: Set the Maximum Bounds of the mean line to 1 to cause it to extend the full height of the Plot Area.

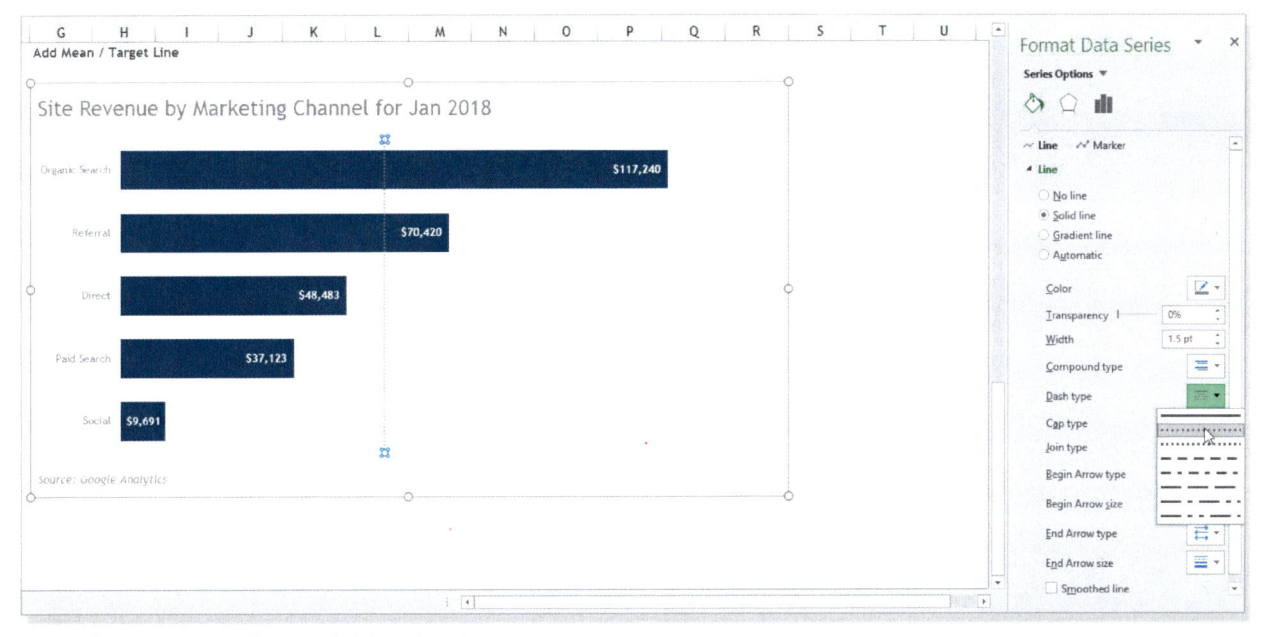

Fig 5.14: Format your mean line to make it less obtrusive.

105

CHAPTER 5 • CREATE CHARTS

Options, select **Value From Cells**. This will open the **Data Label Range** dialog. Select the cell that has your data label and click **OK** to close the **Data Label Range** dialog **(Fig 5.15)**.

Step 10: We don't need the 1 data label at the top of our mean line. To remove it, deselect **Y Value** in the **Label Options** tab of the **Format Data Labels** pane. You can also modify where the data label is positioned in the **Label Position** section of the **Label Options** tab. Excel's default (**Right**) works well for my chart, but you may want to experiment with other positions.

Credit: Shout out to the DataNumen blog for the inspiration to use a scatter plot to get the vertical line: **bit.ly/mdsepc-avg-line**. I always wondered why anyone would ever connect the dots in a scatter plot. Now I know!

Example File
bar-chart.xlsx

Templates
If you'll use these charts again, save them as chart templates. Or download the templates included in the chart templates folder:

+ Bar Chart: mds-bar.crtx
+ Bar Chart with Highlight: mds-bar-highlight.crtx

Note: The mean line chart doesn't save as a template. Only the bar chart survives. The mean line gets ground into a fine powder and blown into the wind. Ergo, you will need to add it manually each time you create this chart. If you want to use the column chart with mean line, at least the mean line survives; you just need to add the data label back in. But the column chart requires less hackery to add a mean line than the bar chart.

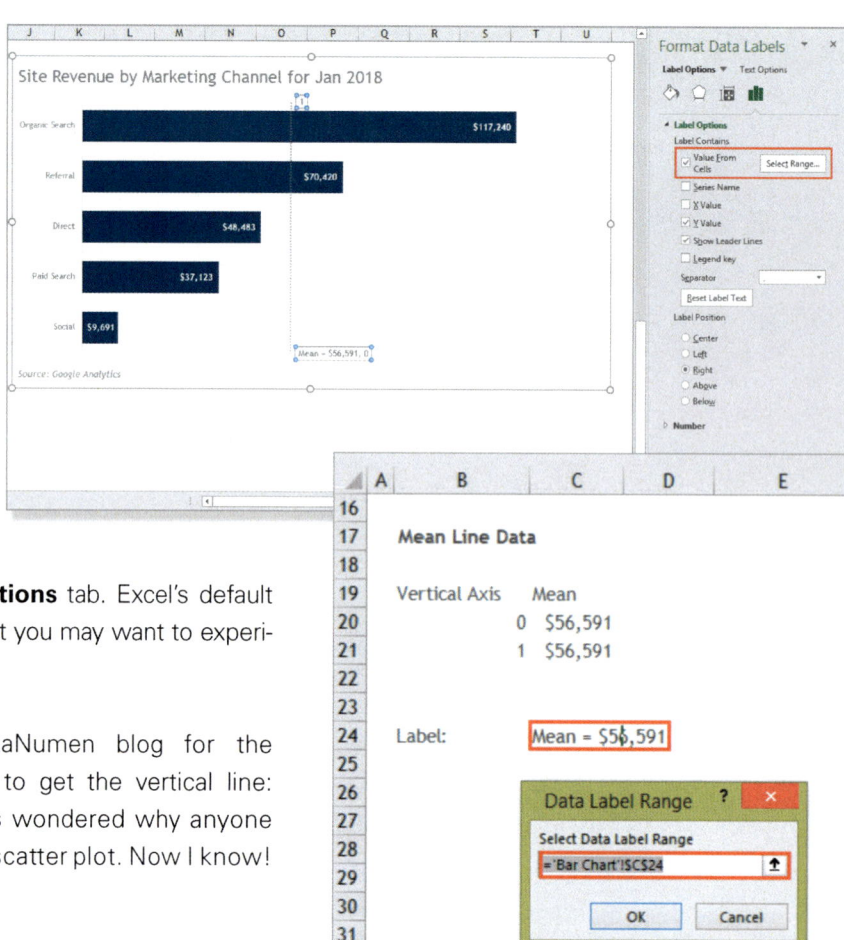

Fig 5.15: Add a value to your mean line using the Value From Cells option.

CHAPTER 5 • CREATE CHARTS

COLUMN CHART

Uses

Column charts are best used to:

- compare data across discrete groups
- visualize categorical data with short labels
- plot data with no time component
- plot data with an aggregated time component (e.g., months, quarters, years)

Tips

- Avoid using vertical column charts for datasets that have long category labels because this requires viewers to tilt their heads to read them (one of my greatest pet peeves).
- For column charts with many columns and a time element, consider labeling every other data point (e.g., Jan, Mar, May, etc.)
- Consider sorting your data to make the data easier to read. Sorting saves viewers' eyes from darting back and forth to figure out the order of your data. Unlike the bar chart, sorting is more intuitive: sorting in descending order puts the longest bar first.
- Consider a chart type with less ink if your data has a high number of categories. That much ink can be visually overwhelming.
- Remove the legend if you're only charting one data series.
- You can use color to highlight an observation by making all the columns except the one you want to highlight gray. Even better: Couple it with an annotation in a text box.

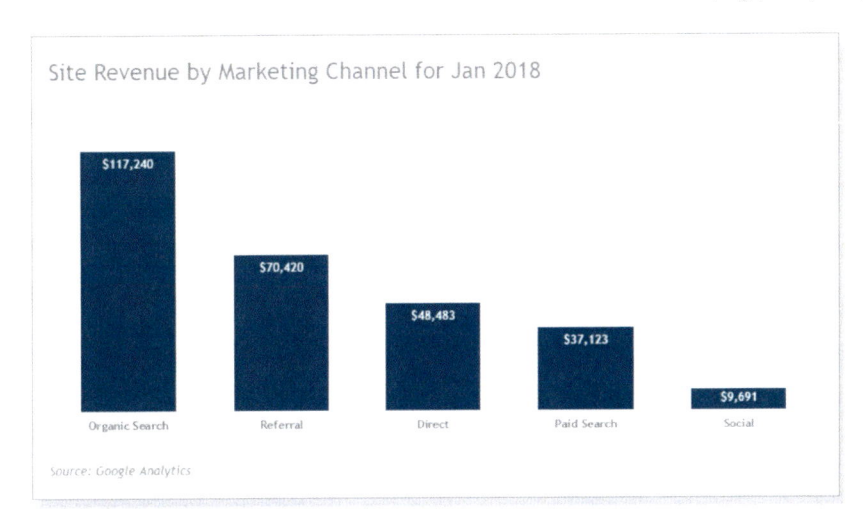

Fig 5.16: Our final chart.

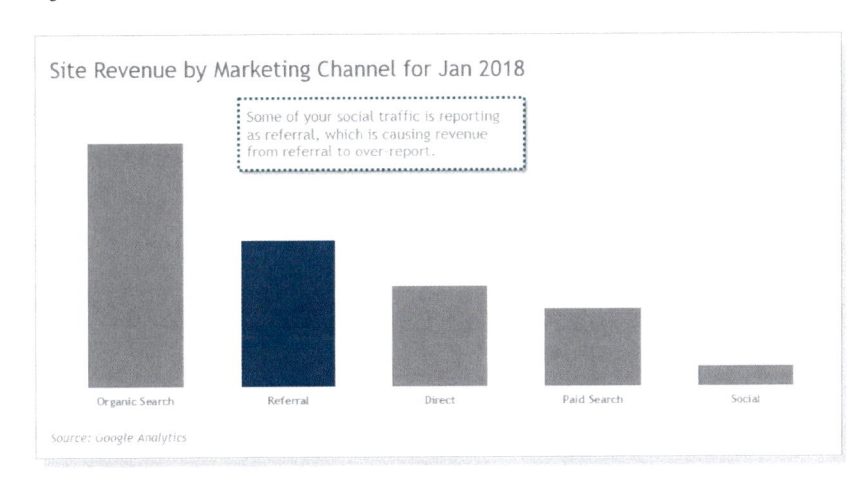

Fig 5.17: Learn how to emphasize a column using color.

LEARN MORE See the Thin Out Axes section of the Design with Elegance chapter to learn how to reduce clutter in your axes.

107

CHAPTER 5 • CREATE CHARTS

💬 **LEARN MORE** See the Highlight with Color variation of the Bar Chart tutorial to learn how to highlight a data series or point (Fig 5.16).

Category / Metric
1 category / 1 metric

Tutorial

Step 1: Select any cell inside your dataset and navigate to **Insert ▸ Charts ▸ Insert Column or Bar Chart ▸ 2-D Column ▸ Clustered Column (Fig 5.18)**.

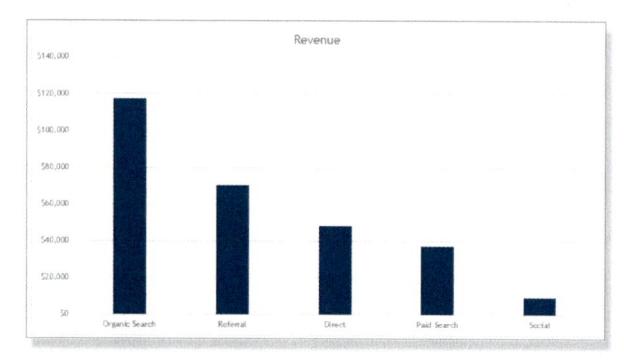

Fig 5.18: Our initial chart.

Step 2: Clean up your chart to taste. I cleaned mine by taking the following steps:

+ made title more descriptive
+ left aligned chart title
+ bumped up title font size
+ made chart title dynamic
+ added annotation for data source
+ removed gridlines
+ made bars wider
+ added data labels
+ removed vertical axis

💬 **LEARN MORE** See the Design with Elegance chapter to learn how to perform these refinements.

Variations

Highlight with Color

You can use color to highlight a particular data point (a single point in a data series) by graying out the other data series and adding one of your darker colors to the data series you want to spotlight. We'll pick up where the main tutorial left off since this is an embellishment to it.

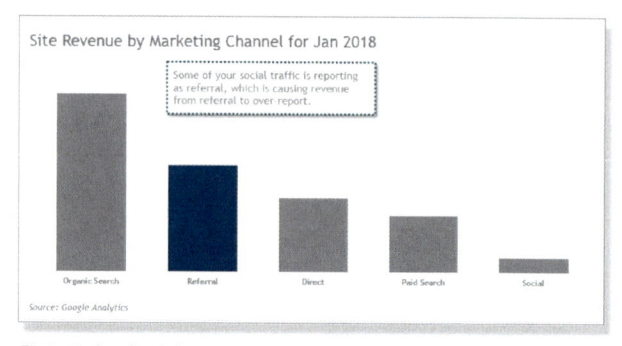

Fig 5.19: Our final chart.

Step 1: Select one bar to select the entire data series, then press Ctrl-1 to open the **Format Data Series** pane. To change the fill color of your data series to gray, open the **Fill & Line** tab and set **Fill Color** to your theme's gray color. Your theme colors appear at the top of the **Fill Color** drop-down menu.

Step 2: Select the bar you want to highlight and set its fill color to one of your theme colors **(Fig 5.20)**.

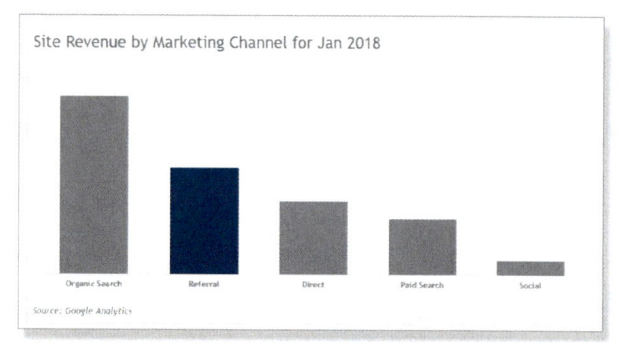

Fig 5.20: Highlight a data point by changing its fill color.

CHAPTER 5 • CREATE CHARTS

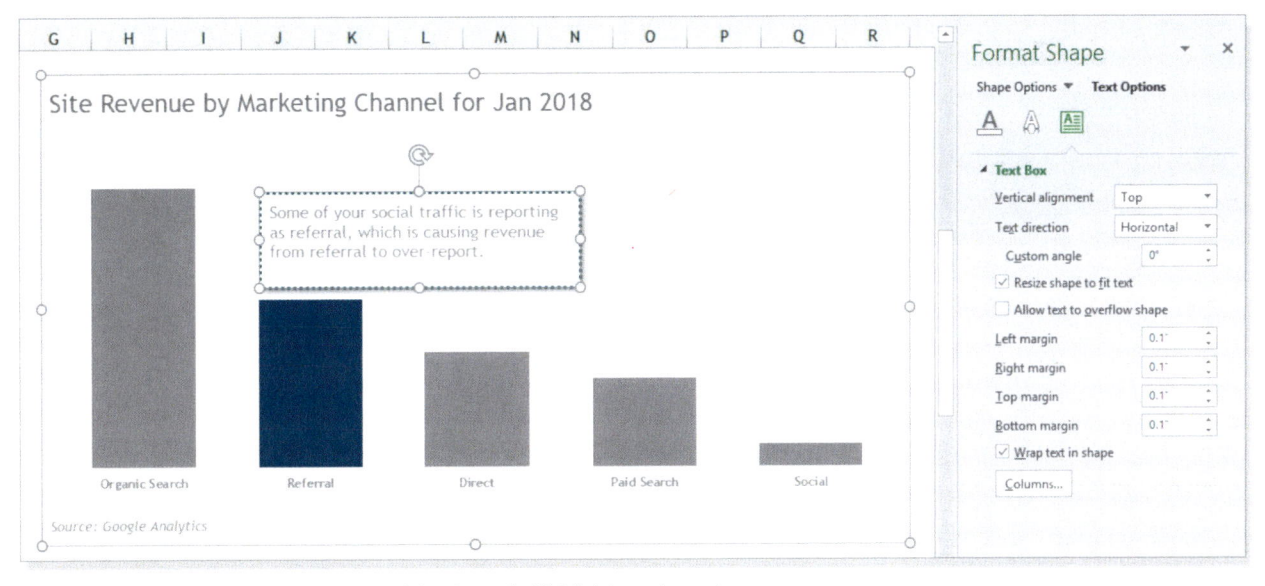

Fig 5.21: Consider adding an annotation to explain why you're highlighting a data point.

Step 3: When you highlight a particular data point, it's a good idea to accompany it with a text box. I added my branded text box **(Fig 5.21)**.

LEARN MORE See the Add Annotations section of the Design with Elegance chapter to learn how to create text boxes.

Add a Mean Line
You can add a mean or target line to your column chart, and it's more intuitive than adding a mean/target line to a bar chart. To add one, take the following steps:

Step 1: Add a column to your data table with the mean or target value **(Fig 5.23)**.

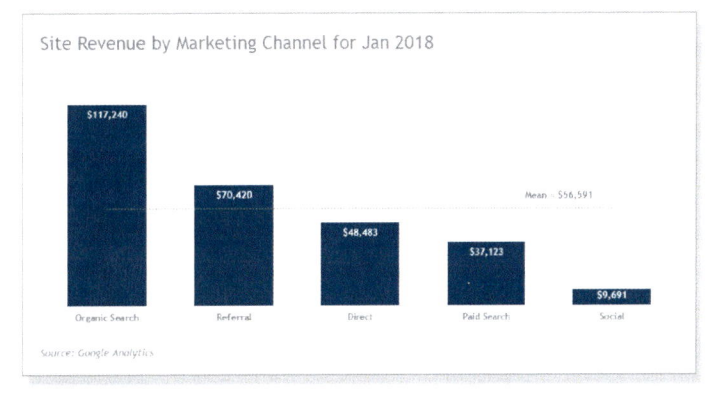

Fig 5.22: Our final chart.

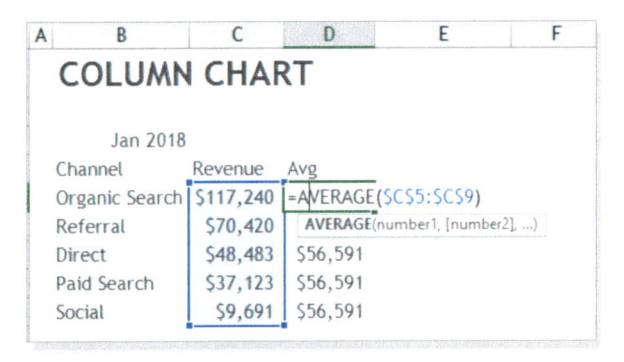

Fig 5.23: Add mean column to data table.

109

CHAPTER 5 • CREATE CHARTS

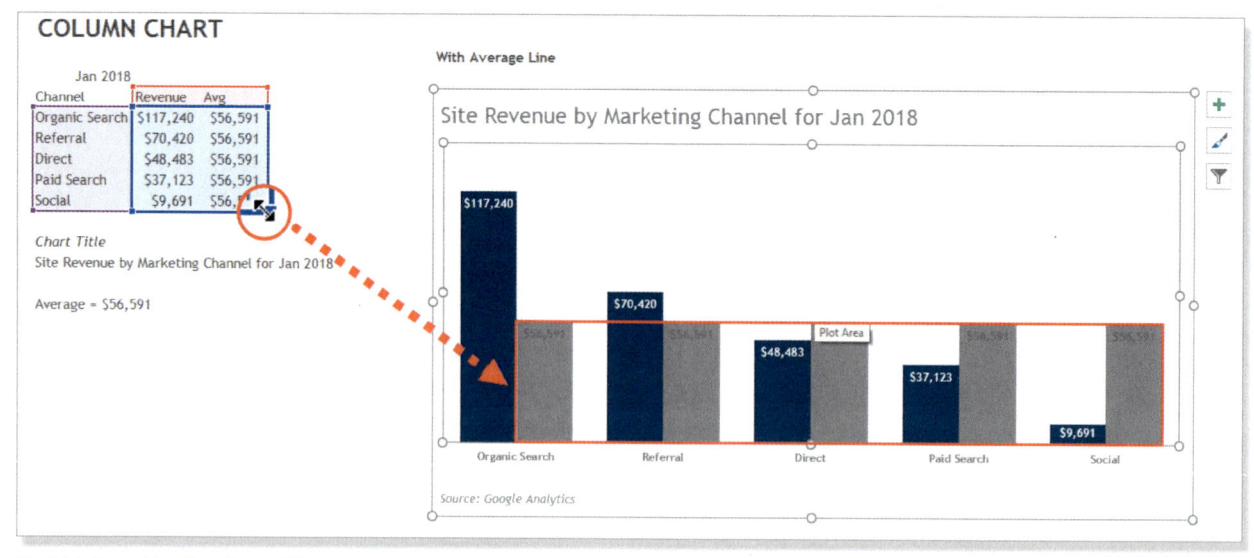

Fig 5.24: Drag sizing handle to add new column to your chart.

Step 2: Click anywhere in the **Plot Area** to highlight the categories and data series in your table, then drag the sizing handle in the bottom-right corner of the data series to the right to add the new column **(Fig 5.24)**. If your handles prove to be uncooperative (they sometimes go on strike) you'll need to update your range using the Select Data Source dialog.

> **LEARN MORE** See the Hack Excel's Charting Engine section of the Design with Elegance chapter to learn how to rock the Select Data Source dialog.

Step 3: Right-click on any one of the new columns in your chart, then choose **Change Series Chart Type** from the contextual menu. In the **Change Chart Type** dialog change your new data series to **Line Chart (Fig 5.25)**.

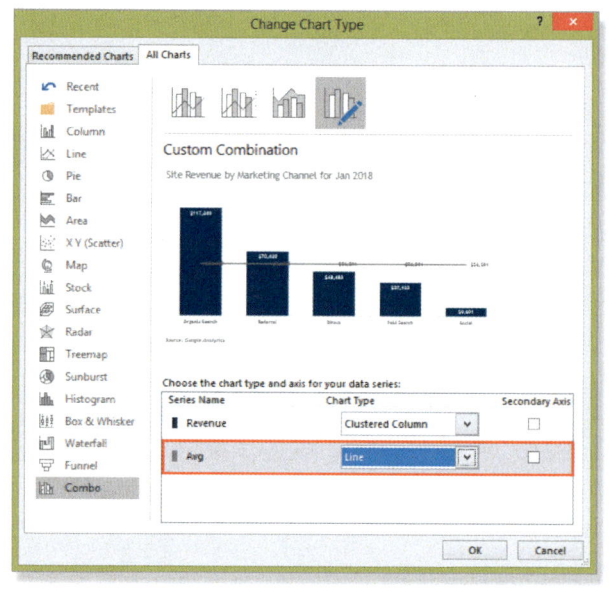

Fig 5.25: Change new data series to a line chart.

Step 4: We need to create a cell with our mean. I like to minimize the need for annotations in my chart, so when I'm adding a mean or target line, I prefer adding text to my data label that identifies what the line represents. Ultimately, my data label will house the mean value. First, let's program the text we want in the data label in a cell. I used this formula:

`="Mean = "&TEXT(B14,"$#,##0")`

> **LEARN MORE** See the Pull in Values section of the Design with Elegance chapter to learn how to use the TEXT function to dynamically update chart elements and the Customize Number Formatting section of the Design with Elegance chapter to learn how to customize your number formats.

110

CHAPTER 5 • CREATE CHARTS

Step 5: We'll add data labels, but in the next step we'll get rid of all but one so don't get too attached. Select the chart and open the **Chart Elements** flyout menu, then select the right-facing arrow next to **Data Labels** and choose **Above (Fig 5.26)**.

Step 6: We don't need a data label for every category; that's excessive. So, to pare your line down to one label, select one of the labels to select all the data labels for that series. Then select each of the labels you don't need and delete them **(Fig 5.27)**. With the remaining label, enter an = sign in the **Formula Bar**, then select the cell that has the text and/or value you want to use for your label **(Fig 5.28)**. Refer to B15 in the example workbook.

Fig 5.26: Add data labels to your mean/target line.

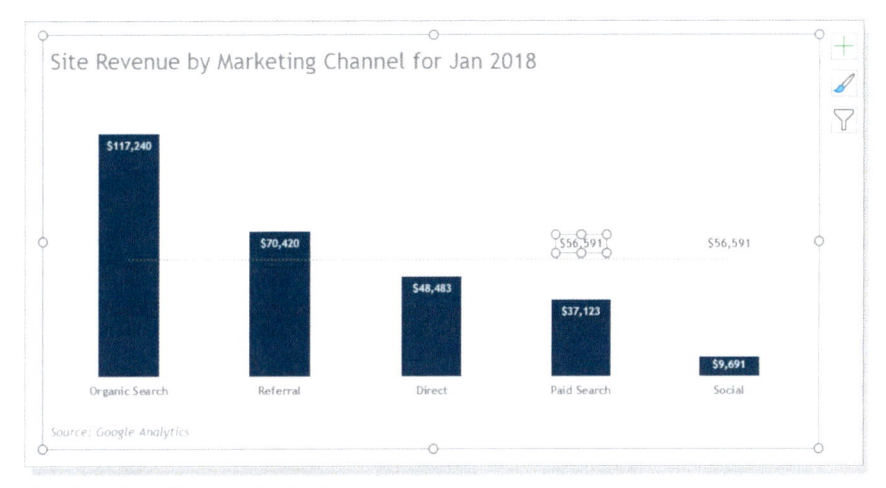

Fig 5.27: Delete all but one data label.

Fig 5.28: Point your last data label to the cell that contains the text you want to explain the line.

111

CHAPTER 5 • CREATE CHARTS

Tip: A data label is safe to delete when it's surrounded by white circles. If you see blue circles, you have all the data labels for the entire series selected, and you won't be able to point it to a cell using the **Formula Bar**.

Step 7: Let's add some finesse to that mean line. It's kind of thick and chunky. Select the line to pull up the **Format Data Series** pane. Select the **Fill & Line** tab, then reduce **Width**. I set mine to 1.5 pt. I then set **Dash type** to **Round Dot (Fig 5.29)**.

Example File
column-chart.xlsx

Templates
If you'll use these charts again, save them as chart templates. Or download the templates included in the chart templates folder:

- Column Chart: mds-column.crtx
- Column Chart with Highlight: mds-column-highlight.crtx
- Column Chart with Mean Line: mds-column-mean.crtx

Note: The data label in the column chart with mean line doesn't come through correctly in a template; it's just too customized. I saved the mean line, but you'll need to add the data label. To add it back, follow steps 5 and 6.

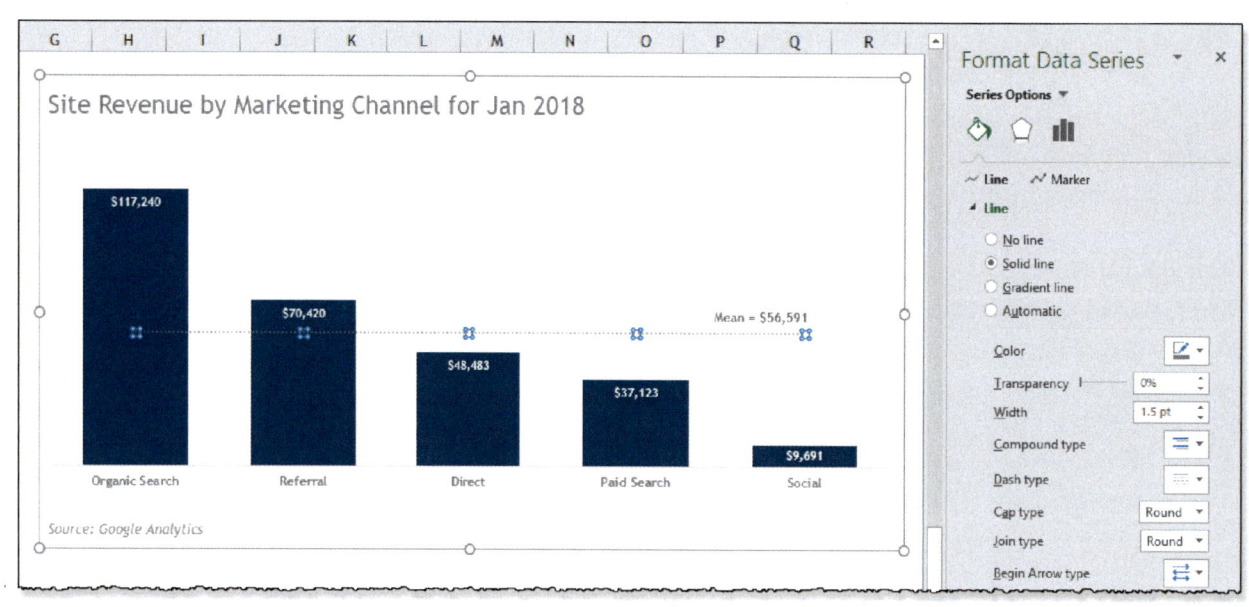

Fig 5.29: Format mean line to make it more elegant.

112

CHAPTER 5 • CREATE CHARTS

PICTOGRAPH

AKA
You may also see pictographs called:

- icon arrays
- ideograms
- ideographs
- isotypes
- pictos
- pictograms
- pictorial charts
- pictorial unit charts
- picture graphs

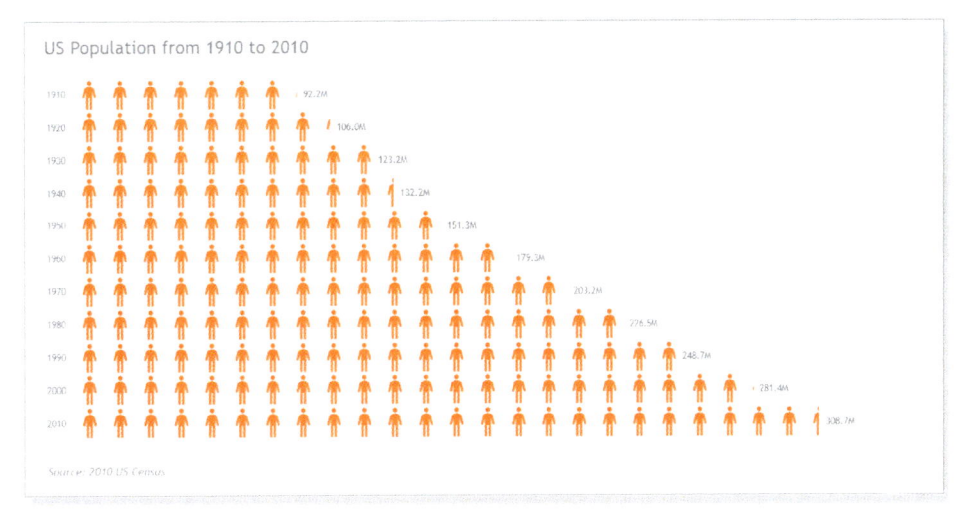

Fig 5.30: Our final chart.

Uses
Pictographs are best used to:

- augment data in infographics
- plot data that lends itself to intuitive use of icons (e.g., data about people, housing industry, money, etc.)
- visualize data that caters to international audiences (icons transcend language)
- plot small datasets

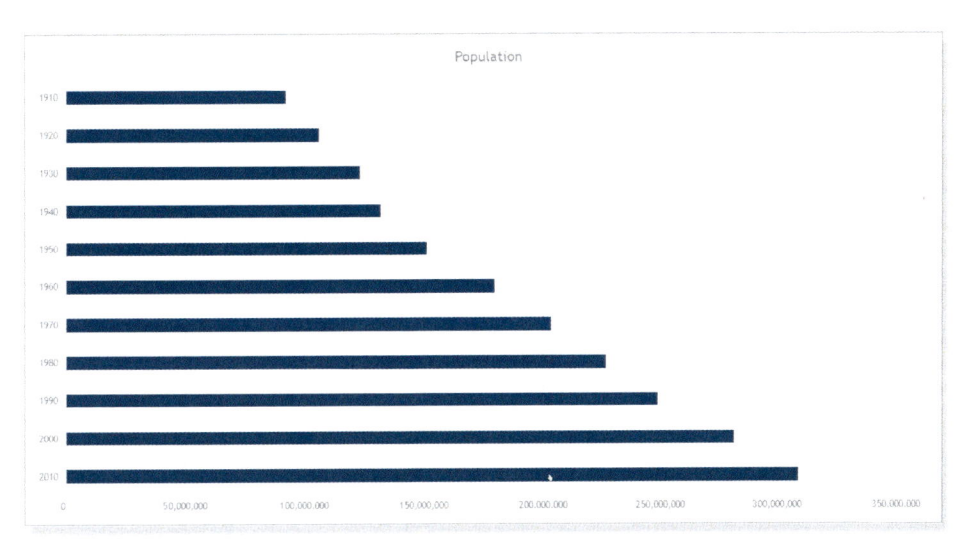

Fig 5.31: Our initial chart.

- Excel's icons are ideal for these types of charts.
- Colorize icons using one of your brand colors.

Category / Metric
1 category / 1 metric

Tutorial
Step 1: Select any cell inside your dataset and navigate to **Insert ▸ Charts ▸ Insert Column or Bar Chart ▸ 2-D Bar ▸ Clustered Bar (Fig 5.31)**.

Tips
- Horizontal bars can be easier to read than vertical bars when you're working with icons.
- Monochromatic icons are less distracting than full-color icons that contain a lot of detail.
- Pictographs can become garish fast, so be careful and take your audience into account.

113

CHAPTER 5 • CREATE CHARTS

Step 2: Sort your chart by right-clicking in the column you want to sort by and choosing **Sort ▸ Sort A to Z** (ascending) or **Sort Z to A** (descending) from the contextual menu. I sorted years in descending order so that the oldest year would be at the top of the stack in my bar chart **(Fig 5.32)**.

Step 3: Remove gridlines.

Step 4: Add data labels to bring your values into your chart. I set mine to **Outside End (Fig 5.33).**

> 💬 **LEARN MORE** See the Add Data Labels section of the Design with Elegance chapter to learn how to add data labels.

Step 5: We no longer need our horizontal axis, so we can remove it by selecting it and pressing the Delete key.

Step 6: Click any cell in your worksheet and drop an image you want to use for your pictograph into your spreadsheet. I picked mine from the icons available in Excel (**Insert ▸ Illustrations ▸ Icons**), but you can paste one from anywhere. I then colorized it to match my branding **(Fig 5.34)**.

> 💬 **LEARN MORE** See the Colorize Icons section of the Design with Elegance chapter to learn how to align icons with your branding.

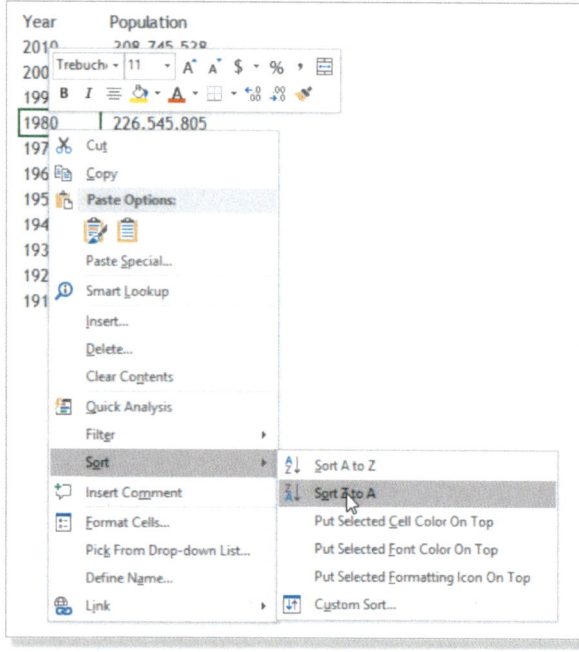

Fig 5.32: Set your sort order.

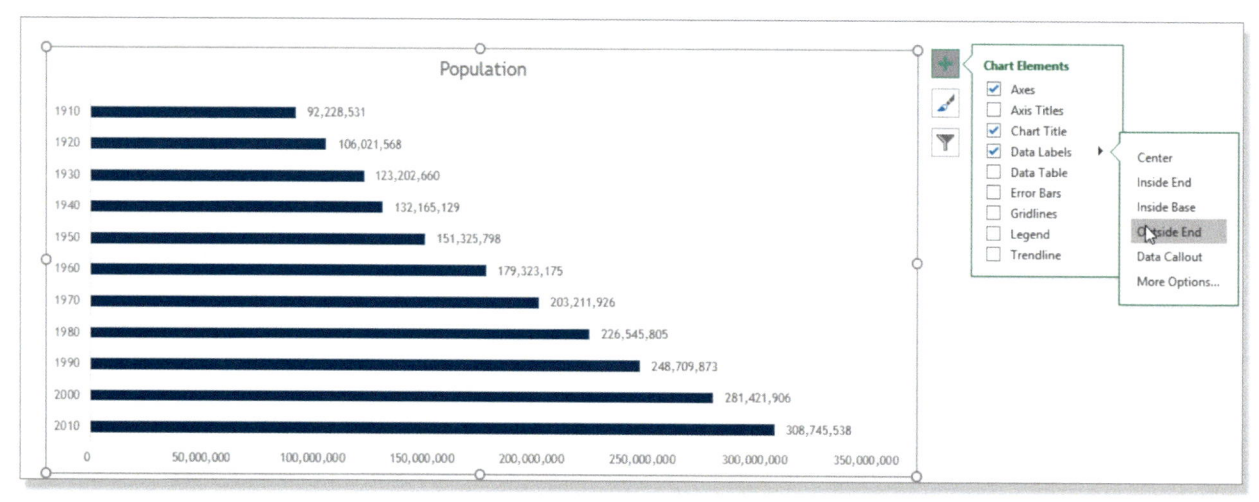

Fig 5.33: Add data labels.

114

CHAPTER 5 • CREATE CHARTS

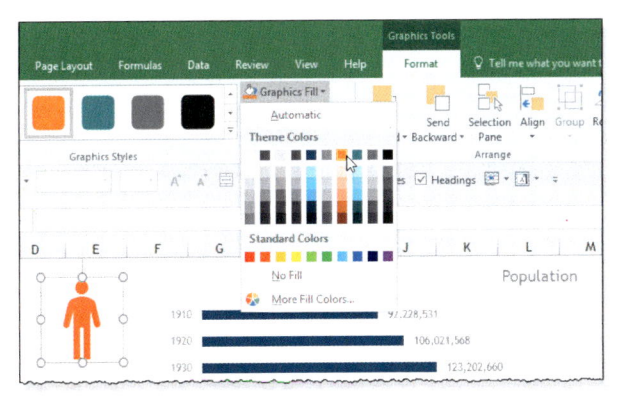

Fig 5.34: Select and colorize the icon you'll use for your pictograph.

Tip: If you bring an icon in from outside Excel, I recommend picking one with a transparent background for maximum flexibility. If you search Google Images, you can restrict your image search to images with transparent backgrounds by clicking **Tools ▶ Color ▶ Transparent (Fig 5.35)**. When you click on an image in Google it will show up in a preview. Right-click on the image in the preview window and choose **Save image as** from the contextual menu. Of course, make sure you have adequate rights to use any image you find on Google. Blah blah blah….

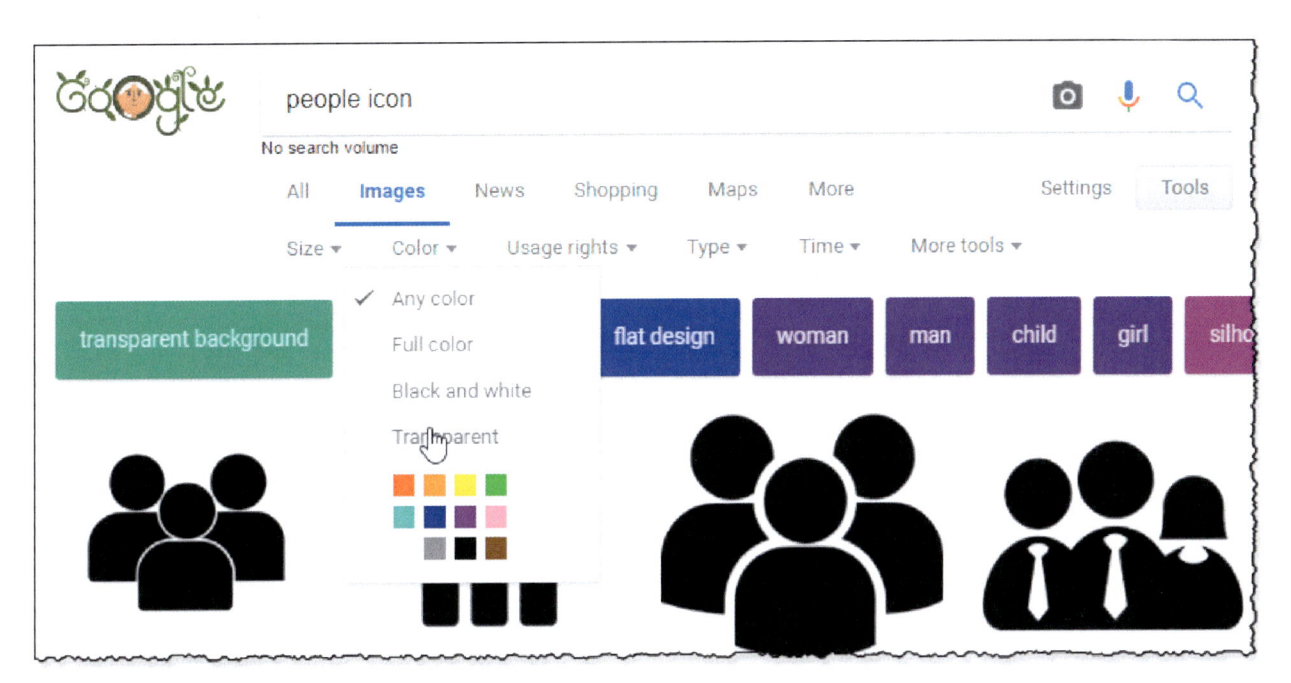

Fig 5.35: Restrict image search on Google to transparent images only.

115

CHAPTER 5 • CREATE CHARTS

Step 7: Copy your icon to the clipboard, then select any bar in your chart to select the entire data series and press Ctrl-1 to open the **Format Data Series** pane. Open the **Fill & Line** tab and set **Fill** to **Picture or texture fill**. Excel will automatically use the icon you stored on the clipboard. Your chart will look like a hot mess, with your icon stretched across each bar. To fix that, select the **Stack** option below the **Transparency** slider **(Fig 5.36)**. (Alternatively, you can choose the **Stack and Scale with** option and set how many units an icon represents.)

Step 8: You might be happy with your pictograph. If so, you're free to go play! I still need to clean mine up some more. If you share my obsessive proclivities, let's keep tweaking! I want my icons to be a little bigger, so I'll reduce the gap width between my bars. With your data series still selected, open the **Series Options** tab and move the **Gap Width** slider to the left. I set mine to 5% **(Fig 5.37)**.

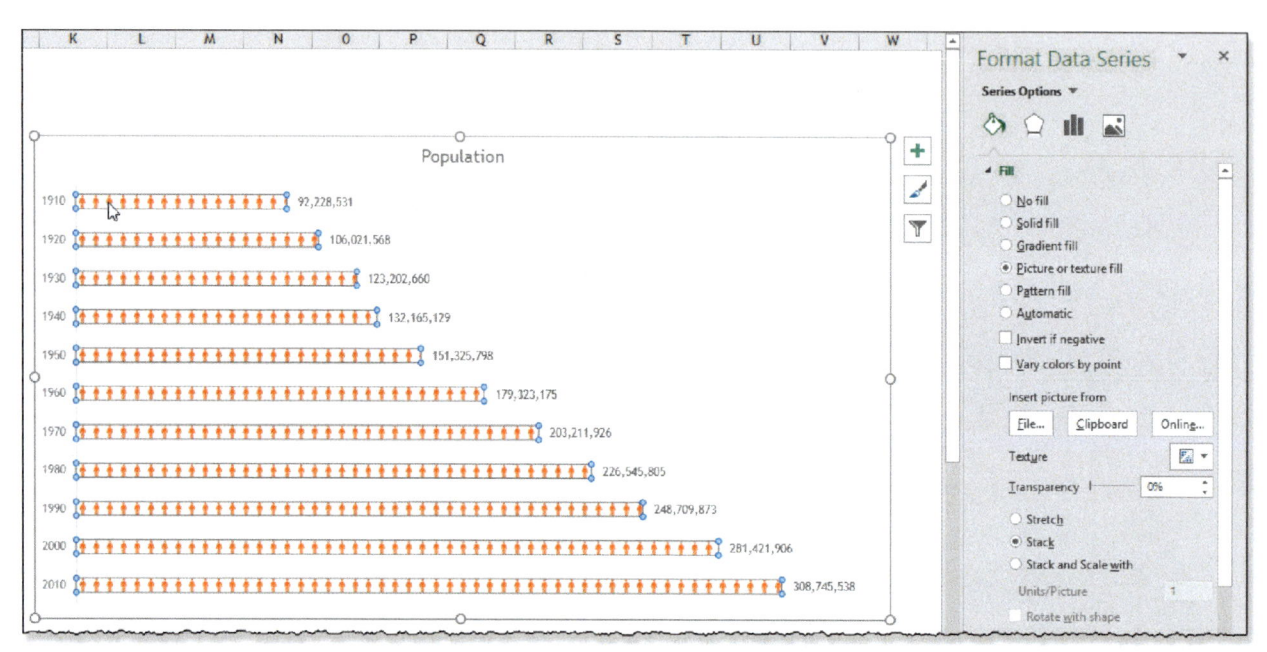

Fig 5.36: Set your data series fill to your icon.

116

CHAPTER 5 • CREATE CHARTS

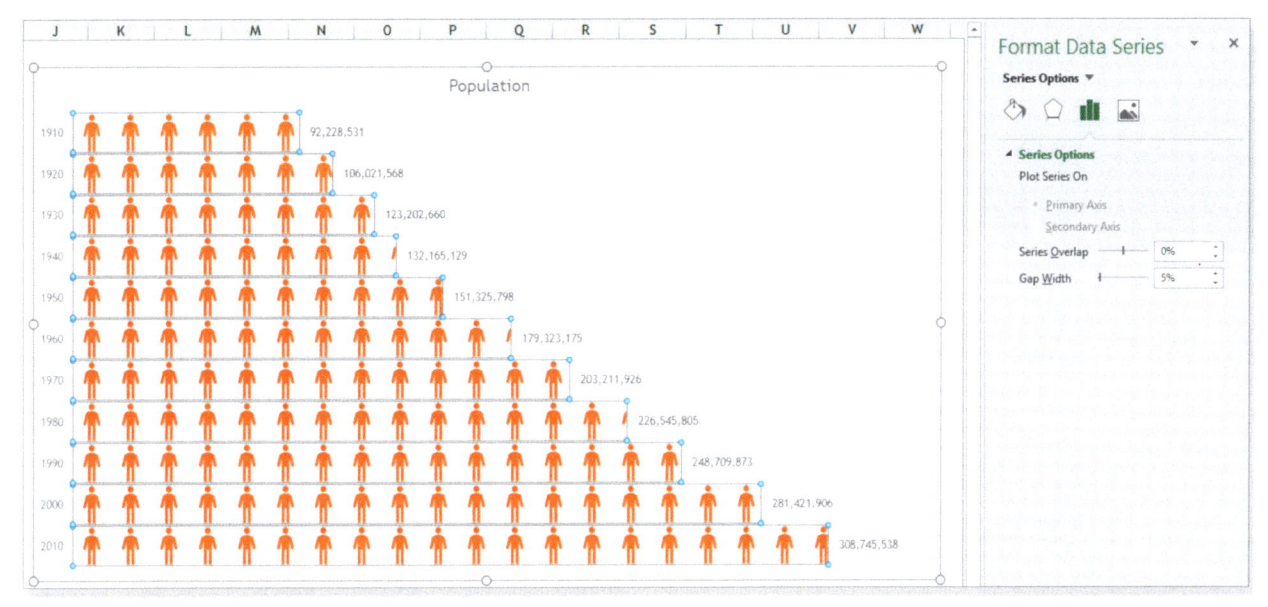

Fig 5.37: Increase the width of your bars to nudge them closer.

Step 9: My numbers are quite large because I'm charting census data, so I also customized the number formatting in my data labels by selecting one to select all of them and open the **Format Data Labels** pane. Under **Numbers**, you can customize your number formatting. I entered the following code into the **Format Code** field, which tells Excel I only want to show total as millions rounded to one decimal, with an "M" to indicate millions **(Fig 5.38)**:

`#.0,,"M";-#.0,,"M";0`

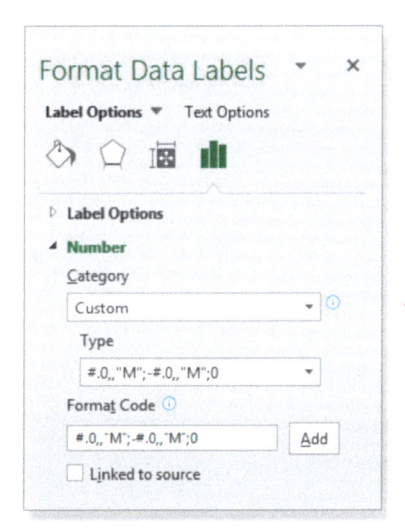

Fig 5.38: Simplify number formatting for values in thousands or millions.

Alternatively, I could have left the M out of my data labels, but then I'd need to add an annotation explaining that my numbers are millions.

LEARN MORE See the Simplify Large Numbers section of the Design with Elegance chapter to learn how to format large numbers.

Step 10: Clean up your chart to taste. I cleaned mine by taking the following steps:

+ made title more descriptive
+ left aligned chart title
+ bumped up title font size
+ added annotation for data source

Example File

pictograph.xlsx

Template

If you'll use this chart again, save it as a chart template. Or download the template included in the chart templates folder: mds-pictograph.crtx.

117

CHAPTER 5 • CREATE CHARTS

CLUSTERED COLUMN CHART

AKA

You may also see clustered column charts called:

- grouped bar/column charts
- paired bar/column charts
- side-by-side bar/column charts

Uses

Clustered column charts are best used to:

- compare data with multiple series across discrete groups
- visualize datasets with a limited number of category items
- compare period-over-period (e.g., week-over-week, month-over-month, quarter-over-quarter, year-over-year, etc.) data

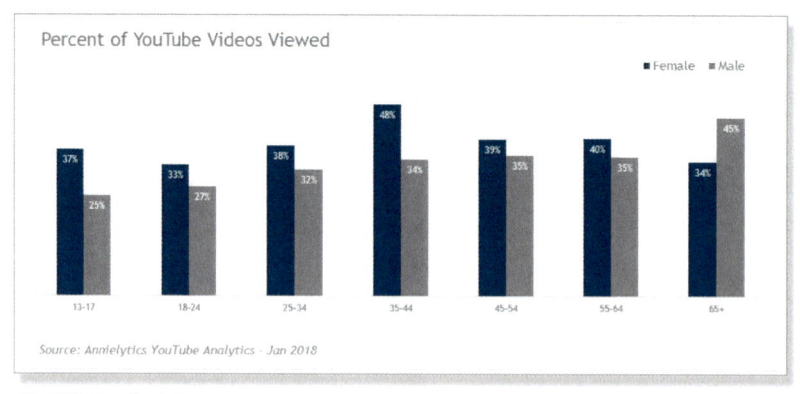

Fig 5.39: Our final chart.

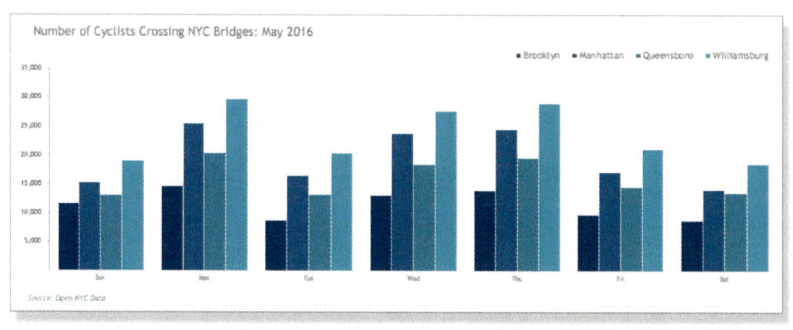

Fig 5.40: Too many categories or series can make clustered column charts difficult to interpret.

Tips

- Don't be afraid to move the legend of your chart. I prefer placing mine below the chart or somewhere inside the Plot Area, as long as it doesn't overlap with the columns.
- If I'm dynamically pulling in new data, like with an API, I avoid placing the legend inside the Plot Area since it may overlap with a column(s) when the data is refreshed.
- Clustered columns allow the direct comparison of multiple data series in a given category item, but it's more difficult to compare data series from item to item, if you have more than two.
- Bringing data into the columns via data labels can make it easier to compare data series across multiple category elements.
- Clustered bar and column charts can get visually complex when you either have too many data series in a category or too many category elements **(Fig 5.40)**. If you only have two data series, you can have more category elements, and if you only have two or three category elements, you may be able to get away with adding more data series before you hit a point of diminishing returns. The goal is to minimize the need for viewers' eyes to dart back and forth to consume the data. It can cause anxiety spikes and bury the story the data is trying to tell.
- The more data series you have, the thicker and chewier your legend is, which can make processing the data tiring. Again, you're causing viewers to look back and forth from your chart to your legend, which is not ideal from a usability standpoint.
- Make sure there is adequate spacing between category elements. White space helps viewers interpret the categorization of your data.

CHAPTER 5 • CREATE CHARTS

- Don't let individual data series bump up against each other. Columns that touch look like histograms, which measure frequencies and are used in statistical analysis.
- If you want to show how each data series contributes to the whole, stacked columns work better.
- You may want to use bars if your labels are longer or your chart looks cluttered with columns because of the space you have available for your chart. Space can be limited when working with dashboards or visualizations that will go into a presentation.
- If you're using tabular data, you may need to use a pivot table to transform your data into a crosstab format. I've included a Pivot Table worksheet in the downloadable workbook to show you how I set mine up.

Category / Metric
2 categories / 1 metric

Tutorial
Step 1: Select any cell inside your dataset and navigate to **Insert ▸ Charts ▸ Insert Column or Bar Chart ▸ 2-D Column ▸ Clustered Column (Fig 5.41)**.

💬 **LEARN MORE** See the Design with Elegance chapter to learn how to perform these refinements.

Step 2: Clean up your chart to taste. I cleaned mine by taking the following steps:

- made title more descriptive
- left aligned chart title
- bumped up title font size
- added annotation for data source
- made source annotation dynamic
- removed gridlines
- added data labels
- moved legend
- removed vertical axis

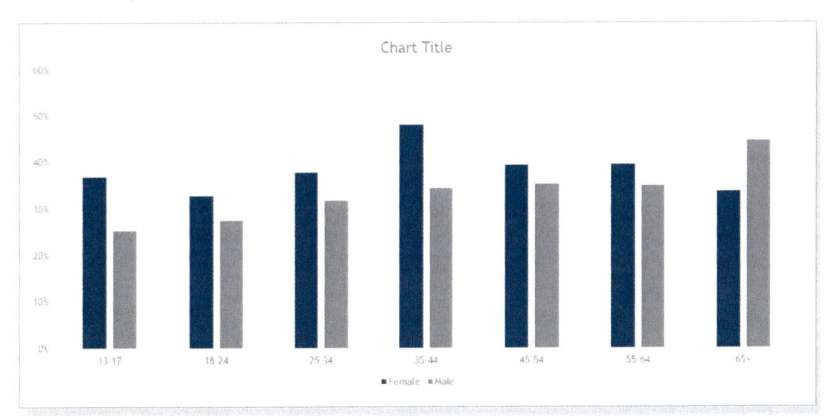

Fig 5.41: Our initial chart.

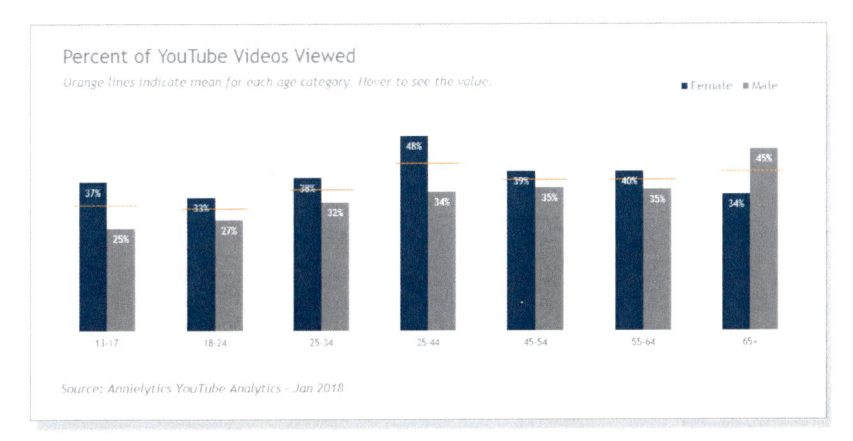

Fig 5.42: Our final chart.

Variations
Add a Mean Line
Hundred-Foot View
To add a mean (or target) line to each of our clusters, we'll add an additional column to our dataset with the mean value. Then we'll add it to our chart and change the chart type to a scatter plot. This will add dots between our columns. Next, we'll add horizontal

119

error bars to this data series and tidy them up. Finally, we'll remove Mean from our legend since it shows as blank (because only its error bars are showing).

Step 1: Add a column to your dataset populated with your average **(Fig 5.43)**.

Fig 5.43: Add a mean column to your dataset.

Step 2: Click anywhere in the **Plot Area** to highlight the category and data series in your dataset, then drag the handle in the bottom-right corner of the data series to the right to add the new column **(Fig 5.44)**.

LEARN MORE See the Hack Excel's Charting Engine section of the Design with Elegance chapter to learn how to add new data series.

Step 3: Right-click one of the new columns and select **Change Series Chart Type**. Change the chart type for your mean line to **Scatter**, under **XY (Scatter)**. If Excel selects **Secondary Axis**, deselect it **(Fig 5.45)**.

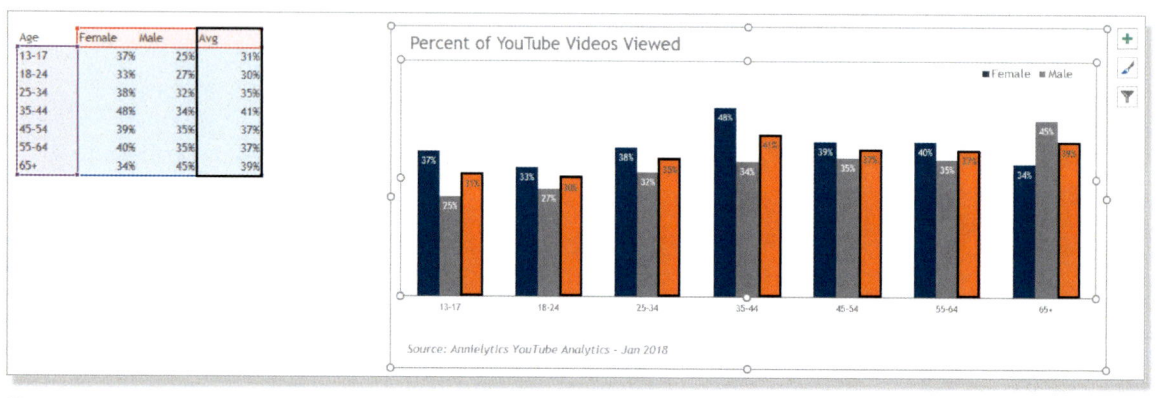

Fig 5.44: Add mean data to your chart.

Step 4: If adding a scatter plot to your chart triggers Excel to add a top and right axis, delete them. Also, adding a new data series pushes your legend onto the shoulder of the road, so you will want to move it to the left **(Fig 5.46)**.

Step 5: Select one of the dots to select the mean data series, then select **Error Bars** from the **Chart Elements** flyout menu to add them to our chart **(Fig 5.47)**.

Step 6: Let's tweak these error bars. First, let's remove the end caps. With the error bars still selected, in the **Error Bar Options** tab, set **End Style** to **No Cap**. Next, set **Error Amount** to **Fixed value**. You'll need to play with this value until your error bars fit your columns. 0.3 worked for my chart **(Fig 5.48)**.

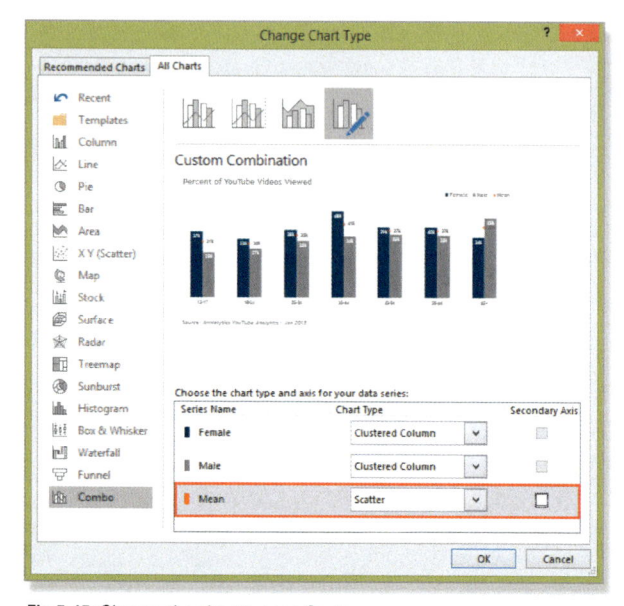

Fig 5.45: Change the chart type to Scatter.

120

CHAPTER 5 • CREATE CHARTS

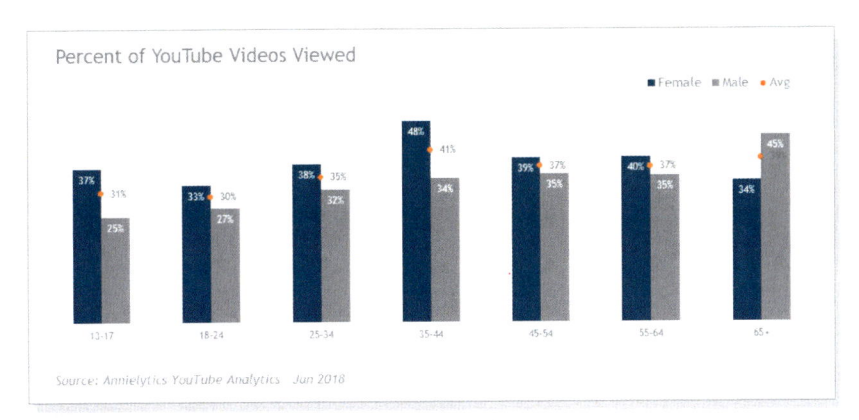

Fig 5.46: Delete unnecessary axes and move your legend.

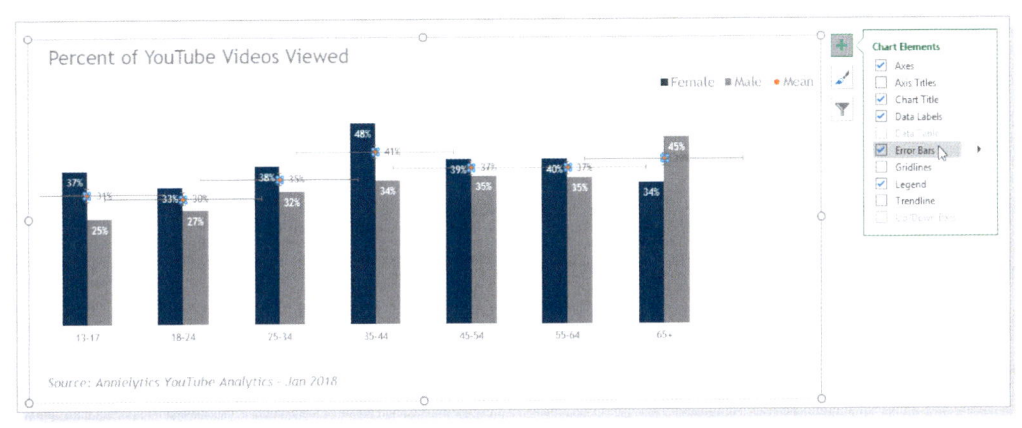

Fig 5.47: Add error bars to your data.

Fig 5.48: Remove end caps and customize Error Amount of your error bars.

121

CHAPTER 5 • CREATE CHARTS

Fig 5.49: Remove markers from error bars.

Step 7: Finally, let's get rid of the markers. To do that, deselect the error bars and select one of the markers to select all of them. In the **Format Data Series** pane, open the **Fill & Line** tab. Click **Marker** at the top of the pane. Open the **Marker Options** section, if it doesn't open automatically. Set **Marker Options** to **None (Fig 5.49)**.

Step 8: I find the data labels distracting, so I removed them by selecting one and pressing the Delete key. A viewer can see the average value by hovering over the center of the error bar. I also chose to remove Mean from the legend by selecting the legend, then right-clicking on it and choosing **Delete** (not **Delete Series**) from the contextual menu **(Fig 5.50)**. Finally, nudge your legend to the right to line it back up with your **Plot Area**.

Tip: You could move the data labels in your columns to the inside base, but then you still run the risk of the data labels for your mean lines being obfuscated by an accompanying column, regardless of where you place them. But feel free to test.

Credit: Shout out to the Peltier Tech Blog for the inspiration to use error bars to show average: **bit.ly/mdsepc-clustered-avg**. I just translated his instructions from 2003 to 2016 and fiddled with some settings.

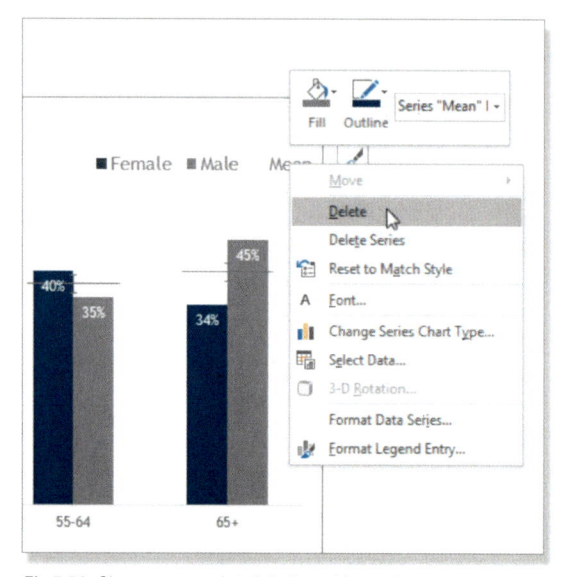

Fig 5.50: Clean up your data labels and legend.

CHAPTER 5 • CREATE CHARTS

Show % Change
Hundred-Foot View

We'll use a pivot table to get each year into its own column. Then we'll use a nested axis to show month and % Δ. To get both Month and %Δ into our horizontal axis, we'll pull both a category (month) and metric (%Δ) into it. Finally, we'll use an IF function to serve up an up-facing arrow if there's a YoY increase, a down-facing arrow if there's a decrease, and a rectangle if there's no change.

Step 1: With the practice data, I exported the data at the month level. This introduces an issue though: Excel doesn't recognize a data point like 201701 (Jan 2017). I had to combine a DATE, LEFT, AND RIGHT function to reconfigure it to a form Excel would recognize **(Fig 5.52)**.

💬 **LEARN MORE** See the Fix Dates section of the Massage Your Data chapter to learn how to fix and format rogue dates.

Step 2: Data doesn't usually come to us with years broken into different columns. It typically comes as one continuous range (e.g., Jan 1, 2017 to Dec 31, 2018), as it should. To get years into their own columns, we need to create a pivot table. Then you'll pull Month into the Rows field, Year into the Columns field, and Sessions (or whatever you want to chart) into the Values field **(Fig 5.53)**.

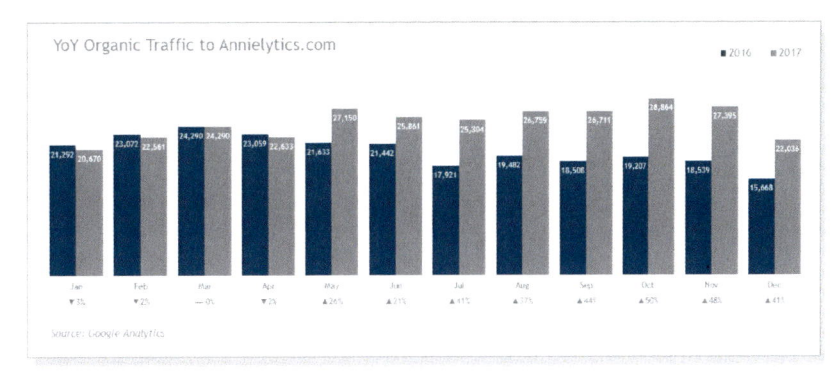

Fig 5.51: Our final chart.

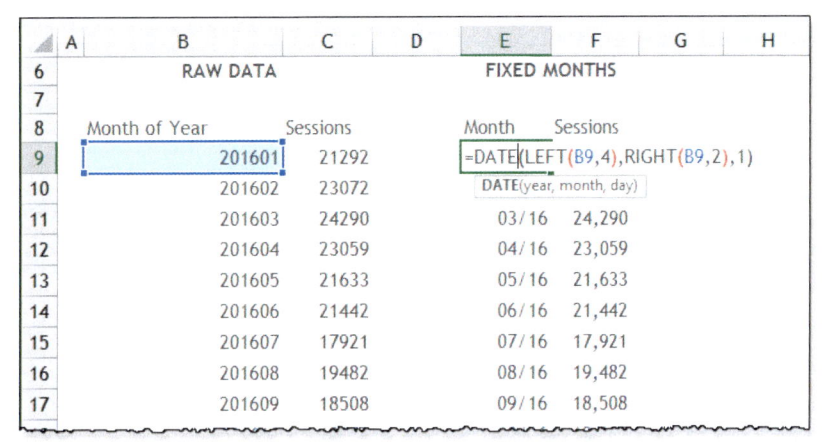

Fig 5.52: Transform Google Analytics' funky month format into dates using the DATE function.

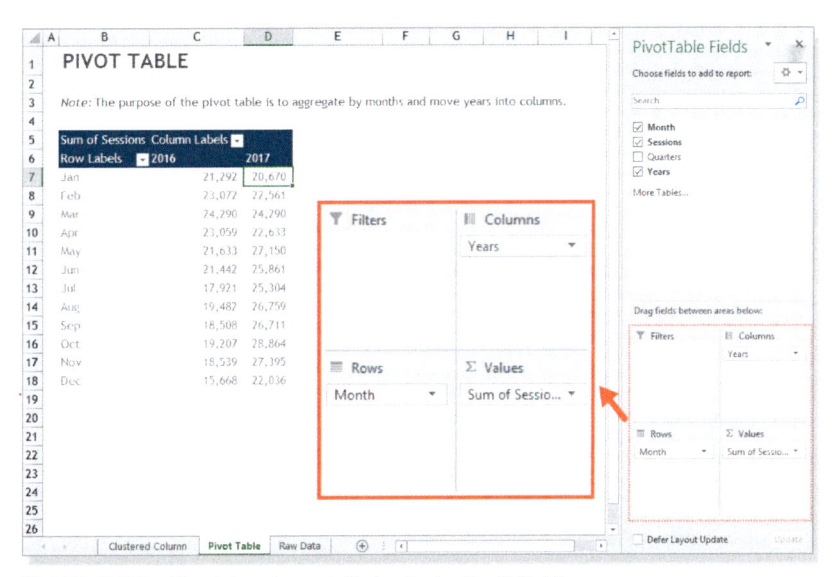

Fig 5.53: Splitting Year into columns is the key to charting YoY data.

123

CHAPTER 5 · CREATE CHARTS

🗨 **LEARN MORE** See the Customize Number Formatting section of the Design with Elegance chapter to learn how to finesse your number formatting.

Month	2016	2017
Jan	21,292	20,670
Feb	23,072	22,561
Mar	24,290	24,290
Apr	23,059	22,633
May	21,633	27,150
Jun	21,442	25,861
Jul	17,921	25,304
Aug	19,482	26,759
Sep	18,508	26,711
Oct	19,207	28,864
Nov	18,539	27,395
Dec	15,668	22,036

Fig 5.54: Copy-and-paste what you need from the pivot table into its own worksheet.

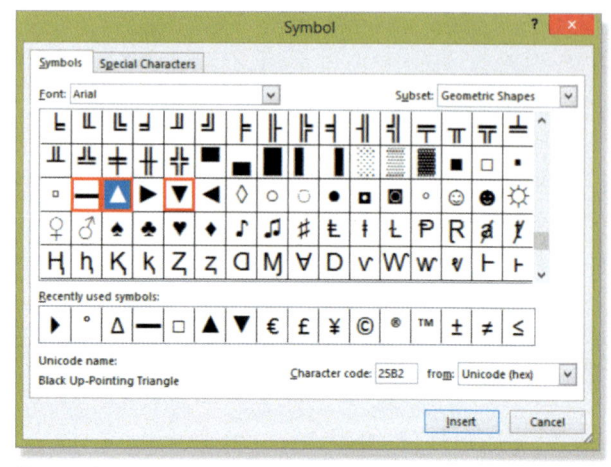

Fig 5.55: Create symbols to use in your chart.

Step 3: Copy-and-paste your massaged data from the table as values from the **Paste Options** button, which can be found under **Paste Values** ▶ **Values & Number Formatting (Fig 5.54)**.

Step 4: We need to create a % Δ column, but first we need to create our triangles that will indicate if our metric is up or down from the year before. To do that, navigate to **Insert** ▶ **Symbols** ▶ **Symbol**. Then set **Font** to **Arial** and **Subset** to **Geometric Shapes**. Click **Black Down-Pointing Arrow** and click **Insert** to insert it in your cell. Rinse and repeat with **Black Up-Pointing Arrow** and **Black Rectangle (Fig 5.55)**.

Tip: To save yourself some scrolling, use the filter to select your font and subset faster.

Step 5: If you want to use the delta symbol in your column header, set **Font** to **(normal text)** and **Subset** to **Greek and Coptic** from the **Symbol** dialog **(Fig 5.56)**.

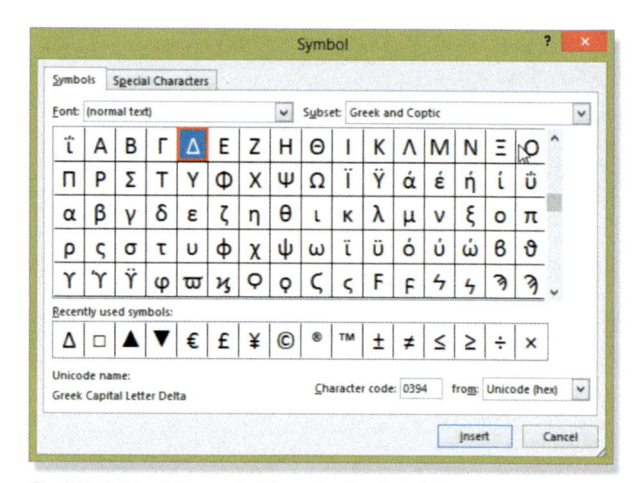

Fig 5.56: Add a delta symbol from the Symbol dialog.

Note: The (normal text) option is at the top of the Font filter list but doesn't come up if you start to type it like the other options.

124

CHAPTER 5 · CREATE CHARTS

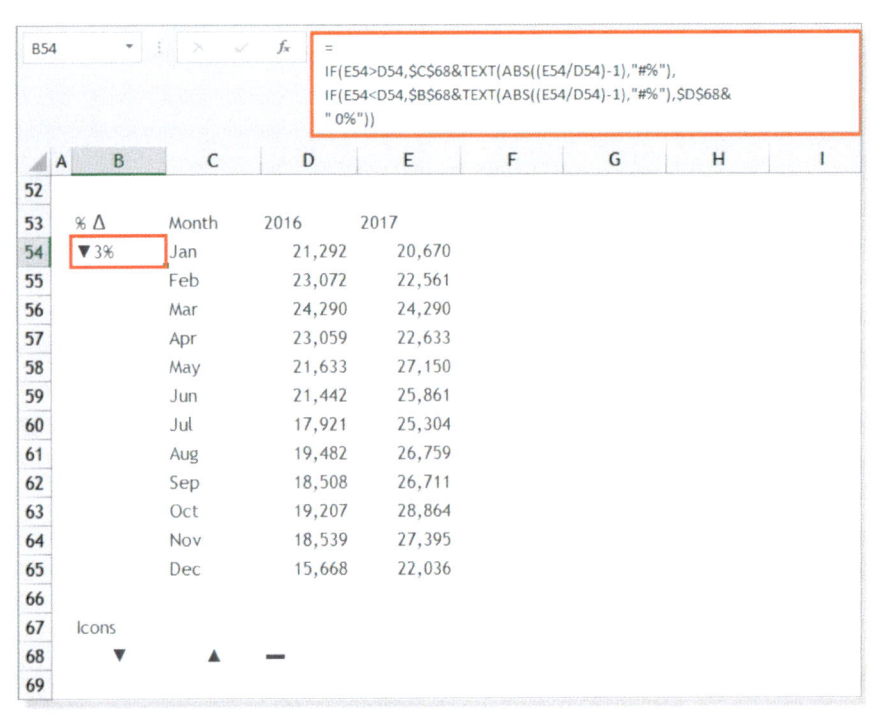

Fig 5.57: Use a nested IF function to dynamically pull arrows into your chart.

Step 6: We'll need a nested IF function to generate an up-facing arrow when YoY was a gain, a down-facing arrow when there was a loss, and a rectangle when there was no change **(Fig 5.57)**. The first line says if Jan 2017 is greater than Jan 2016, concatenate the up-facing arrow with the absolute of the percent difference. Deconstructing the formula further, E54/D54 – 1 calculates the % difference, and the ABS function removes the negative formatting. Finally, the #% sets the number formatting. You can, of course, tweak the number formatting to meet the needs of your data.

Tip: I find that breaking the formula up with line breaks (using Alt-Enter) helps you find issues, such as missing parentheses and commas. You will need to drag the bottom edge of the Formula Bar down to view the entire formula.

Step 7: Double-click the bottom-right corner of the cell with your formula to apply it to the rest of the column using Excel's auto-fill functionality. Or you can drag the fill handle if you're more comfortable with that.

125

CHAPTER 5 • CREATE CHARTS

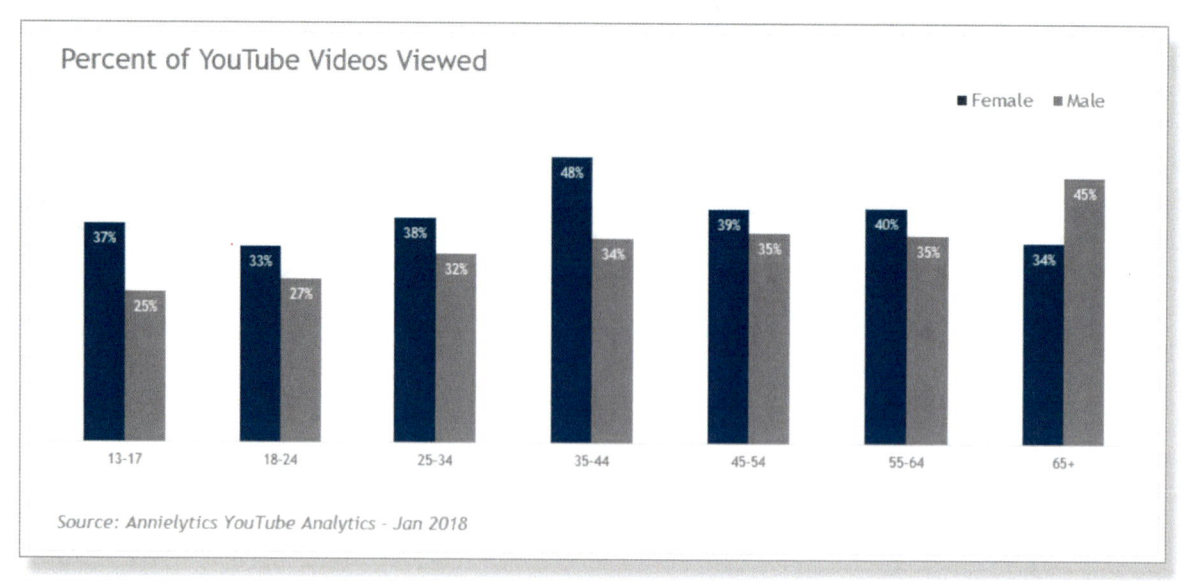

Fig 5.58: Start with your original clustered column chart.

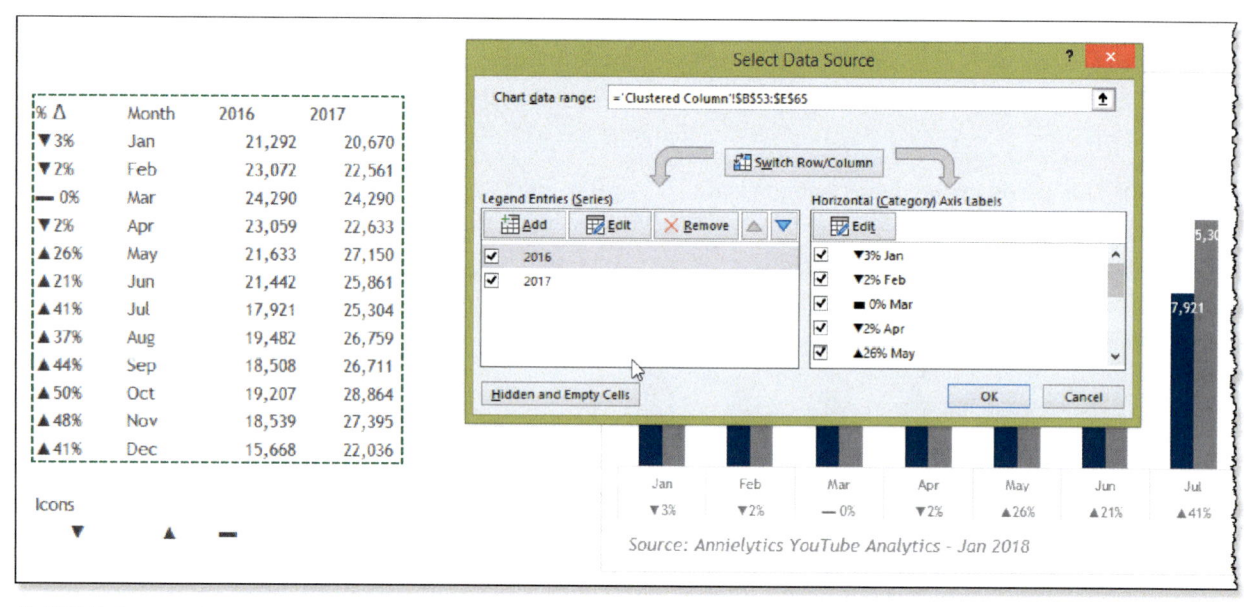

Fig 5.59: Update your data range.

Step 8: We'll pick up where we left off with our original chart **(Fig 5.58)**. Alternatively, you could create a new chart using the chart template.

Step 9: Select your chart, then right-click anywhere inside and choose **Select Data** from the contextual menu. Set **Chart data range** to your new dataset by clicking-and-dragging over it **(Fig 5.59)**.

126

Step 10: If you click the **Edit** button inside the **Horizontal (Category) Axis Labels** field, you should see the marching ants surround your first two columns **(Fig 5.60)**. If you don't, update the field with your first two columns, sans your column headers.

Step 11: I've found that the data labels can sometimes be off-centered inside the bars. If this happens with your chart, you can play with the margins by selecting a data label, which will open the **Format Data Labels** pane. Open the **Size & Properties** tab, then under **Alignment,** set **Left Margin** to 0 **(Fig 5.61)**. Ymmv.

Step 12: Clean up your chart to taste. I cleaned mine by taking the following steps:

- updated chart title
- updated annotation with new data source
- made bars wider
- reduced font size of data labels

Credit: Shout out to the My Online Training Hub blog for the inspiration to use the arrows and accompanying formula: **bit.ly/mdsepc-chart-labels**.

Example File
clustered-column-chart.xlsx

Fig 5.60: Our chart has two columns in the Horizontal (Category) axis labels field.

Template
The only chart in this section that would save as a template was the main clustered column chart. If you'll use it again, save it as a chart template. Or download the template included in the chart templates folder: mds-clustered-column.crtx.

You'll need to create the variation visualizations from scratch or save them in a file where you keep all your charts that won't save as templates.

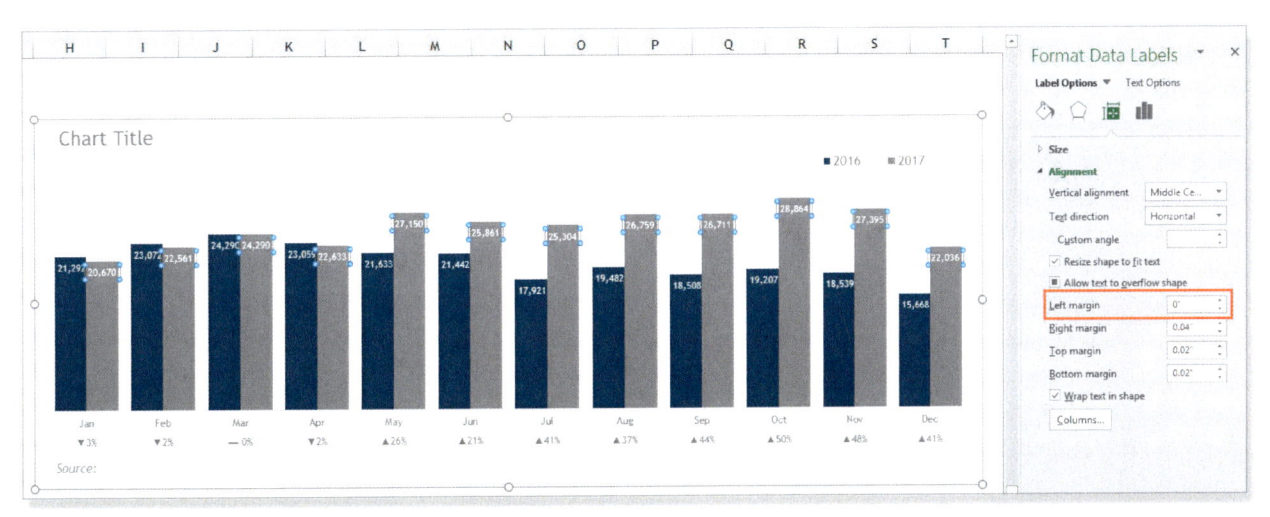

Fig 5.61: You may need to adjust the margins in your data labels for aesthetics.

CHAPTER 5 • CREATE CHARTS

STACKED COLUMN CHART

Uses

Stacked column charts are best used to:

* demonstrate how each data series contributes to the whole
* compare data across discrete groups
* visualize a sum of changes over time
* plot survey data to show the number of responses of each choice (e.g., very satisfied to very unsatisfied) for each question

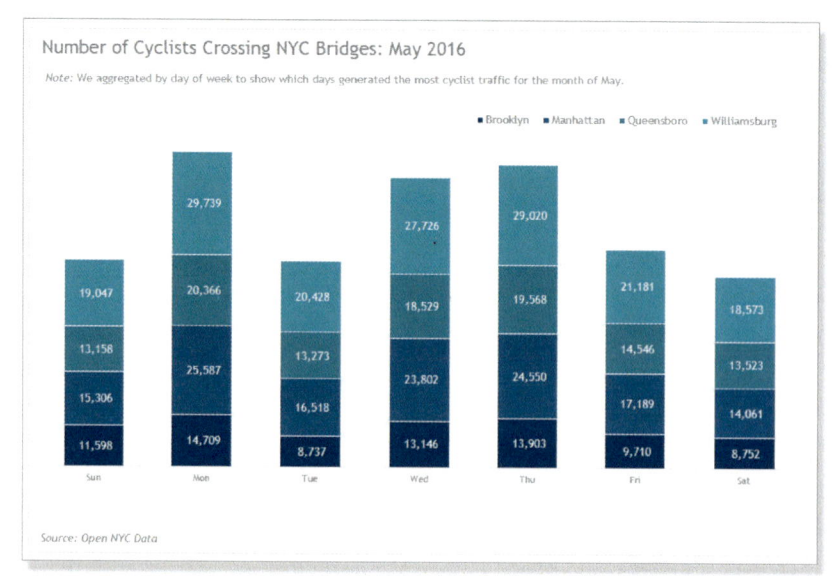

Fig 5.62: Our final chart.

Tips

* Keep in mind that only the data series closest to the axis (x for bars charts and y for columns) is easy to compare across category elements or dates because it has a floor.
* Bringing the data into your chart with data labels can aid in comparing data series across category items.
* If you want to draw attention to a particular series, consider graying out the other data series with different tints of gray and only applying color (one of your branded colors!) to the series you're making queen for the day.
* You might be tempted to employ gridlines to help viewers interpret the relative sizes of your series, but they will typically just make an already busy chart even busier and add more noise than signal.
* If you want to be able to compare the totals of your different columns, consider adding them to the chart over using axis labels and gridlines. I demonstrate an easy hack for adding totals to stacked columns in the Add Column Totals variation of the tutorial in this section.

* Consider using different tints of a single color over discrete colors for each series.
* Stacked columns aren't good for comparing the individual elements within a single bar/column. The individual elements in a bar show the contribution of each data point to the whole (that particular bar/column). The comparison piece comes into play by comparing the stacked bars/columns to each other **(Fig 5.63)**.
* If your dataset includes a time element, columns work better than bars because it's more natural to read left to right when dealing with data over time.
* I've seen stacked column charts coupled with a line chart using a combination chart. I don't recommend combining these because trying to cram that much data into a chart can cause analysis fatigue. If you can't avoid it, consider using tints of gray for your data series (i.e., the individual sections of your bars) to draw attention to the metric you want to highlight with the line (e.g., MoM or YoY trends).

- If you're using tabular data, you may need to use a pivot table to transform your data into a cross-tab format. I've included a Pivot Table worksheet in the downloadable workbook to show you how I set mine up. Click anywhere in the pivot table to open the PivotTable Fields pane. From there you can easily see how I constructed it.

Category / Metric
2 categories / 1 metric

Tutorial

Step 1: The data I used for the example file was poorly organized and structured. You will sometimes need to unpivot data like this before you can use it **(Fig 5.64)**.

LEARN MORE See the Unpivot Data section of the Massage Your Data chapter to learn how to restructure your data.

Fig 5.63: A stacked column chart contains part-to-whole and comparison elements.

Fig 5.64: Sometimes you need to unpivot your data before you can work with it.

CHAPTER 5 • CREATE CHARTS

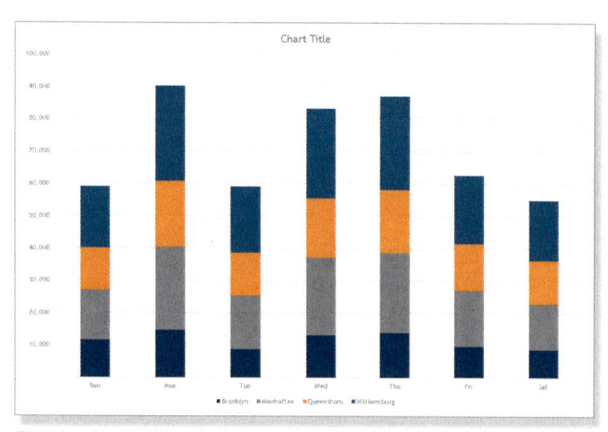

Fig 5.65: Our initial chart.

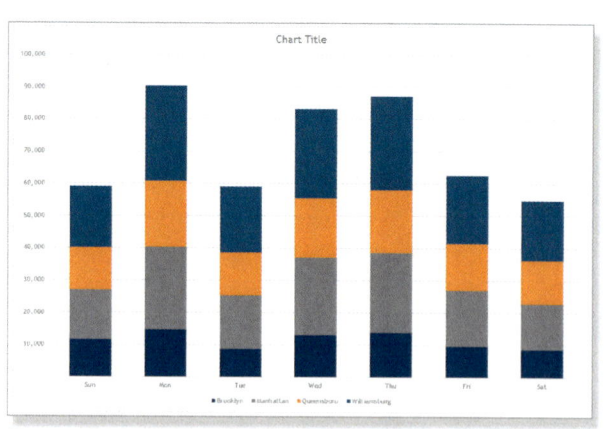

Fig 5.66: Fatten up your columns with Gap Width.

Step 2: Select any cell inside your dataset and navigate to **Insert ▸ Charts ▸ Insert Column or Bar Chart ▸ 2-D Column ▸ Stacked Column (Fig 5.65)**.

Step 3: I'm not crazy about all the wasted white space in Excel's default column chart, so I fattened up my columns by setting **Gap Width** to 80% **(Fig 5.66)**.

> 💬 **LEARN MORE** See the Adjust Chart Series Spacing section of the Design with Elegance chapter to learn how to customize the space between columns.

Step 4: Clean up your chart to taste. I cleaned mine by taking the following steps:

- made title more descriptive
- left aligned chart title
- bumped up title font size
- added annotation for data source
- removed gridlines
- added data labels
- bumped up legend font size
- moved legend
- removed vertical axis
- added border to data series
- changed series fill from discrete colors to tints

> 💬 **LEARN MORE** See the Design with Elegance chapter to learn how to perform these refinements.

Variation
Add Column Totals

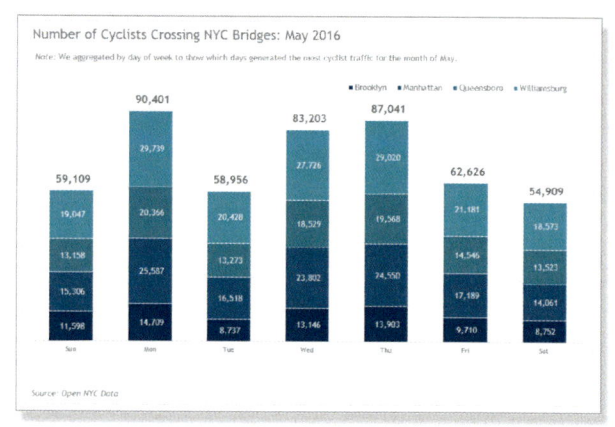

Fig 5.67: Our final chart.

If you're using a stacked column chart, there's a very simple hack to add column totals to your chart. If you're using a stacked bar chart, best of luck! I tried to crack the code and aborted the mission moments before tossing my laptop through the nearest window.

Step 1: Add a total column to your dataset **(Fig 5.68)**. The easiest way to generate sums is to select the cell directly to the right (or below, if you're summing a column) and press Alt-=. Then double-click the

130

bottom-right corner of the cell to send it down your column or drag the fill handle to apply the formula to your column.

Step 2: Select your chart anywhere in the **Plot Area** and drag the bottom-right corner of the blue sizing handle to the right to add the total column to your chart **(Fig 5.69)**. This will jack up your chart, but it will only be a temporary flash of insanity! If you can't drag the handle because Excel is having an off day, you'll need to crack open the Select Data Source dialog and add your series that way.

💬 **LEARN MORE** See the Hack Excel's Charting Engine section of the Design with Elegance chapter to learn how to rock the Select Data Source dialog.

Fig 5.68: Use Alt-= to generate a SUM function on the fly.

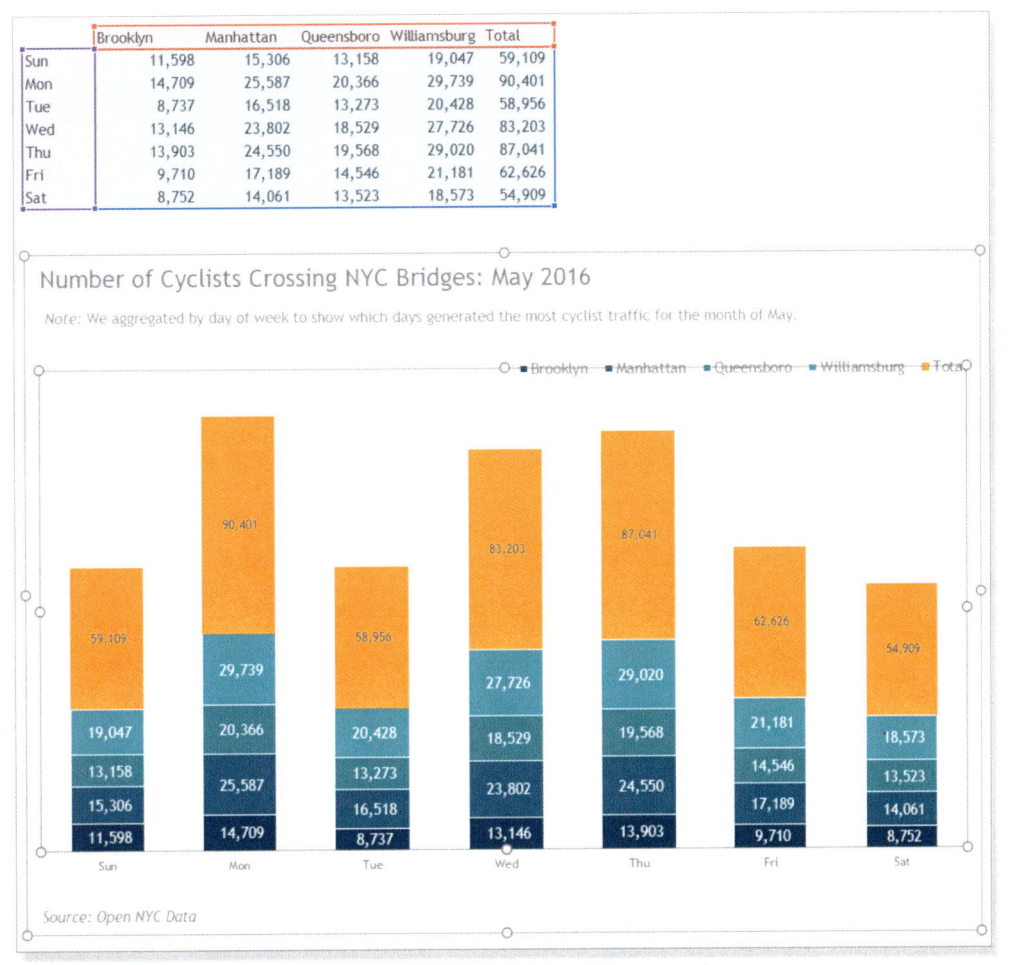

Fig 5.69: Dragging sizing handles is usually the easiest way to add a series to your chart.

CHAPTER 5 · CREATE CHARTS

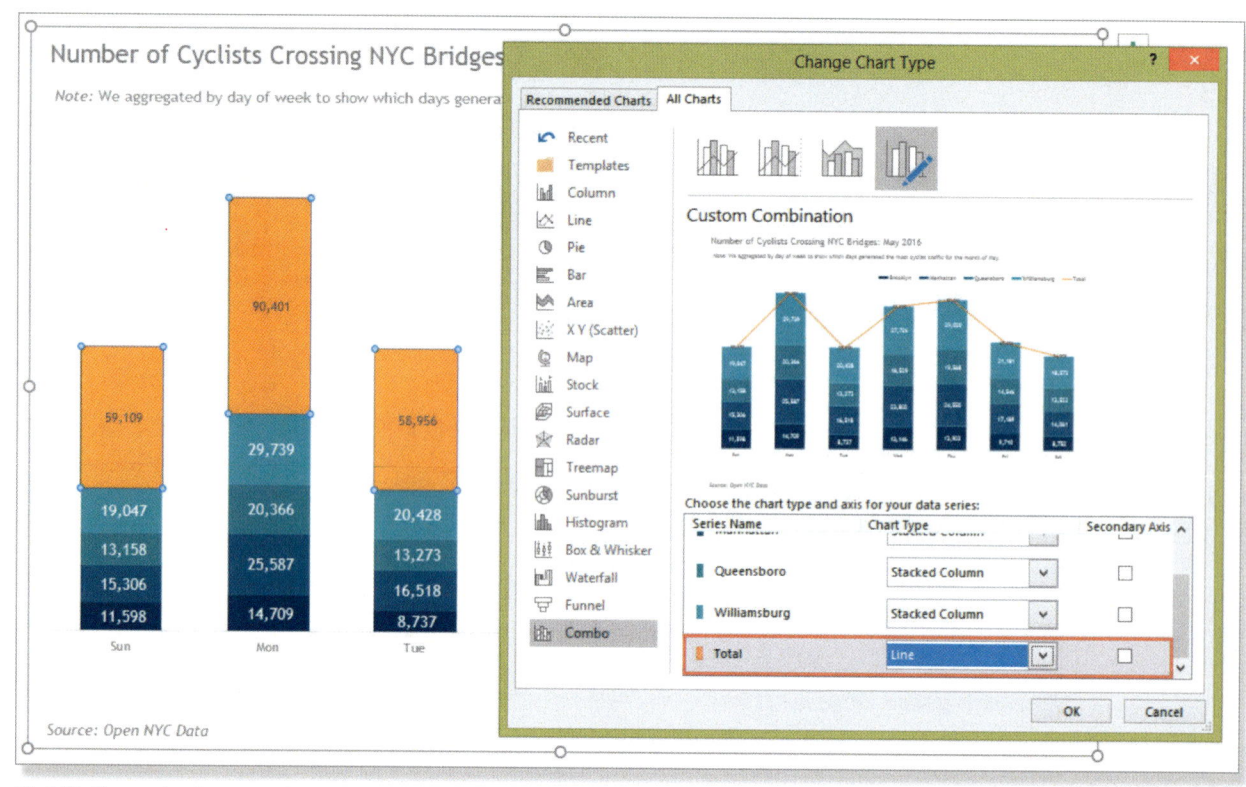

Fig 5.70: Change the chart type of your total series to Line.

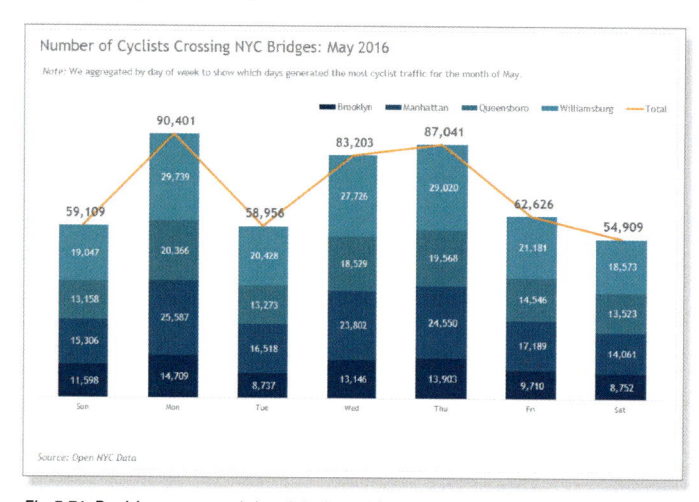

Fig 5.71: Position your total data labels and format to taste.

Step 3: Select any one of your newly minted total bars to select the entire data series, then right-click and choose **Change Series Chart Type** or navigate to **Chart Tools ▶ Design ▶ Type ▶ Change Chart Type**. In the **Change Chart Type** dialog, set your Total data series to **Line** and click **OK**. If Excel tries to put it on a secondary axis, shut that down. You want your total on the same axis as the rest of your data **(Fig 5.70)**.

Step 4: Select one of the data labels to select all of them. Press Ctrl-1 to open the **Format Data Labels** pane and open the **Label Options** tab. Under the **Label Options** section, set **Label Position** to **Above** to move them above the line. While they're still selected, format them to taste. I set mine to **Bold** and set **Font Size** to 12 to make them stand out more than the other data labels **(Fig 5.71)**.

132

Step 5: That line is like the kid who blocks your face in the roller coaster pic with his antics! We can't make that kid disappear, but we can get rid of the line. Its only purpose is to provide a carrier for data labels. (Sneaky, I know!) To hide it from view, select it, then set **Line** to **No line** under the **Fill & Line** tab **(Fig 5.72)**. Bye, obnoxious camera hog!

Step 6: That line's not going down without a fight. You'll see evidence of your nefarious visualization practices in the legend. No problem! Select the legend, then click Total and press Delete on your keyboard to send it sleeping with the fishies.

Example File
stacked-column.xlsx

Templates
If you'll use these charts again, save them as chart templates. Or download the templates included in the chart templates folder:

Stacked column chart: mds-stacked-column.crtx

Stacked column with totals chart:
mds-stacked-column-total.crtx

Number of Cyclists Crossing NYC Bridges: May 2016

Note: We aggregated by day of week to show which days generated the most cyclist traffic for the month of May.

■ Brooklyn ■ Manhattan ■ Queensboro ■ Williamsburg Total

Day	Brooklyn	Manhattan	Queensboro	Williamsburg	Total
Sun	11,598	15,306	13,158	19,047	59,109
Mon	14,709	25,587	20,366	29,739	90,401
Tue	8,737	16,518	13,273	20,428	58,956
Wed	13,146	23,802	18,529	27,726	83,203
Thu	13,903	24,550	19,568	29,020	87,041
Fri	9,710	17,189	14,546	21,181	62,626
Sat	8,752	14,061	13,523	18,573	54,909

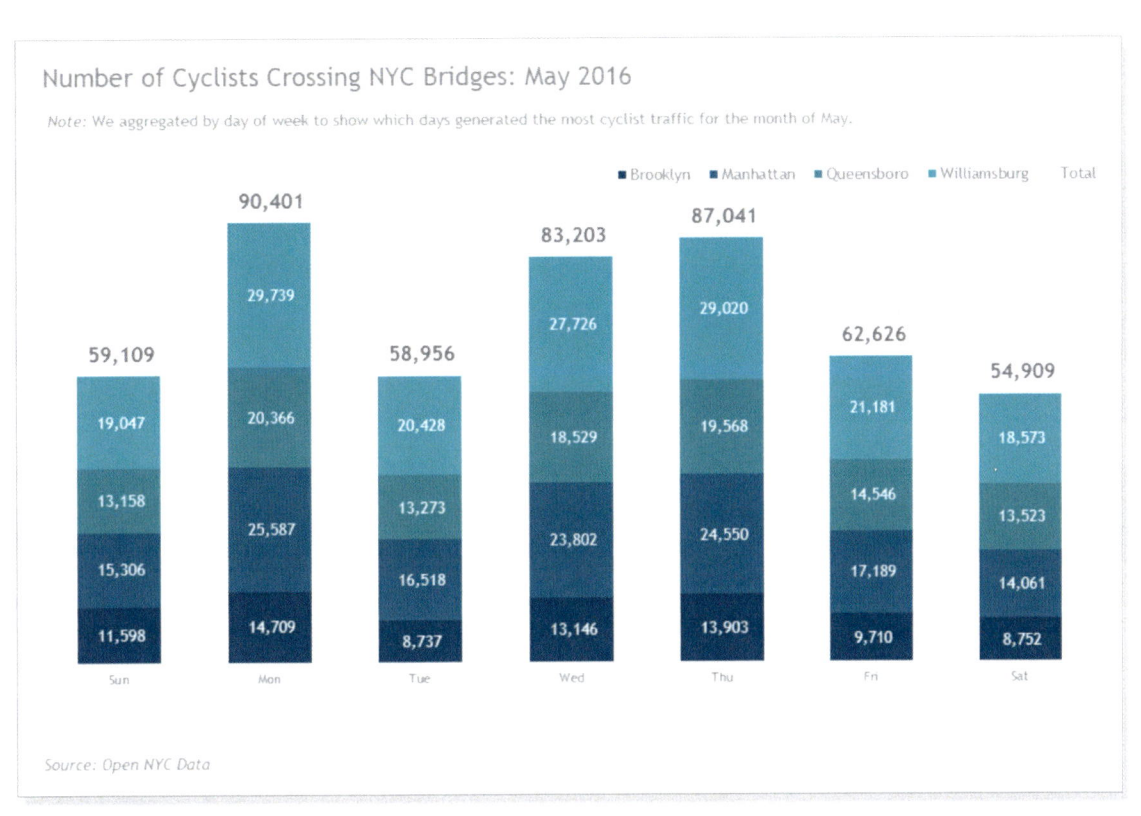

Source: Open NYC Data

Fig 5.72: Hide the line from view.

CHAPTER 5 • CREATE CHARTS

CLUSTERED STACKED COLUMN CHART

AKA
You may also see clustered stacked column charts called grouped stacked column (or bar) charts.

Uses
Clustered stacked column charts are best used to:

- compare period-over-period (e.g., week-over-week, month-over-month, quarter-over-quarter, or year-over-year) data with segmentation
- compare data that easily lends itself to segmentation, such as regional performance

Tips
- Keep category elements limited—a handful at most.
- Format axis to show intuitive grouping.
- Avoid combining categories into one field (e.g., year and quarter).
- Bringing data into your chart will make comparing performance among your categories easier.
- Skinny columns need not apply. Fatten them up.
- If you're using tabular data, you may need to use a pivot table to transform your data into a cross-tab format. I've included a Pivot Table worksheet in the downloadable workbook to show you how I set mine up. Click anywhere in the pivot table to open the PivotTable Fields pane. From there you can easily see how I constructed it.

Category / Metric
3 categories / 1 metric

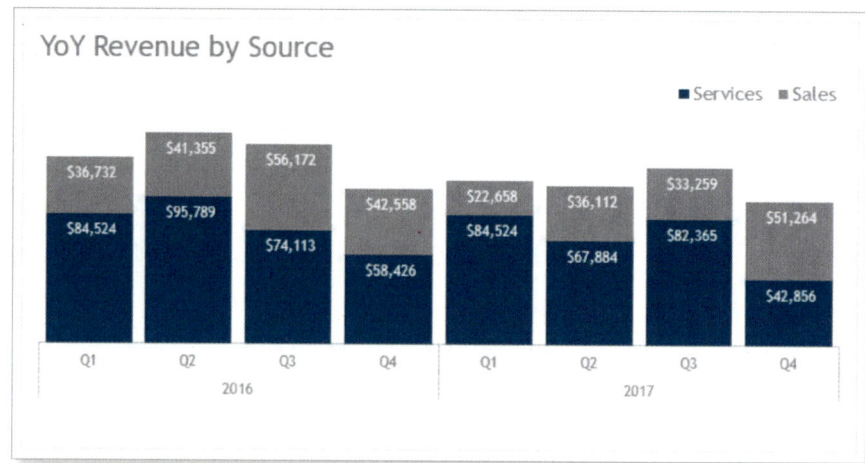

Fig 5.73: Our final chart.

Tutorial
Hundred-Foot View
Excel doesn't allow you to create clustered stacked column charts by default, so we'll need to hack the x axis with line formatting. This border will help give the impression that our top-level category (years for the example chart) is clustered, even though all of our columns will be equidistant.

Step 1: Before we can start charting, we need to get our data structured to give us two categories in our horizontal axis. I've seen bloggers recommend manually massaging your data into the right format, but we can use a pivot table to save a lot of manual effort. In the example dataset, I pulled Revenue Source into the **Columns** field, Year and Quarter into the **Rows** field, and Revenue into the **Values** field **(Fig 5.74)**.

Step 2: My pivot table defaulted to the compact layout. In this layout, the quarters and years are in the same column. We will need them in separate columns, with Q1 in the same row as the year. To accomplish this, navigate to **PivotTable Tools ▸ Design ▸ Layout ▸ Report Layout ▸ Show in Tabular Form**. Excel may

134

CHAPTER 5 • CREATE CHARTS

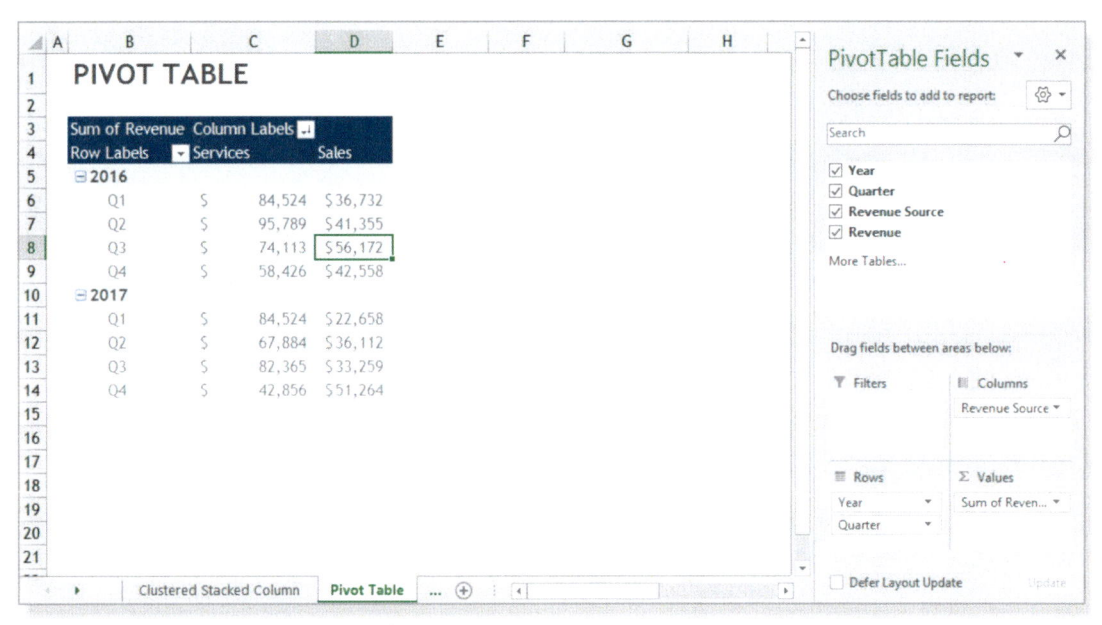

Fig 5.74: Restructure your data with a pivot table.

Fig 5.75: Structure pivot table in tabular form and turn off all totals.

try to be helpful and add subtotals. If so, you'll need to remove them by setting **Subtotals** to **Don't Show Subtotals**—also accessible from the **Layout** group **(Fig 5.75)**.

Step 3: Copy-and-paste your data into another sheet, then click-and-drag over your dataset and navigate to **Insert ▸ Charts ▸ Insert Column or Bar Chart ▸ 2-D Column ▸ Stacked Column (Fig 5.76)**.

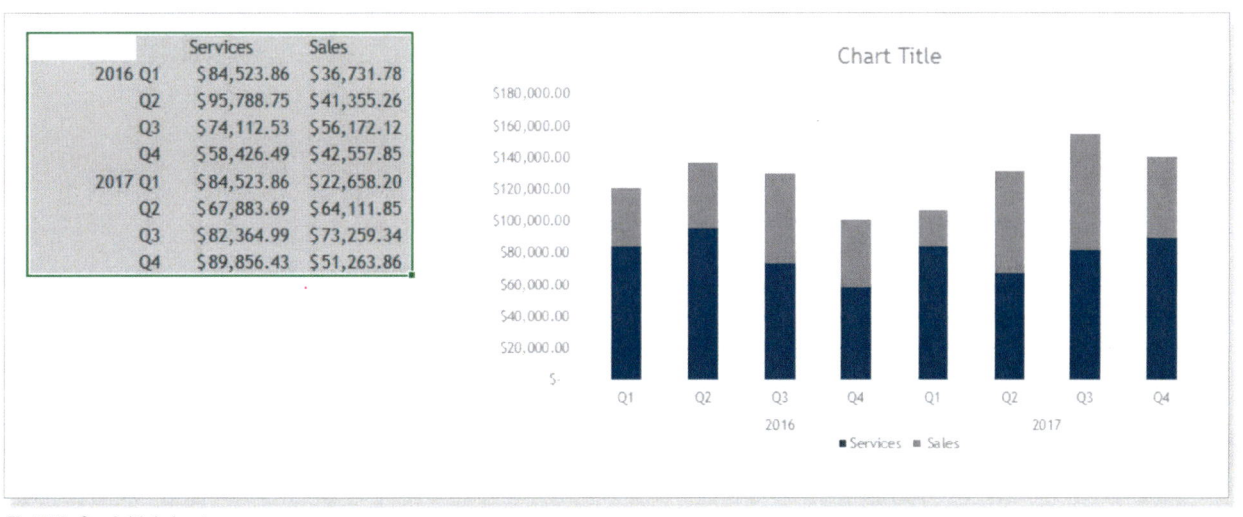

Fig 5.76: Our initial chart.

135

CHAPTER 5 · CREATE CHARTS

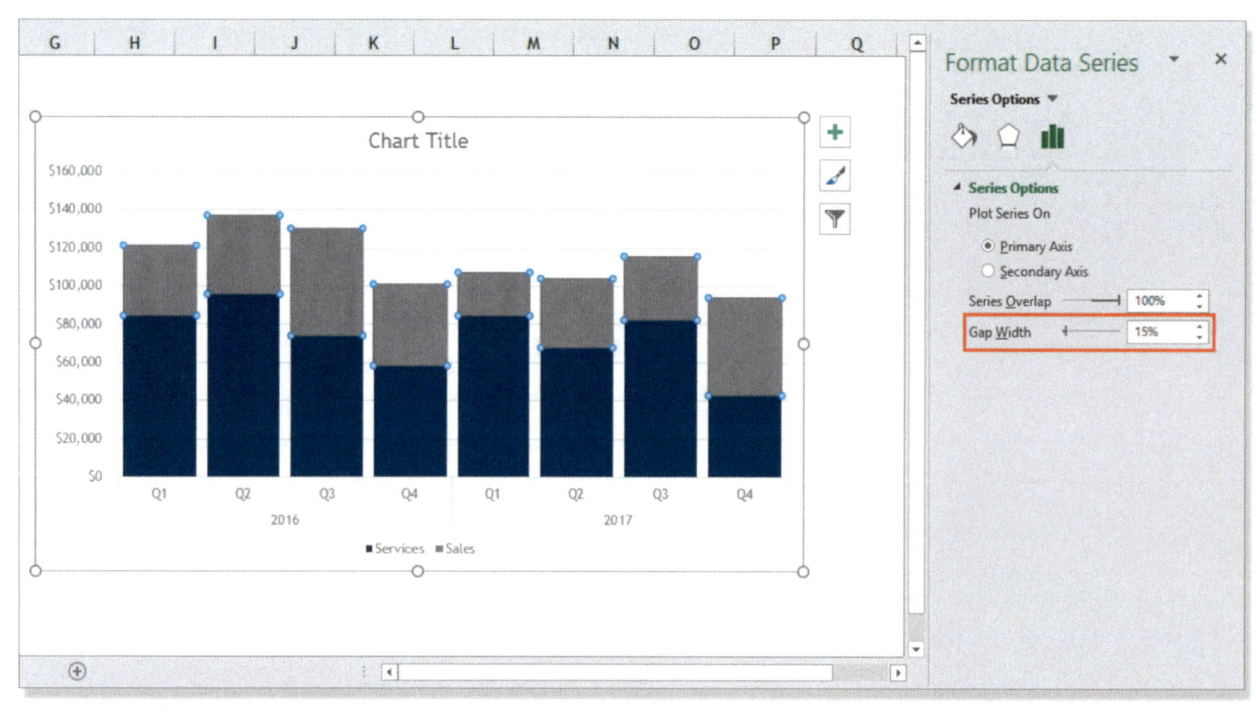

Fig 5.77: Reduce Gap Width to nudge the columns closer together.

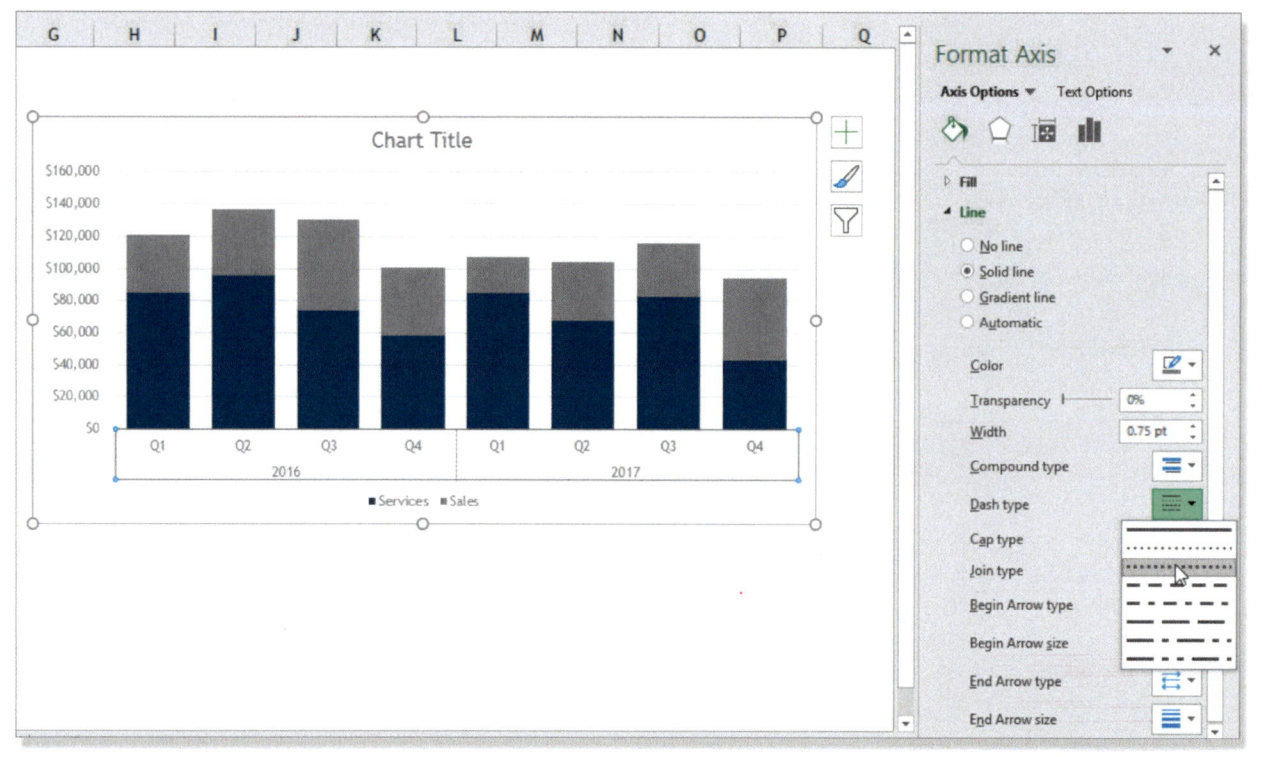

Fig 5.78: I modified the color and dash type for my axis border.

Step 4: I want to nudge our columns in each cluster closer together. To do that we need to reduce **Gap Width**. I set mine to 15% **(Fig 5.77)**.

LEARN MORE See the Adjust Chart Series Spacing section of the Design with Elegance chapter to learn how to customize the space between columns.

Step 5: There's not a clear demarcation between the different years. I would recommend darkening the borders between years in your axis—or whatever your category items are. I also opted for a dotted line because I didn't like how harsh the darker color looked, but I'm a princess when it comes to color. To modify the color of the axis border, select any item in the horizontal axis, which will open the **Format Axis** pane. Open the **Fill & Line** tab, and set **Line** to **Solid line** and color to a darker gray. I also set **Dash Type** to **Square Dot (Fig 5.78)**.

Step 6: Clean up your chart to taste. I cleaned mine by taking the following steps:

- made title more descriptive
- left aligned chart title
- bumped up title font size
- removed gridlines
- bumped up legend font size
- moved legend
- added data labels
- removed vertical axis

LEARN MORE See the Design with Elegance chapter to learn how to perform these refinements.

Example File

clustered-stacked-column-chart.xlsx

Template

If you'll use this chart again, save it as a chart template. Or download the template included in the chart templates folder: mds-clustered-stacked-column.crtx.

CHAPTER 5 • CREATE CHARTS

DIVERGING BAR CHART

AKA

You may also see diverging bar charts called:

- back-to-back graphs
- bidirectional charts
- bikini charts
- butterfly charts
- converging stacked bar charts
- deviation bar charts
- population distribution charts
- split axis graphs
- tornado chart/diagrams
- two-sided bar charts

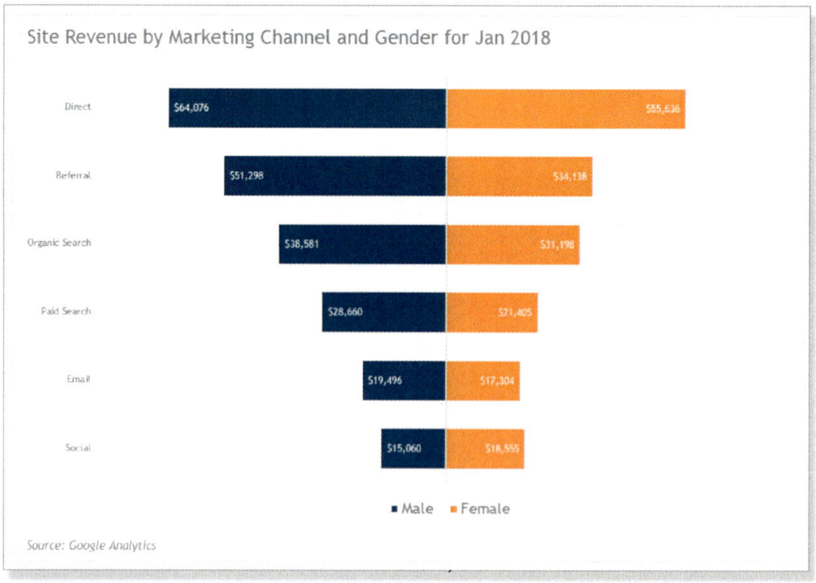

Fig 5.79: Our final chart.

Uses

Diverging bar charts are best used to:

- compare opposites, e.g., male vs. female, print vs. online assets, agree vs. disagree, impact of active vs. sedentary lifestyles
- compare the relative importance of variables

Tips

- Sorting your data can make these charts easier to read.
- Column charts don't work as well for diverging data, in my opinion. They tend to be harder to interpret.
- Although the use of red and green can be intuitive when charting opposites, it can be problematic for colorblind readers.

💬 **LEARN MORE** See the Design for Colorblind Users section of the Design with Elegance chapter to learn how to design with accessibility in mind.

Category / Metric

2 categories / 1 metric

Tutorial

Hundred-Foot View

Excel doesn't give us the option to create diverging bar charts, so we'll need to hack a bar chart by making one of our data series negative (by multiplying the values by -1). Then we'll use number formatting to hide evidence of our sleight of hand by removing negative signs.

Step 1: To get bars to populate on the other side of the tracks (to wit, the left side of the y axis), we need to multiply one of our data columns by -1 to make the values negative. To do this, drop a -1 into a cell, copy it, and select the data column you want to position to the left of the y axis by clicking-and-dragging or selecting the first cell and pressing Ctrl-Shift-down arrow. Then right-click and choose **Paste Special ▸ Paste Special**. At the risk of stating the obvious, this will open the **Paste Special** dialog. Under **Operation**, select **Multiply** and click **OK (Fig 5.80)**.

Step 2: We don't want anyone to know our data is negative. We want it to appear positive—like when you're at a work event when you'd rather be at home

138

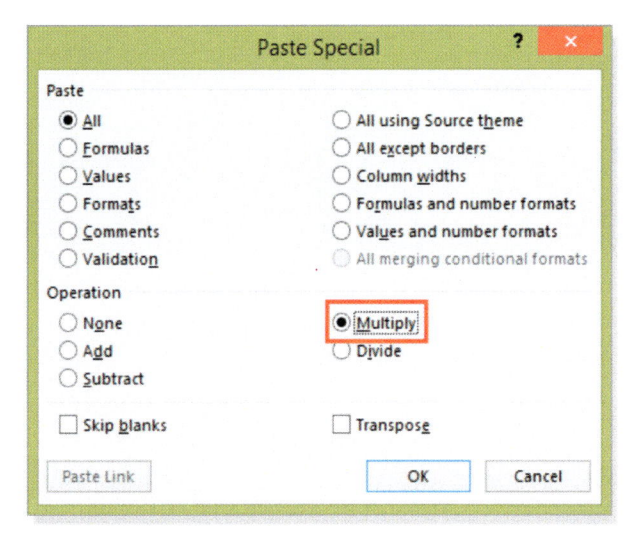

Fig 5.80: Set data you want on the left side of the chart to negative values by multiplying by -1.

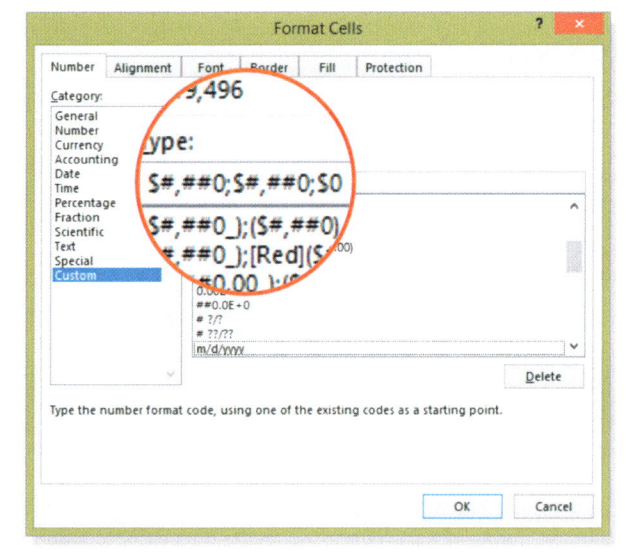

Fig 5.81: Use custom number formatting to make negative numbers look positive.

watching *The Handmaid's Tale* in your pajamas. To change the appearance of negative values, we need to use custom number formatting **(Fig 5.81)**. Since my dataset uses currency data and I don't want to show cents, I set mine using currency symbols and no decimals:

$#,##0;$#,##0;$0

💬 **LEARN MORE** See the Customize Number Formatting section of the Design with Elegance chapter to learn how to finesse your number formatting.

Step 3: I like to order my bar charts so that the bars are in ascending or descending order, depending on the story I want to tell and the data I want to emphasize. To do this, you'll need to create a total column and sort it to preference. Just keep in mind that, with bar charts, if you want your bars to get smaller as you work your way south in the chart, your total column will need to be sorted in ascending order—and vice versa. To sort, right-click in any cell in your total column and **Sort** ▶ **Sort Smallest to Largest** or **Sort Largest to Smallest**. If you don't have a total column before you convert one column to negative values and you need to use a calculated column, you'll need to account for that in your formula, as I did in the example file by multiplying my negative values by -1 **(Fig 5.82)**. Oh, what a tangled web we weave when first we practice to deceive. *sigh*

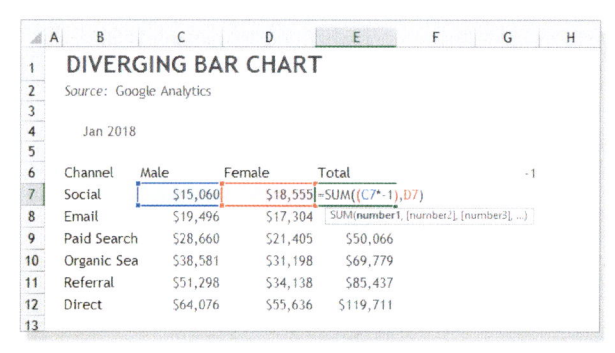

Fig 5.82: Your total formula will need to account for the negative values.

CHAPTER 5 • CREATE CHARTS

139

Step 4: We're finally ready to start building our chart. First, select everything but the total column, then navigate to **Insert ▸ Charts ▸ Insert Column or Bar Chart ▸ 2-D Bar ▸ Clustered Bar (Fig 5.83)**. We have our work cut out for us! *rolls up sleeves*

Step 5: Let's get the axis labels out of the way. To do that, select the axis to open the **Format Axis** pane. In the **Axis Options** tab, under **Labels**, set **Label Position** to **Low (Fig 5.84)**. This moves your axis labels to the left edge of your **Plot Area**.

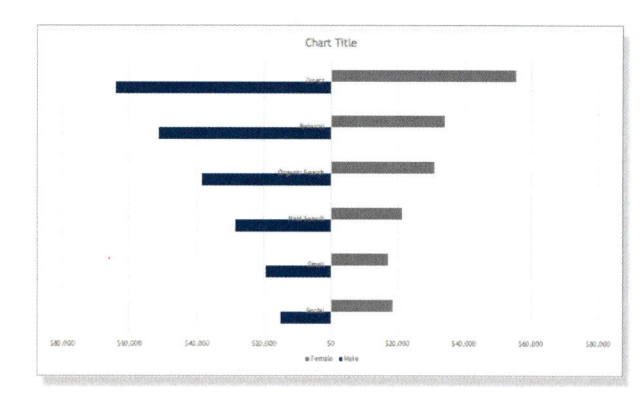

Fig 5.83: Our initial chart.

Step 6: Next, let's line up our data series. Select any bar to select its data series and open the **Format Data Series** pane. In the **Series Options** tab, set **Series Overlap** to 100% **(Fig 5.85)**. If these were on the same side of the axis, one data series would be on top of the other. But, in this case, no data gets snuffed out because they're on opposite sides of the vertical axis.

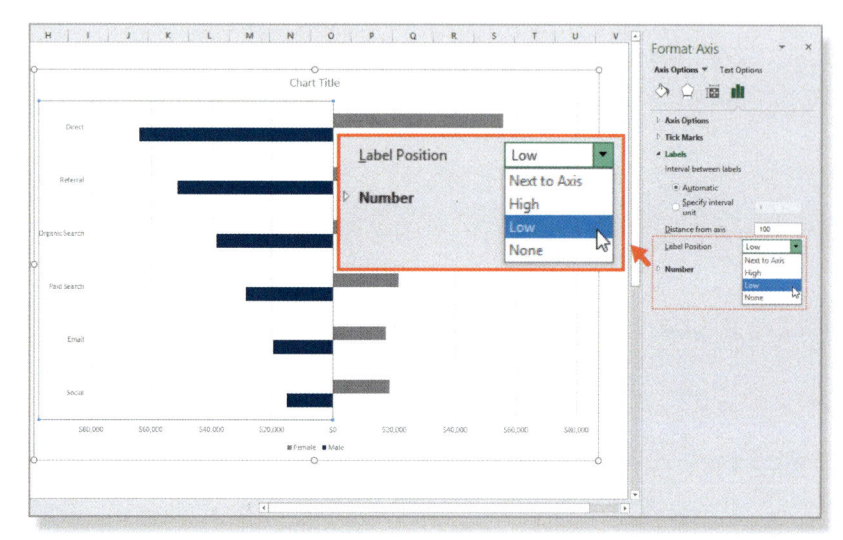

Fig 5.84: Move axis labels out of the way.

Step 7: My template uses gray as a secondary color, which I highly recommend doing. It complements most color palettes and is good for auxiliary chart elements. This isn't a good use of gray because it could potentially communicate that the values in blue are more important. For my chart, I changed the gray bars to orange, which is one of my theme colors. If you need to change the color of one of your data series, with the **Format Data Series** pane still open, navigate to the **Fill & Line** tab, then set **Fill Color** to your preferred (branded!) color **(Fig 5.86)**. Rinse and repeat with your other series.

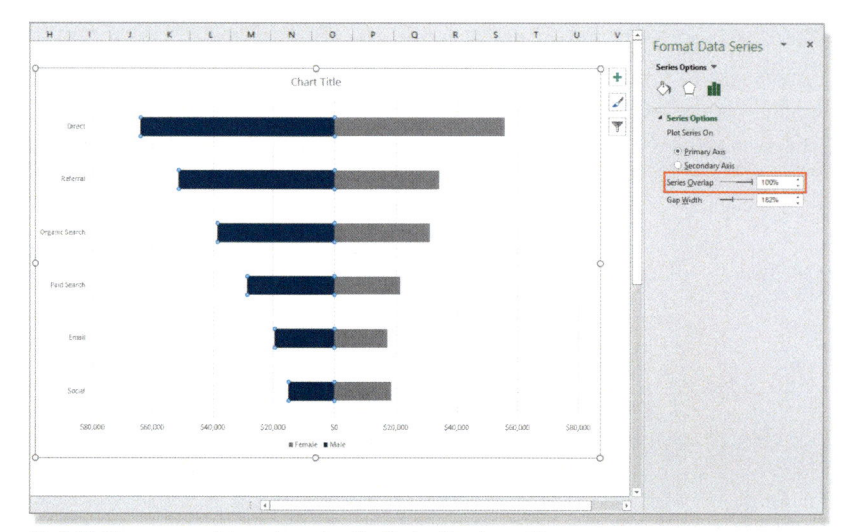

Fig 5.85: Line up your data series using the Series Overlap setting.

CHAPTER 5 • CREATE CHARTS

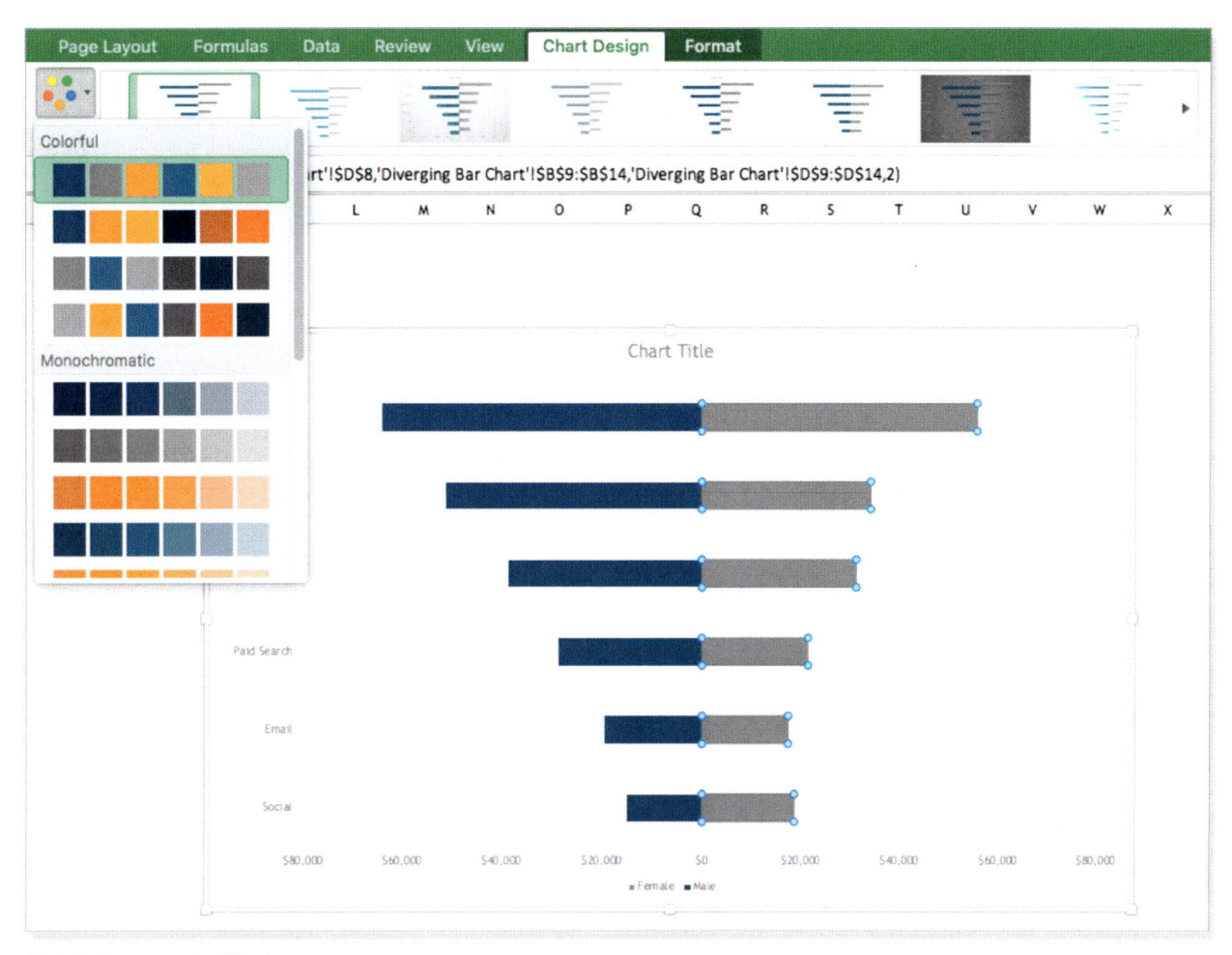

Fig 5.86: Set your series fill colors

Step 8: I'm a huge fan of data labels because they save your viewers processing power, with not needing to look back and forth between your data series and the value axis. They're elegant, efficient, and exact. I set mine to **Inside End**. You might need to change the text color to white to make them pop. Once you do this, you won't need your gridlines or horizontal axis labels, so you can send them packing **(Fig 5.87)**.

LEARN MORE See the Add Data Labels section of the Design with Elegance chapter to learn how to add data labels.

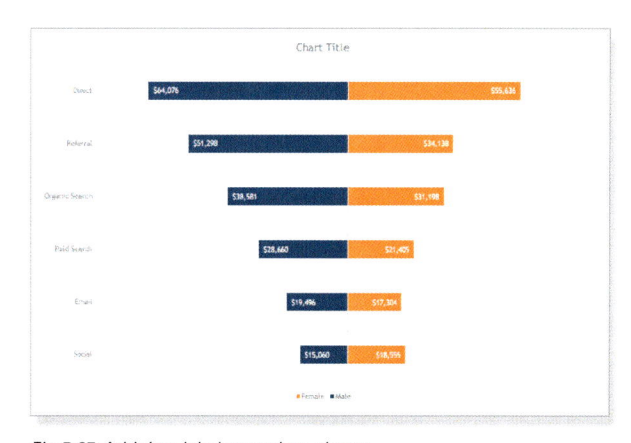

Fig 5.87: Add data labels to reduce clutter.

141

CHAPTER 5 • CREATE CHARTS

Step 9: It's counter-intuitive for your legend items to appear in the opposite order of your data series. To remedy this, right-click anywhere inside your chart and choose **Select Data** from the contextual menu. In the **Select Data Source** dialog, under **Legend Entries (Series)**, select the top entry and click the down-facing arrow to move it down (or click the bottom entry and click the up-facing arrow to move it up). Then click **OK (Fig 5.88)**.

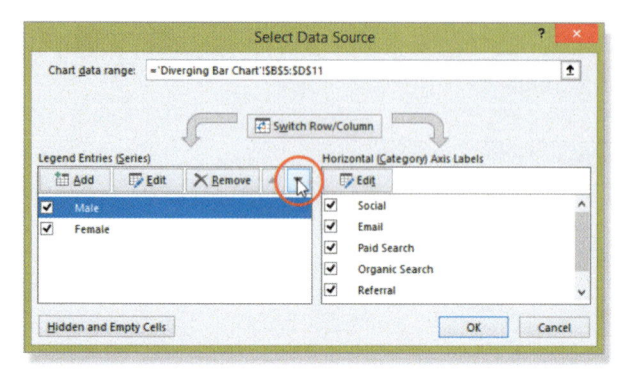

Fig 5.88: Smash the patriarchy! Lol j/k......

Step 10: Clean up your chart to taste. I cleaned mine by taking the following steps:

- made title more descriptive
- left aligned chart title
- bumped up title font sizes
- added annotation for data source
- bumped up legend font
- made bars wider
- moved legend
- added border to data series
- centered legend with axis

> 💬 **LEARN MORE** See the Design with Elegance chapter to learn how to perform these refinements.

Variation
Center Label

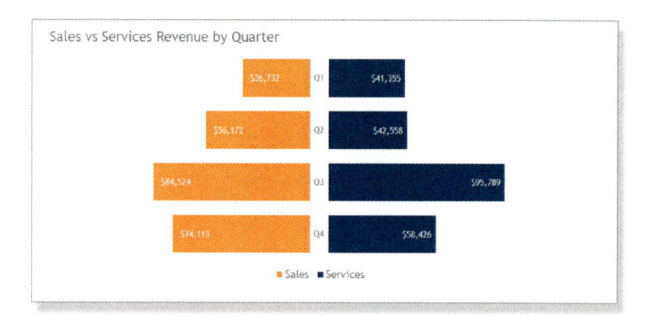

Fig 5.89: Our final chart.

Hundred-Foot View

This is a super hacked chart, and its use case is very limited. The only way it works is if your labels are equal in pixel width—or at least very close, as you'll see in the example file. Excel doesn't allow you to center your axis like this. What we'll do to achieve the parting of the columns is add two new columns to our table: one to make room for our centered labels and another to calculate that sum of that value and whatever value we assigned to the left side of our vertical axis (Sales in the example workbook). For example, if we made $50,000 from sales revenue and we need $10,000 of a fake data point to give us enough room for our axis labels, that total value will be $60,000. Technically, -$60,000 but let's not get pedantic. Finally, we'll point the data labels at our real sales data instead of the fake data the bars to the left are built on. So stealth!

Okay, let's pick up where the last tutorial left off.

Step 1: Add a column to your data table to make room for your labels. -10,000 worked for the sample dataset. Then create a new column that sums your original value and the value in this column. I renamed my original Sales column so that it would be clear which one I'd be using in my chart. My Old Sales column is the only one I won't be charting **(Fig 5.90)**.

CHAPTER 5 • CREATE CHARTS

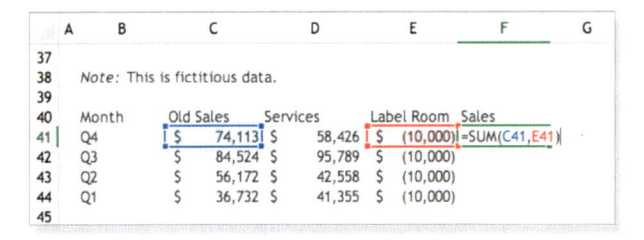

Fig 5.90: Add columns to make room for your future axis labels.

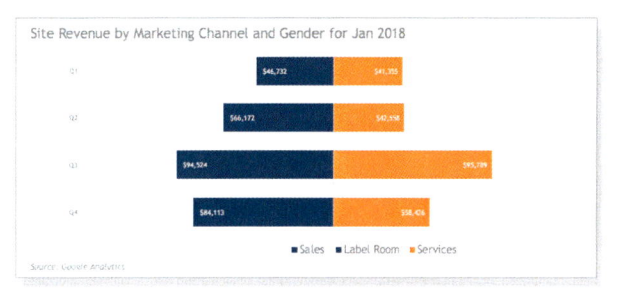

Fig 5.91: I had to add my orange fill and data labels back in.

Fig 5.92: Nudge your labels toward the axis.

Step 2: Right-click anywhere inside the **Plot Area** and choose **Select Data** from the contextual menu to open the **Select Data Source** dialog. Reset **Chart data range** by clicking-and-dragging over your updated data range. If your columns are non-contiguous, like mine, you'll need to press-and-hold the Ctrl key as you add your columns. Click **OK** to close the dialog. You may lose your data labels and any customized fill colors. (At least I did.) If so, add them back in for only the data series that will be visible **(Fig 5.91)**. In other words, not for your Label Room series.

💬 **LEARN MORE** See the Hack Excel's Charting Engine section of the Design with Elegance chapter to learn how to rock the Select Data Source dialog.

Step 3: Select your axis labels and press Ctrl-1 to open the **Format Axis** pane. In the **Axis Options** tab, under **Labels**, set **Distance from axis** to 0 and **Label Position** to **Next to Axis**. This will nudge our labels up against the axis **(Fig 5.92)**.

143

CHAPTER 5 • CREATE CHARTS

Step 4: Let's change up the fill color in our Label Room data series. You'll want it to blend with your **Plot Area** fill color. Mine is white, so I changed mine to white **(Fig 5.93)**.

Step 5: Next we'll ditch the border on the axis. First, select the axis, then set **Line** to **No line** under the **Fill & Line** tab **(Fig 5.94)**.

Step 6: Our data labels in the bars to the left of the axis point to #fakedata, so we'll need to point them to our original data column (Old Sales in the example file). From the **Label Options** tab, under **Label Options**, select **Value from Cells**, then click the **Select Range** button. Click-and-drag over the data in your original column. Deselect **Value (Fig 5.95)**.

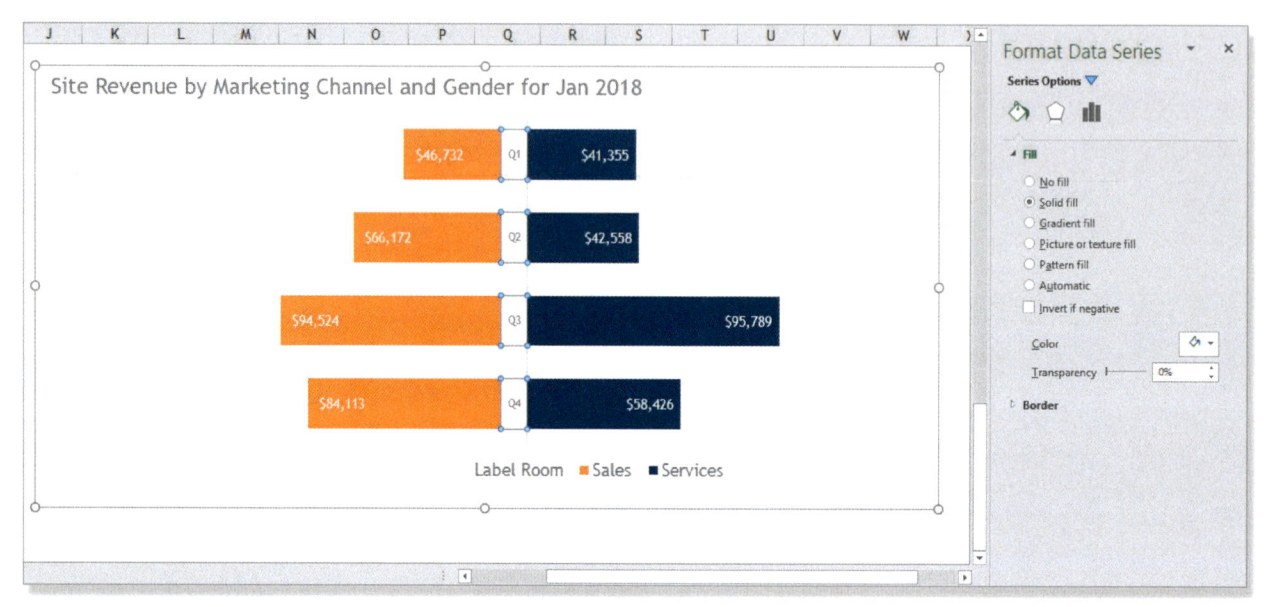

Fig 5.93: Set your Label Room data series fill to blend with your Plot Area fill color.

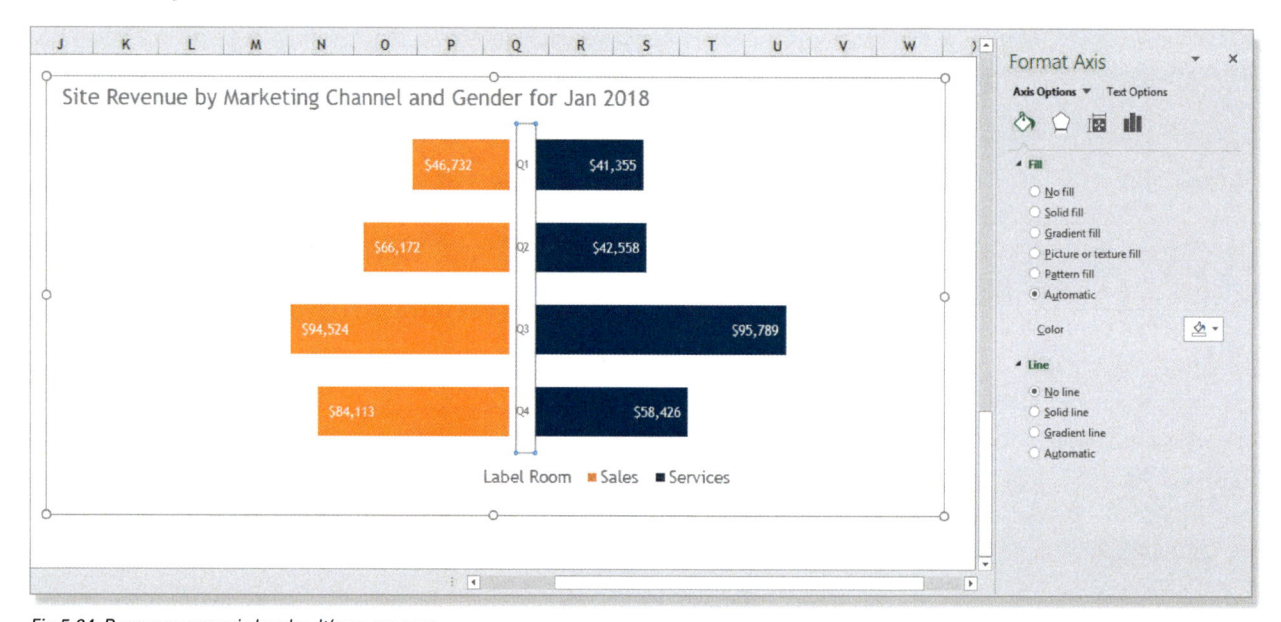

Fig 5.94: Remove your axis border. It's an eye sore.

144

CHAPTER 5 • CREATE CHARTS

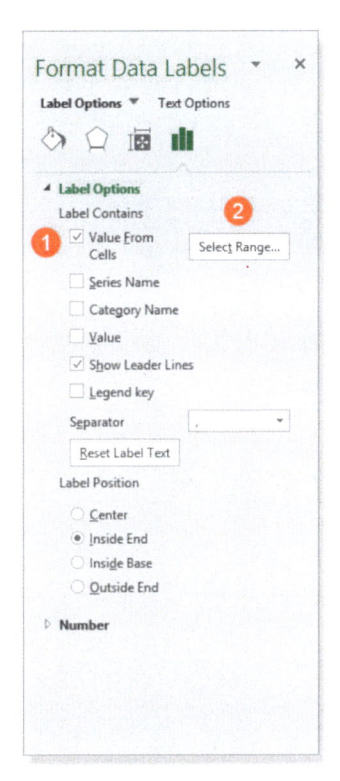

Fig 5.95: Excel allows you to use a different range for your labels than the one you used to plot the series.

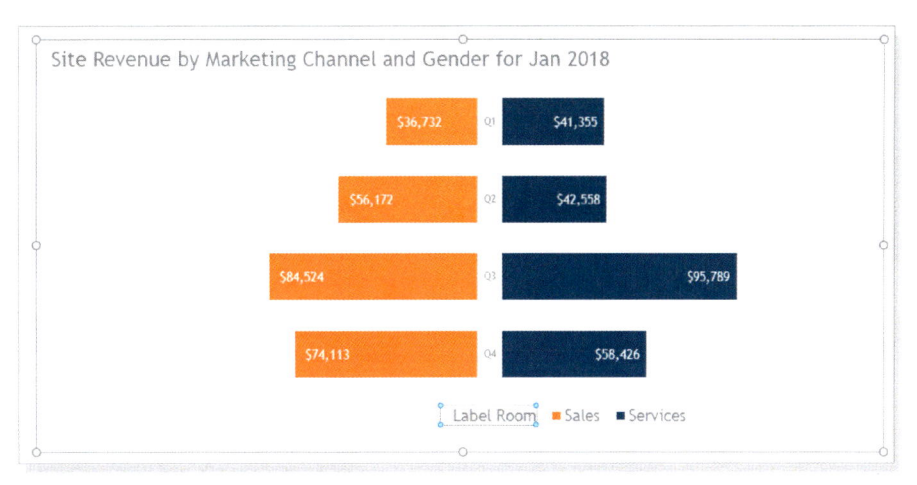

Fig 5.96: Isolate Label Room, then delete it.

Step 7: We don't need Label Room in the legend. Select the legend, then select Label Room to isolate it. Then press the Delete key on your keyboard **(Fig 5.96)**.

Tip: You'll know you've isolated a single legend item when it's surrounded by blue circles.

Step 8: Finally, switch out your chart title and call it a day.

Example File

diverging-bar-chart.xlsx

Template

The variation won't save as a template; it's too customized. If you'll use the main diverging bar chart again, save it as chart template. Or download the template included in the chart templates folder: mds-diverging-bar.crtx.

Note: If you use this chart template, you may need to reverse the order of the data series in the legend and tweak the theme colors Excel pulls in, if you're not happy with the defaults.

145

CHAPTER 5 • CREATE CHARTS

DIVERGING STACKED BAR CHART

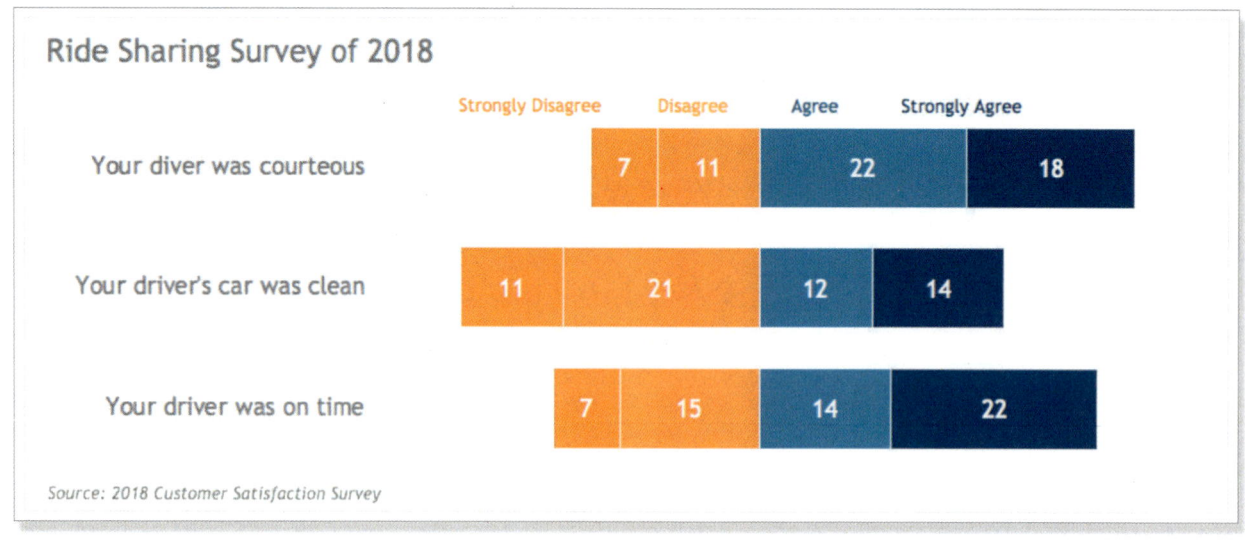

Ride Sharing Survey of 2018

	Strongly Disagree	Disagree	Agree	Strongly Agree
Your diver was courteous	7	11	22	18
Your driver's car was clean	11	21	12	14
Your driver was on time	7	15	14	22

Source: 2018 Customer Satisfaction Survey

Fig 5.97: Our final chart.

AKA
You may also see diverging bar charts called:

- stacked back-to-back graphs
- stacked bidirectional charts
- stacked bikini charts
- stacked butterfly charts
- stacked converging stacked bar charts
- stacked deviation bar charts
- stacked population distribution charts
- stacked split axis graphs
- stacked tornado chart/diagrams
- stacked two-sided bar charts

Uses
Diverging stacked bar charts are best used to:

- visualize rating scales, such as Likert scales
- visualize survey results where positive responses are to the right side of the vertical axis and negative responses are to the left

Tips
- If you're charting survey data, your dataset would need to have three categories: name (or some kind of identifier), question, and answer. Then the metric would be the count of responses. But you don't plot the ID; you just tally up the participants' responses.
- You can't chart survey data with different types of questions. Instead, you need to ask the same type of question (e.g., make a statement and ask recipients to rate how much they agree or disagree) or chart levels of positivity in responses. You should also use the same number possible responses for each question for best results.
- There are no laws against replacing your legend with text boxes. You can also use icons, as you'll see in the Dumbbell Plot tutorial. However, if you do this, you'll want to first save your chart template since anything you hardcode into your chart will become an element in every chart you build from that template.

Category / Metric
1–2 categories / 1–2 categories

146

CHAPTER 5 • CREATE CHARTS

Tutorial

Hundred-Foot View

Excel doesn't give us the option to create diverging stacked bar charts, so we'll need to hack a stacked bar chart by making two of our data series negative (by multiplying the values by -1). Then we'll use number formatting to remove negative signs from our axis.

Step 1: Make negative responses—the responses you want to show up on the left side of the y axis—negative, following steps 1 and 2 in the Diverging Bar Chart tutorial **(Fig 5.98)**.

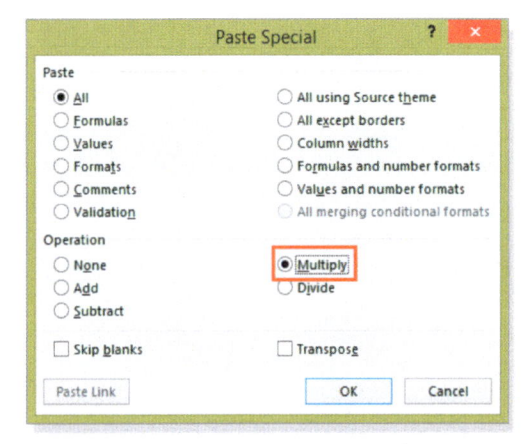

Fig 5.98: Set data you want on the left side of the chart to negative values.

Step 2: Select any cell in your dataset and navigate to **Insert ▶ Insert Column or Bar Chart ▶ 2-D Bar ▶ Stacked Bar (Fig 5.99)**. Brace yourself. It's unsightly.

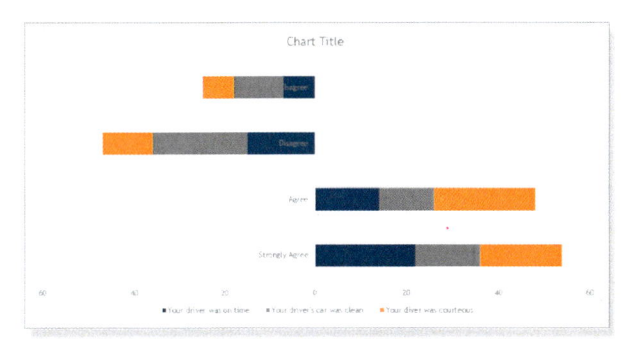

Fig 5.99: Our initial chart.

Step 3: One of the [many] problems with our initial chart is that we need the responses in the legend. In the **Select Data Source** dialog, they should be in the **Legend Entries (Series)** section, and the questions should be in the **Horizontal (Category) Axis Labels** section. To remedy this, we'll need to switch them by clicking the **Switch Row/Column** button. Click **OK** to close the dialog. In the example dataset, you'll see the response options (Strongly Agree to Strongly Disagree) in the Series section after the switcheroo **(Fig 5.100)**.

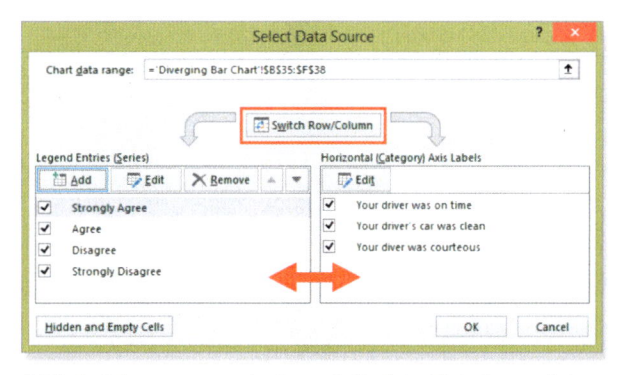

5.100: Switch your rows and columns in the Select Data Source dialog.

💬 LEARN MORE See the Hack Excel's Charting Engine section of the Design with Elegance chapter to learn how to rock the Select Data Source dialog.

Step 4: Sometimes your series get out of order in the legend. If this happens to your data, back into the **Select Data Source** dialog you go! In the **Legend Entries (Series)** field, use the up- or down-facing arrows to reshuffle your data series **(Fig 5.101)**.

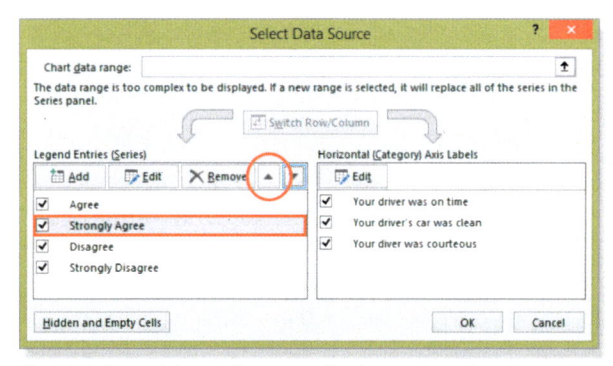

Fig 5.101: You might need to manually change the order of your data series. Stay calm and visualize on.

CHAPTER 5 • CREATE CHARTS

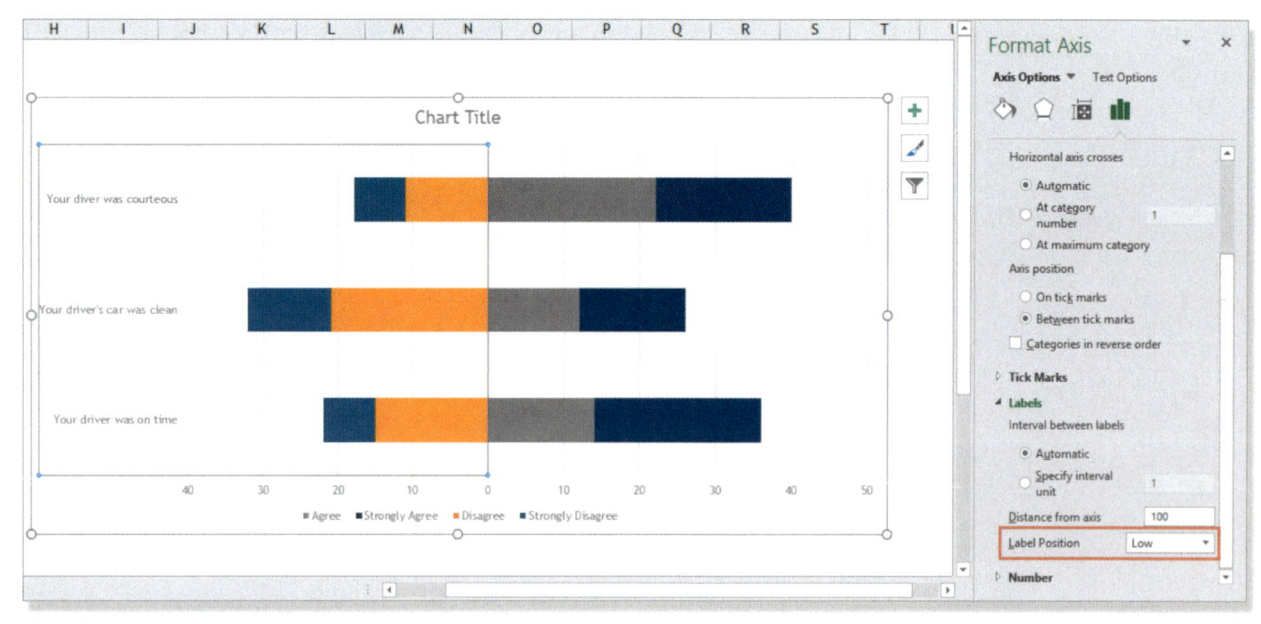

Fig 5.102: Sorry, guys. Snuggle time is over.

Step 5: Let's move the vertical axis labels away from the axis, so our labels don't overlap with our bars. First, select the axis and press Ctrl-1 to open the **Format Axis** pane. In the **Axis Options** tab, under **Labels**, set **Label Position** to **Low (Fig 5.102)**.

Note: Don't worry about the legend being out of order. We'll be replacing it with a younger model.

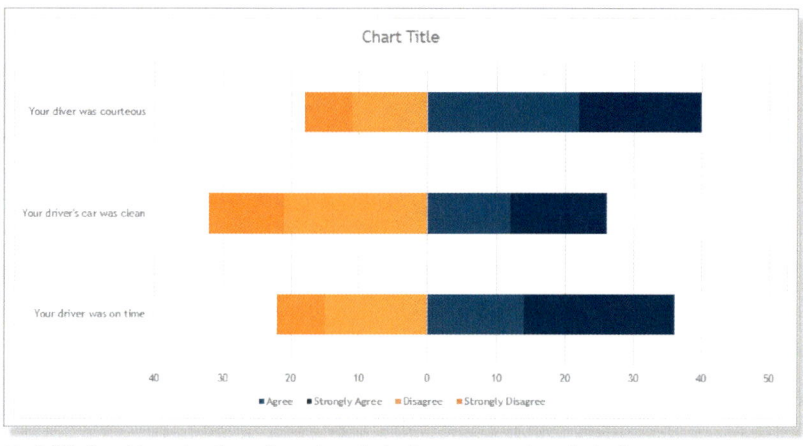

Fig 5.103: Consider using tints of your branded colors over non-branded colors.

Step 6: I don't recommend using different colors in a chart to show survey results. Instead, I recommend using tints of your branded (theme) colors. I want the darker color to be associated with the strong values—i.e., Strongly Agree/Disagree—so we'll apply our new tints to the data series closer to the y axis. First, select one of these bars (data points) to select the entire data series, which will open the **Format Data Series** pane. Open the **Fill & Line** tab and set **Fill Color** to your preferred color **(Fig 5.103)**. Our chart is starting to shape up! If you need to change any of your other series, rinse and repeat this step until they're formatted correctly.

> 💬 **LEARN MORE** See the Stick to Branded Colors section of the Design with Elegance chapter to learn more about restricting your designs to branded colors.

Step 7: No matter what I tried, I couldn't get the legend to follow the same order as the data series. I think I just pushed Excel a little too hard, so I had to fake a legend with text boxes. I made the font colors matchy matchy for additional clarity **(Fig 5.104)**. There's a fine line between clear and garish. Tweaking font colors dances on that line. However, creating colored rectangles would require grouping each rectangle with its corresponding label and then aligning everything. That's a little extra, even for me!

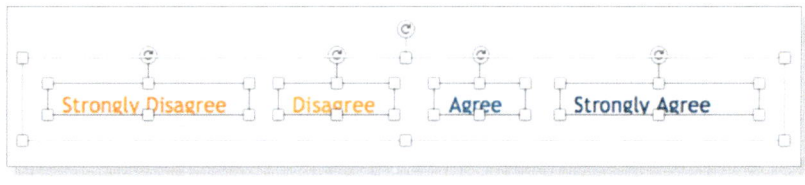

Fig 5.104: You may need to manually create a legend out of text boxes for this chart.

LEARN MORE See the Align Strategically section of the Design with Elegance chapter to learn how to align objects in your workbook.

Step 8: Clean up your chart to taste. I cleaned mine by taking the following steps:
- made title more descriptive
- left aligned chart title
- bumped up title font size
- added annotation for data source
- removed gridlines
- made bars wider
- added data labels
- added border to data series

LEARN MORE See the Design with Elegance chapter to learn how to perform these refinements.

Example File
diverging-stacked-bar-chart.xlsx

Template
If you'll use this chart again, save it as a chart template. Or download the template included in the chart templates folder: mds-diverging-stacked-bar.crtx.

Note: If you use this chart template, you may need to reverse the order of the data series in the legend and tweak the theme colors Excel pulls in, if you're not happy with the defaults.

CHAPTER 5 • CREATE CHARTS

FLOATING BAR CHART

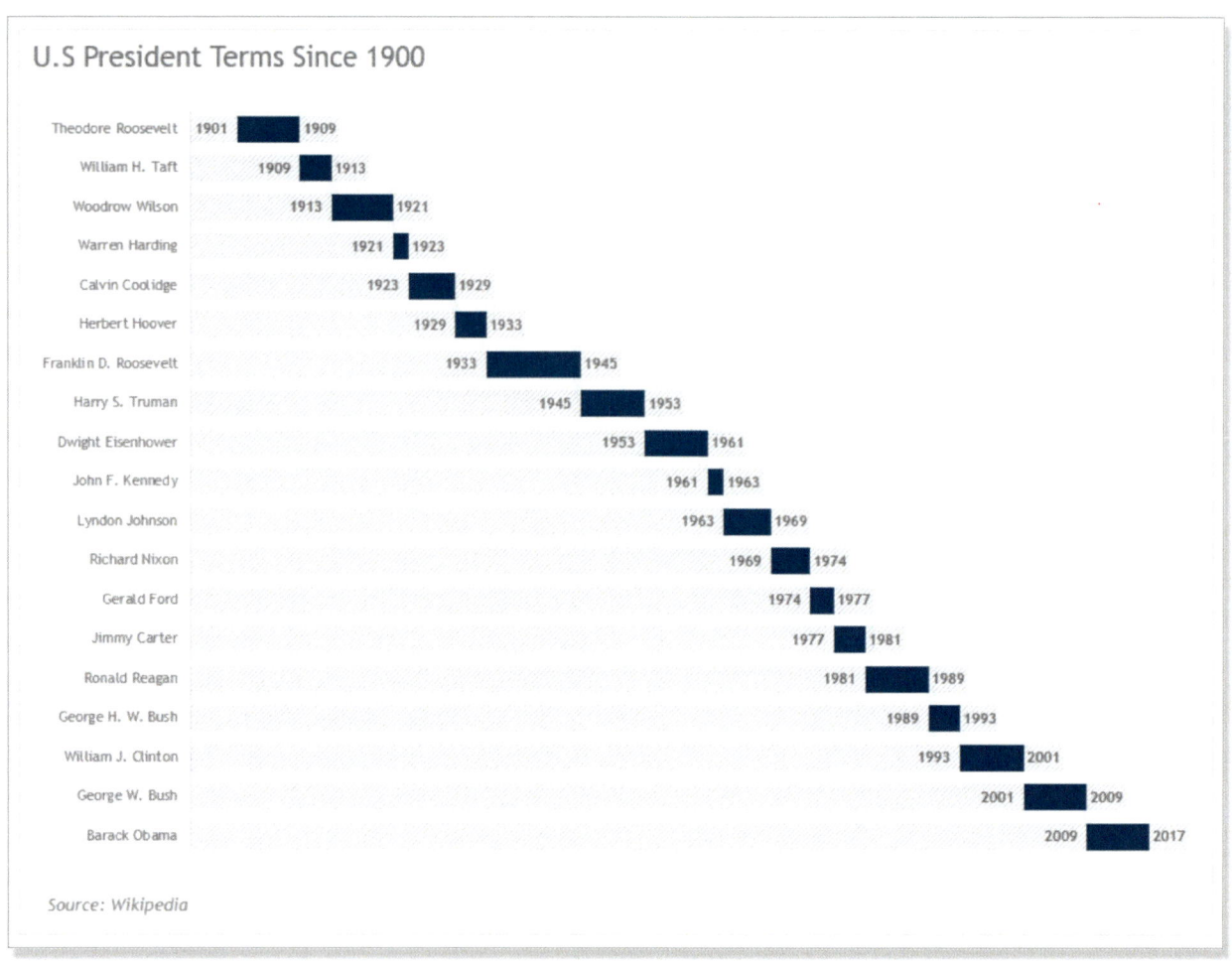

U.S President Terms Since 1900

President	Start	End
Theodore Roosevelt	1901	1909
William H. Taft	1909	1913
Woodrow Wilson	1913	1921
Warren Harding	1921	1923
Calvin Coolidge	1923	1929
Herbert Hoover	1929	1933
Franklin D. Roosevelt	1933	1945
Harry S. Truman	1945	1953
Dwight Eisenhower	1953	1961
John F. Kennedy	1961	1963
Lyndon Johnson	1963	1969
Richard Nixon	1969	1974
Gerald Ford	1974	1977
Jimmy Carter	1977	1981
Ronald Reagan	1981	1989
George H. W. Bush	1989	1993
William J. Clinton	1993	2001
George W. Bush	2001	2009
Barack Obama	2009	2017

Source: Wikipedia

Fig 5.105: Our final chart.

AKA
You may also see floating bar charts called:
- column graphs
- difference graphs
- high-low graphs
- range bars
- span charts

Uses
Floating bar charts are best used to:
- compare data across discrete groups
- visualize ranges between a minimum and maximum value, such as temperatures, river flow rates, oxygen levels, etc.
- provide a visual overview of varying start times, durations, and overlaps for a number of processes

Tips

- I frequently see floating bar charts created with gridlines to help the viewer see where each bar starts and stops. But, as with most visualizations, gridlines can detract from the elegance of the floating bars.
- If you want to highlight a key event in the middle of the range, you could add a mean/target line as demonstrated in the Add a Mean Line variation of the Bar Chart tutorial or the Add a Mean Line variation of the Column Chart tutorial.
- Floating bars are good for comparing ranges among category elements, but this is still aggregated data. You won't be able to view skew or kurtosis in your dataset, which you will learn more about in the Box & Whisker Chart tutorial. If I can convince you to kick off your analyses of aggregated data with a box chart, I will feel better about all my life decisions to this point.
- Gantt charts are just floating bar charts for project management. (Mind blown, I know.)
- If one series picks up where another lets off, you might want to consider using a Pareto chart. It's a floating column chart with panache.

LEARN MORE See the Pareto Chart section of this chapter to learn how to use a Pareto chart.

Category / Metric
1 category / 2 metrics

Tutorial
Hundred-Foot View
A floating bar chart is a stacked bar chart that typically has data series with no fill or outline. Because of the number of bars and their distance from the vertical axis, I added a subtle background pattern to make it easier to see which bar went with which president.

Each bar is made of data from three data series (columns in my dataset):

- *Start:* This data series extends from the y axis to the start year of each president's term from the 20th and 21st centuries.
- *Years in Office:* Franklin Roosevelt, for example, was inaugurated in 1933 and served 12 years, so this second bar extends from 1933 to 1945. This data series is the belle of the ball, and it's where we want users' eyes to focus. This column is a calculated column that subtracts Start from End.
- *Additional Bar:* This bar is a carrier mule for an additional data label. Its only purpose is to extend long enough to make a place for the data label at the end of the Years in Office bar. I added a matching pattern to it as the first bar because it looked weird to have it floating out in space like a free electron. But if your dataset doesn't have as many categories as mine—with bars floating farther and farther from the y axis—you won't need to add a fill to this series. Instead of using its value (5) in the data label, we'll point Excel to the End column.

Step 1: We need to add a couple helper columns. The first column will calculate the difference between the end and start of each term. This will set the length of our visible bars, which will represent the length of each president's term **(Fig 5.106)**.

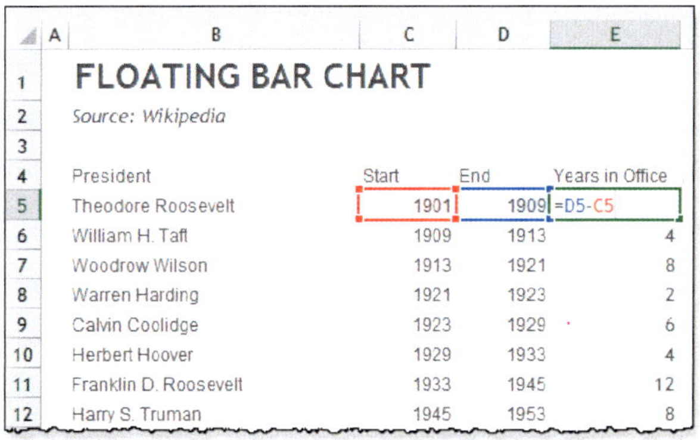

	A	B	C	D	E
1		**FLOATING BAR CHART**			
2		Source: Wikipedia			
3					
4		President	Start	End	Years in Office
5		Theodore Roosevelt	1901	1909	=D5-C5
6		William H. Taft	1909	1913	4
7		Woodrow Wilson	1913	1921	8
8		Warren Harding	1921	1923	2
9		Calvin Coolidge	1923	1929	6
10		Herbert Hoover	1929	1933	4
11		Franklin D. Roosevelt	1933	1945	12
12		Harry S. Truman	1945	1953	8

Fig 5.106: You may need a helper column to use later in your chart.

CHAPTER 5 • CREATE CHARTS

Fig 5.107: You will need a helper column to house your hacktastic data labels.

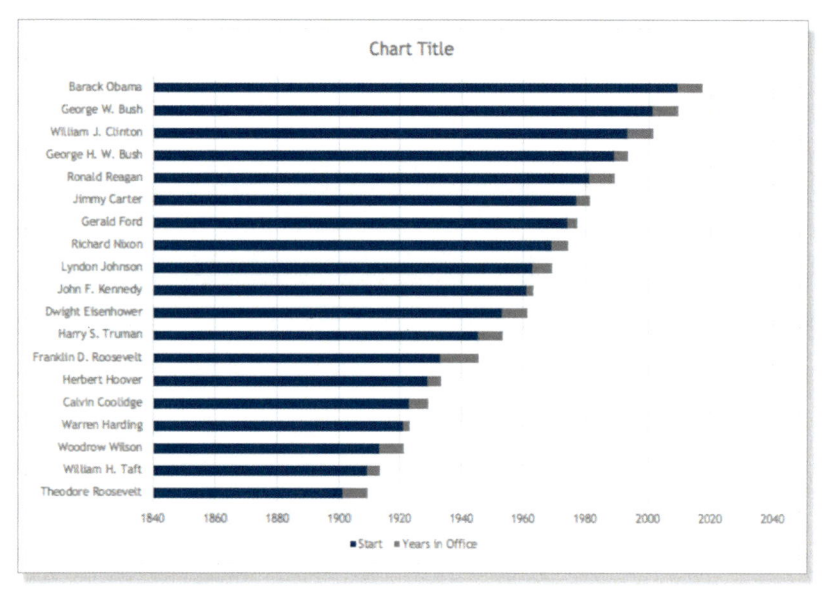

Fig 5.108: Our initial chart.

Step 2: The second helper column will give us a bar that will bear us a data label. I had to add this column to get a data label with the year the president's term ended because Excel doesn't give you the option to add a data label outside the data series with a stacked bar. I set this value to 5 **(Fig 5.107)**.

Step 3: The only column we won't need in our chart from the data table is the End column, so click-and-drag over the first two columns, then hold down the Ctrl key and click-and-drag over the last two columns to add them to your selection. Finally, navigate to **Insert ▸ Charts ▸ Insert Column or Bar Chart ▸ 2-D Bar ▸ Stacked Bar (Fig 5.108)**.

Step 4: I'd prefer that the presidents appear in chronological order, with Roosevelt at the top of the stack in our chart. But because bar charts are charted in the reverse order that they appear in the data table, we'll need to reverse the sort. To set your sort, right-click any cell in the column you want to sort by and choose your sort option from the **Sort** flyout menu. I sorted mine in descending order.

CHAPTER 5 • CREATE CHARTS

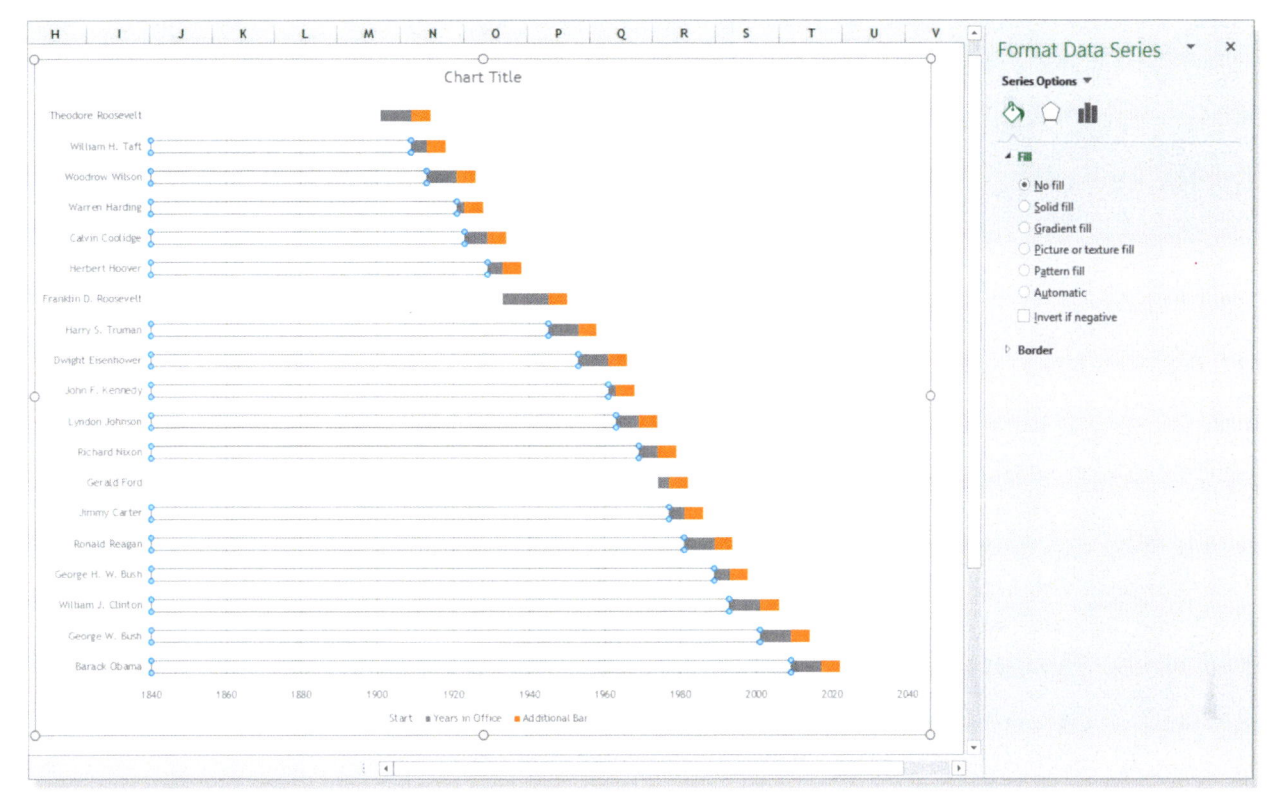

Fig 5.109: Strip the data series closest to the y axis of its fill and line.

Step 5: Select the data series closest to the vertical axis and press Ctrl-1 to open the **Format Data Series** pane. Open the **Fill & Line** tab, then set **Fill** to **No fill** and **Border** to **No line (Fig 5.109).** I circle back and add a pattern fill to this data series later, but you usually won't want to add these pixels to your visualizations. Sometimes sacrifices need to be made for the sake of clarity.

Step 6: Repeat step 5 with the data series farthest from the vertical axis so that only the data series depicting years in office remains visible **(Fig 5.110)**.

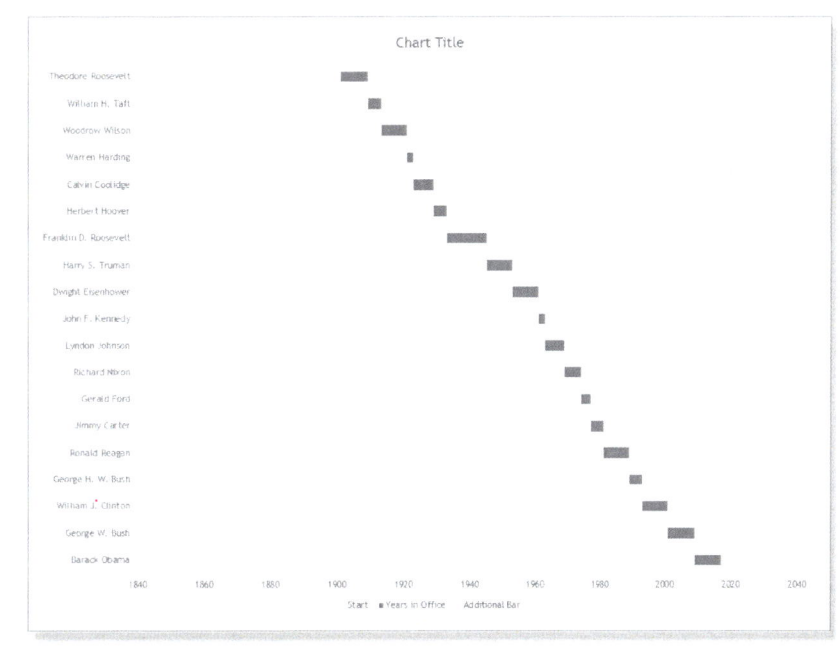

Fig 5.110: Our bars are finally levitating.

153

CHAPTER 5 • CREATE CHARTS

Step 7: We can trim down the horizontal axis. We just need it to start at 1900 and end at 2017. However, we'll need to allow for a little extra space since we'll have data labels on both sides of our visible bar. So we'll set the minimum value to 1895 and maximum to 2020. To set the minimum and maximum values, select the axis to open the **Format Axis** pane, then open the **Axis Options** tab. Under **Bounds**, set **Minimum** to 1895 and **Maximum** to 2020—or whatever makes sense for your data **(Fig 5.111)**.

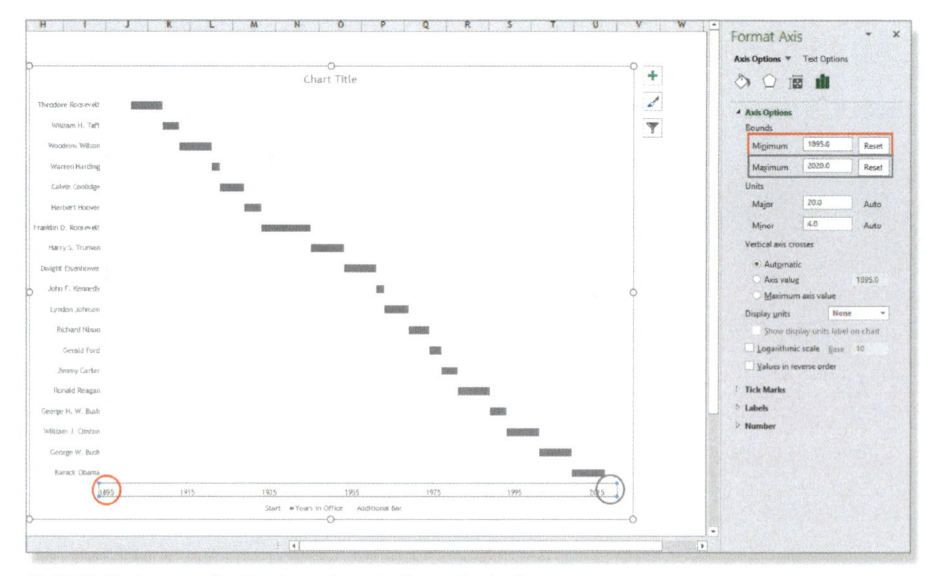

Fig 5.111: Nudge your floating bars closer to the vertical axis.

Step 8: I don't want gray for my fill color, so I set it to my branded blue color. To do this, select one of the bars to select the data series and open the **Format Data Series** pane. Open the **Fill & Line** tab. Set **Fill** to **Solid fill** and **Fill Color** to your preferred (branded) color.

Step 9: The bars are a little on the thin side, so I made them wider by setting **Gap Width** to 50% **(Fig 5.112)**.

> **LEARN MORE** See the Adjust Chart Series Spacing section of the Design with Elegance chapter to learn how to customize the space between columns.

Step 10: I want to add the start and end years as data labels, but this will take some trickery because we'll need to add the start year label to the inside end position of the far-left "invisible" bar and the end year to the center of the far-right bar. Let's begin with the start year. Select one of the bars to select the entire series. From the **Chart Elements** flyout menu, click the right-facing arrow next to **Data Labels** to open the flyout menu and select **Inside End (Fig 5.113)**.

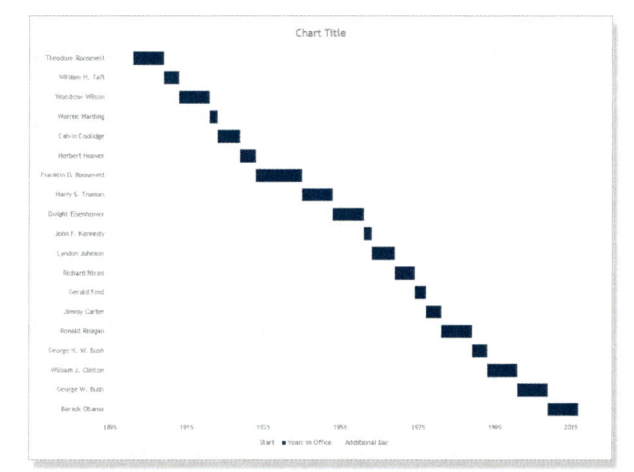

Fig 5.112: Reduce Gap Width to thicken up your bars.

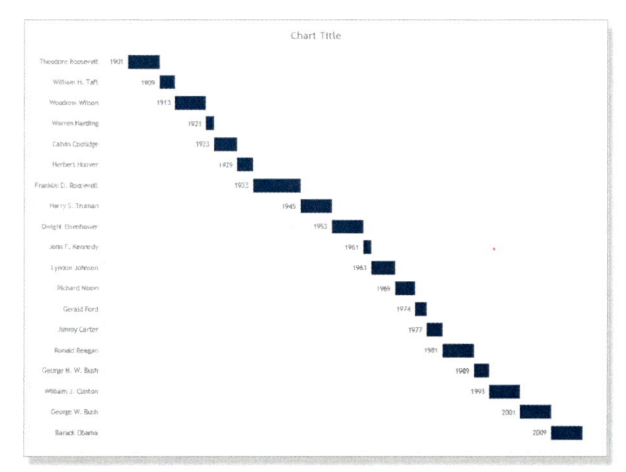

Fig 5.113: Add start date to first data series.

CHAPTER 5 • CREATE CHARTS

Step 11: Rinse and repeat with the Additional Bar data series—except pull the end year into the data label instead of the default value. To do this, open the **Label Options** tab and set **Label Contains** to **Value From Cells (Fig 5.114a)**. This will summon the **Data Label Range** dialog. Click-and-drag over the end years—sans the column header—and click **OK (Fig 5.114b)**. Finally, deselect **Value** and **Show Leader Lines** to evict our random value. You can set **Label Position** to **Center** or **Inside Base**, whichever works best for your data **(Fig 5.114c)**. (**Inside End** gives you no margin between the visible bar and the label.) *Tip:* You only see leader lines if you drag your data labels away from their default position, so deselecting them in this case is optional.

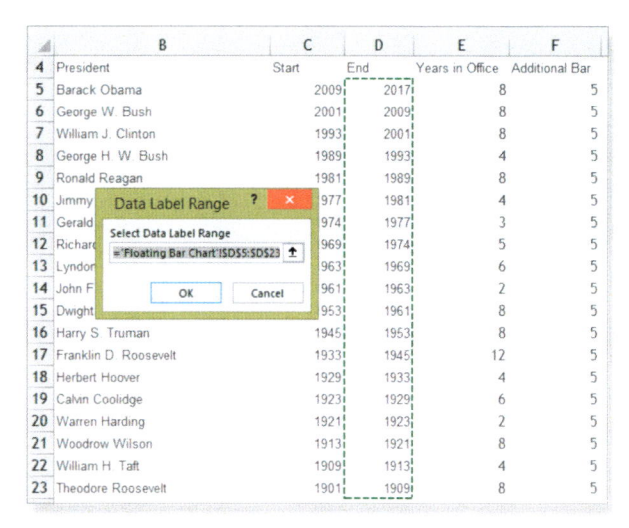

Fig 5.114b: Add end year to bars.

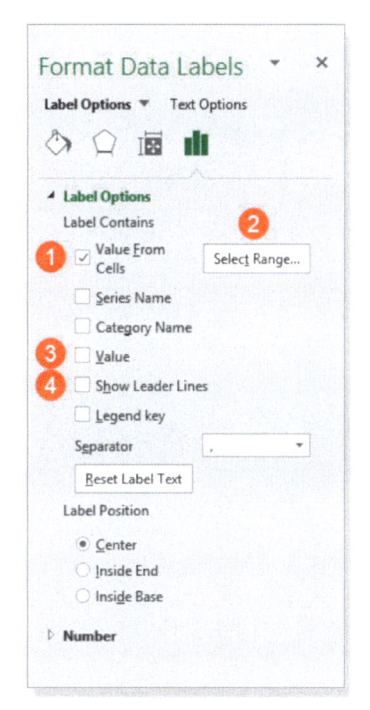

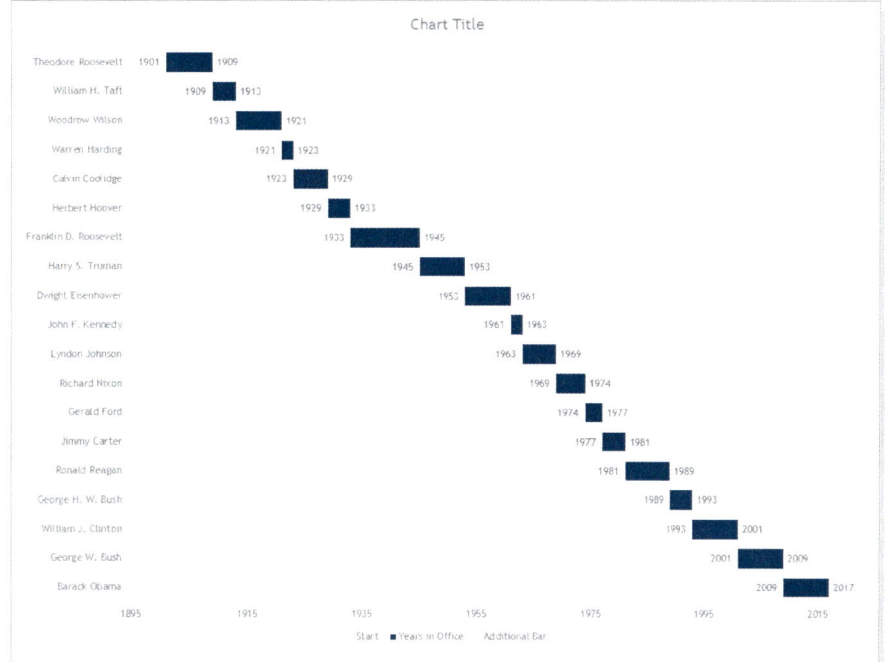

Fig 5.114a: Select Value From Cells.

Fig 5.114a: Tada! You've hacked Excel to add start and end years!

155

CHAPTER 5 • CREATE CHARTS

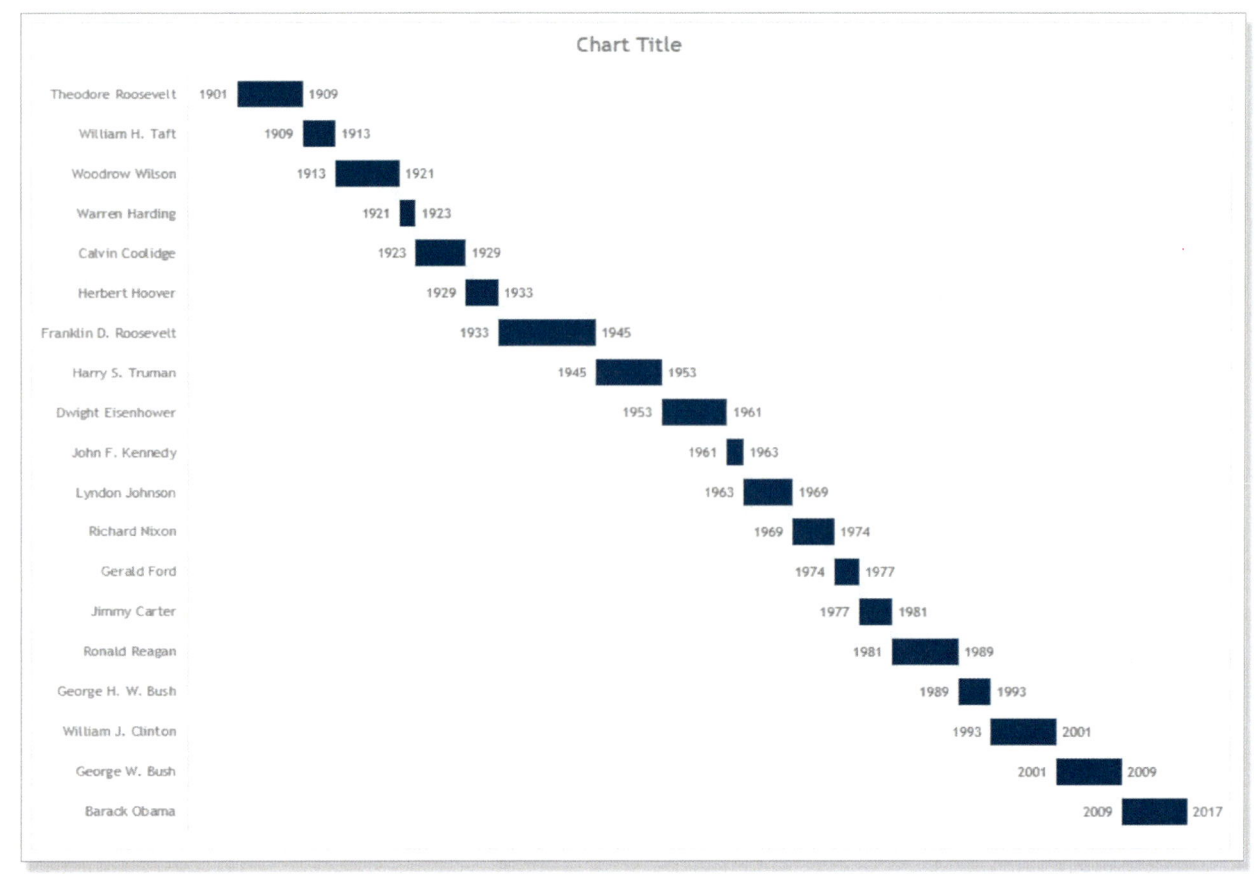

Fig 5.115: Bye, chart junk!

Step 12: Now we can delete the gridlines, horizontal axis, and legend **(Fig 5.115)**.

Step 13: Our chart is shaping up, but the bars toward the bottom of the chart are so far away from the y axis, it's difficult to associate the bar with the corresponding president. To help guide viewers' eyes, let's add a pattern fill to the now-hidden data series on the left. Select the data series to open the **Format Data Series** pane. Open the **Fill & Line** tab and set **Fill** to **Pattern Fill** and choose a swatch. I chose **Diagonal stripes: Light upward** pattern (third row down, second swatch over) and set **Foreground**. I used light gray. Then I used the same settings for the other hidden data series for balance **(Fig 5.116)**.

Step 14: Clean up your chart to taste. I cleaned mine by taking the following steps:

+ made title more descriptive
+ left aligned chart title
+ bumped up title font size
+ added annotation for data source
+ boldfaced data labels

💬 **LEARN MORE** See the Design with Elegance chapter to learn how to perform these refinements.

Example File

floating-bar-chart.xlsx

156

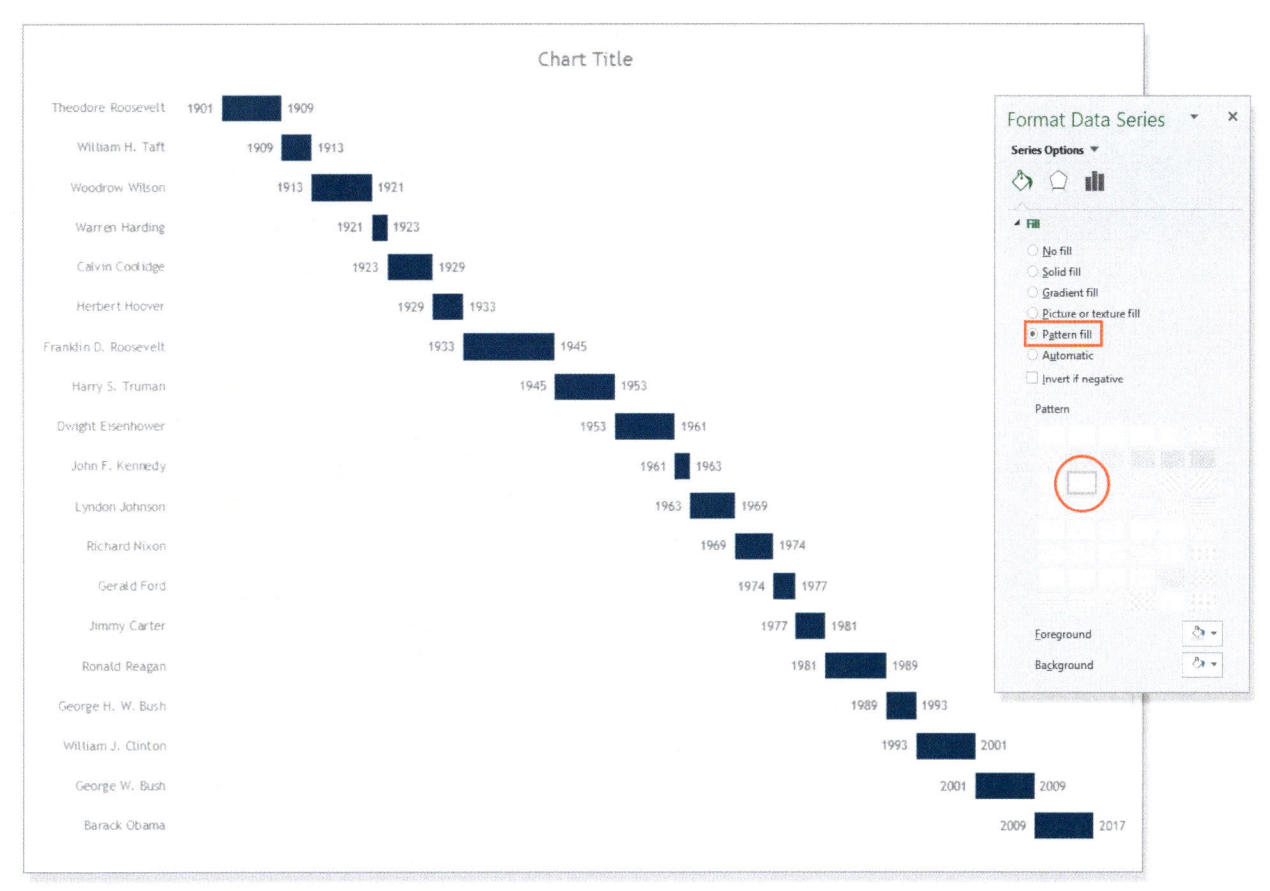

Fig 5.116: You can add shading to improve readability of your chart.

Template

If you'll use this chart again, save it as a chart template. Or download the template included in the chart templates folder: mds-floating-bar.crtx.

Warning: If you need to customize the minimum and maximum values in an axis, set it back to the default before saving as a template. Otherwise, every chart you create from your template will be hard-coded with your custom values.

Note: The data label to the right of the visible bar needs to be set up again since it points to a data range. Repeat step 10 to set it back up.

CHAPTER 5 • CREATE CHARTS

DOT PLOT

Avg Page Load Times for Top 10 Landing Pages — Jan 2018

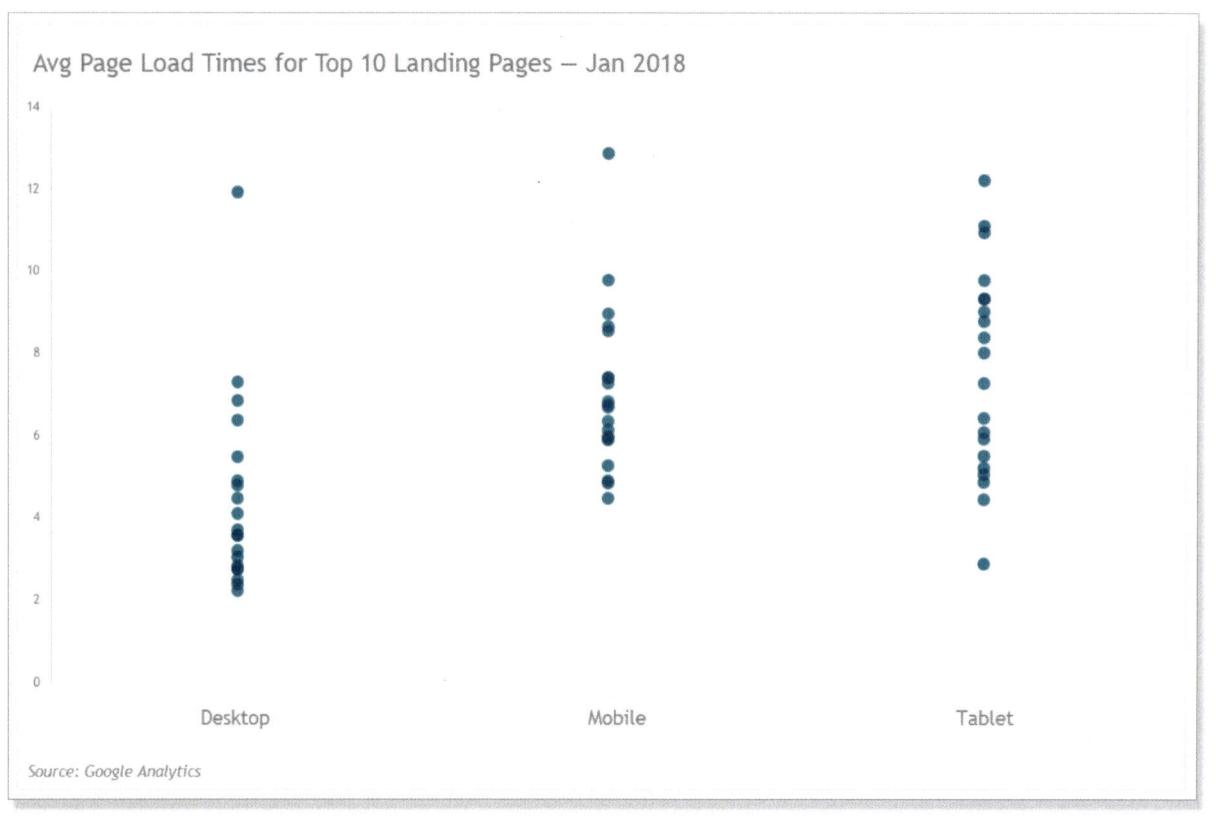

Source: Google Analytics

Fig 5.117: Our final chart.

AKA
You may also see dot plots called:
+ overplotted dot plots
+ strip dot plots
+ technical dot plots

Uses
Dot plots are best used to:
+ visualize small datasets with categories
+ demonstrate the distribution of categorical data
+ identify gaps and outliers

Tips
+ The dot plot is also a distribution chart but is weak on the distribution piece (box plots are much better). So I classified it as a comparison chart. Ymmv.
+ If you have a large number of data points, you may need to add transparency to your markers to help viewers see the dots at the bottom of the stack.
+ You can customize the marker type, size, and color. But don't get crazy! (Said in my best Bon Qui Qui voice: **bit.ly/mdsepc-dont-get-crazy**.)
+ Excel will want to give each of your category elements a different color. Since you'll be deleting the legend, this may cause confusion. You don't want viewers to spend unnecessary time trying to interpret what the colors mean.
+ Axes and legends aren't sacrosanct. Don't be afraid to create custom elements using text boxes.

158

CHAPTER 5 · CREATE CHARTS

- We will be rocking the Select Data Source dialog so hard with this chart type Microsoft never intended for us to create in Excel.

💬 **LEARN MORE** See the Hack Excel's Charting Engine section of the Design with Elegance chapter to learn how to rock the Select Data Source dialog.

- With a dot plot you have two categories and one metric, but only one of the categories is identified in the chart. With the example data, we gathered the page load times for a site's top 10 landing pages and segmented them by device category.

- If you have two metrics and one category and you want dots, consider a scatter plot with categories.

Category / Metric
2 categories / 1 metric**

Tutorial
Hundred-Foot View
Excel doesn't provide a dot plot by default, so we'll use a scatter plot. To line our dots up along our horizontal axis, we'll assign pretend categories to our data. The x value will determine where each of the data series is positioned along the x axis, and the y value will determine where they fall along the vertical axis. To ease the pain of overlap, we'll increase the transparency of our dots.

Step 1: The first thing we need to do is get our data into a crosstab format. If you're following along with the sample data, we'll do that by dragging Landing Page into the **Rows** field, Device Category into **Columns**, and Avg. Page Load Time into **Values (Fig 5.118)**.

Step 2: Copy-and-paste your data into another worksheet. Choose **Value & Number Formatting** from the **Paste Options** button that appears in the bottom-right corner of your pasted range. I also capitalized my device categories, but that's purely a style preference.

Step 3: Add a column to the left of each of these columns by right-clicking the column headings and choosing **Insert** from the contextual menu **(Fig 5.119)**. If you're following along with the example data, you'll be adding three columns total. In this supplemental table, we'll need to assign an x value for each of our

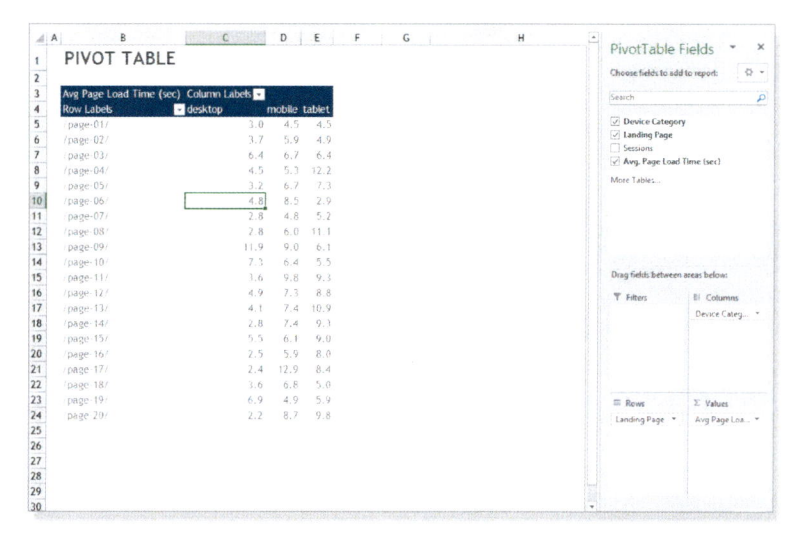

Fig 5.118: Massage your data into a crosstab format.

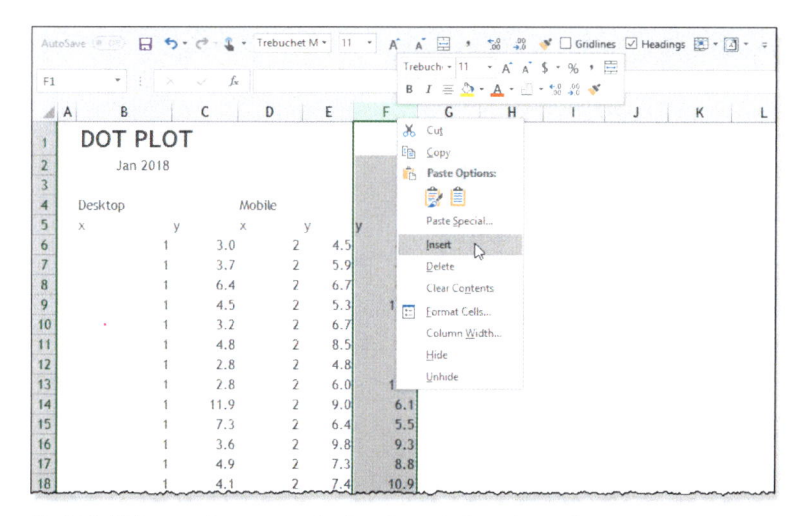

Fig 5.119: Make room for your contrived x columns by inserting a column to the left of each y column.

**Technically, you will have three categories, but the third one is contrived and won't come from your dataset.

159

CHAPTER 5 • CREATE CHARTS

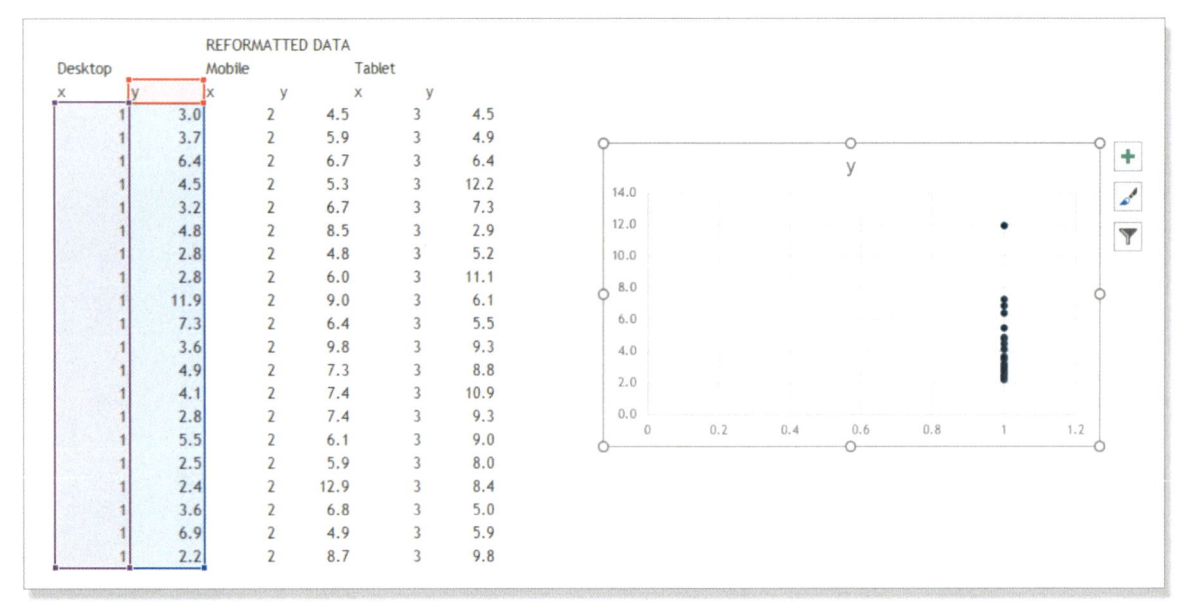

Fig 5.120: Our initial chart.

category elements. The x value determines where these points will appear along our horizontal axis. I'll assign Desktop the value of 1, Mobile the value of 2, and Tablet the value of 3. It doesn't matter what those values are, as long as they're equidistant from each other. They could have been 5, 10, and 15 or 100, 200, and 300.

Also, we don't need the actual landing pages because, with a dot plot, we're only interested in the distribution of the metric segmented by our category (e.g., page load time segmented by device category).

Step 4: Click-and-drag over the x and y columns for Desktop, then navigate to **Insert ▸ Charts ▸ Insert Scatter (X, Y) or Bubble Chart ▸ Scatter ▸ Scatter (Fig 5.120)**.

Tip: Press Ctrl-Shift-down arrow to select your entire range without scrolling.

Step 5: Right-click on the chart and choose **Select Data** from the contextual menu. In the **Select Data Source** dialog, click the **Edit** button in the **Legend Entries (Series)** field. In the **Edit Series** dialog set **Series name** by selecting Desktop. Series X and Y values are already set **(Fig 5.121)**.

> 💬 **LEARN MORE**   See the Hack Excel's Charting Engine section of the Design with Elegance chapter to learn how to rock the Select Data Source dialog.

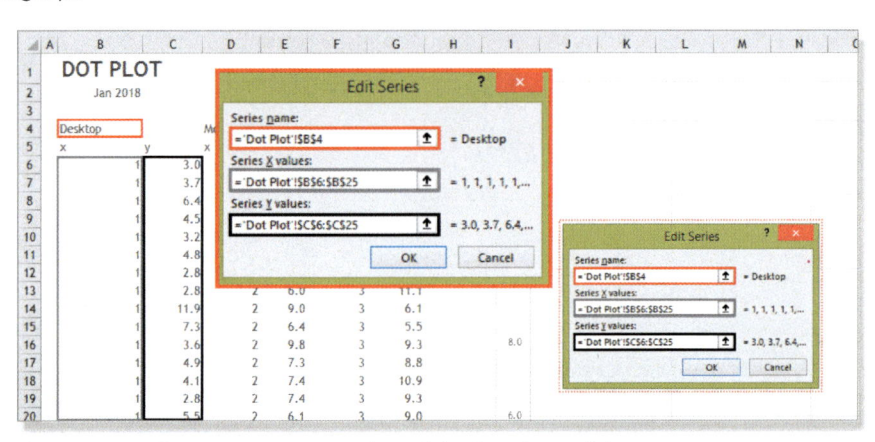

Fig 5.121: Assign Name to your data series in the Select Data Source dialog.

160

CHAPTER 5 • CREATE CHARTS

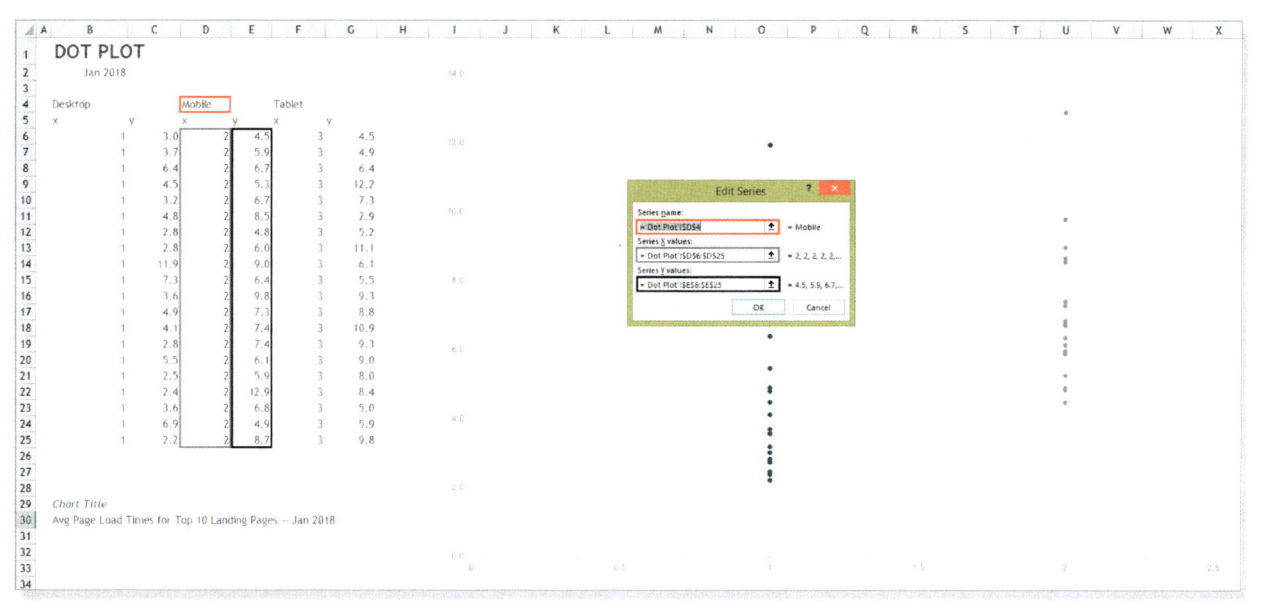

Fig 5.122: Adding a data series from scratch like whuhh!

Step 6: With the **Select Data Source** dialog open, click the **Add** button to add your next data series. This time we're assigning **Series name**, **Series X values**, and **Series Y values** for the Mobile device category manually **(Fig 5.122)**.

Step 7: Rinse and repeat with Tablet **(Fig 5.123)**. Not a bad start.

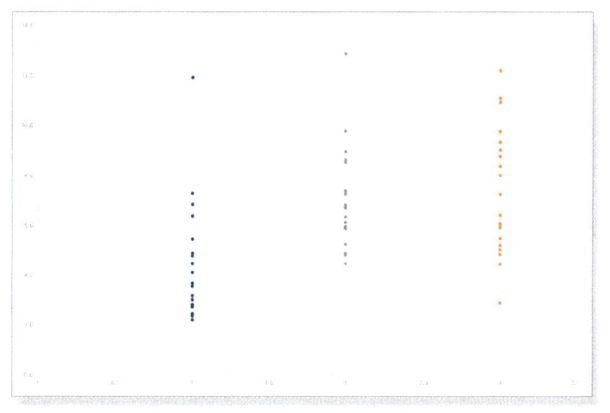

Fig 5.123: Our initial chart.

Step 8: Let's add a legend to our chart, so we can easily see which data series is from Desktop, Mobile, and Tablet. First, select the chart and open the **Chart Elements** flyout menu, then select the right-facing arrow next to **Legend** and choose **Bottom (Fig 5.124)**.

Fig 5.124: Add a legend (which we'll remove later).

161

CHAPTER 5 · CREATE CHARTS

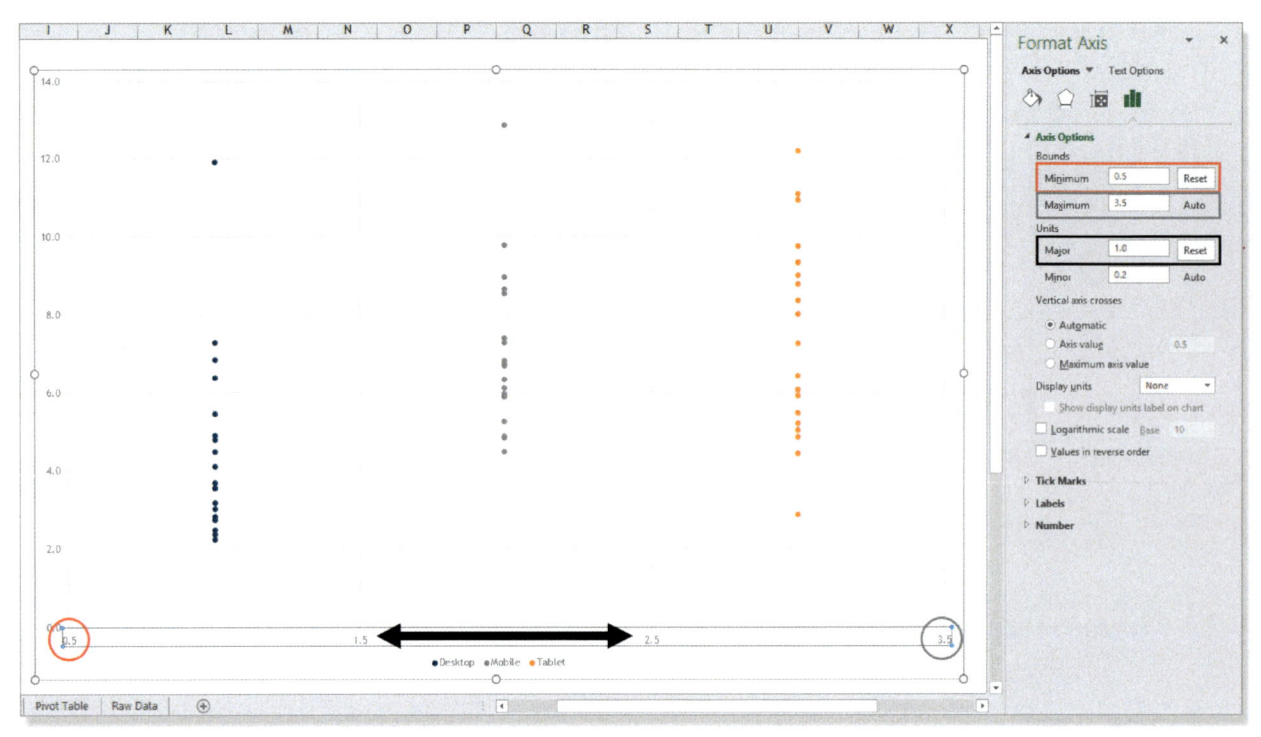

Fig 5.125: Position your dot plots along the horizontal axis.

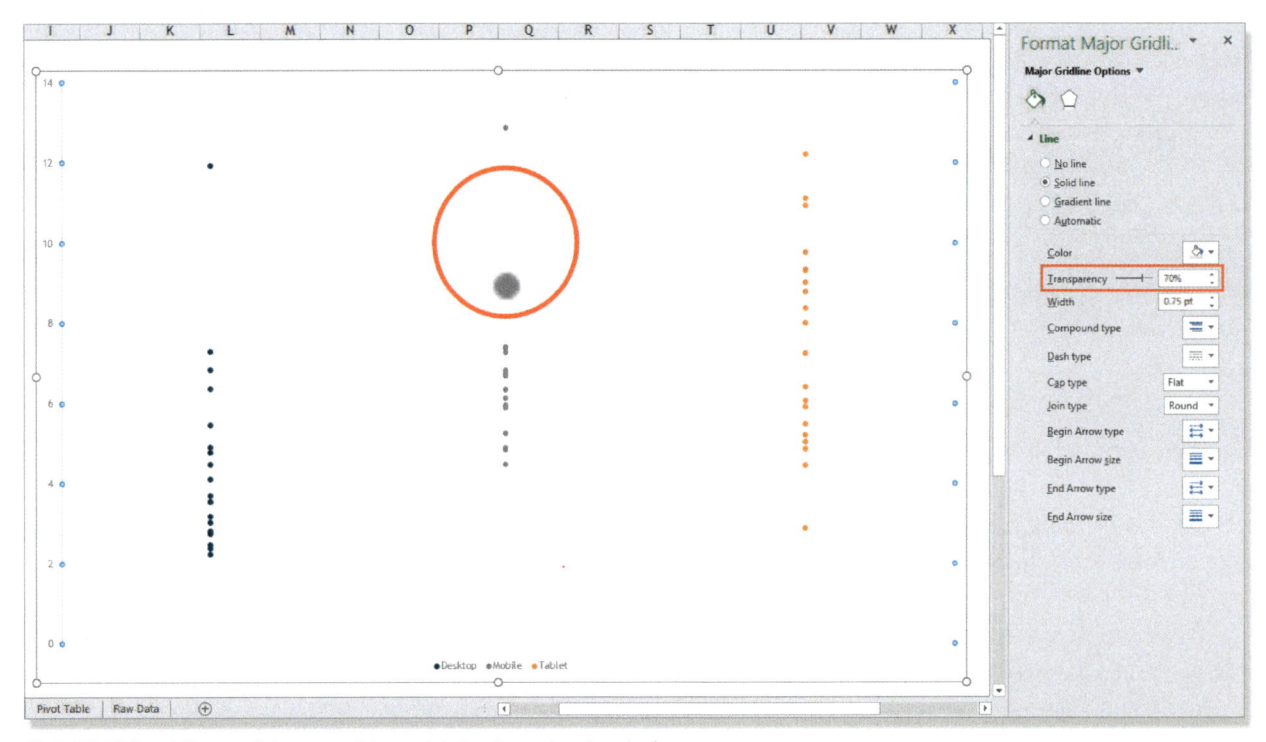

Fig 5.126: Make gridlines as light as possible to minimize distracting chart junk.

162

Step 9: Our horizontal axis is pretty useless, but before we delete it, we'll customize it to nudge the first series (Desktop in the example file) closer to the vertical axis. This is for aesthetics alone. To start, select the horizontal axis and press Ctrl-1 to open the **Format Axis** pane, then open the **Axis Options** tab. Under **Bounds**, set **Minimum** to 0.5 and **Maximum** to x.5, with x being the number of data series in your chart (3.5 for the example workbook). Under **Units**, set **Major** to 1 to center your series nicely in the plot—or the value that separates your x values **(Fig 5.125)**. Now you can delete your horizontal axis.

Step 10: Our vertical axis has an unnecessary decimal place because our data includes a decimal place. By default, Excel mimics the number formatting of your data in the axis. We'll customize the number axis formatting to reduce this chart junk. Select the vertical axis to open the **Format Axis** pane. In the **Axis Options** tab, under the **Number** section, set **Decimal places** to 0.

Step 11: We still have too much chart junk, so we'll delete the vertical gridlines and lighten the horizontal ones. To lighten them, select one to select all of them and open the **Format Major Gridlines** pane. Open the **Fill & Line** tab and either set **Fill Color** to a lighter color and/or dial up **Transparency** using the slider. I increased it to 70% **(Fig 5.126)**.

Step 12: Let's bump the marker size up a bit. Select a marker in your first data series to select all of them and open the **Format Data Series** pane. Open the **Fill & Line** tab and select **Marker** to format yours to taste **(Fig 5.127)**. I used the following settings:

- **Marker Options**: **Built-in**
- **Type**: Circle
- **Size**: 9
- **Fill**: Solid fill
- **Fill Color**: blue
- **Transparency**: 30%
- **Border**: No line

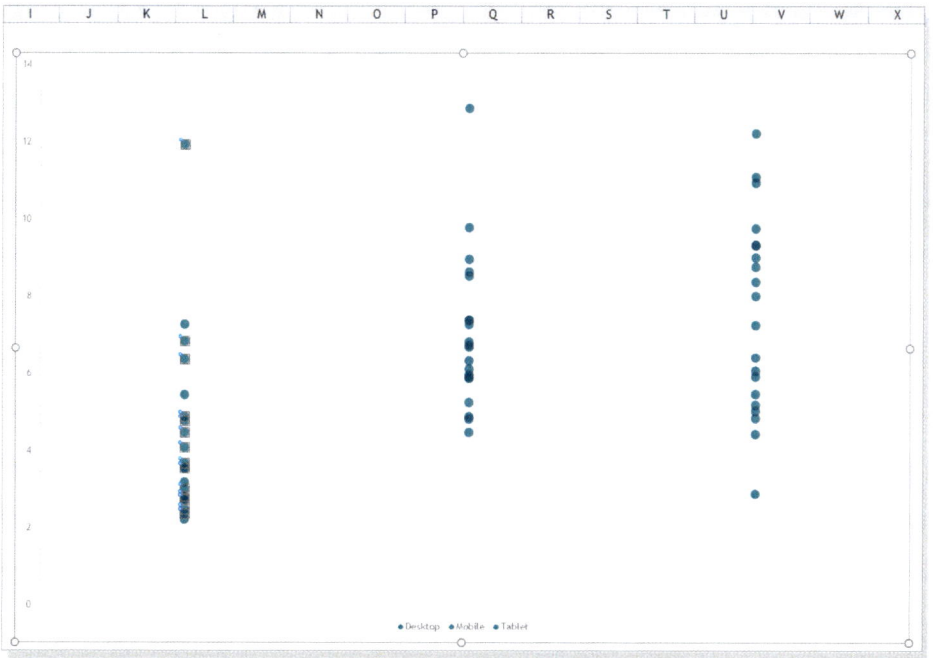

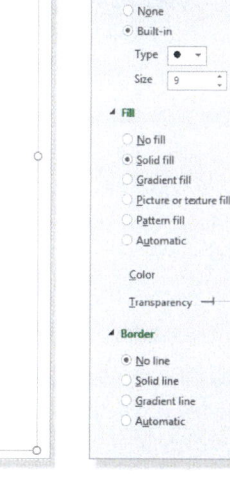

Fig 5.127: Adding transparency to your markers helps mitigate overplotting.

CHAPTER 5 • CREATE CHARTS

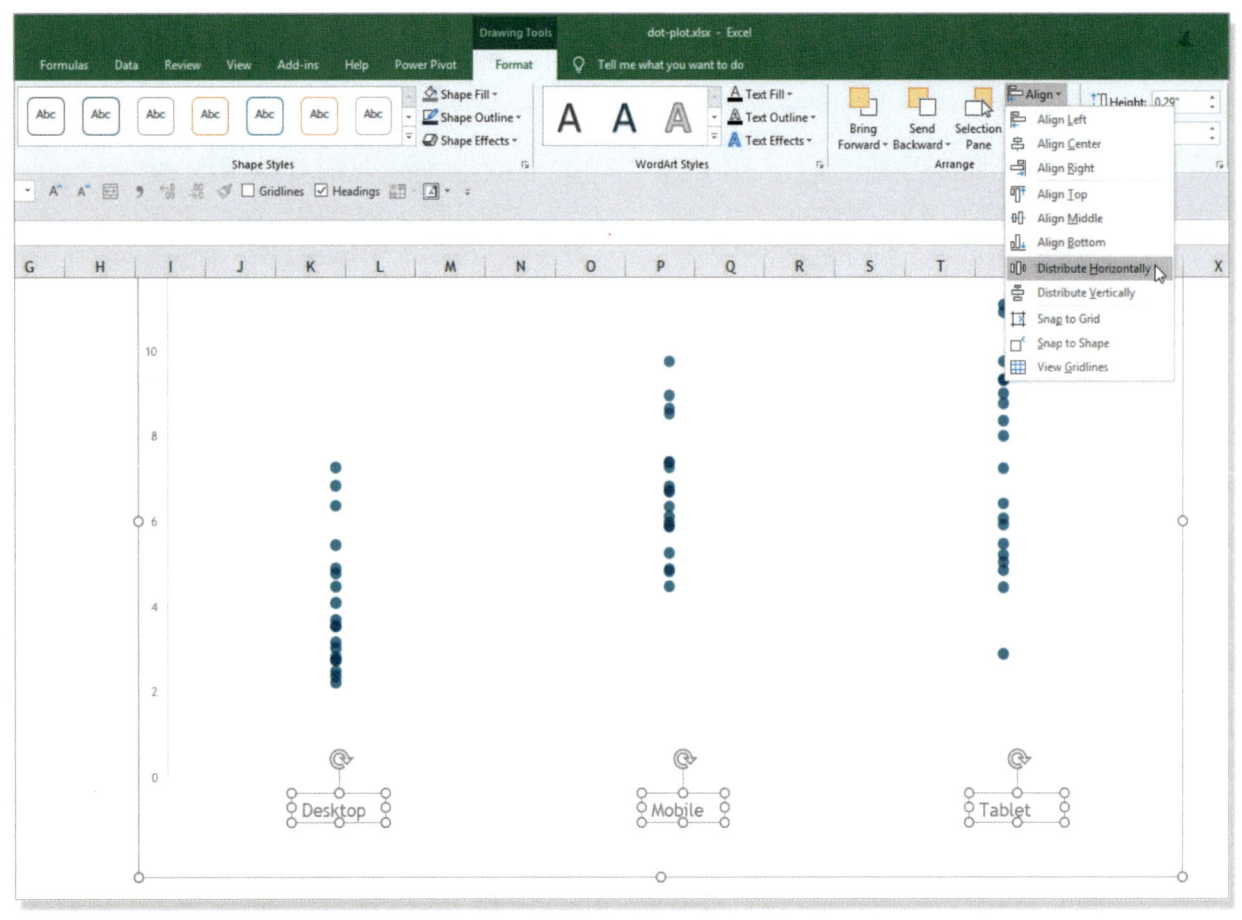

Fig 5.128: Align and evenly distribute text boxes when manually creating axis labels.

If you prefer to keep a border, make it the same color as your fill. Repeat this process with your other data series. I also recommend using the same fill color for each of your series. I addressed why in the Tips section.

Step 13: We no longer need the legend, so you can delete it. We'll use text boxes to create a custom legend.

Step 14: Pull the **Plot Area** up by dragging the bottom-center sizing handle to make room for the text boxes. With your chart still selected, navigate to **Insert ▶ Text ▶ Text Box** and drag out your first text box inside the **Chart Area** but outside the **Plot Area**.

Create one for each of your data series. Align them by holding down the Shift key as you select each one, then navigate to **Drawing Tools ▶ Format ▶ Arrange ▶ Align ▶ Align Top**. You can also select **Distribute Horizontally** under **Align** to ensure they're distributed equally **(Fig 5.128)**. You can also group them by selecting **Group ▶ Group** under the **Arrange** section.

💬 **LEARN MORE** See the Align Strategically section of the Design with Elegance chapter to learn how to align objects in your workbook.

164

CHAPTER 5 • CREATE CHARTS

Step 15: Clean up your chart to taste. I cleaned mine by taking the following steps:

- made title more descriptive
- left aligned chart title
- bumped up title font size
- added annotation for data source

LEARN MORE See the Design with Elegance chapter to learn how to perform these refinements.

Credit: Shout out to the Peltier Tech Blog for the inspiration to use faux categories to line up dots: **bit.ly/mdsepc-dot-plot**.

Variation
Jitter Plot
Hundred-Foot View

If crowding in your dot plot is significant, you can add jitter to your categorical values to give them some breathing room. We'll use a tweaked-out RAND function to generate very small positive and negative values, which we'll sum with our category numbers. This will scatter our dots slightly along the line they adhered to in the dot plot chart.

Step 1: We'll pick up where we left off with the dot plot's dataset **(Fig 5.130)**.

Fig 5.129: Our final chart.

Fig 5.130: The jitter plot table looks exactly like our strip dot plot table at first.

165

CHAPTER 5 • CREATE CHARTS

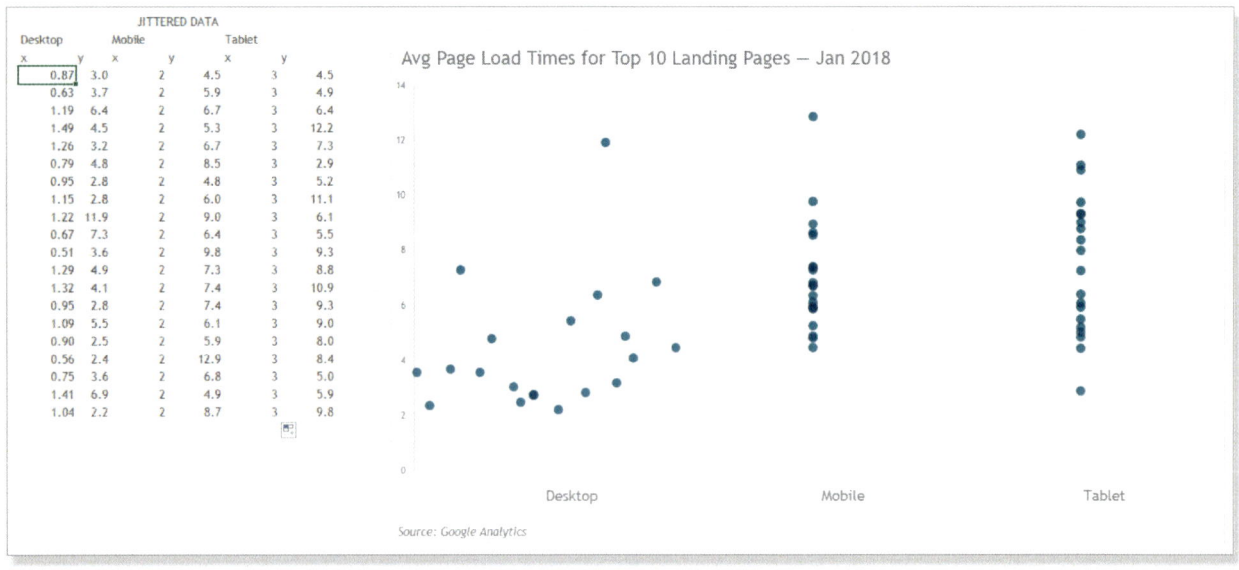

Fig 5.131: OMG this jitter is way too caffeinated!!! How many shots were in that formula anyway?!?

Step 2: We need to add some jitter around our category numbers (1, 2, 3). This will cause our dots to dance all around these three vertical lines. We'll use the RAND function to accomplish this jitter. It's used to generate numbers between 0 and 1. If you added RAND-generated values to a set of numbers, they would add values between 0 and 1 to each of your data points. That would effectively nudge our dots to the right of the line they're currently lined up along but not to the left. We want our dots scattered in both directions. With that goal in mind, we'll subtract 0.5 from RAND to generate random values between -0.5 and 0.5. The only problem with this approach is, when I enter the formula below, the jitter is a little too AMPED UP…like it drank a triple shot latte on an empty stomach **(Fig 5.131)**. I used the formula below.

= B6+(RAND()-0.5)

Step 3: Think back to math class circa sixth grade. How can we make a number much smaller? The best way is to multiply it by a fraction or divide it by another number—whichever is more intuitive for you. This takes some experimentation. I tweaked the formula until I achieved an acceptable amount of jitter. In the example file, 8 was the winning denominator **(Fig 5.132)**. I used the formula below.

=B6+(RAND()-0.5)/8

Step 4: To copy your final formula to the other category columns, we can't select the column and drag the fill handle to the right, like we usually do, because we'd run roughshod over our other y columns. Instead, copy the x column that contains your formula, then paste it over each of the other x columns. Because we didn't lock down our cell references, B6 will become D6 in the Mobile x column and F6 in the Tablet x column. If you get errors, select the cells with error indicators (green triangles in the upper-left corner), and choose **Ignore Error** from the menu that appears to the left **(Fig 5.133)**.

CHAPTER 5 • CREATE CHARTS

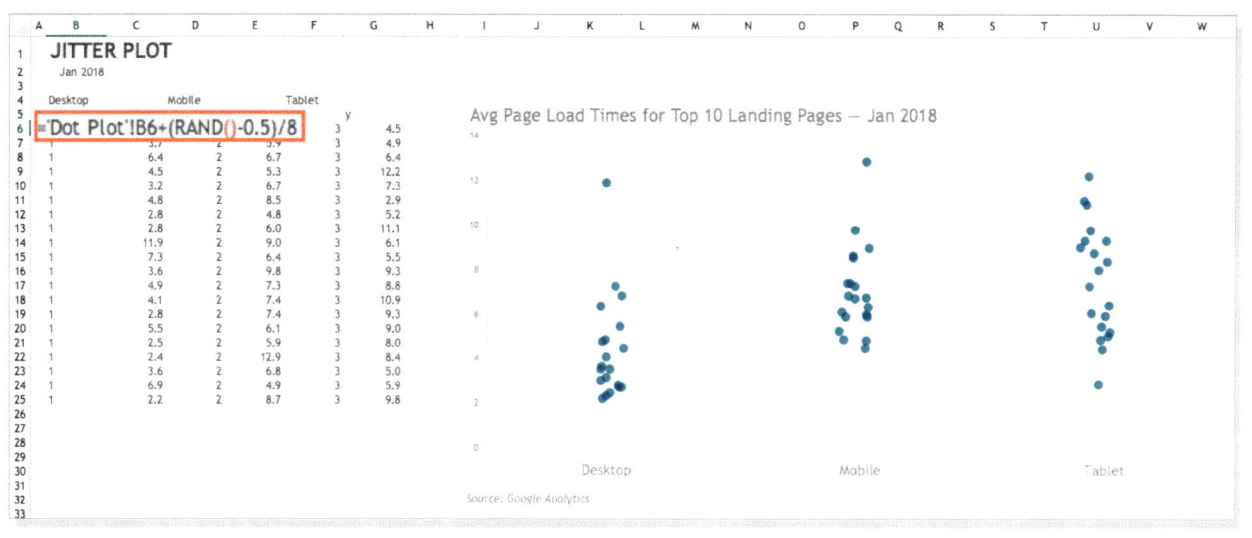

Fig 5.132: I reduced the jitter by dividing my original formula by 8.

Fig 5.133: Calm down, Excel.

Step 5: If anyone will see your table, you can get rid of the decimal by selecting your data (hold down the Ctrl key to select non-contiguous ranges) and navigating to **Home ▶ Number ▶ Decrease Decimal**.

Credit: Shout out to Eddie Smith for the inspiration to use the RAND function to add random variance to the category values: **bit.ly/mdsepc-jitter-plot**.

Example File
dot-plot.xlsx

Template
Because this chart is so highly customized and needs to be built from the ground up, you can't save it as a chart template. The best thing you can do is save a file in a custom folder of all your crazy charts that you can't save as chart templates and create a copy of it, then switch out your data with your new data. It will at least save you a few steps.

167

CHAPTER 5 · CREATE CHARTS

CYCLE PLOT

Interest for Game of Thrones Over Time
Jan 2011 - Dec 2017

Source: Google Trends

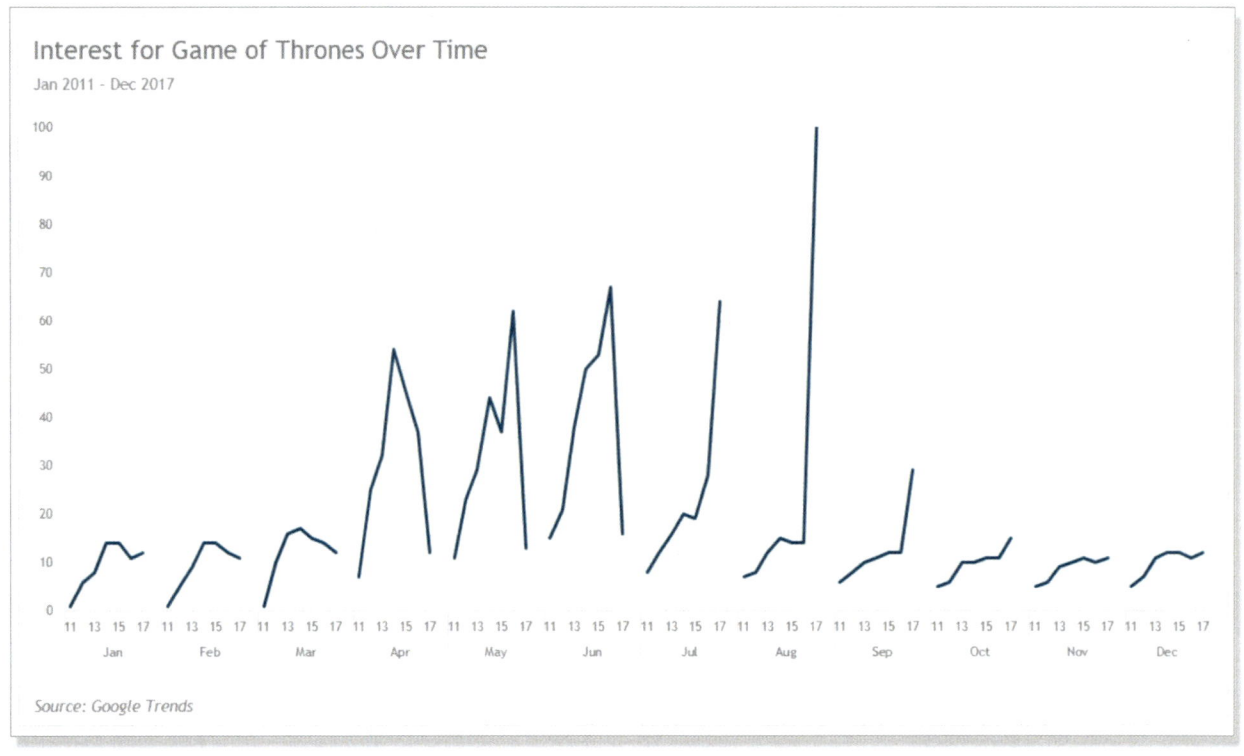

Fig 5.134: Our final chart.

AKA
You may also see cycle plots called:

- month plots
- seasonal subseries plots
- subseries plots

Uses
Cycle plots are best used to:

- compare period-over-period data (e.g., week-over-week, month-over-month, quarter-over-quarter, or year-over-year)
- compare data from different seasons (e.g., compare Halloween and Christmas sales seasons)
- detect outliers that seasonality doesn't account for
- prepare data for modeling and forecasting

Tips
- A pivot table will be your best friend in getting your data into the right format, especially separating date values, such as months and years. I've included a Pivot Table worksheet in the downloadable workbook to show you how I set mine up. Click anywhere in the pivot table to open the PivotTable Fields pane. From there you can easily see how I constructed it.
- If you want to make individual data points stand out more, add markers. You can make them less obtrusive by using the same color as your lines and reducing their size. If you need to choose between clean and clear, go with clear every time.
- You can add a mean or target line to your individual plots.
- Keep your individual series the same color.

Category / Metric
2 categories / 1 metric

168

CHAPTER 5 • CREATE CHARTS

Tutorial

Hundred-Foot View

A cycle chart seems complicated but is actually pretty straightforward. It's basically a line chart with nested axis labels and breaks between our lines. We'll generate those breaks by adding empty rows to our dataset.

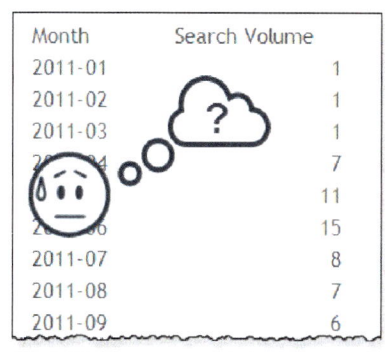

Fig 5.135: Pre-aggregated dates frequently need to be reformatted.

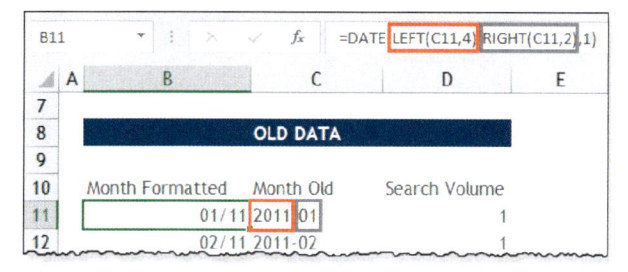

Fig 5.136: The DATE function allows you to restructure your data so Excel into a recognizable date format.

Step 1: Before I could start visualizing my data, I had to reformat the month data from Google Trends because Excel doesn't recognize yyyy-mm as a date **(Fig 5.135)**. The DATE function enables you to reconfigure these dates into a format Excel will recognize **(Fig 5.136)**.

> **LEARN MORE** See the Fix Dates section of the Massage Your Data chapter to learn how to rock the DATE function and the Customize Number Formatting section of the Design with Elegance chapter to learn how to format your values in a date format that works best for your data.

Step 2: When I was knee deep in data, testing this chart, one issue that nearly drove me to drink (in the Irish sense) was crowding in the axis. Because this is such a highly customized chart, the normal editing ability you have with the chart axis is lost forever. When I threw my question out to the Twitterverse, David Rekuc (**@DaveRekuc**) came through with a suggestion to convert my dates to text using the TEXT function. This minimized the data drama because Excel isn't such a princess about charting text as it is with dates. To that end, we'll use a TEXT function to convert our dates to text **(Fig 5.137)**.

Fig 5.137: Ultimately, processing your dates as text causes less drama downstream.

CHAPTER 5 • CREATE CHARTS

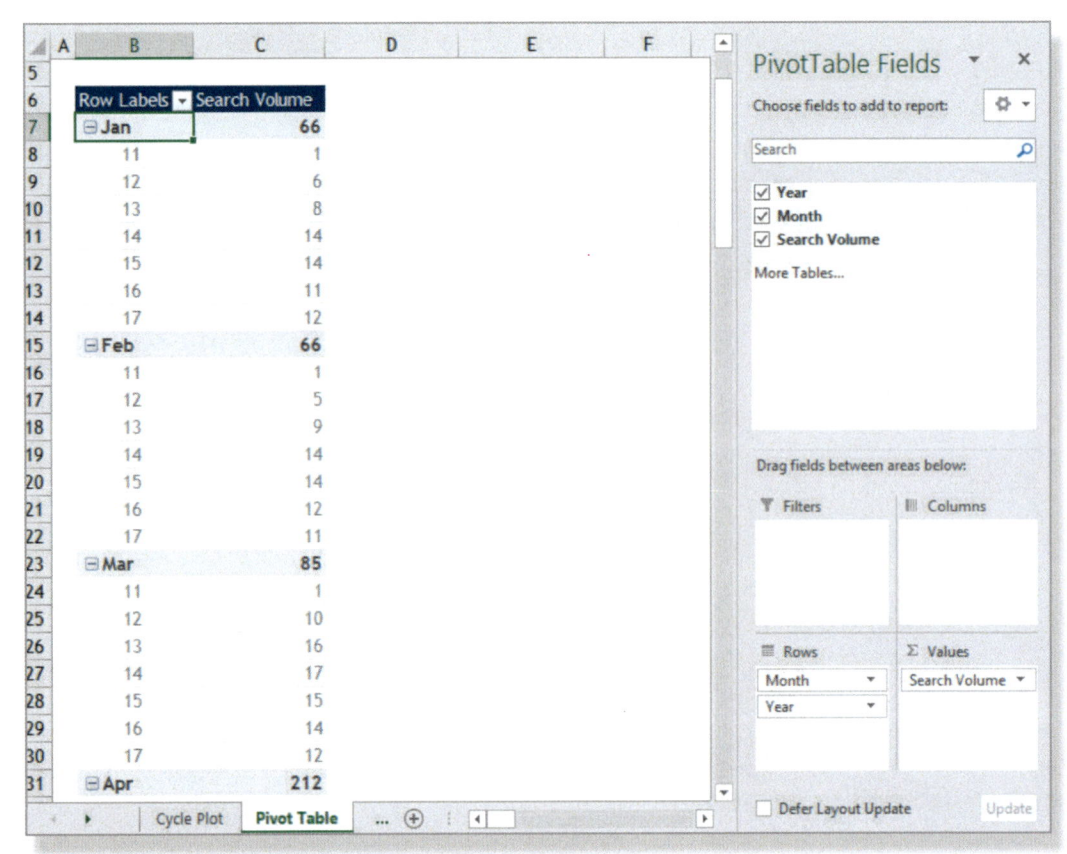

Fig 5.138: Position month above year so that we can segment month by year.

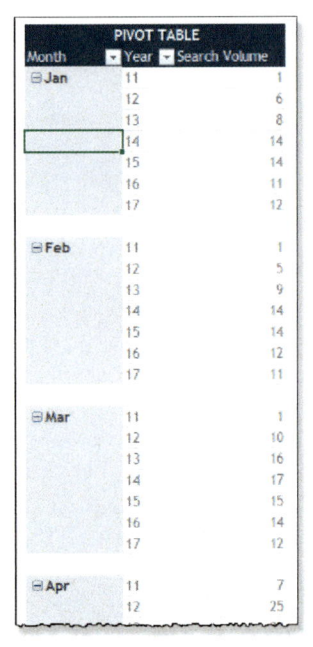

Fig 5.139: The tabular format will set you up for a nested horizontal axis.

Tip: If you're comfortable with embedding formulas, you can combine steps 1 and 2 by embedding your DATE function inside the TEXT function: `=TEXT(DATE(LEFT(C8,4),RIGHT(C8,2),1),"yy")`. I used a helper column to make it easier for neophytes to understand. Learn more about using helper columns in my post on the Search Engine Land website: **bit.ly/mdsepc-helper-columns**.

Step 3: We'll need a pivot table to get the nested axis labels. We'll start by dragging Month into the **Rows** field, then pull Year below it. This is the reverse of our typical stack order, but with a slope chart we're grouping by month, not year. Next, pull Search Volume (or whatever metric you're working with) into the **Values** field **(Fig 5.138)**.

💬 **LEARN MORE** See the Pivot Tables section of the Massage Your Data chapter to learn how to rock pivot tables.

170

CHAPTER 5 • CREATE CHARTS

Step 4: If your pivot table adds grand totals, you'll want to turn them off under **PivotTable Tools ▶ Design ▶ Layout ▶ Grand Totals ▶ Off for Rows and Columns**. Same with subtotals (**Layout ▶ Subtotals ▶ Do Not Show Subtotals**).

Step 5: To create separate lines for each month, we'll need a blank row between each of our months. You can achieve this by enabling blank lines: **PivotTable Tools ▶ Design ▶ Layout ▶ Blank Rows ▶ Insert Blank Line after Each Item**.

Step 6: Change your report layout from Compact (Excel's default layout) to Tabular: **Tools ▶ Design ▶ Layout ▶ Report Layout ▶ Show in Tabular Form**. This will set you up for a nested (or bilevel) horizontal axis **(Fig 5.139)**.

Step 7: To get rid of the *Sum of* in your column header, double-click the cell to open the **Value Field Settings** dialog, then delete *Sum of* (or *Average of* or whatever aggregation unit you chose). Excel won't allow you to enter the same header you used in your original dataset, so you'll need to get around this by modifying your header in some way. The easiest way is to just add a space to the end.

Tip: Alternatively, you can edit the header in cell or in the **Formula Bar**.

Step 8: Copy-and-paste your massaged data from the pivot table as values from the **Paste Options** button (**Paste Values ▶ Values & Number Formatting**). This is the raw data we'll be charting. Excel no longer recognizes your years as values, but that's fine. It doesn't matter anymore.

Step 9: Click-and-drag to select all your data (or select the cell in the upper-left corner, hold down the Shift key, and select the cell in the bottom-right corner to select the entire table), then navigate to **Insert ▶ Charts ▶ Insert Line or Area Chart ▶ 2-D Line ▶ Line (Fig 5.140)**.

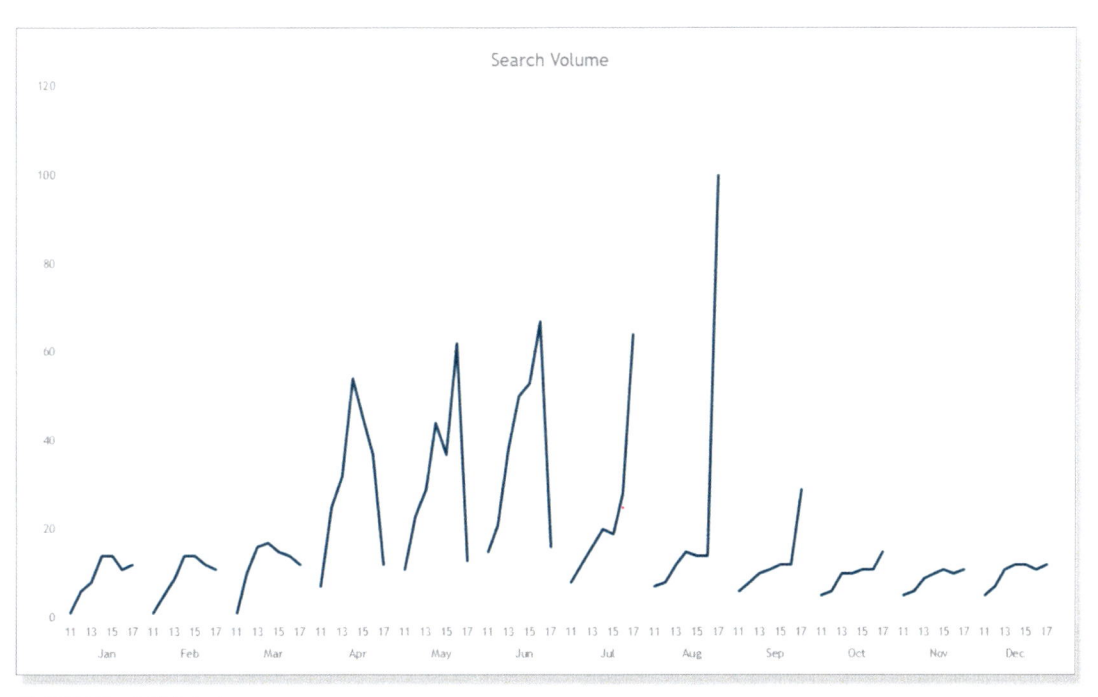

Fig 5.140: Our initial chart.

171

CHAPTER 5 • CREATE CHARTS

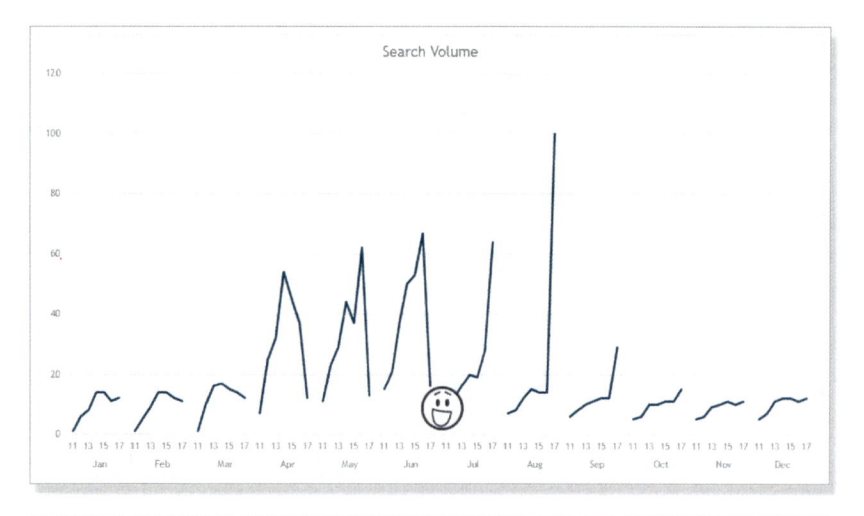

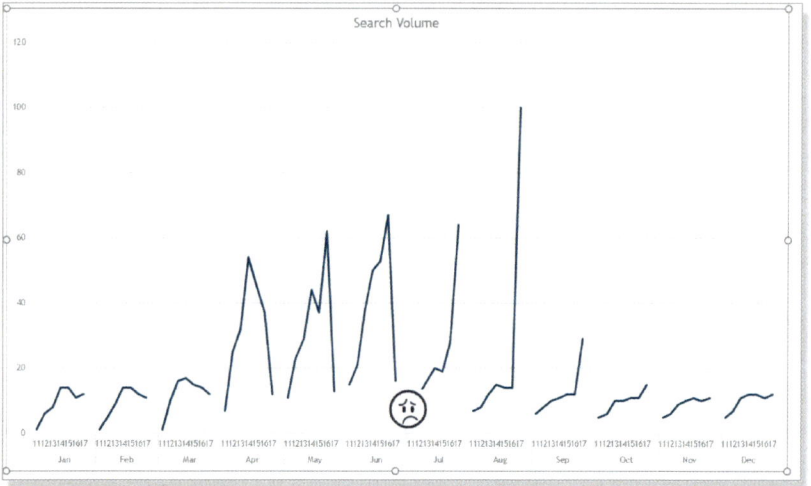

Fig 5.141: Excel: "I swear, move that Chart Area one more millimeter... Okay, Karen, you did this to yourself."

💬 **LEARN MORE** See the Design with Elegance chapter to learn how to perform these refinements.

Step 10: I want a space between my years to prevent crowding in the axis, but the option to set the interval between axis labels doesn't work with a nested axis. You can drop a 2 in there, but Excel will be unaffected. Therefore, the key is to not make your chart too wide. If you do, your axis labels will be more crowded than your neighborhood *Walmart* on Black Friday **(Fig 5.141)**.

Step 11: Clean up your chart to taste. I cleaned mine by taking the following steps:

* made title more descriptive
* left aligned chart title
* bumped up title font size
* added annotation for data source
* bumped up transparency of gridlines

Example File
cycle-plot.xlsx

Template
Because this chart is so highly customized and needs to be built from the ground up, you can't save it as a chart template. The best thing you can do is save a file in a custom folder of all your crazy charts that you can't save as chart templates and create a copy of it, then switch out your data with your new data. It will at least save you a few steps.

CHAPTER 5 • CREATE CHARTS

SLOPE CHART

AKA
You may also see slope charts called slopegraphs.

Uses
Slope charts are best used to:

- compare opposites or competitors (e.g., females vs. males, drug vs. placebo, teams, students in different programs or schools, etc.)
- demonstrate overall change between two time periods
- visualize two endpoints where the in between points are less important
- demonstrate injustices between groups who should be equally matched

Tips

- If you only have one category, you can plot two metrics, but if you have two categories, you can only plot one metric.
- Consider bringing labels for your data series into the chart so you can delete your legend.
- Consider ditching your gridlines.
- Lining up labels is difficult in the chart, and you can't select them and align them like you can objects. If you need to align yours, you can insert a thin line, then line the labels up along the line.

Category / Metric
1+ categories / 1+ metrics

Tutorial
Hundred-Foot View

As with many of the charts in this book, Excel doesn't offer the ability to create a slope chart by default. Not being swayed by

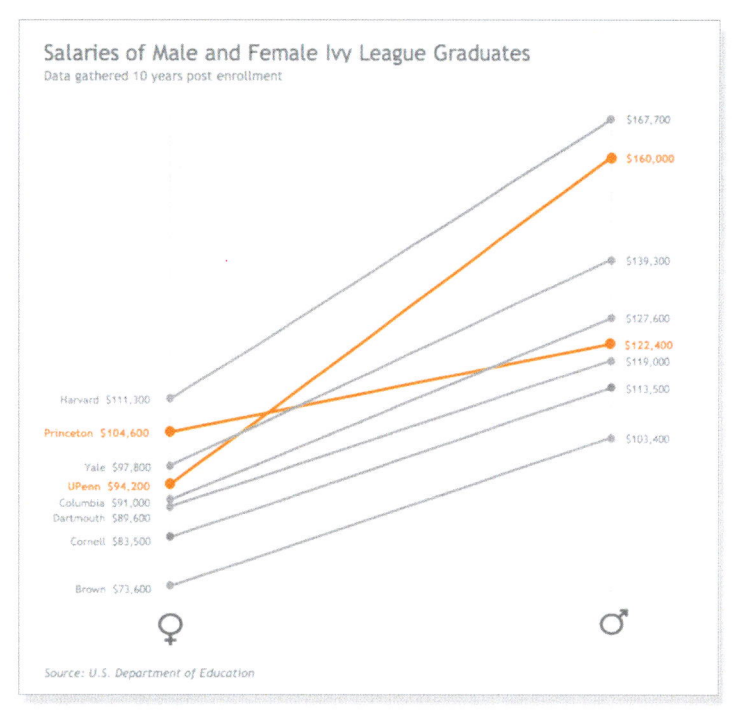

Fig 5.142: Our final chart.

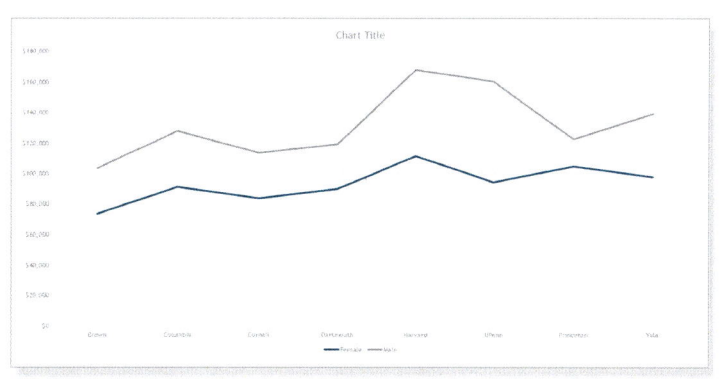

Fig 5.143: Our initial chart.

that, we'll take a line chart with only two data series and tilt it on its axis by switching the rows and columns. Next, we'll add markers to our lines to add some panache and maximize the functionality of our data labels. This will enable us to dump the legend. Finally, we'll use an obscure axis option to give our lines something to hold onto, so they don't look like they're floating.

Step 1: Select any cell in your dataset and navigate to **Insert ▸ Charts ▸ Insert Line or Area Chart ▸ 2-D Line ▸ Line (Fig 5.143).**

CHAPTER 5 • CREATE CHARTS

Step 2: We'll need to flip this chart by switching the categories and data series. To start, right-click on the chart and choose **Select Data** from the contextual menu. In the **Select Data Source** dialog, click the **Switch Row/Column** button. This will move the schools into the legend and gender into the axis labels. You also want your chart to follow more of a portrait orientation, as opposed to the more traditional landscape orientation **(Fig 5.144)**.

> 💬 **LEARN MORE** See the Hack Excel's Charting Engine section of the Design with Elegance chapter to learn how to rock the Select Data Source dialog.

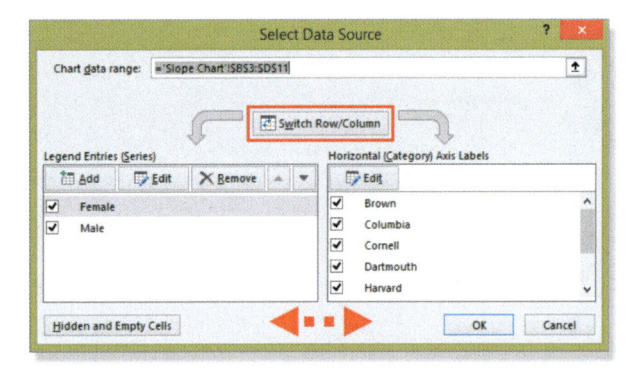

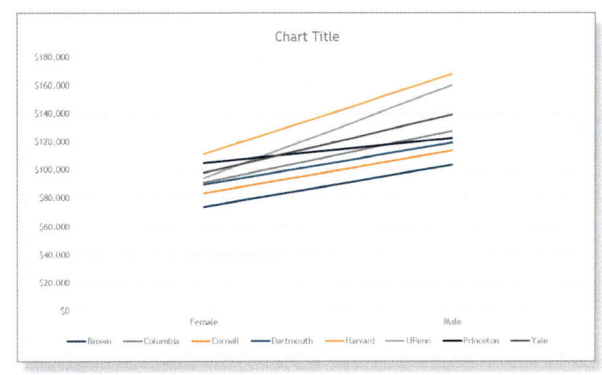

Fig 5.144: Switch your rows and columns to create a rudimentary slope chart.

> 💬 **LEARN MORE** See the Add Data Labels section of the Design with Elegance chapter to learn how to use data labels.

Step 3: Next, we'll format our lines. To start, select one and press Ctrl-1 to open the **Format Data Series** pane. Open the **Fill & Line** tab and adjust the line color and width **(Fig 5.145)**. I used the following settings:

- **Line**: Solid line
- **Line Color**: gray
- **Width**: 2.75 pt

Step 4: For dramatic effect, I set all my lines to gray, except the line showing the greatest difference between male and female salaries (University of Pennsylvania) and the line showing the least difference (Princeton). Those I set to orange **(Fig 5.146)**.

Note: Being the perennial tinkerer, I later bumped **Width** up to 3 pt on these accent lines to cause viewers' eyes to focus on those data series.

Step 5: If you want to add a marker, select **Marker** at the top of the **Fill & Line** tab **(Fig 5.147)**. I used the following settings:

- **Marker Options**: Built-in
- **Type**: Circle
- **Size**: 7 (9 for orange markers)
- **Fill**: Solid fill
- **Fill Color**: gray (orange for accent lines)
- **Border**: No line

Rinse and repeat for each line.

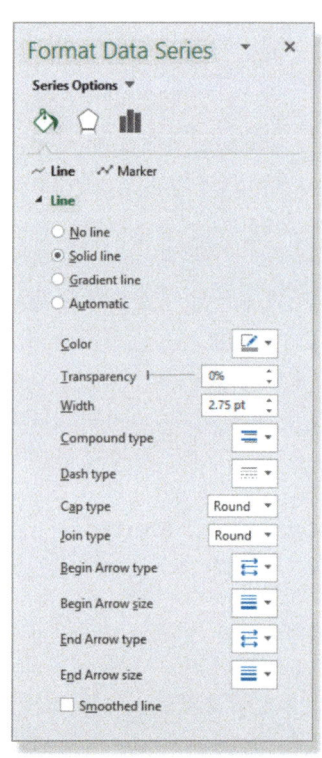

Fig 5.145: Format your lines to taste.

174

CHAPTER 5 • CREATE CHARTS

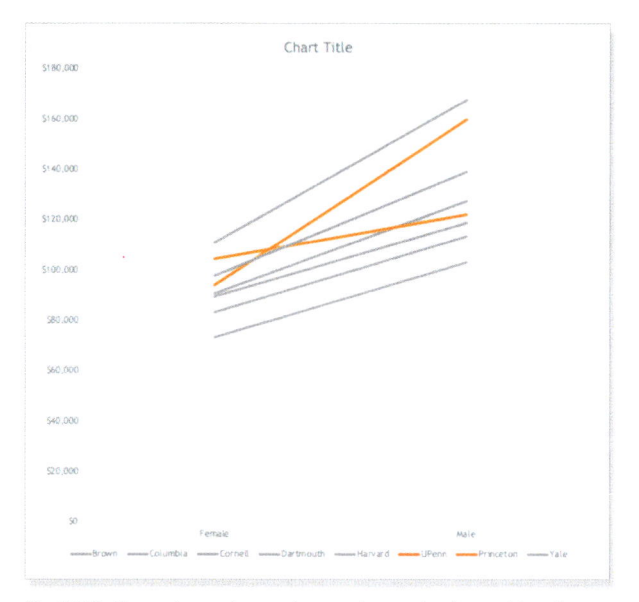

Fig 5.146: Draw viewers' attention to data series by making them a different color.

Fig 5.147: I like adding stylized markers to my slope lines for accent.

Step 6: We'll use data labels so that we can ditch the legend. I try to avoid making viewers' eyes wander too far to see the legend. In this chart, we want the label on the right to sit to the right of each line and the label on the left to sit to the left. To do this, select the first line. From the **Chart Elements** flyout menu, click the right-facing arrow next to **Data Labels** to open its flyout menu and choose **Right (Fig 5.148)**. Rinse and repeat for each line.

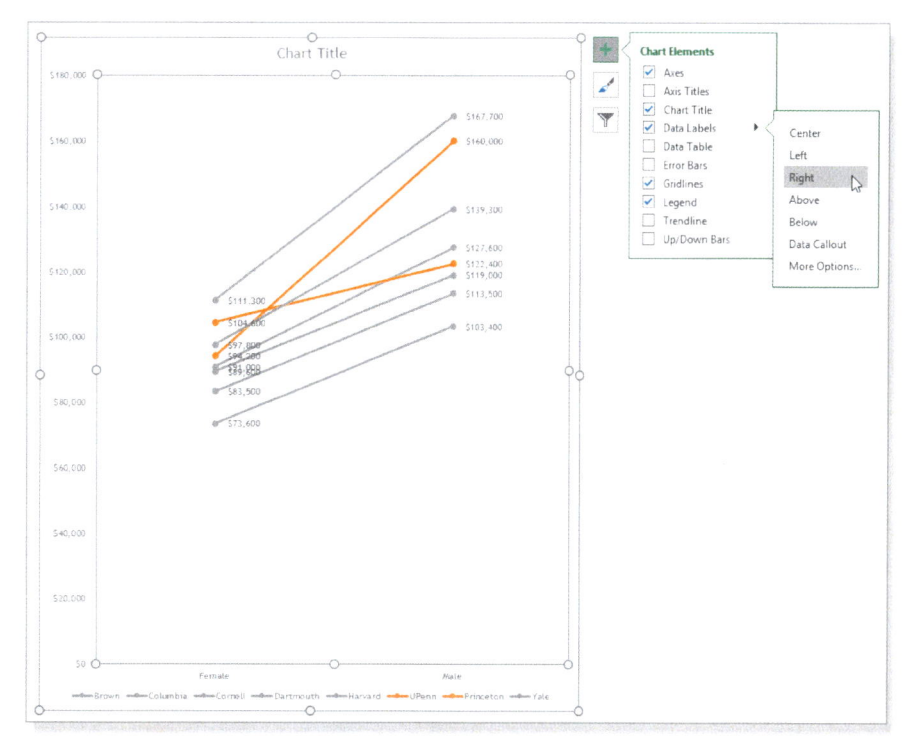

Fig 5.148: Add data labels to the right of the lines.

175

Step 7: We need to flip our data labels to the left on the left side of each line. Unfortunately, you can't select multiple data labels to edit them en masse, so we'll need to edit them individually. To reposition the first data label on the left, you'll need to isolate it, which can be a little tricky. First, click it, which will select both data labels for that line and open the **Format Data Labels** pane.

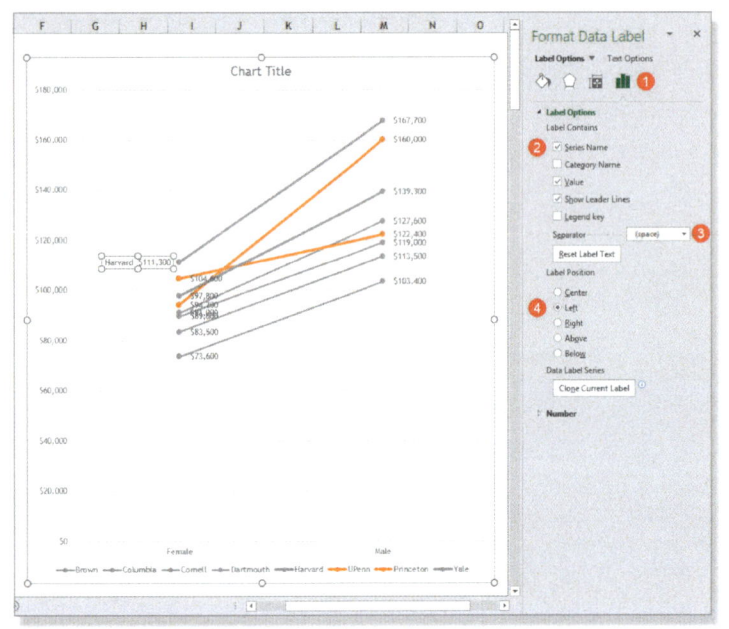

Fig 5.149: Move data label to the left side of your data series.

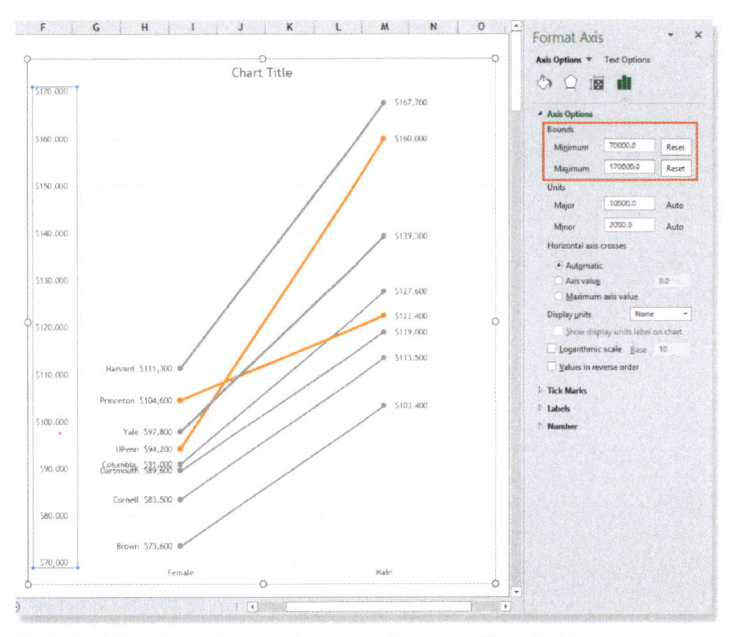

Fig 5.151: Adjust the minimum and maximum bounds to fill the Plot Area.

Then click it again to isolate it and open the **Format Data Label** pane. Open the **Label Options** tab and select **Series Name** under **Label Options.** Next, set **Separator** to **(space)** to separate the series name and value with a space. Excel defaults to a comma, which I'm not keen on. You may want to experiment with different separators to see which one you prefer. Finally, set **Label Position** to **Left (Fig 5.149)**.

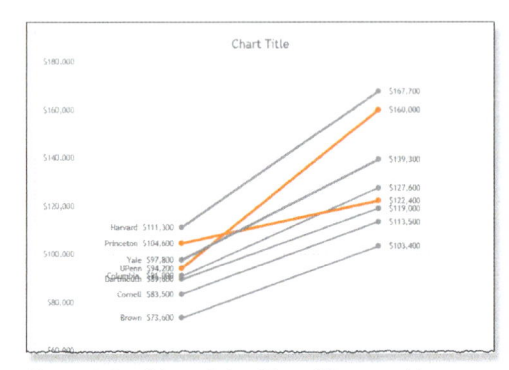

Fig 5.150: Break it up, Columbia and Dartmouth!

Step 8: Rinse and repeat until all lines are formatted. You might deal with crowding, as I did in the example file, where Columbia and Dartmouth are going all *Orange is the New Black* on us **(Fig 5.150)**. We'll deal with that in the next step.

Step 9: Now let's address the crowding. It's rare that I ever suggest doing this, but with a slope chart you don't need the vertical axis to start at 0, so we can adjust the maximum and minimum values to give our data series breathing room. Select the vertical axis to open the **Format Axis** pane and open the **Axis Options** tab, if it doesn't open automatically. Under **Bounds**, adjust **Minimum** and **Maximum** to straddle your smallest and largest values. I set mine to 70,000 and 170,000, respectively **(Fig 5.151)**.

CHAPTER 5 • CREATE CHARTS

Step 10: Now we don't need the vertical axis, legend, or gridlines, so let's zap them **(Fig 5.152)**.

Step 11: Our lines look like they're levitating, so let's give them lines to hold onto. First, we'll need to use a setting that will trash our chart for a New York minute, but it will allow us to add vertical lines using the **Major Gridlines** setting. To start, select the horizontal axis to open the **Format Axis** pane. Open the **Axis Options** tab, if it doesn't open automatically. Under **Axis Options**, set **Axis position** to **On tick marks (Fig 5.153)**.

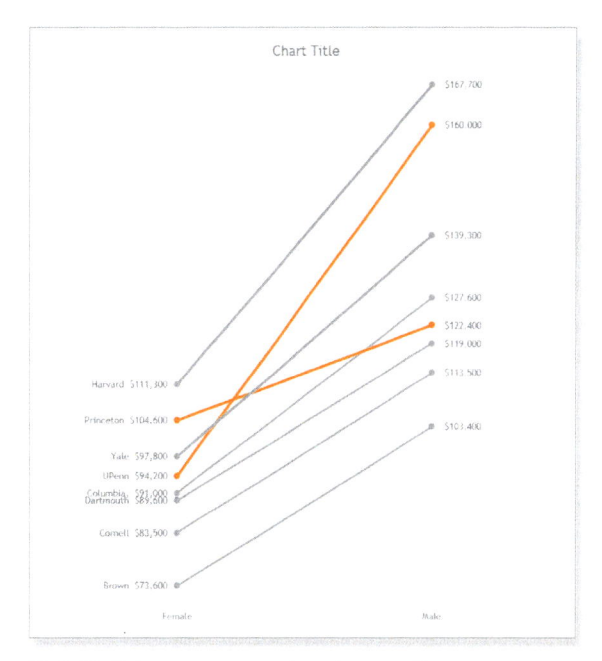

Fig 5.152: Delete unnecessary chart elements, like axes and gridlines.

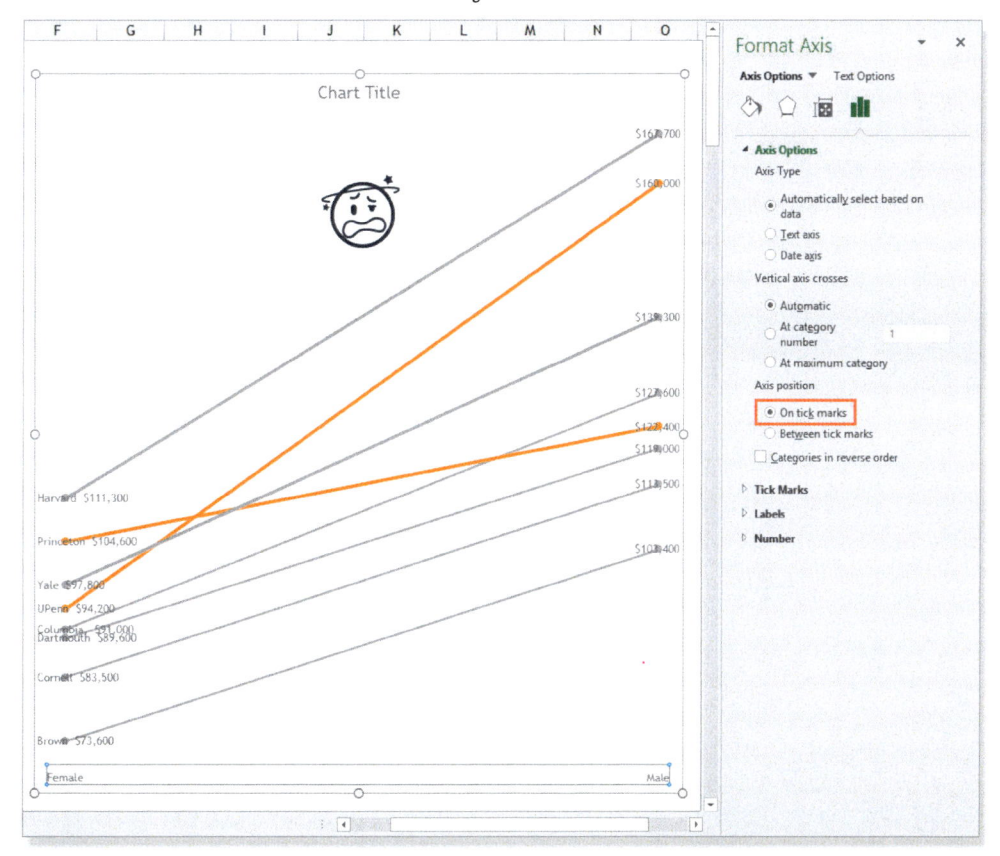

Fig 5.153: Set Axis Position to On tick marks. And try not to freak out. All will be made right.

177

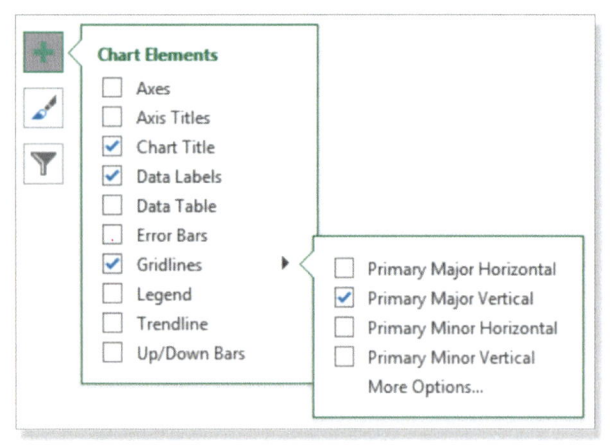

Fig 5.154: Add vertical gridlines to your chart.

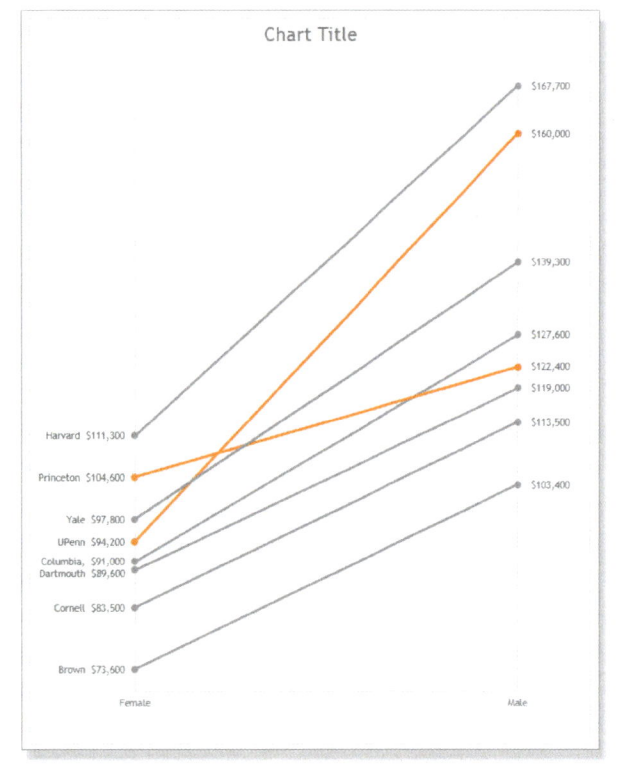

Fig 5.155: Now our lines have something to affix themselves to.

Step 12: Select anywhere in the chart and open the **Chart Elements** flyout menu. Click the right-facing arrow next to **Gridlines** and select **Primary Major Vertical** from the flyout menu **(Fig 5.154)**. Then click anywhere in the **Plot Area** and drag the handles on both sides toward the center to give your data labels room to stretch **(Fig 5.155)**.

Disclosure: Did I cheat and bump Dartmouth down for this screenshot? It depends upon what the meaning of the word *cheat* is. #90sheadlines

Step 13: Clean up your chart to taste. I cleaned mine by taking the following steps:

- made title more descriptive
- left aligned chart title
- bumped up title font size
- added subtitle
- added annotation for data source
- increased line width and marker size for orange lines
- set font color for the two highlighted labels to match line color
- replaced the x axis with male and female icons
- tweaked positioning of data labels to further reduce crowding

Credit: Shout out to the Evergreen Data blog for the inspiration for this hack: **bit.ly/mdsepc-slopegraph**.

Example File

slope-chart.xlsx

Template

Because this chart is so highly customized and needs to be built from the ground up, you can't save it as a chart template. The best thing you can do is save a file in a custom folder of all your crazy charts that you can't save as chart templates and create a copy of it, then switch out your data with your new data. It will at least save you a few steps.

LOLLIPOP GRAPH

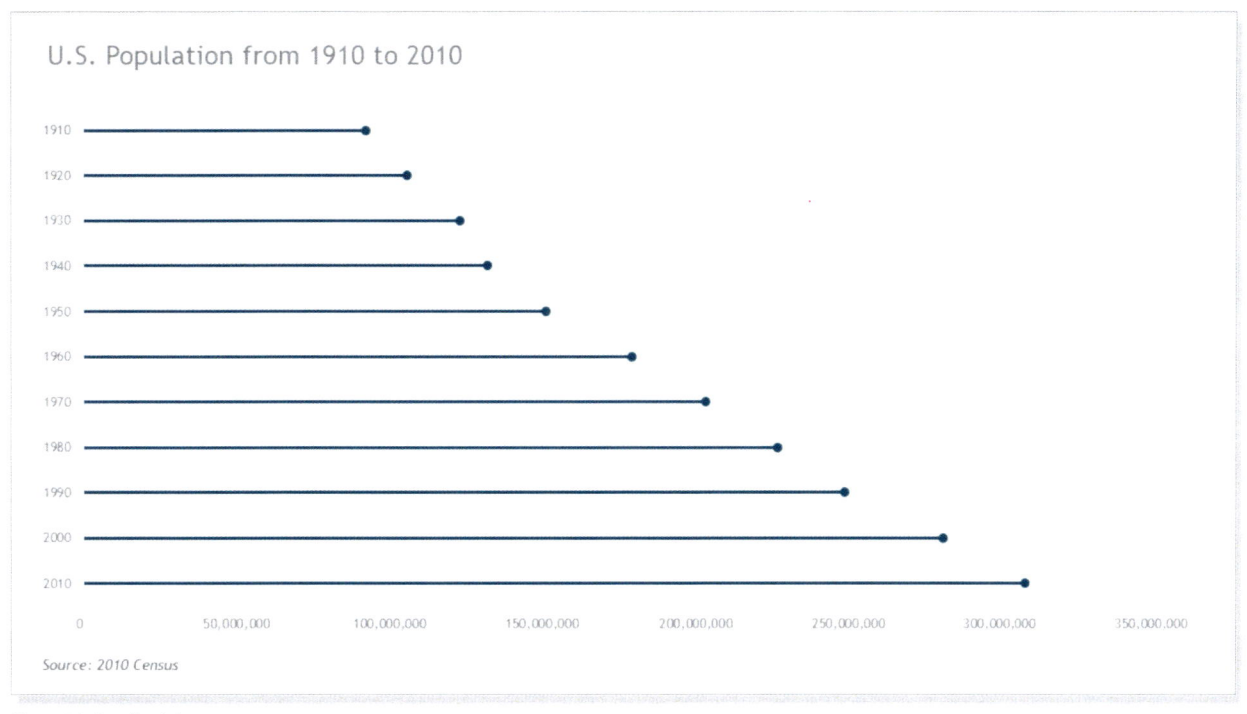

U.S. Population from 1910 to 2010

[Lollipop chart showing U.S. population from 1910 to 2010, with values ranging from approximately 92 million in 1910 to approximately 309 million in 2010]

Source: 2010 Census

Fig 5.156: Our final chart.

Uses

Lollipop graphs are best used to:

- compare data across discrete groups
- plot datasets with long categories labels because of its horizontal orientation
- visualize data with no time component
- plot datasets with many category elements
- present data that has very similar values (easier to see differentiations with less ink)

Tips

- Consider sorting your data to make it easier to read because it saves viewers' eyes from darting back and forth to figure out the order of your data. Sort your data in ascending order to get your bars in descending order.
- Look for opportunities to simplify your bar charts by removing gridlines, simplifying your axes, and bringing data into the bars.

- Remove the legend for datasets with one series.
- Make one of your bars a different color if you want to highlight an observation. Even better: Couple it with an annotation in a text box.
- Be careful with the proportions of the width of the stick to the radius of the disc. You don't want the disc to be too small or it just looks like an end cap, but you also don't want it to look like a bobblehead.

Category / Metric

1 category / 1 metric

Tutorial

Hundred-Foot View

In this chart we'll hack error bars to get our lollipops. Why error bars? Very simply because we can add spheres to them, and we can't do that with regular, run-of-the-mill bars. If you've ever wondered why you would ever use an arrowhead the shape of a circle, here you go! In short, our bars will be nothing more than a host organism for our error bars, and we'll remove their formatting to make them look invisible. It will be a bit of a dog-and-pony show to get those spheres, but the final product will [hopefully] be worth it.

Step 1: Select any cell inside your dataset and navigate to **Insert ▶ Charts ▶ Insert Column or Bar Chart ▶ 2-D Bar ▶ Clustered Bar (Fig 5.157)**.

Step 2: Let's add error bars to our chart. Select the chart and open the **Chart Elements** flyout menu. Select the right-facing arrow next to **Error Bars** and choose **More Options** from the flyout menu to open the **Format Error Bars** pane **(Fig 5.158)**. Select the **Error Bar Options** tab and modify the following settings:

- **Direction**: **Minus** (This will cause the error bar to extend from the end of the bar back to the y axis.)
- **End Style**: **No Cap**
- **Error Amount**: **Percentage – 100%** (This will ensure the error bar extends the full length of the bar.)

Step 3: To bring our error bars out of the shadows, we'll drain the color from our regular bars. To do this, select one of the bars to select the entire data series and open the **Format Data Series** pane. Open the **Fill & Line** tab, if it doesn't open automatically, and set **Fill** to **No fill**. This will leave us with a stack of pick-up sticks along a vertical axis **(Fig 5.159)**.

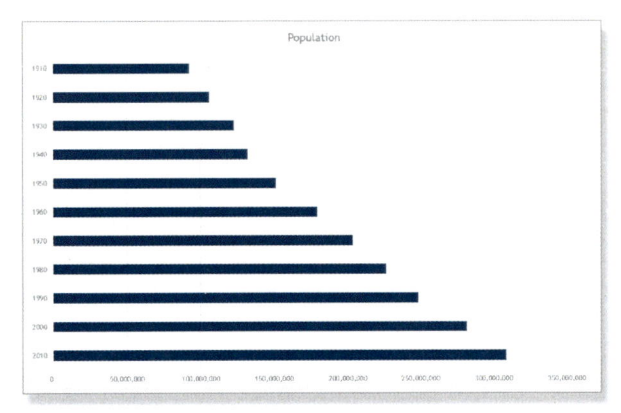

Fig 5.157: Our initial chart.

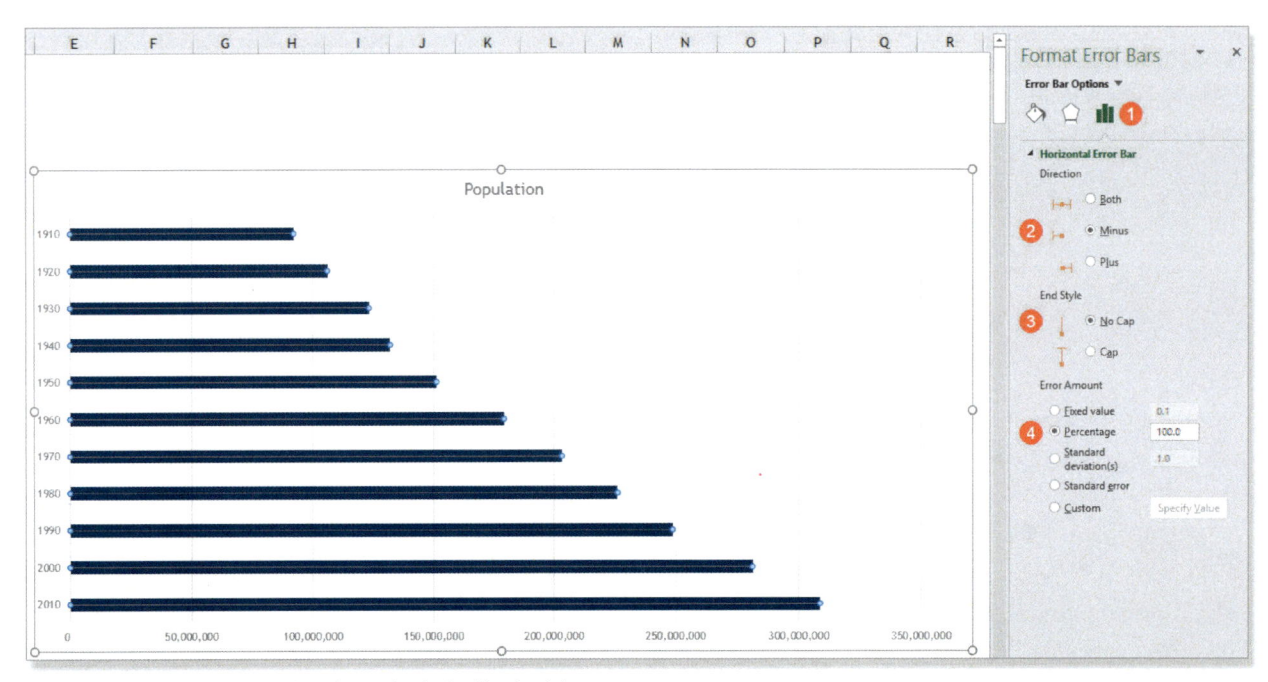

Fig 5.158: Add error bars to your bar chart to let the hacking begin!

CHAPTER 5 • CREATE CHARTS

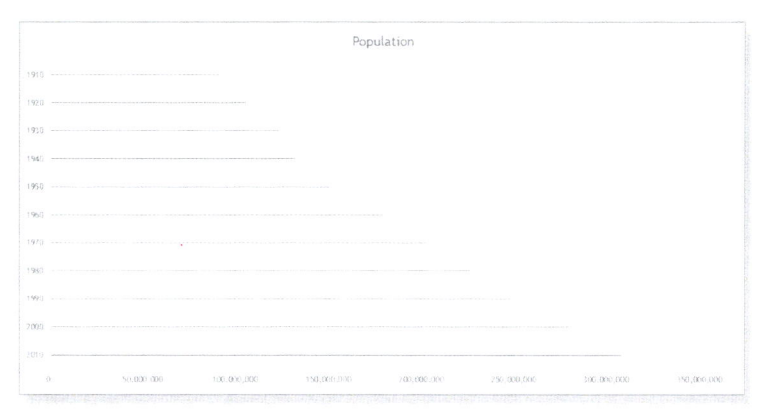

Fig 5.159: Hocus pocus, alakazam!

Step 4: We want to beef up our bars since they look a little emaciated. Select one of the error bars to select all of them and open the **Format Error Bars** pane. Open the **Fill & Line** tab and set **Width** to a setting of your choice. I used 2.25 pt. And I set **Outline color** to blue.

Step 5: While still in the **Fill & Line** tab, let's add arrowheads to our error bars **(Fig 5.160)**. Because we set our error bars to Minus in step 2, they will need to start at the end of the bar and extend back to the y axis. So, to add spheres to what *appears* to be the end of each bar, we will need to add them to the beginning **(Fig 5.161)**. I used the following settings:

+ **Begin Arrow type**: Oval Arrow
+ **Begin Arrow size**: Arrow L Size 5 (the arrow in the middle)
+ **End Arrow type**: No arrow

Tip: If your bars seem a little stretched out, you may need to drag the sizing handle on the right side of your **Chart Area** to the left to reduce the width of your chart.

Step 6: Clean up your chart to taste. I cleaned mine by taking the following steps:

+ made title more descriptive
+ left aligned chart title
+ bumped up title font size
+ added annotation for data source
+ bumped up transparency of gridlines

Fig 5.160: Add an arrowhead to the beginning of your error bars.

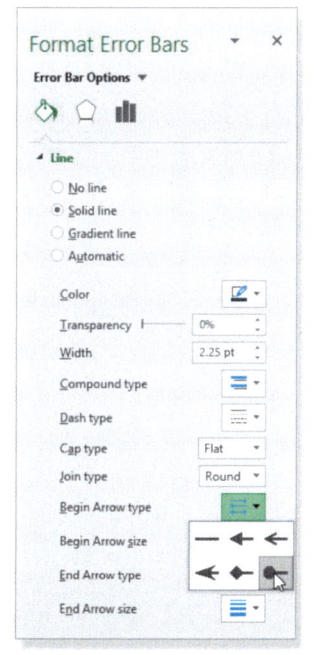

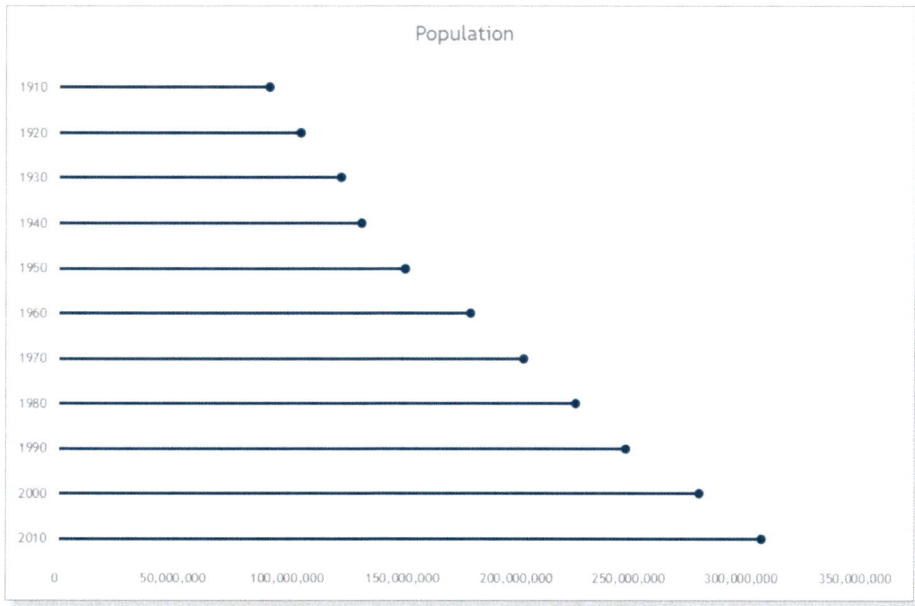

Fig 5.161: If you thought round arrowheads on error bars were useless, I present you Exhibit A!

181

CHAPTER 5 · CREATE CHARTS

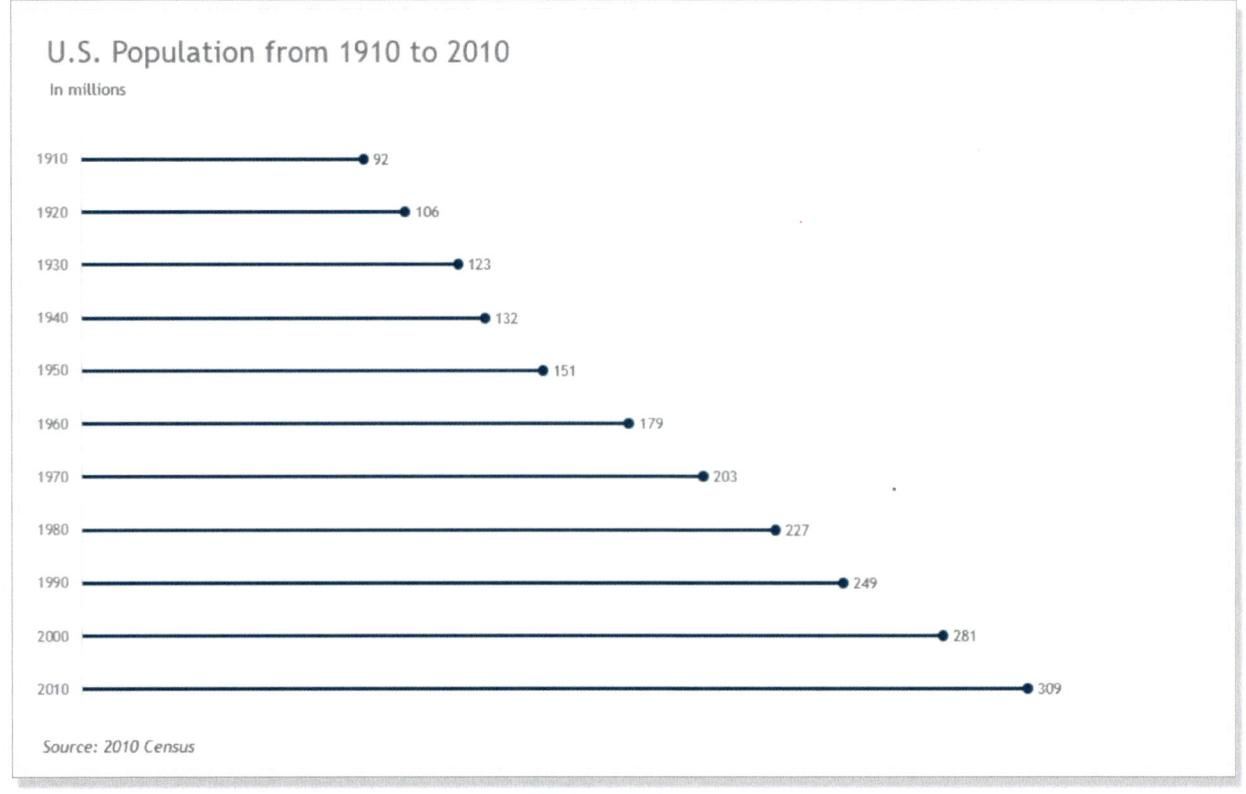

U.S. Population from 1910 to 2010
In millions

Year	Population
1910	92
1920	106
1930	123
1940	132
1950	151
1960	179
1970	203
1980	227
1990	249
2000	281
2010	309

Source: 2010 Census

Fig 5.162: Data labels may add more noise than they're worth with lollipop graphs.

💬 LEARN MORE See the Design with Elegance chapter to learn how to perform these refinements.

Credit: Shout out to the PolicyViz blog for the brilliant idea to use error bars with custom arrowheads to get the rounded tips: **bit.ly/mdsepc-lollipop**.

Variation
Add Data Labels
You could add data labels to the end of your bars **(Fig 5.162)**. In my example, I also simplified the number formatting by using the code #,, and adding a text box noting that the values are expressed in millions. Personally, I don't prefer this format because I think it distracts from the elegance of the lollipops, but you do you, boo!

Example File
lollipop-graph.xlsx

Templates
If you'll use these charts again, save them as chart templates. Or download the templates included in the chart templates folder:

Lollipop graph: mds-lollipop.crtx

Lollipop graph with data labels: mds-lollipop-data-labels.crtx

DUMBBELL PLOT

AKA

You may also see dumbbell plots called

- barbell plots
- connected dot plots
- DNA charts

Uses

Dumbbell plots are best used to:

- demonstrate changes between two points in time or between two conditions (e.g., health metrics before and after a drug study or performance after the implementation of a new protocol)
- compare opposites or competitors (e.g., females vs. males, drug vs. placebo, team A vs. team B, etc.)

Tips

- Some say you can display up to four dots in a connected dot plot. Personally, I don't like more than three—to demonstrate minimum, maximum, and median home values, for example.
- To enhance understanding, order your categories in some intuitive way. In the example file, I sorted my data in descending order using the difference between salaries for men and women for each Ivy League school to draw attention to the schools with the greatest salary differentials between men and women. (Looking at you, UPenn!)
- Bringing category labels into your chart and ditching the legend may make your chart easier to read.
- Since you're not dealing with error bars like you do with lollipop graphs, you can adjust the size of the dots. Don't be afraid to add some weight to those dumbbells. (Sorry.)
- If you want to include data labels (as I did in the example chart), I recommend pulling the data label inside the dots to avoid adding a bunch of chart junk that will distract from the overall impact of a dumbbell plot.
- Overall, I'm not a fan of axes that don't start at 0. However, with a dumbbell chart, it's not essential for your dumbbells to start at 0, as you'll see in the tutorial.
- If you wanted to emphasize dumbbells that indicate a decrease, you could assign one color for lines with a positive slope and another for lines with a negative slope. Just avoid green and red for the sake of colorblind users.

LEARN MORE See the Design for Colorblind Users section of the Design with Elegance chapter to learn how to design with accessibility in mind.

- Personally, I find charts with horizontal dumbbells to be more aesthetically pleasing and easier to interpret than vertical dumbbells.

Category / Metric

1 category / 2+ metrics

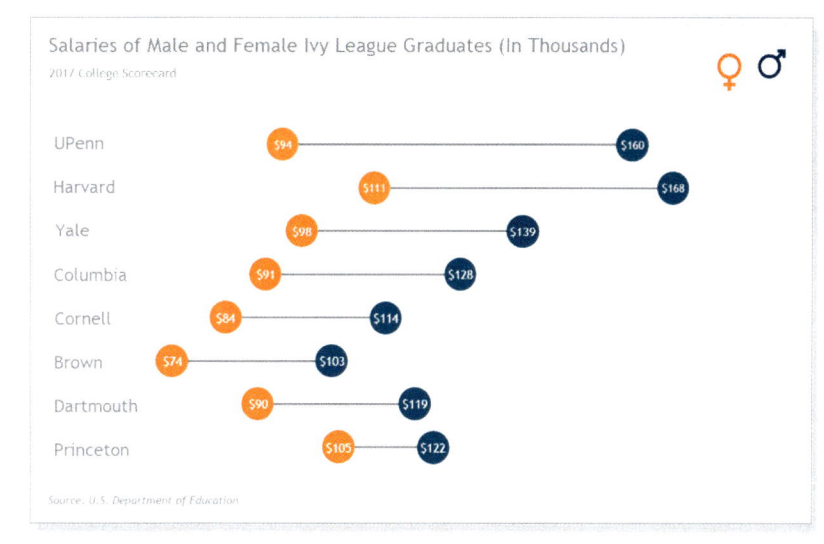

Fig 5.163: Our final chart.

Tutorial

Hundred-Foot View

We'll use a scatter plot with connected dots to build out our dumbbells. But for a scatter plot you need x and y values. (See the Scatter Plot tutorial to learn more.) So, in the same way we assigned x values in the Dot Plot tutorial, we'll give each data series—in the case of the example dataset, Ivy League schools—a contrived y value to line them up along horizontal lines. Excel lets us get crazy with the marker size of the dots in a scatter plot, so we'll take full advantage of that. Then we'll use data labels to add our values to our dumbbells, so we can remove axis labels for both the vertical and horizontal axes. Finally, we'll use text boxes to give each of our data series a label.

Step 1: Before I could plot my data, I needed to sort it. I wanted to emphasize the school with the greatest differential in salary between men and women, so I calculated the delta and sorted by that column **(Fig 5.164)**. I sorted them in ascending order to place the school with the highest pay differential (i.e., the University of Pennsylvania) at the top of the stack in the chart.

Step 2: We need to assign y values to each of our data series. One easy way to increment values down a column is to enter 1 in a cell and 2 in the cell directly below (E4:E5 of the Dumbbell Plot worksheet in the example dataset). Then select both cells by clicking-and-dragging over them and double-click the fill handle in the

EXTRACTED DATA

UNITID	INSTNM	MN_EARN_WNE_MALE0_P	MN_EARN_WNE_MALE1_P	Δ
186131	Princeton University	$ 104,600	$ 122,400	=E4-D4
182670	Dartmouth College	$ 89,600	$ 119,000	$29,400
217156	Brown University	$ 73,600	$ 103,400	$29,800
190415	Cornell University	$ 83,500	$ 113,500	$30,000
190150	Columbia University in the City of New York	$ 91,000	$ 127,600	$36,600
130794	Yale University	$ 97,800	$ 139,300	$41,500
166027	Harvard University	$ 111,300	$ 167,700	$56,400
215062	University of Pennsylvania	$ 94,200	$ 160,000	$65,800

Fig 5.164: I calculated the delta and sorted by it to emphasize the schools with the largest pay gaps.

DUMBBELL PLOT

Institution	Female	Male	Female Y	Male Y
Princeton	$ 104,600	$ 122,400	1	
Dartmouth	$ 89,600	$ 119,000	2	
Brown	$ 73,600	$ 103,400	3	
Cornell	$ 83,500	$ 113,500	4	
Columbia	$ 91,000	$ 127,600	5	
Yale	$ 97,800	$ 139,300	6	
Harvard	$ 111,300	$ 167,700	7	
UPenn	$ 94,200	$ 160,000	8	

Fig 5.165: Assign y values to each of your data series.

Fig 5.166: Our initial...chart?

bottom-right corner of the range to fill the rest of the column with values 3 through 8. While the range is still selected, drag the fill handle to the right to fill the Male Y column with a copy of the values from the Female Y column—or copy-and-paste **(Fig 5.165)**. Notice how University of Pennsylvania is assigned the y value of 8. This will place it at the top of the chart, with the rest of the schools lining up below it.

Tip: It will be easier to build your data series if your x and y value columns are side by side (i.e., x|x|y|y), instead of alternating them (i.e., x|y|x|y), as we did in the Dot Plot tutorial.

Step 3: We'll need to build our chart manually, and it would be more work to select a cell in our dataset (as we normally do) because this data is whack. Instead, select a cell outside your data range and navigate to **Insert ▸ Charts ▸ Insert Scatter (X, Y) or Bubble Chart ▸ Scatter ▸ Scatter with Smooth Lines and Markers (Fig 5.166)**.

Step 4: Right-click anywhere inside the chart and choose **Select Data** from the contextual menu. In the **Select Data Source** dialog, under **Legend Entries (Series)**, click the **Add** button **(Fig 5.167)**. We're fixin' to add our first data series. (Shout out to the Southerners out there! I got y'all!)

Step 5: In the **Edit Series** dialog, set each of the ranges by clicking inside each of the fields and selecting the ranges as shown in the accompanying screenshot **(Fig 5.168)**. See why it's easier if you keep your x and y columns together?

Tip: Remember to delete the ={1} that you'll find loitering in the **Series y values** field.

> 💬 **LEARN MORE** See the Hack Excel's Charting Engine section of the Design with Elegance chapter to learn how to rock the Select Data Source dialog.

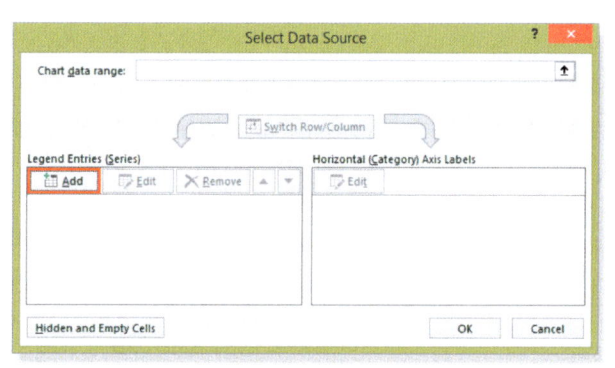

Fig 5.167: It all starts with the Select Data Source dialog.

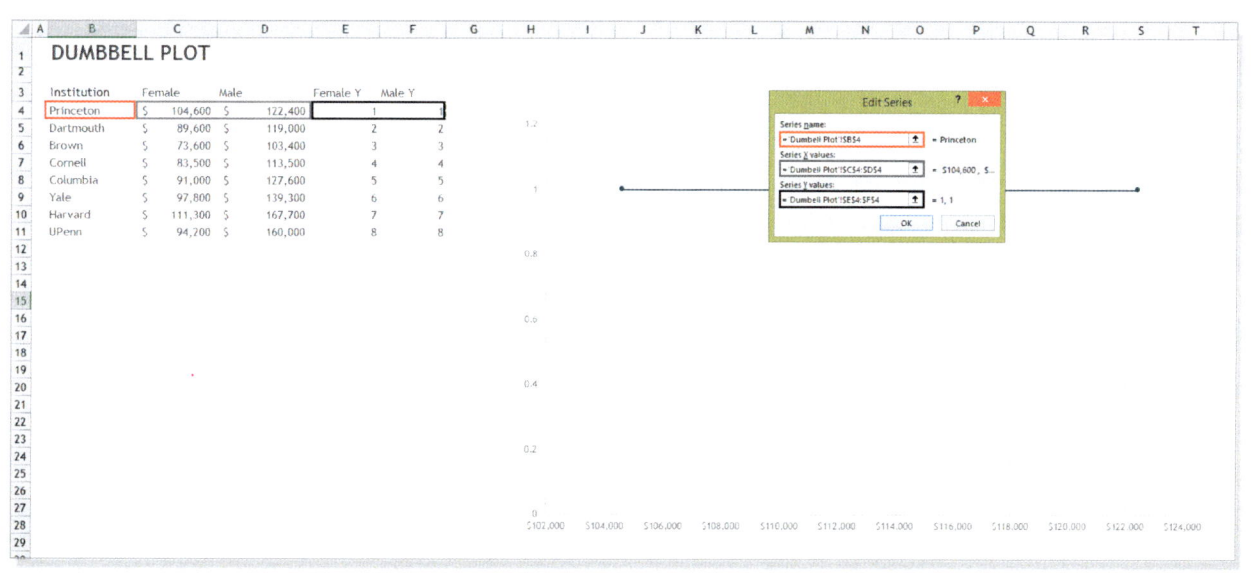

Fig 5.168: Assign your series values in the Edit Series dialog.

185

CHAPTER 5 · CREATE CHARTS

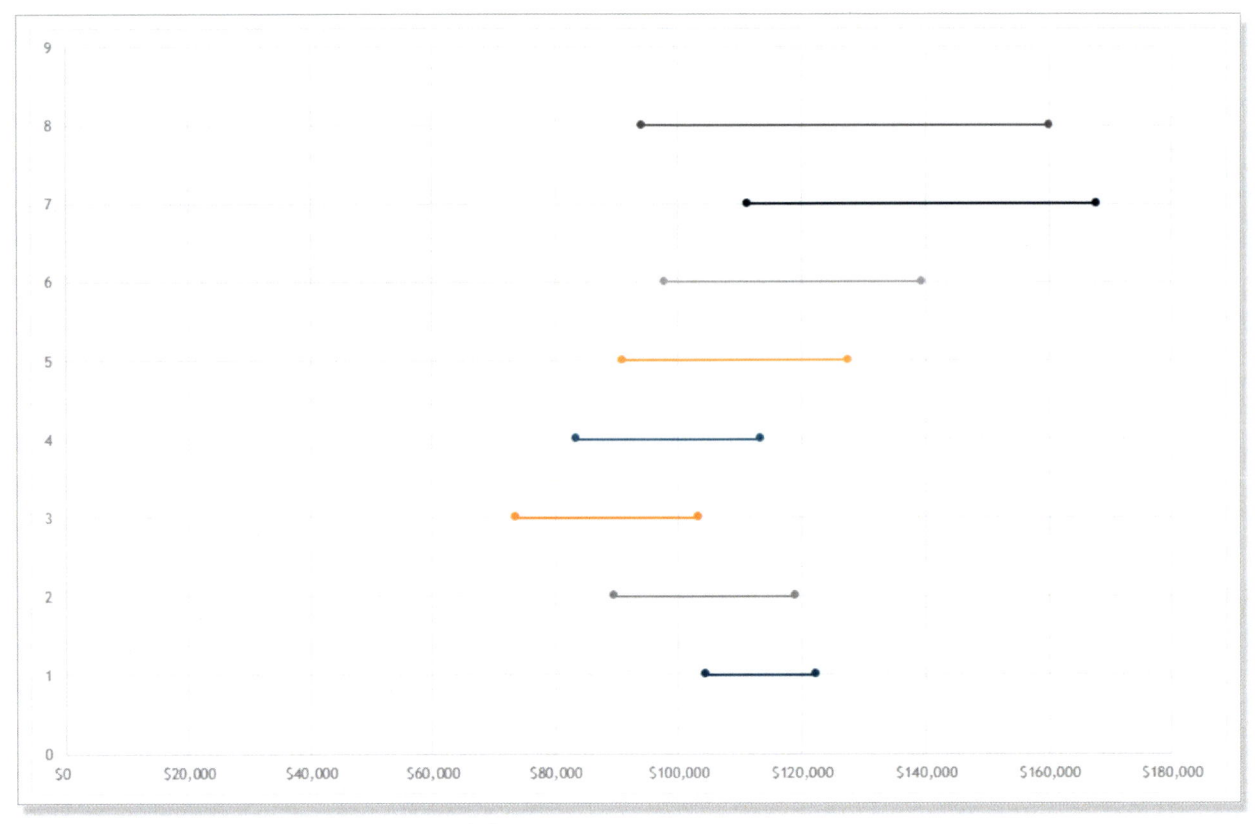

Fig 5.169: Our rudimentary dumbbells are starting to take form.

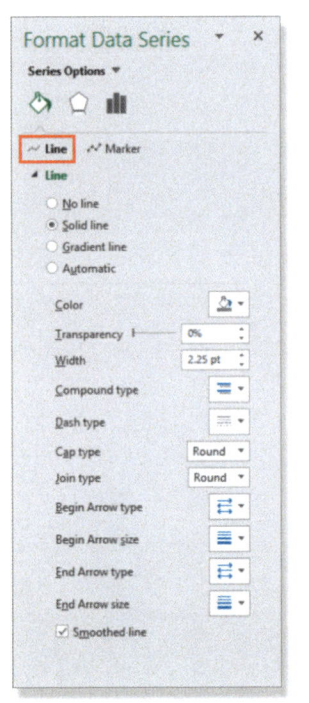

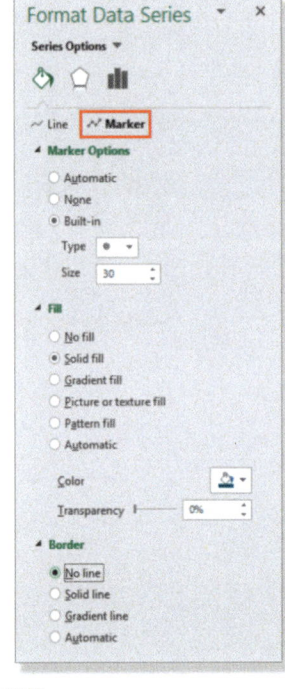

Fig 5.170: Format your data points to taste.

Step 6: Rinse and repeat for each of your data series **(Fig 5.169)**.

Step 7: Now we'll format our lines and markers using our branded colors. To start the formatting process, select one of your dumbbells, then press Ctrl-1 to open the **Format Data Series** pane **(Fig 5.170)**. I used the settings below:

Line

- **Line Color**: gray
- **Width**: 2.25 pt

Markers

- **Marker Options**: **Built-in**
- **Type**: **Circle**
- **Size**: 30
- **Fill**: **Solid fill**
- **Fill Color**: blue
- **Border**: **No line**

186

CHAPTER 5 • CREATE CHARTS

Step 8: Once you've formatted your data series, select one of the dots to modify its fill color. I selected the female data point. This will open the **Format Data Point** pane. Set **Fill Color** to another one of your branded colors. I used orange. This way, if women from a particular school had the higher salaries (lol, if only!), you would see the orange dot to the right of the blue dot in the example file. Sadly, that is not the case for any of the schools included in this analysis.

Step 9: Rinse and repeat for each of your data series **(Fig 5.171)**.

Step 10: We don't need the y axis since that's #fakedata. Select it and delete it. And because we'll be pulling data labels into the markers, we won't need the x axis either. However, before we delete it, we'll modify the minimum value. We don't need it to start at 0; it creates an unnecessary (and unsightly) gap between our dumbbells and the vertical axis. We don't want to set it to match our smallest value, though, because we'll need space for the text boxes we'll use for our categories. After some experimentation, I found 50,000 to be a good value. If you need to modify this value in your dataset, select your axis to open the **Format Axis** pane. Open the **Axis Options** tab and set **Minimum Bounds** to a value that makes sense for your data—again, only if you have excessive space **(Fig 5.172)**. Now delete your x axis. #buhbai

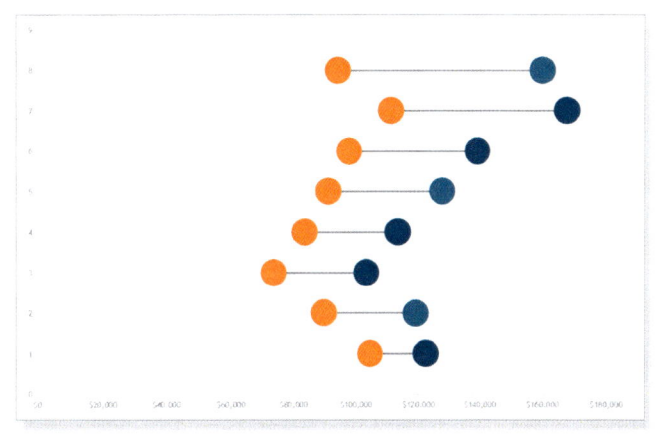

Fig 5.171: Format each of your data series to taste.

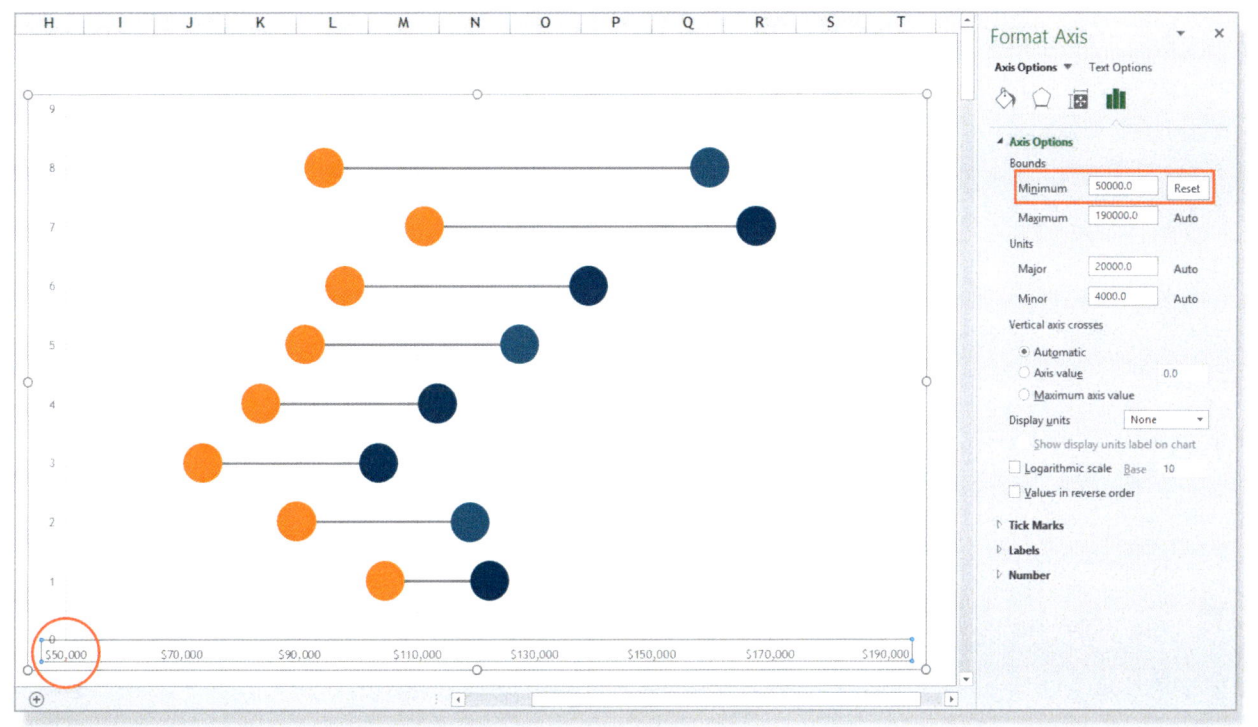

Fig 5.172: Modify x axis minimum and maximum bounds to remove unnecessary white space.

187

CHAPTER 5 · CREATE CHARTS

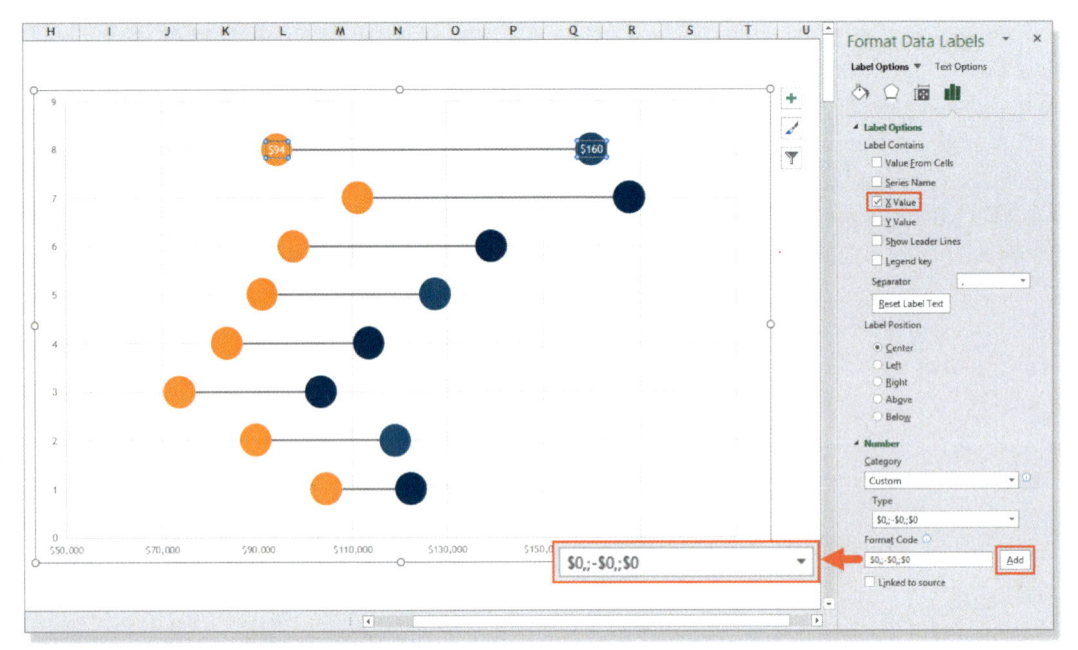

Fig 5.173: Format your data labels to taste.

Step 11: Add data labels. I set **Font Color** to white and **Font Size** to 10. With the labels still selected, open the **Label Options** tab. Under the **Label Options** section, set **Label Contains** to **Series X Value** to pull in the **X values** we set in the **Edit Series** dialog in step 5 (i.e., female and male salaries in the example file). I also applied custom number formatting to express salary in thousands **(Fig 5.173)**.

> 💬 **LEARN MORE** See the Add Data Labels section of the Design with Elegance chapter to learn how to add data labels and the Simplify Large Numbers section to learn how to format large numbers.

Tip: If you want to save yourself some steps, you can bring your categories into your data labels **(Fig 5.174)**. Mynda Treacy did this in her dumbbell tutorial on the My Online Training Hub blog: **bit.ly/mdsepc-mynda-dumbbell**.

Step 12: Now you can delete both of your axes **(Fig 5.175)**. But don't delete your gridlines yet. We'll use them in the next step.

Fig 5.174: Alternatively, use data labels to label categories.

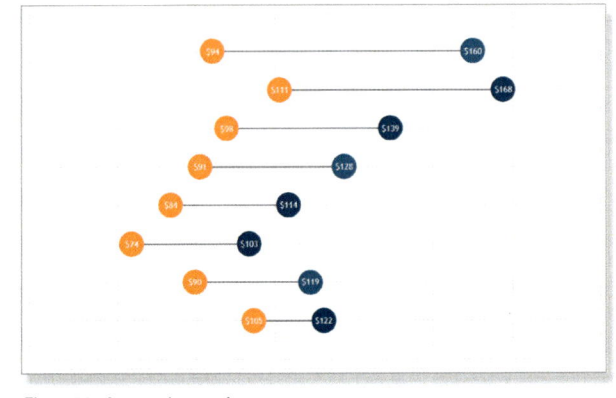

Fig 5.175: Au revoir, axes!

CHAPTER 5 • CREATE CHARTS

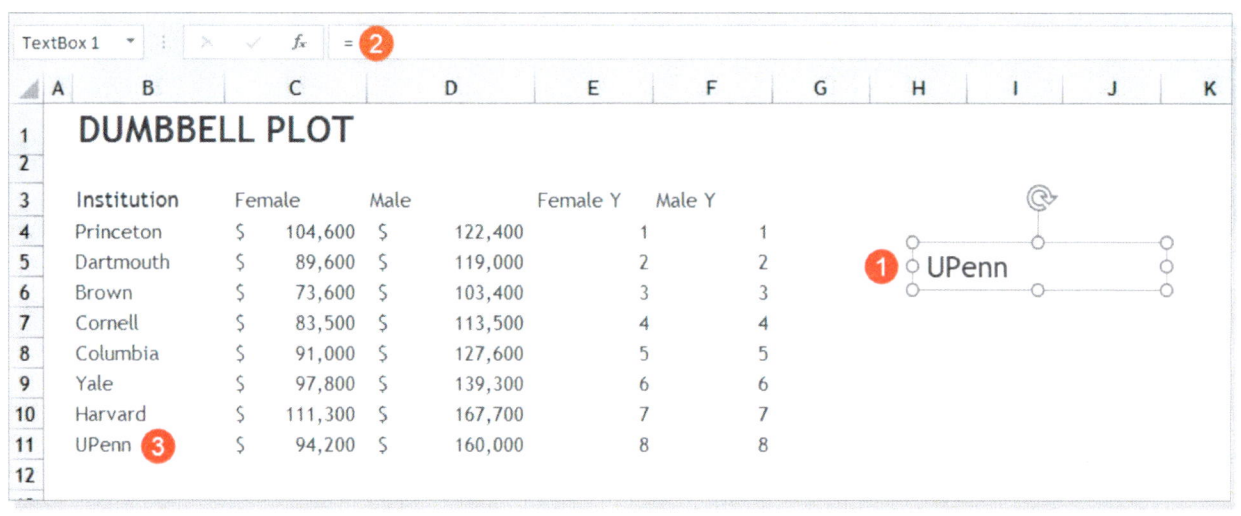

Fig 5.176: Use a formula to pull values into your text boxes.

Step 13: Now we're ready to create our text boxes. I used a formula to pull the first value from my table into the text box **(Fig 5.176)**. Then I formatted the text box.

Step 14: Copy-and-paste your text boxes and update their formulas. Then align them with your horizontal gridlines. To ensure they're aligned, use the **Align** settings. In the **Align** menu, I recommend choosing **Align Left** and **Distribute Vertically**. Then you can group them (another option under the **Arrange** tools) and move them where you want them **(Fig 5.177)**.

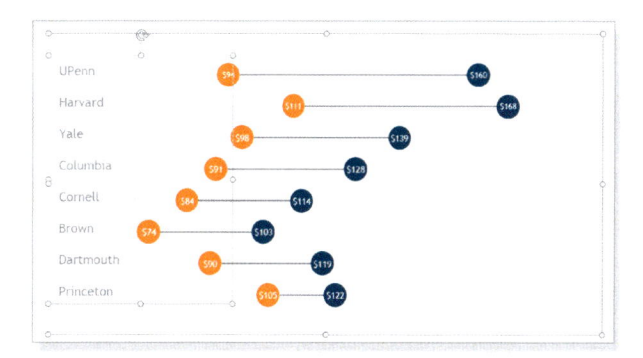

Fig 5.177: You can't be too detailed when aligning text boxes in Excel.

Step 15: Clean up your chart to taste. I cleaned mine by taking the following steps:

- added chart title
- left aligned chart title
- added subtitle
- added annotation for data source
- removed gridlines
- added custom legend using colorized icons
- grouped legend with chart

💬 LEARN MORE See the Design with Elegance chapter to learn how to perform these refinements—and particularly the Colorize Icons section to learn how to align icons with your branding.

Credit: Shout out to the Evergreen Data blog for the inspiration to use text boxes for the category labels: **bit.ly/mdsepc-dumbbell**.

Example File
dumbbell-plot.xlsx

Template
Because this chart is so highly customized and needs to be built from the ground up, you can't save it as a chart template. The best thing you can do is save a file in a custom folder of all your crazy charts that you can't save as chart templates and create a copy of it, then switch out your data with your new data. It will at least save you a few steps.

189

CHAPTER 5 · CREATE CHARTS

COMBINATION CHART

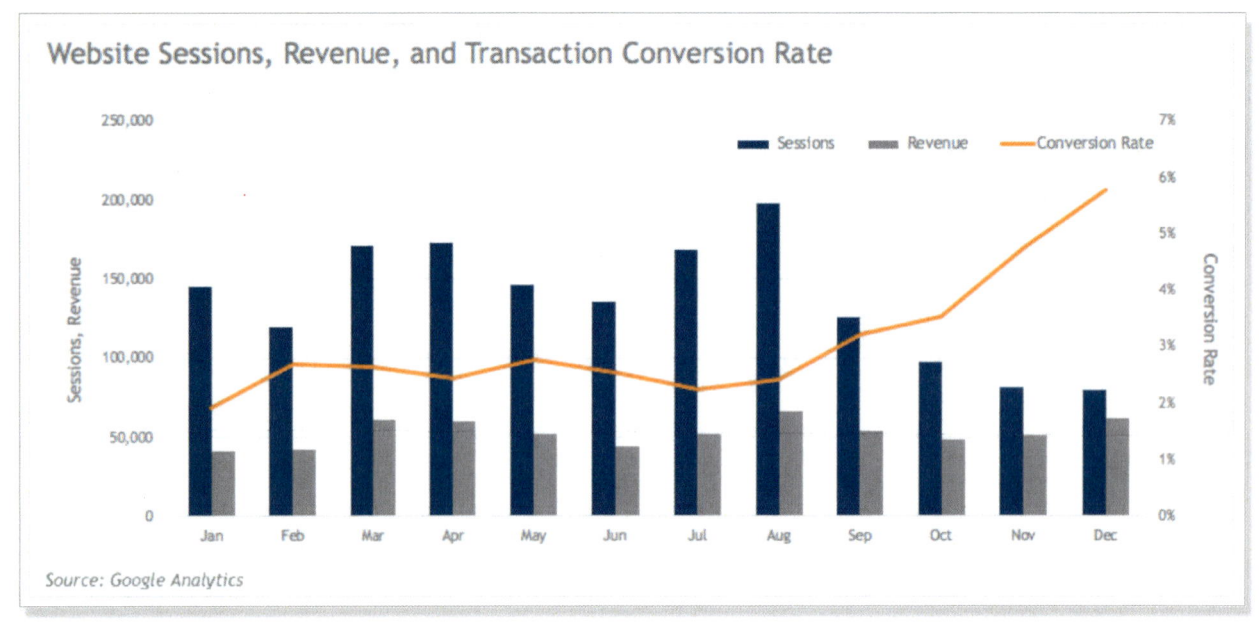

Website Sessions, Revenue, and Transaction Conversion Rate

Source: Google Analytics

Fig 5.178: Our final chart.

AKA
You may also see combination charts called multi-series dual y combination charts.

Uses
Combination charts are best used to:

- compare data across discrete groups
- visualize categorical data with short labels
- view correlation among data ranges with very different values (e.g., whole numbers and percentages)
- plot data with series that differ by orders of magnitude (e.g., number of customers vs. lifetime value)

Tips
- If your data is exported as individual dates and you want to aggregate them by month (or quarter), the easiest way to do that is with a pivot table. I've included a Pivot Table worksheet in the downloadable workbook to show you how I set mine up. Click anywhere in the pivot table to open the PivotTable Fields pane. From there you can easily see how I constructed it.

  LEARN MORE See the Pivot Tables section of the Massage Your Data chapter to learn how to rock pivot tables.

- If the data values for your individual series are similar in scale, consider having them share an axis for quick comparisons.
- To keep the visualization clean, try to mix chart types that differ significantly—for example, column with line or column with area. Mixing chart types that are too similar, such as line and area, can be confusing to interpret and kinda ugly.
- If you want a bar element, you need to use a [vertically oriented] column chart over a [horizontally oriented] bar chart because you can't mix bar charts with line or area charts—hacks like the bar chart with mean line excepted.

190

- If you have two data series with values that are very different from each other in magnitude, I highly recommend putting the data series with the smaller values on the secondary axis. In most cases, it's odd to have the smaller data series in the left vertical axis.

- If you use a secondary axis, add axis titles to both vertical axes. (PLEASE.)

- When you add an axis title to a secondary axis, Excel orients it in the same direction as the primary axis. This is a matter of preference, but I rotate it so that the bottom edges of the labels are closest to the axes.

- If you use a line chart and there aren't clear notches—because the data points form a straight(ish) line—you may want to add markers to your line. If it's not necessary to see the individual data points, I wouldn't add them though as they tend to add unnecessary pixels to your chart.

- Avoid placing your legend to the right side of your chart. I recommend placing it above, below or inside the chart. Positioning it to the right can throw off the sense of balance among the chart elements, in my opinion.

Category / Metric
1+ categories / 2+ metrics

Tutorial
Step 1: Select any cell inside your dataset and navigate to **Insert ▶ Charts ▶ Insert Combo Chart ▶ Clustered Column–Line on Secondary Axis (Fig 5.179)**.

Tip: If you don't get the chart types you want, right-click anywhere inside the chart and choose **Change Chart Type**. In the **Change Chart Type** dialog, set **Chart Type** for each of your data series. You can also opt to pop a data series onto a secondary axis here. You'll see a preview of your chart inside the dialog.

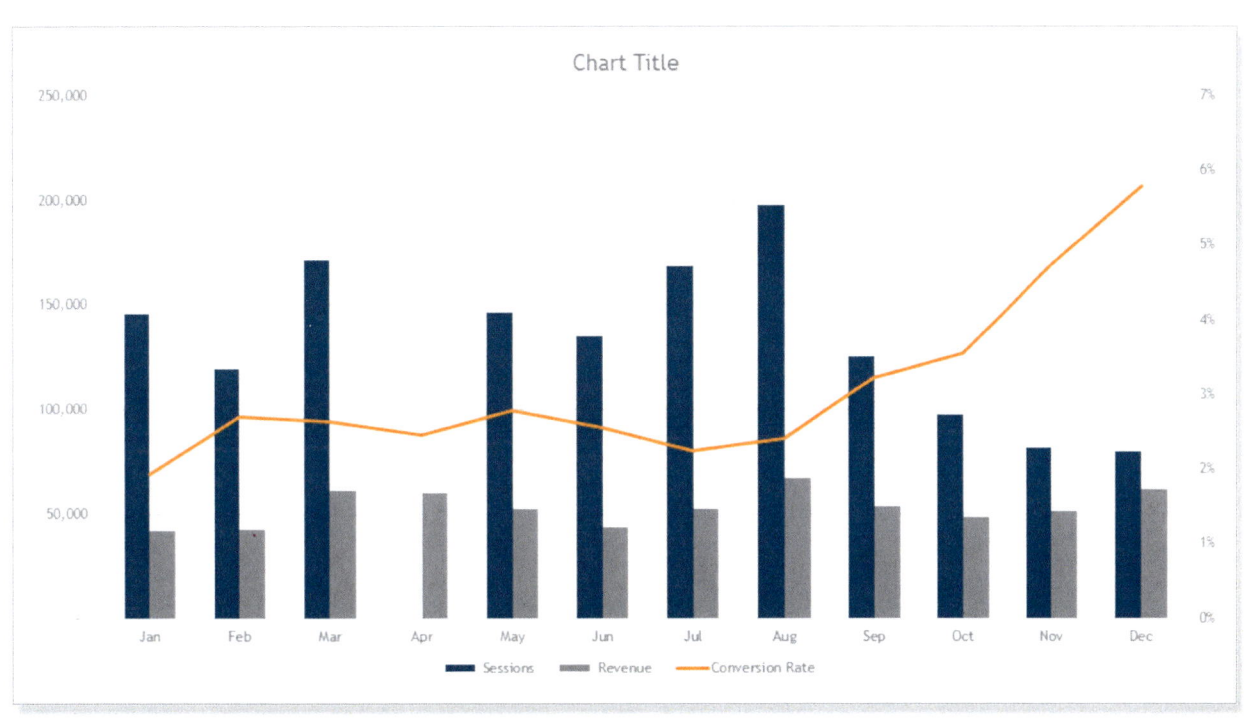

Fig 5.179: Combination charts are now default charts in Excel 2016.

Step 2: Next, we'll add axis titles by selecting the chart and opening the **Chart Elements** flyout menu. Click the right-facing arrow next to **Axis Titles** and choose **Primary Vertical** and **Secondary Vertical** from the flyout menu **(Fig 5.180)**. Change the text by either [carefully] double-clicking inside the text box or selecting the text box and entering your text in the **Formula Bar**.

Step 3: I don't like the default orientation of the secondary axis title. I prefer it to mirror the left axis title. If you agree, select the axis title, then press Ctrl-1 to open the **Format Axis Title** pane. Open the **Size & Properties** tab and set **Text direction** to **Rotate all text 90° (Fig 5.181)**.

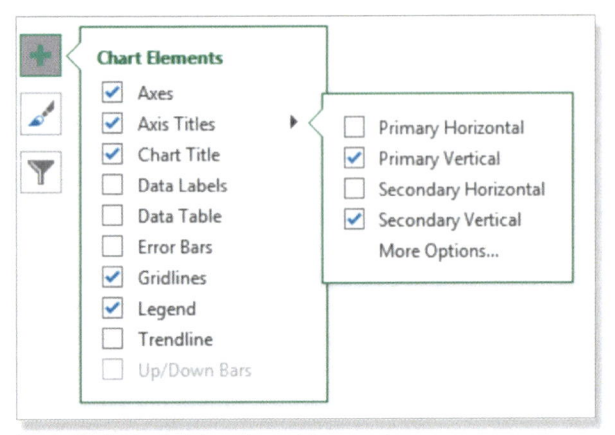

Fig 5.180: Add axis titles when adding a series to a secondary axis.

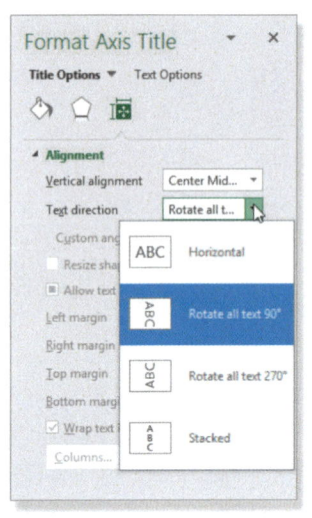

Fig 5.181: Flip that axis title like a hotcake.

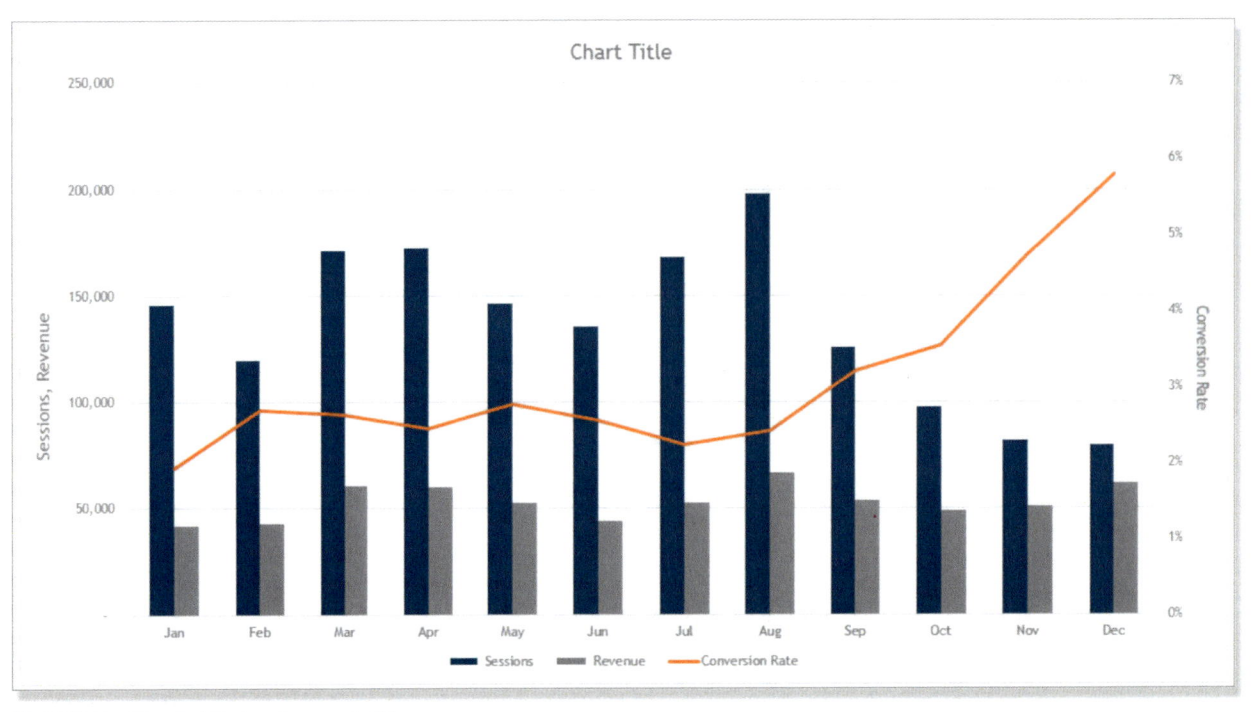

Fig 5.182: Don't be afraid to tweak your axis titles.

CHAPTER 5 · CREATE CHARTS

Step 4: I also like to bump up the text size and drag the title boxes farther away from the axes. If you hold down the Shift key when you drag them, you'll move them along a horizontal plane. This will prevent your axis titles from getting misaligned **(Fig 5.182)**. You can type into these boxes just like any other text box: Double-click to select the text and type over it or select the text box and type into the **Formula Bar**.

Step 5: Clean up your chart to taste. I cleaned mine by taking the following steps:

- made title more descriptive
- left aligned chart title
- bumped up title font size
- added annotation for data source
- bumped up transparency of gridlines
- moved legend

LEARN MORE See the Design with Elegance chapter to learn how to perform these refinements.

Variation
Use Area Chart

Sometimes it makes more sense to use an area graph instead of a line graph. This can be a great way to emphasize that data series. In this case you would follow all the steps listed above, but set the data series you want to go on the secondary axis to Area instead of Line.

Example File
combination-chart.xlsx

Template
If you'll use these charts again, save them as chart templates. Or download the templates included in the chart templates folder:

Combination chart with line:
mds-combination-line.crtx

Combination chart with area:
mds-combination-area.crtx

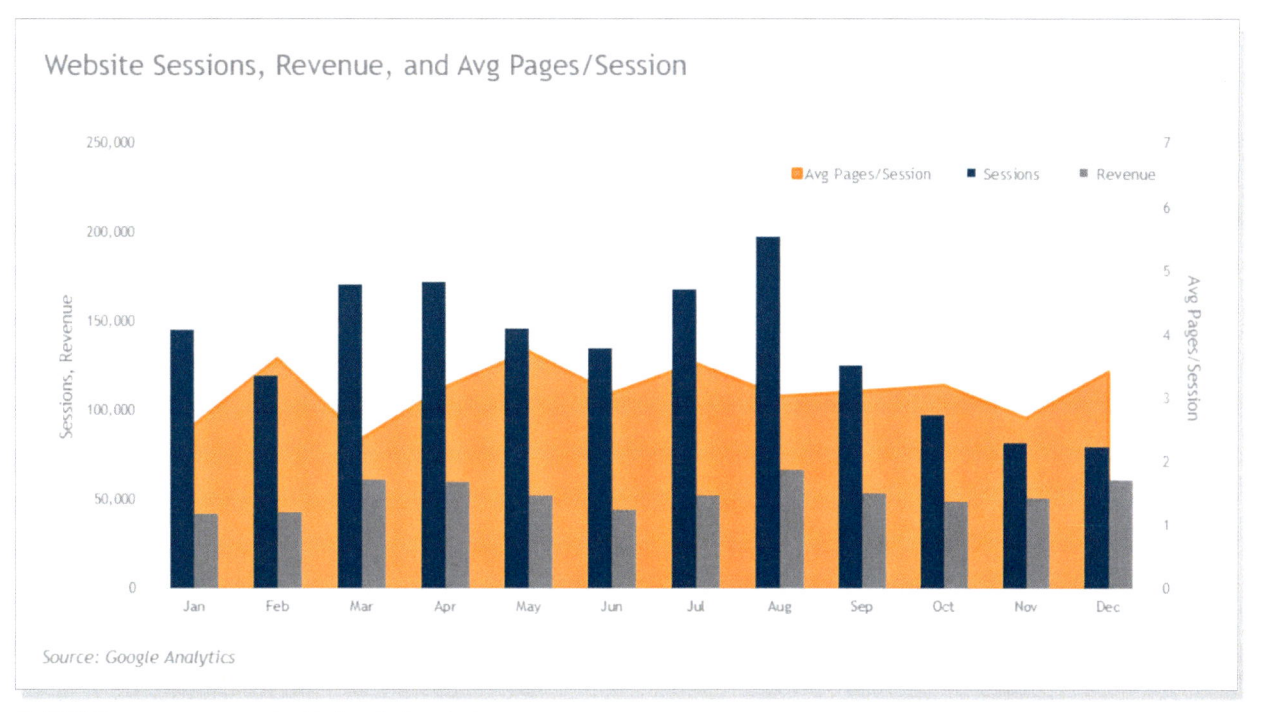

Fig 5.183: You can use an area chart in lieu of a line chart.

193

CHAPTER 5 • CREATE CHARTS

SMALL MULTIPLES

Fig 5.184: Our final chart.

AKA
You may also see small multiples called:

- grid charts
- lattice charts
- panel charts
- trellis charts

Uses
Small multiples are best used to:

- view multiple slices from the same dataset
- compare trends
- plot data from the same time period(s)
- plot data that would be crowded in other visualizations
- perform correlation analysis for multiple potential causes of a phenomenon
- visualize data with GPS coordinates (using maps)
- augment infographics

Tips

- I only include 0 for categories because technically you could use histograms or scatter plots in small multiples. However, if you do, make sure your audience is very comfortable with these chart types.
- If your data is exported as individual dates and you want to aggregate them by month (or quarter), the easiest way to do that is with a pivot table. I've included a Pivot Table worksheet in the downloadable workbook to show you how I set mine up. Click anywhere in the pivot table to open the PivotTable Fields pane.
- If data across visualizations is similar in scale, consider having them share an axis for quick comparisons.
- Avoid gorging charts with too much data. Small multiples are best used to compare trends across slices of your data.
- Use intuitive and descriptive chart titles.
- The more panels your small multiples contain, the simpler the charts should be.
- Keep text in the multiples to a minimum.
- If data contains a location component, consider using maps.

> 💬 **LEARN MORE** See the Maps section of this chapter to learn more.

- Aim for an alignment that is both aesthetic and makes sense. The way you group your charts into columns and rows can suggest grouping unintentionally.
- If your charts use different fill colors, place legend as close to the chart as possible.
- Avoid unnecessary color. If you have different metrics demonstrated across multiple charts, there's no need to use a new color for each visualization. Our brains look for the significance of details like color, which can cause unnecessary processing time and frustration.

CHAPTER 5 • CREATE CHARTS

- You may need to thin out the axes of your charts to reduce ink in your visualizations.
- Gridlines are especially distracting in small multiples. If your data needs them, panels might not be the best choice.
- Data labels need not apply. No room.
- Less is more for small multiples. Minimalism is mission critical or the size of the charts will just compound and magnify the noise.

Category / Metric
0+ categories / 1+ metrics

Tutorial

Hundred-Foot View

As with many of the charts in this book, Excel doesn't offers the ability to create small multiples by default. But what we can do is create a number of charts that Excel offer out of the box, and use a shape (e.g., rectangle) to group them.

For the example data, I used data from my daughter's **Vinspire.me** Facebook page, **www.facebook.com/ VinspireStyling**. I was called young man through 9th grade, but somehow she received a deeply recessive gene that drives her to want to resuscitate beauty from a different era.

But I digress.

Data from social networks is ideal for small multiples because their analytics tools typically provide us with different slices—or facets—from the same dataset. Some analysts will say that this data doesn't qualify for small multiples because it is too dissimilar in scale. New likes would be lost in translation once I added reach and impressions. However, I purposely chose data with different scales because I don't subscribe to the philosophy that data must share the same scale (i.e., value axis) to use small multiples. Feel free to email me at noreply@makingdatasexy .com if you disagree. #lol

Step 1: As we've seen before, if you're working with data that's broken down by day, you'll need to use a pivot table to aggregate your data into larger chunks, such as months or quarters. I fed my pivot table dates from Jan 1, 2017 to Dec 31, 2017, and Excel automatically aggregated it into months **(Fig 5.185)**.

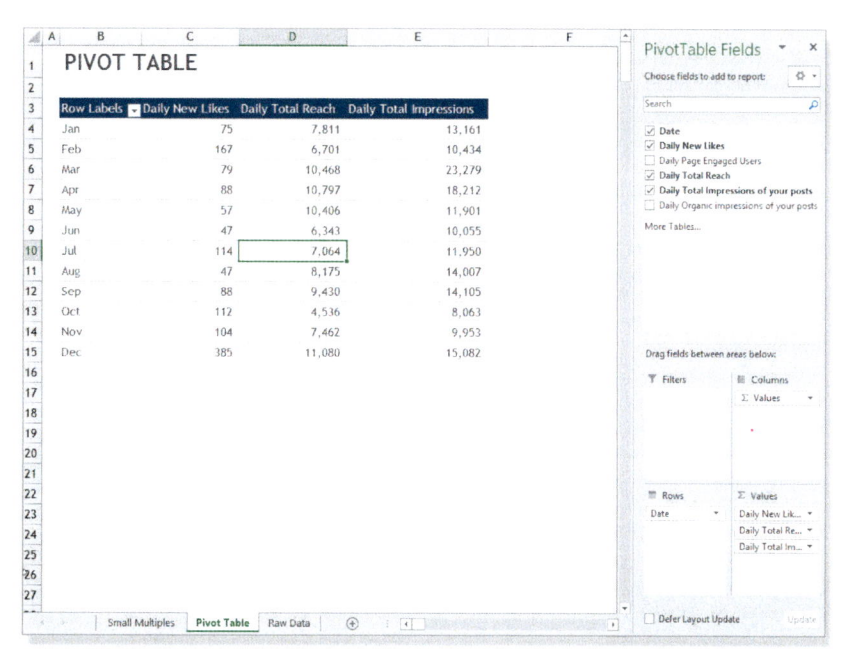

Fig 5.185: It all starts with a pivot table.

195

Step 2: Copy-and-paste your massaged data from the pivot table as values from the **Paste Options** button (**Paste Values** ▸ **Values & Number Formatting**). Click-and-drag over the data you want and create a chart of your choice from it. I'll start with Daily New Likes, and I'll use line charts for my small multiples **(Fig 5.186)**.

Step 3: You will most likely need to thin out at least one of your axes. For my chart, I don't need the level of detail my x or y axis provides, but we'll start with the y axis. Select it, then press Ctrl-1 to open the **Format Axis** pane, then open the **Axis Options** tab. I will typically double the **Major Units** value **(Fig 5.187)**.

Step 4: Now let's address the x axis. If I'm plotting any kind of continuous data—such as data with a time element—I will usually thin the axis. This simple step goes a long way toward simplifying and decluttering your chart. Select it to open the **Format Axis** pane. Then, under the **Labels** section of the **Axis Options** tab, set **Interval between labels** to **Specify interval unit**. If you set this value to 2 and you're plotting months, Excel will skip a month **(Fig 5.188)**. If you set it 3, you'll see the first month of the quarter.

> 💬 **LEARN MORE** See the Thin Out Axes section of the Design with Elegance chapter to learn how to reduce clutter in your axes.

Note: If you're plotting quarters, I recommend beginning your dataset with the first month of your fiscal year, regardless of the default for the tool you're exporting data from.

Step 5: Dump the gridlines.

Step 6: After creating each of your charts, you may want to add them to a container. Even if you don't format the container, containing them in some way will make it easier to adjust them as a single unit. I created a rectangle (**Insert** ▸ **Illustrations** ▸ **Shapes** ▸ **Rectangles** ▸ **Rectangle**), then selected it to open the **Format Shape** pane. If you want to follow along, open the **Fill & Line** tab and set **Fill** to **Pattern fill** **(Fig 5.189)**. If you use a pattern fill, I recommend using a subtle pattern with one of your branded colors. I used Dotted 90% and blue. If you'd like to give it a border, set **Line** to **Solid line** and **Width** to taste. I used .75 pt.

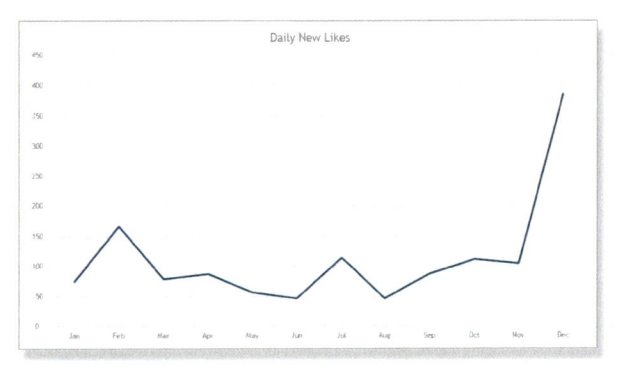

Fig 5.186: Our initial chart.

Fig 5.187: You will most likely need to thin your y axis.

CHAPTER 5 · CREATE CHARTS

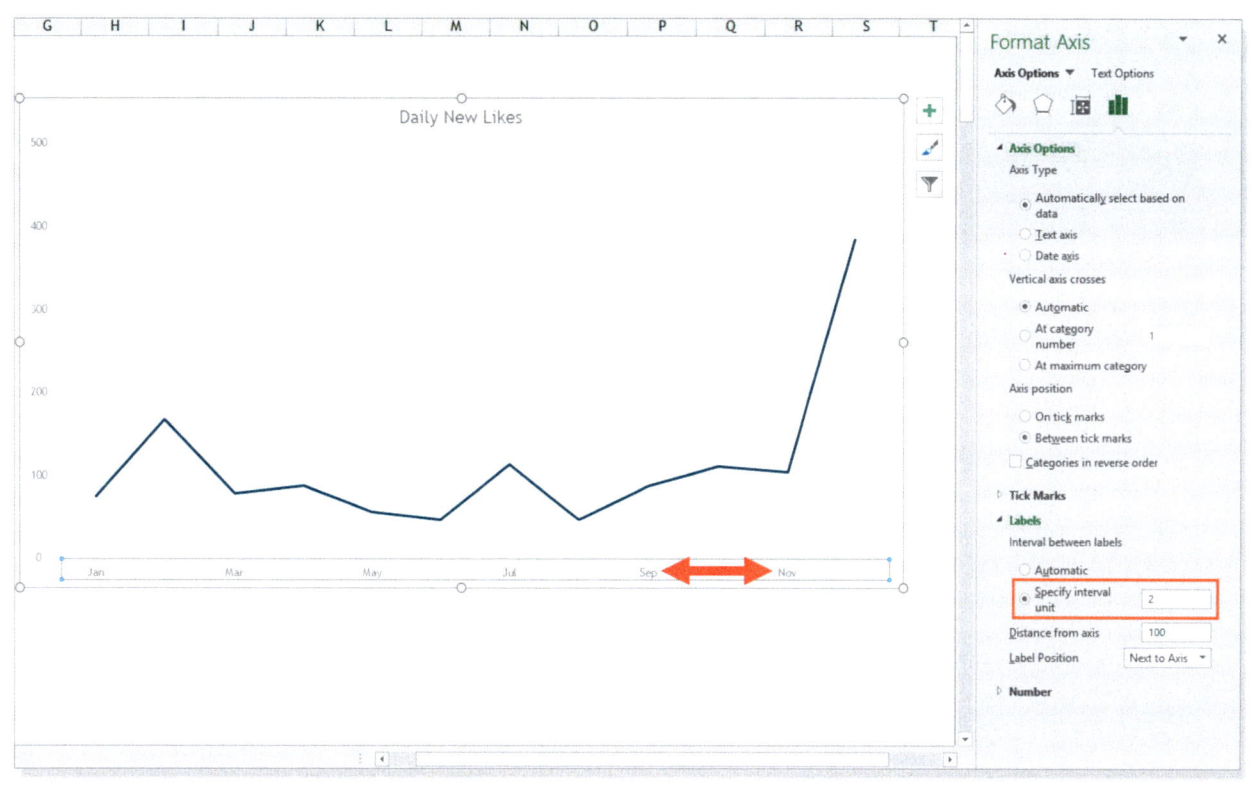

Fig 5.188: Thin your x axis by customizing the interval of your labels.

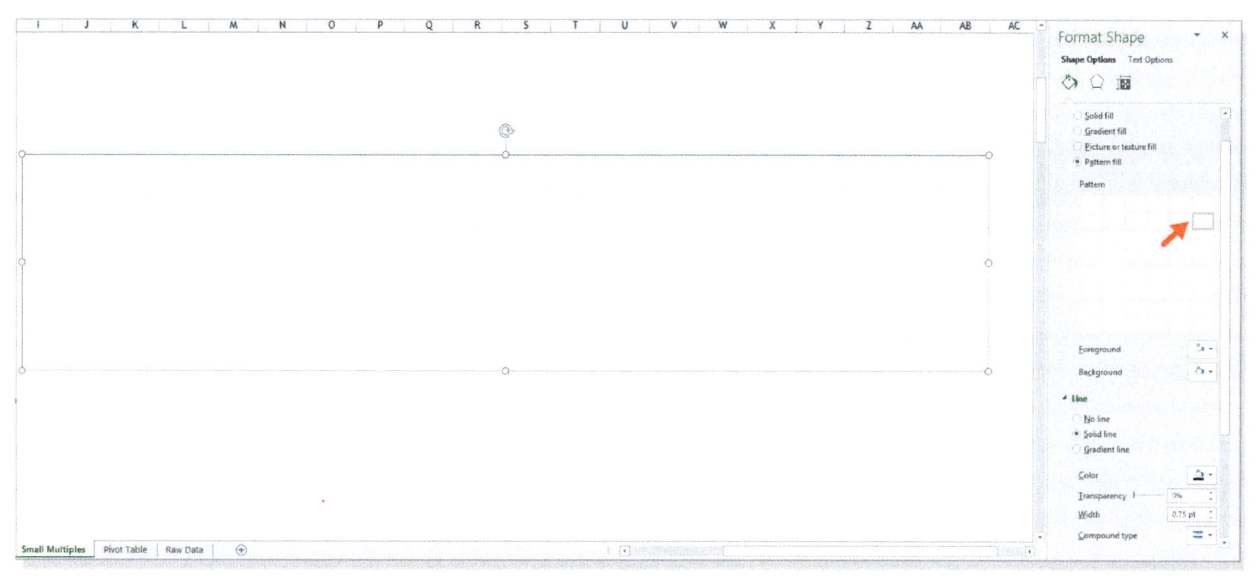

Fig 5.189: You may want to create a container for your charts.

197

CHAPTER 5 • CREATE CHARTS

Fig 5.190: Drop your charts into your container.

Tip: If you think you'll use this shape again, open your global template, paste it in, delete it, and overwrite your template file.

💬 **LEARN MORE** See the Create a Branded Template section of the Set It and Forget It chapter to learn how to create branded themes.

Step 7: Select each of your charts while pressing-and-holding the Ctrl key, then copy-and-paste them into your container. They will congregate in the top-left corner of the container **(Fig 5.190)**.

Step 8: Drag the charts where you want them placed. Then select all of them, along with the container, and navigate to **Drawing Tools ▶ Format ▶ Arrange ▶ Align ▶ Align Middle**.

Step 9: Finally, we'll distribute the charts. With the Shift or Ctrl key pressed, click the container to deselect it. Then, from the **Align** drop-down, select **Distribute Horizontally**. With your charts distributed and still selected, from the **Arrange** area of the **Format** tab, select **Group ▶ Group**. (Or use the contextual menu you get to by right-clicking on the object.) Now add the container back into your selection and choose **Align Center** to make sure the group of charts is aligned in the center of your container. You may want to group the container and charts to move them as one unit.

💬 **LEARN MORE** See the Align Strategically section of the Design with Elegance chapter to learn how to align objects in your workbook.

Variations
Popular charts included in small multiples include:
- Bars
- Columns
- Maps
- Scatter plots
- Gauges
- Pies

Warning: If you'll use pie charts, make sure they are simple. Pie charts can be difficult to compare.

Example File
small-multples.xlsx

Template
Because this chart is so highly customized and needs to be built from the ground up, you can't save it as a chart template. The best thing you can do is save a file in a custom folder of all your crazy charts that you can't save as chart templates and create a copy of it, then switch out your data with your new data. Alternatively, you could save your simplified chart as a chart template. Either of these options could at least save you a few steps.

SPARKLINES

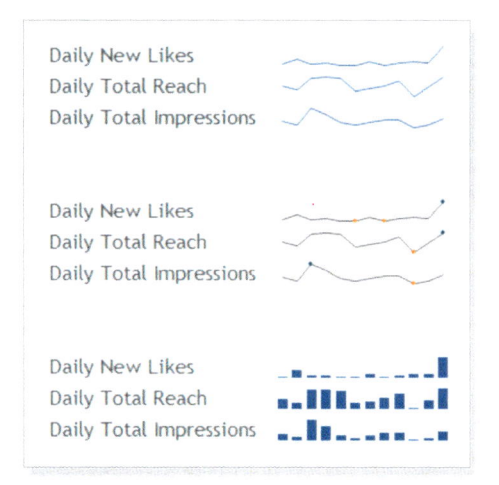

Fig 5.191: Our final chart.

AKA

You may also see sparklines called:

+ spark charts
+ in-cell charts
+ inline charts

Uses

Sparklines are best used to:

+ visualize data within a cell, such as within a table
+ plot data over time
+ compare trends, e.g., weather, sales, stocks, etc.
+ embellish text since they can be embedded inline in text editors like Microsoft Word

Tips

+ If your data is exported as individual dates and you want to aggregate them by month (or quarter), the easiest way to do that is with a pivot table. I've included a Pivot Table worksheet in the downloadable workbook to show you how I set mine up. Click anywhere in the pivot table to open the PivotTable Fields pane. From there you can easily see how I constructed it.
+ Be careful not to try to jam pack too much information into a sparkline.
+ Because a sparkline isn't an object—it's a cell background—you can add other items like text to the cell. But be careful; you can junk up your sparklines with this kind of layering.
+ You can use special formatting to show the highest and/or lowest point in a series. If this is an insight that's important to your viewer(s), I would recommend adding it. I also recommend using a different color for the high and low points and using your branded colors over red and green.
+ If your data includes negative values and you want to use a line chart (even though the Win/Loss sparkline type is clearer, in my opinion), you may want to add an axis.
+ The Win/Loss sparkline is ideal for any kind of sports data (obvs).
+ Excel offers you options for how to deal with empty cells. I usually go with the 0 option.

LEARN MORE See the Handling Blanks section of the Massage Your Data chapter to learn how to chart blanks cells.

+ If you find yourself wanting or needing more detail than sparklines afford, consider using small multiples instead. See the Small Multiples tutorial to learn how.
+ You can create sparklines by either selecting the cells that will house them first and then the data (the process the tutorial below uses) or by selecting your data first and then the sparkline type (**Insert ▶ Sparklines**).
+ If you want to delete a sparkline from a cell, you can't just select the cell and press delete. You must clear the sparkline. You can do this using the contextual menu by selecting the cell(s) and right-clicking and then navigating to **Sparklines ▶ Clear Selected Sparklines**. If you're more comfortable with the ribbon, select the cell(s) and navigate

to **Sparkline Tools ▸ Design ▸ Group ▸ Clear ▸ Clear Selected Sparklines**.

Category / Metric
1 category / 1 metric

Tutorial
Hundred-Foot View
To illustrate the difference between small multiples and sparklines, I used the same data source: my daughter's **Vinspire.me** Facebook page, **www.facebook.com/VinspireStyling**. You can see how much more Spartan the data presentation is with a sparkline over a panel approach.

Step 1: If you'll be creating multiple sparklines, you'll want to include your headers in contiguous cells because sparklines don't include labels. You can do that by either copying-and-pasting them into the cells adjacent to your sparklines or referencing them with a formula **(Fig 5.192)**.

Step 2: Select the cell you want to add the sparkline to and navigate to **Insert ▸ Sparklines ▸ Line, Column, or Win/Loss**. We'll start with a line sparkline, so select **Line**, which will open the **Create Sparklines** dialog **(Fig 5.193)**.

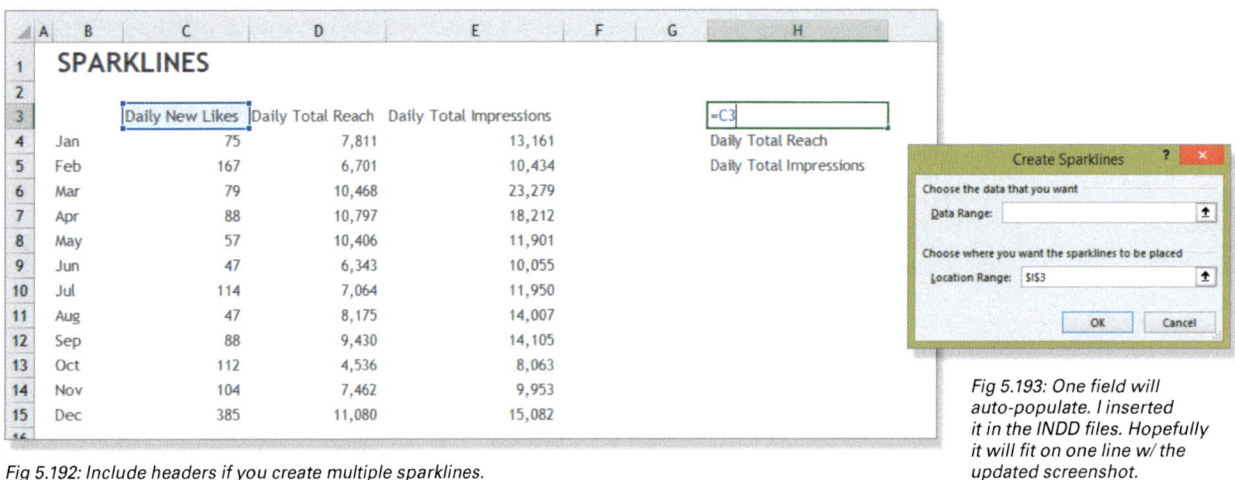

Fig 5.192: Include headers if you create multiple sparklines.

Fig 5.193: One field will auto-populate. I inserted it in the INDD files. Hopefully it will fit on one line w/ the updated screenshot.

Fig 5.194: Although this cuts against the grain, don't include category in your data range.

CHAPTER 5 • CREATE CHARTS

Step 3: In the **Create Sparklines** dialog, click-and-drag over your data range **(Fig 5.194)**.

Note: If you started by first selecting your data range, it will be populated in the dialog, and you will need to set **Location Range**.

Step 4: Choose your sparkline embellishments, such as formatting for high/low points, negative points, etc. **(Fig 5.195)**. Rinse and repeat to create new sparklines or copy-and-paste your sparkline and update the cell reference by right-clicking on the cell with the new sparkline and choosing **Sparklines ▸ Edit Single Sparkline's Data** to open the **Edit Sparkline Data** dialog.

Tip: If you have a table and want to quickly add a column with sparklines to represent each of your rows, you can employ the **Quick Analysis** button. To summon it, click-and-drag over your data (no headers). It will appear in the bottom-right corner of your selection. Select **Sparklines** along the top of the menu and your sparkline style **(Fig 5.196)**.

Example File
sparklines.xlsx

Template
Since sparklines aren't objects (like traditional charts) they can't be saved as templates.

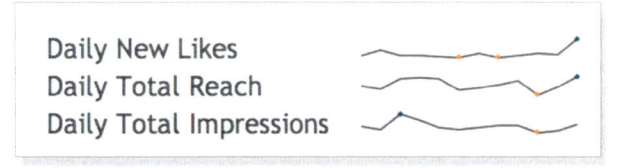

Fig 5.195: Add embellishments to your sparklines, if they help tell your data's story.

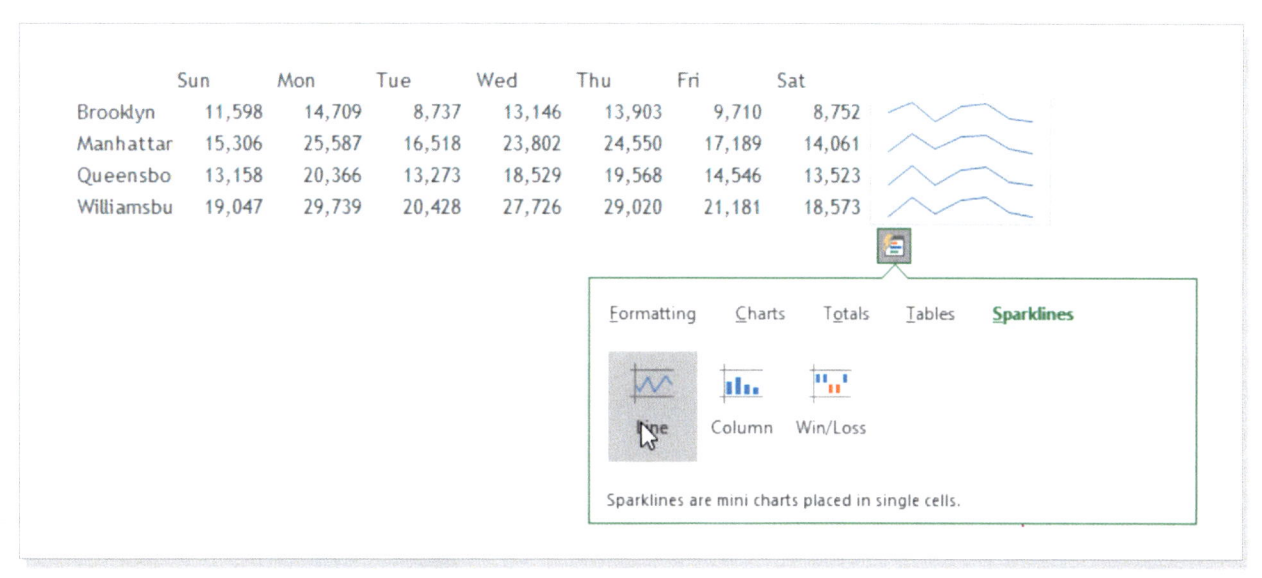

Fig 5.196: Add sparklines quickly to a table with Excel's Quick Analysis feature.

201

TREND

LINE CHART

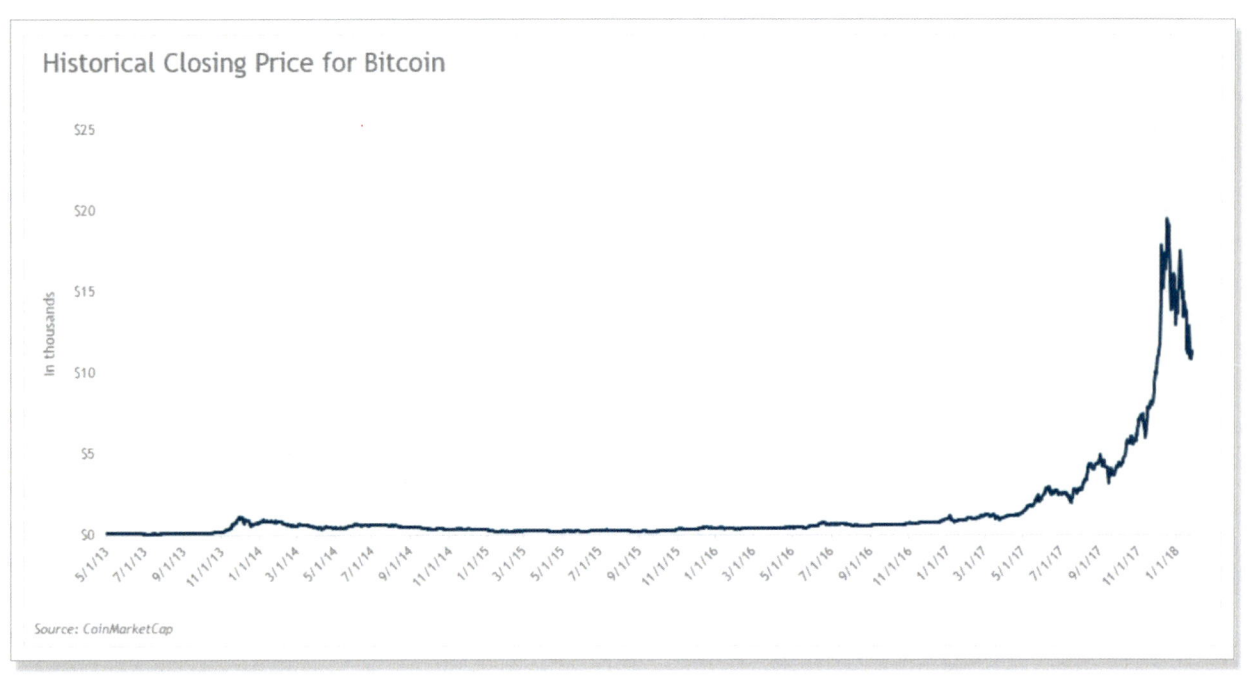

Historical Closing Price for Bitcoin

Source: CoinMarketCap

Fig 5.197: Our final chart.

Uses

Line charts are best used to:

- plot data over time
- compare multiple data series over time
- compare trends
- compare continuous variables, such as distance, revenue, incidences of an event (e.g., disease)

Tips

- Gridlines are almost always necessary with line charts, but you should consider increasing the transparency and/or changing them to a lighter color to reduce their intensity.
- If you use markers, keep them as subtle as possible to reduce the potential for distraction. If you really need them, you may want to consider using a column chart instead.

- You can add a straight line that intersects the line(s) to note a significant event (such as the redesign of a website or date a test began). If you add one, I recommend keeping the line thin and making it dotted or broken to keep it from becoming too cumbersome. You can also attach a data label to your line to document what the line marks. Alternatively, you could use a text box, but it won't move with your line, if your data updates.

- You can add a mean or target line to your chart following the technique in the Add a Mean Line variation of the Column Chart tutorial. (It's an entirely different technique from the bar chart.)

- I don't recommend stacking lines using the Stacked Line chart (**Insert ▸ Charts ▸ Insert Line or Area Chart ▸ Stacked Line**). It's difficult to interpret. If you want to stack your lines to demonstrate how

CHAPTER 5 • CREATE CHARTS

each data series contributes to the whole, I recommend using the Stacked Area chart (**Insert** ▸ **Charts** ▸ **Insert Line or Area Chart** ▸ **2-D Area** ▸ **Stacked Area**).

Category / Metric
1 category / 1+ metrics

Tutorial
Step 1: Select any cell inside your dataset and navigate to **Insert** ▸ **Charts** ▸ **Insert Line or Area Chart** ▸ **2-D Line** ▸ **Line (Fig 5.198)**.

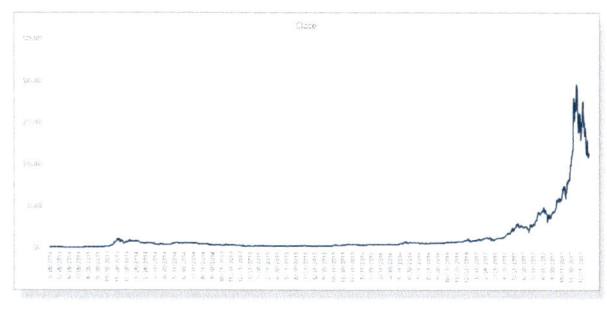

Fig 5.198: Our initial chart.

Step 2: If your horizontal axis is as thick as mine is, you can thin it out by selecting it and pressing Ctrl-1 to open the **Format Axis** pane. When Excel detects your horizontal axis consists of dates, you will see date-oriented options in the **Axis Options** tab. To reduce the number of dates in the axis by half, set **Major Units** to 2 **(Fig 5.199)**.

Step 3: Clean up your chart to taste. I cleaned mine by taking the following steps:

✦ made title more descriptive
✦ bumped up title font size
✦ added annotation for data source
✦ bumped up transparency of gridlines
✦ simplified number formatting in vertical axis
✦ added axis title indicating the simplification

💬 **LEARN MORE** See the Design with Elegance chapter to learn how to perform these refinements.

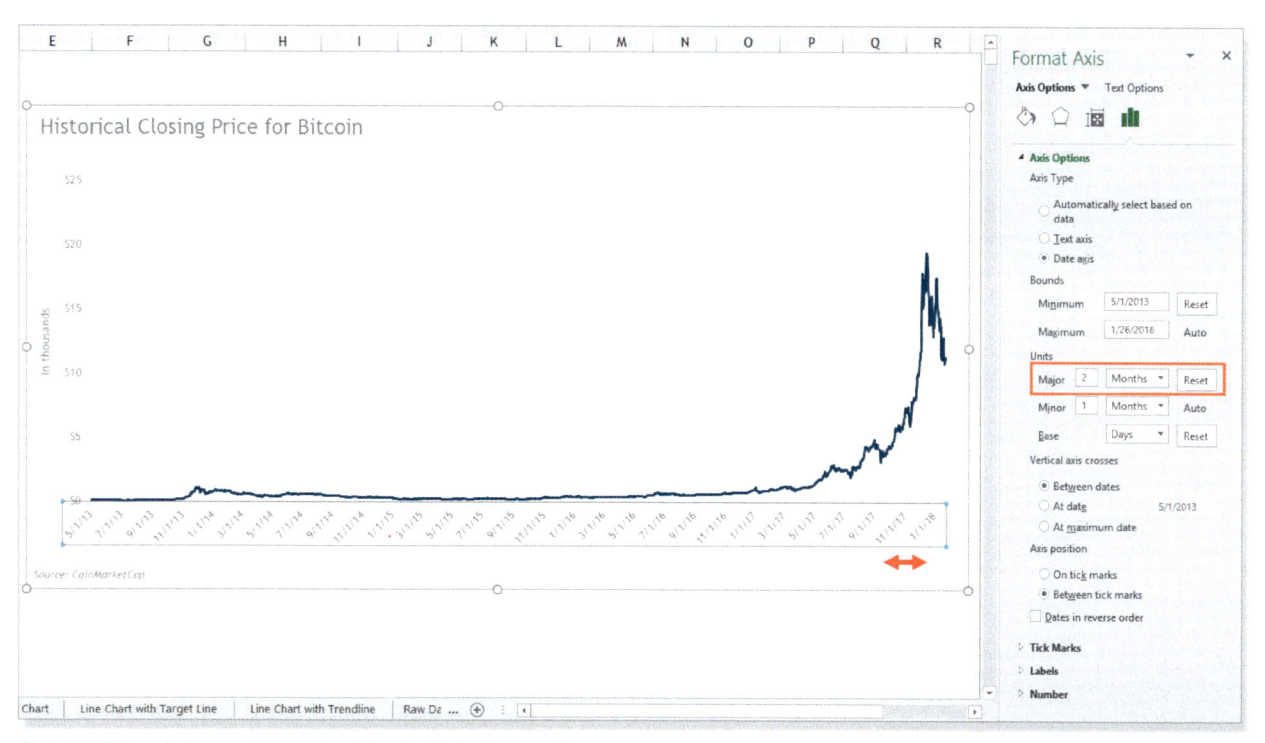

Fig 5.199: Thin out a busy axis by customizing the Major Units setting.

203

Variations
Add Sizzle

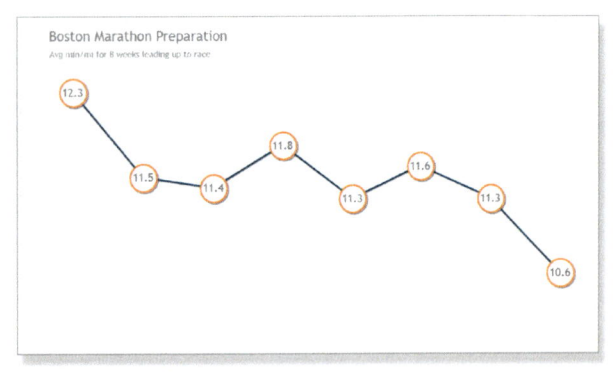

Fig 5.200: Our final chart.

Hundred-Foot View

Infographics have engendered an upending in the way people approach data visualization. Once you're comfortable with the many options Excel gives you to augment your charts, it quite easy to put even a fairly humdrum chart like the line chat in Stilettos.

One very popular option is to add circles to the data points. Supes simple. But if you're going do that, you don't want an axis outing your line chart as a basic bro. This got the gerbils running in their cage, and I thought to myself, *Hmm…I wonder if I could add data labels to the markers*. And whudduyaknow? You sure can! Marked that as a no brainer and sent my axis divorce papers. #itsnotme #itsyou #sobasic

The only thing I didn't like is it seemed a little flat. My internal editor was like, *Why is your right hand hovering over that drop shadow?!? Don't even think about it. You've staked your reputation on this book!* But I ignored her and added an ever-so-slight (barely noticeable, really) drop shadow to just the markers. Go ahead and judge me. I already got your money!

Step 1: Select any cell inside your dataset and navigate to **Insert ▸ Charts ▸ Insert Line or Area Chart ▸ 2-D Line ▸ Line (Fig 5.201)**.

Step 2: Excel freaks out a bit when you use a category that contains numbers, dates excepted. It's like, "Abort mission! Repeat: Abort mission!" and adds whatever is in your far-left column to both the axis and the plot. Usually, removing the header for that column resolves the issue, and Excel's blood pressure returns to a safer pace. If this happens to you and removing the label doesn't resolve the issue,

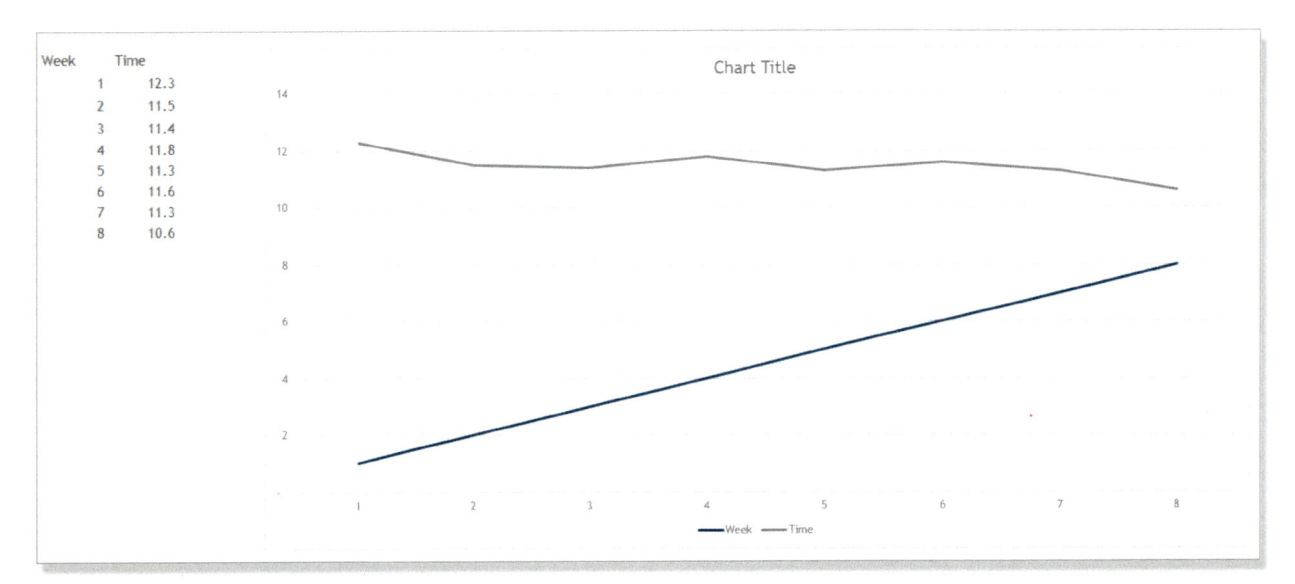

Fig 5.201: Houston, we have a problem.

right-click on the chart and choose **Select Data** from the contextual menu. In the **Horizontal (Category) Axis Labels** section of the **Select Data Source** dialog you can see that Excel has rightly interpreted my leftmost column as the source for my horizontal axis labels, but it still added it as a series. Why? Your guess is as good as mine! But if you just select it and click the **Remove** button, then click **OK**, you'll be back in business **(Fig 5.202)**! Excel will also update your chart title to the column header for your data series.

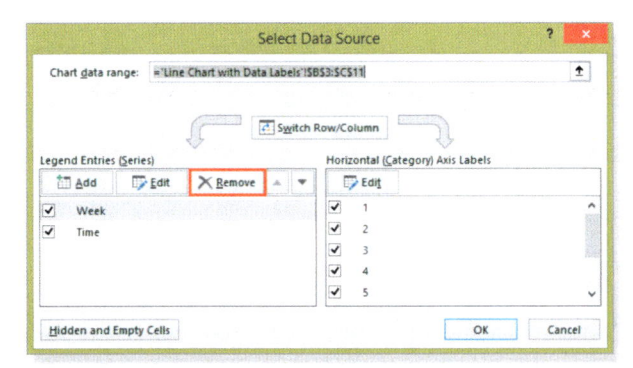

Fig 5.202: Excel freaks out when you use numbers in your category column.

Step 3: Get rid of everything: gridlines, legend, and both axes **(Fig 5.203)**.

Step 4: Next, we'll add some markers. Select the line, then press Ctrl-1 to open the **Format Data Series** pane. Open the **Fill & Line** tab and select **Marker (Fig 5.204)**. I used the following settings:

+ **Marker Options**: Built-in
+ **Type**: Circle
+ **Size** 36 pt

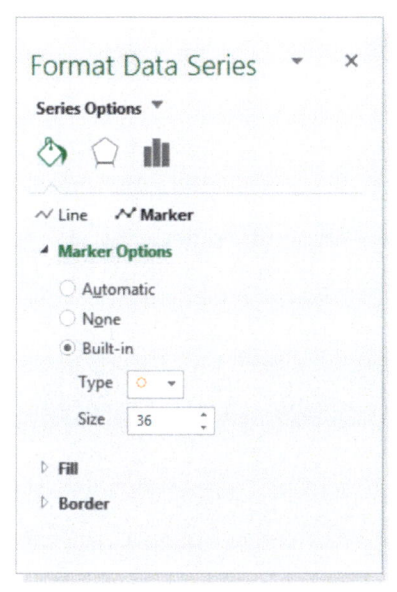

Fig 5.204: Add circle markers to data points.

Fig 5.203: Completely useless line charts FTW!

205

CHAPTER 5 · CREATE CHARTS

Step 5: Now we'll add fill and a border to our markers **(Fig 5.205)**. I used the following settings:

+ **Fill**: **Solid fill**
+ **Fill Color**: white
+ **Border**: **Solid line**
+ **Outline color**: orange
+ **Width**: 2.5 pt

Step 6: Moving our attention to our lines, click **Line** at the top of the **Line & Fill** tab **(Fig 5.206)**. I used the following settings:

+ **Line**: **Solid line**
+ **Outline color**: blue
+ **Width**: 2.5 pt

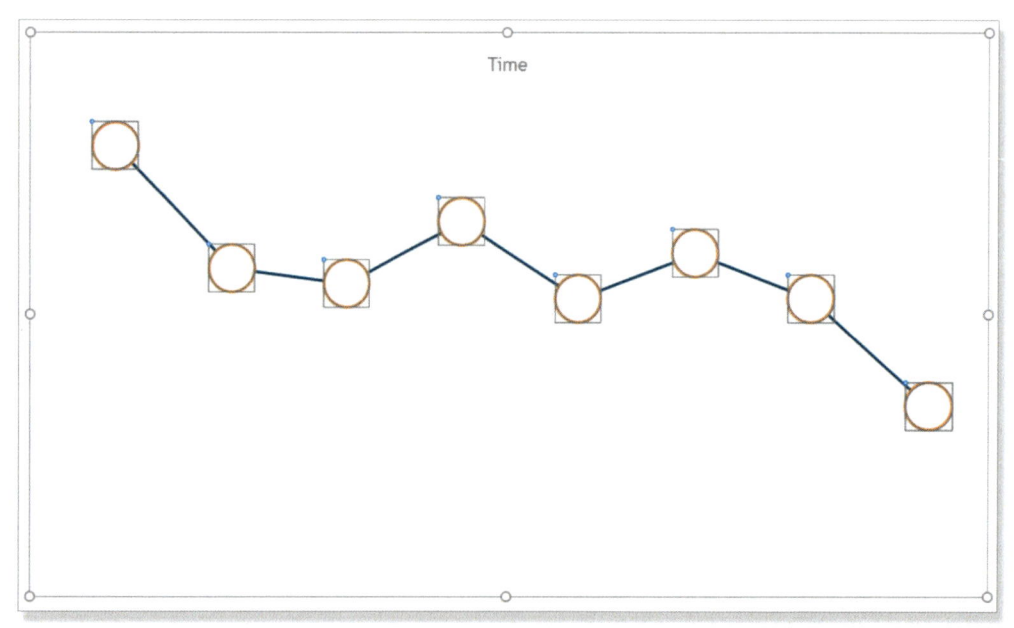

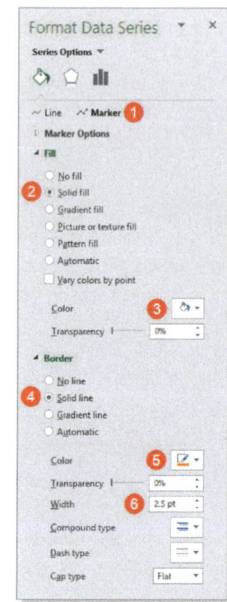

Fig 5.205: Bump up the Width of your markers and add a fill and border to get them infographic ready.

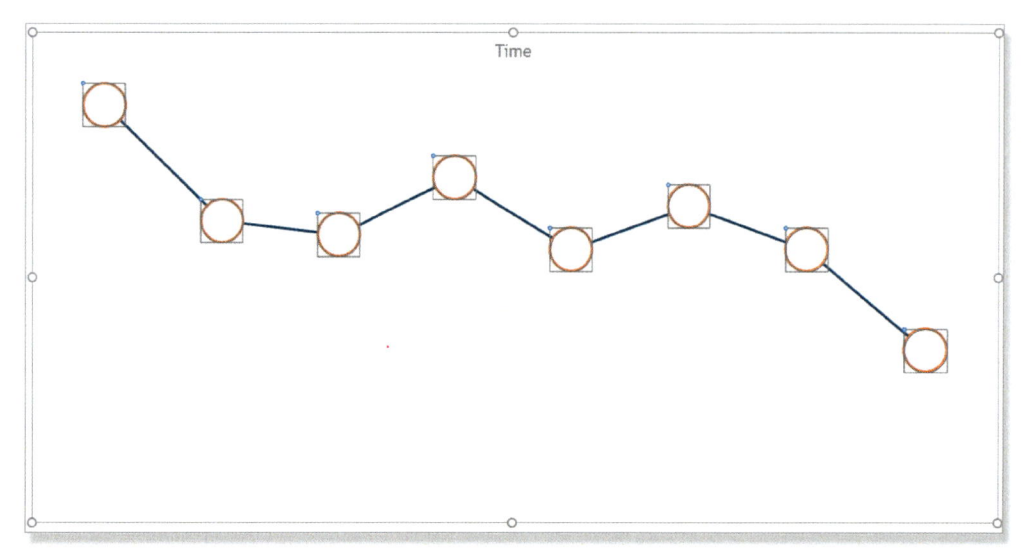

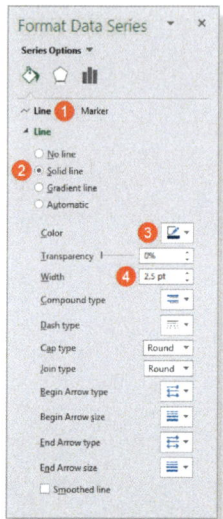

Fig 5.206: Format your line.

206

Step 7: We can't have a line chart with all form and no function, so let's add some data labels to our markers. With the line still selected, open the **Chart Elements** flyout menu and click the right-facing arrow next to **Data Labels** to open the flyout menu. Select **More Options** to open the **Format Data Labels** pane. Open the **Label Options** tab. Set **Label Position** to **Center**. Format text size to taste. I bumped **Font Size** up to 14 pt **(Fig 5.207)**.

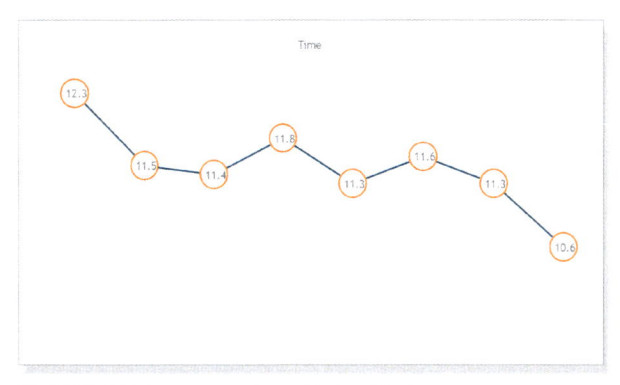

Fig 5.207: Add data to your chart with data labels.

Step 8: The alignment of the numbers in the data labels is off. Some number codes can cause this misalignment. To adjust it, open the **Label Options** tab and open the **Number** section. You can try different number codes from the **Type** menu or write custom code and drop it into the **Format Code** field **(Fig 5.208)**. I went with the latter option and used the code below:

0.0;-0.0;0

> 💬 **LEARN MORE** See the Fix Alignment section of the Design with Elegance chapter to learn how to align numbers.

Step 9: Okay, let #shadowgate begin. Call me crazy, but our infographic line chart just seems to lack dimension. (Probably because, as a two-dimensional chart, it does.) I decided to add a subtle drop shadow to the line and markers. To play,

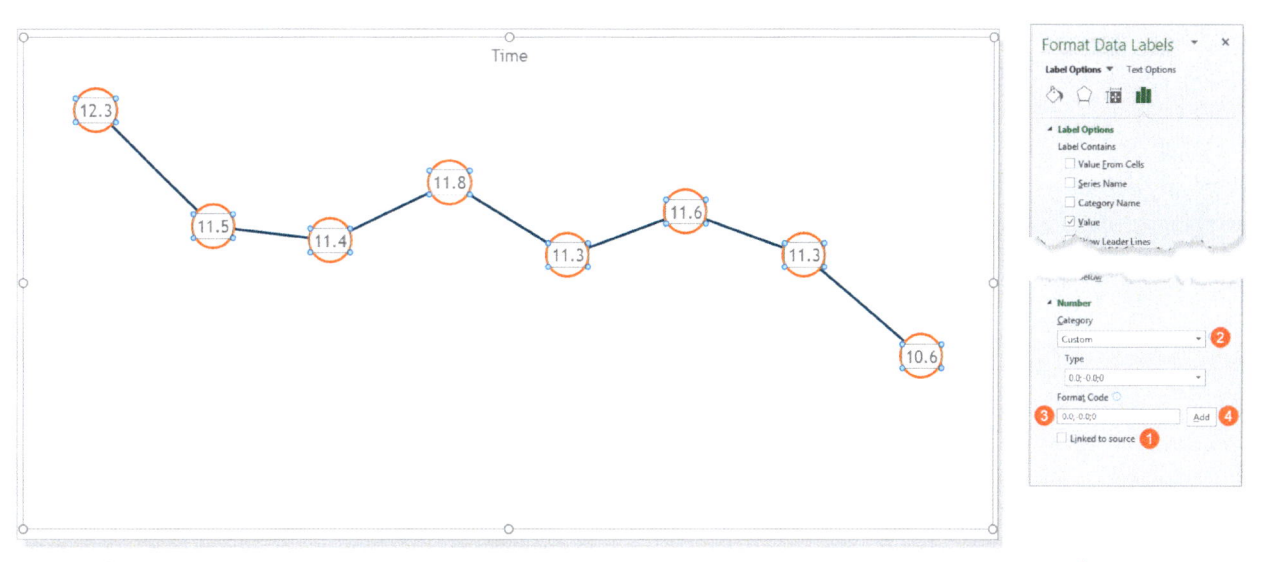

Fig 5.208: Customize your number formatting to center text in data labels.

CHAPTER 5 • CREATE CHARTS

select your line, which will open the **Format Data Series** pane. Open the **Effects** tab. Under **Shadow**, set **Distance** to a small number. I used 1 pt. I also set **Blur** to 1.2 pt **(Fig 5.209)**.

Step 10: Clean up your chart to taste. I cleaned mine by taking the following steps:

* made title more descriptive
* left aligned chart title
* bumped up title font size
* added subtitle
* tweaked the Minimum Bounds of my vertical axis (to minimize excessive white space in the chart)

💬 **LEARN MORE** See the Design with Elegance chapter to learn how to perform these refinements.

Fig 5.209: Let's add just a touch of shadow to our line and markers.

Add a Target Line
Hundred-Foot View

Sometimes you want to add a target or mean line to your line chart to add insight and provide context for your data. Unlike adding a target or mean line to a bar chart (see the Add a Mean Line variation of the Bar Chart tutorial to see the black magic you need to employ), augmenting a line chart with one—or several in my case—is pretty straightforward. You just need to add a column for each of your values. Since the target or mean values are static, we'll get a horizontal line that stretches across our chart. Then all we need to do is customize it to make it more finessed and less obtrusive.

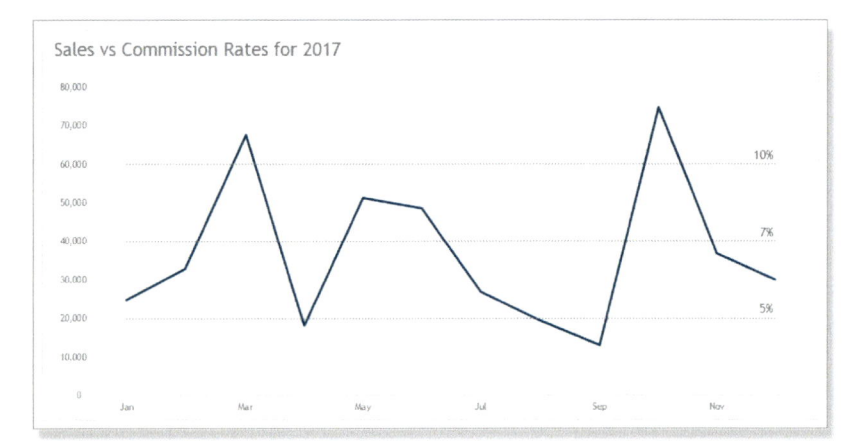

Fig 5.210: Our final chart.

Warning: If you want to save yourself major headaches, make sure you don't have a header in your first column. If you do, you could waste quite a bit of time with cleanup **(Fig 5.211)**!

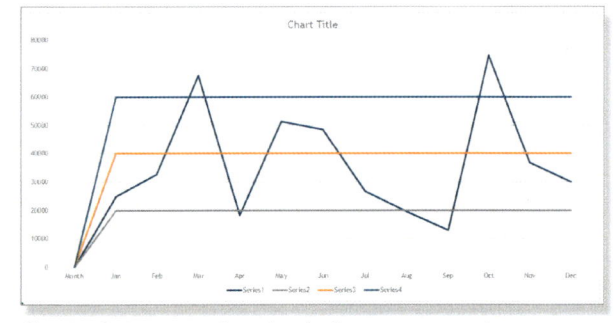

Fig 5.211: Delete your column header for your category data.

208

CHAPTER 5 • CREATE CHARTS

Step 1: Select any cell inside your dataset and navigate to **Insert ▶ Charts ▶ Insert Line or Area Chart ▶ 2-D Line ▶ Line (Fig 5.212)**.

Step 2: Because I added three target lines to my chart, I took a calculated risk and made an executive decision to remove gridlines altogether. If you're only adding one target line, you may choose to keep gridlines. If you do, I'd recommend lightening them by increasing the transparency.

Step 3: It's not necessary for my target lines to each get their own color. So, if you have more than one, I recommend using one color—preferably a subtle one that doesn't make you feel like you're getting punched in the face by a Care Bear. To get started, select one of your lines and press Ctrl-1 to open the **Format Data Series** pane. Open the **Fill & Line** tab to set your line formatting. Under **Line**, I used the following settings:

+ **Outline color**: gray

+ **Width**: 1.25 pt

+ **Dash Type**: **Round Dot**

Rinse and repeat with any other target lines you have, if you have multiple target lines like the chart in the example file **(Fig 5.213)**.

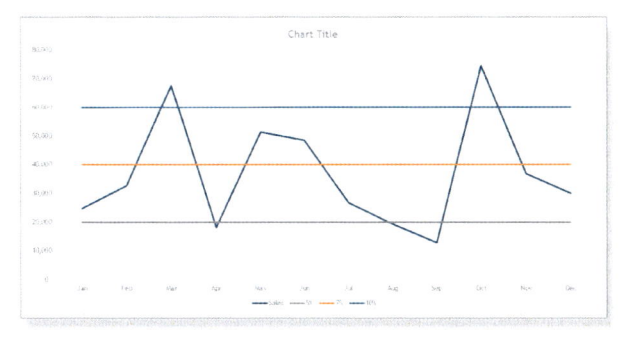

Fig 5.212: Our initial chart.

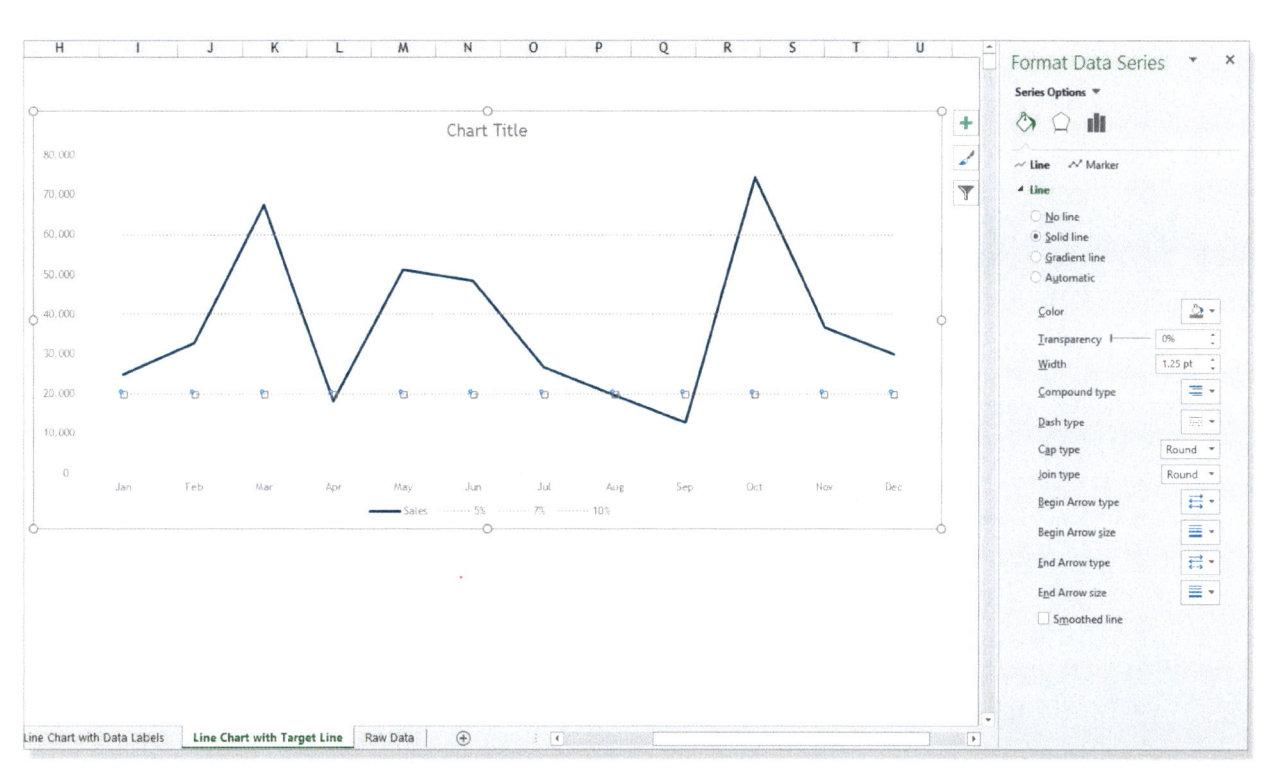

Fig 5.213: I have a gravitational pull toward target lines that are light gray and dotted.

209

CHAPTER 5 • CREATE CHARTS

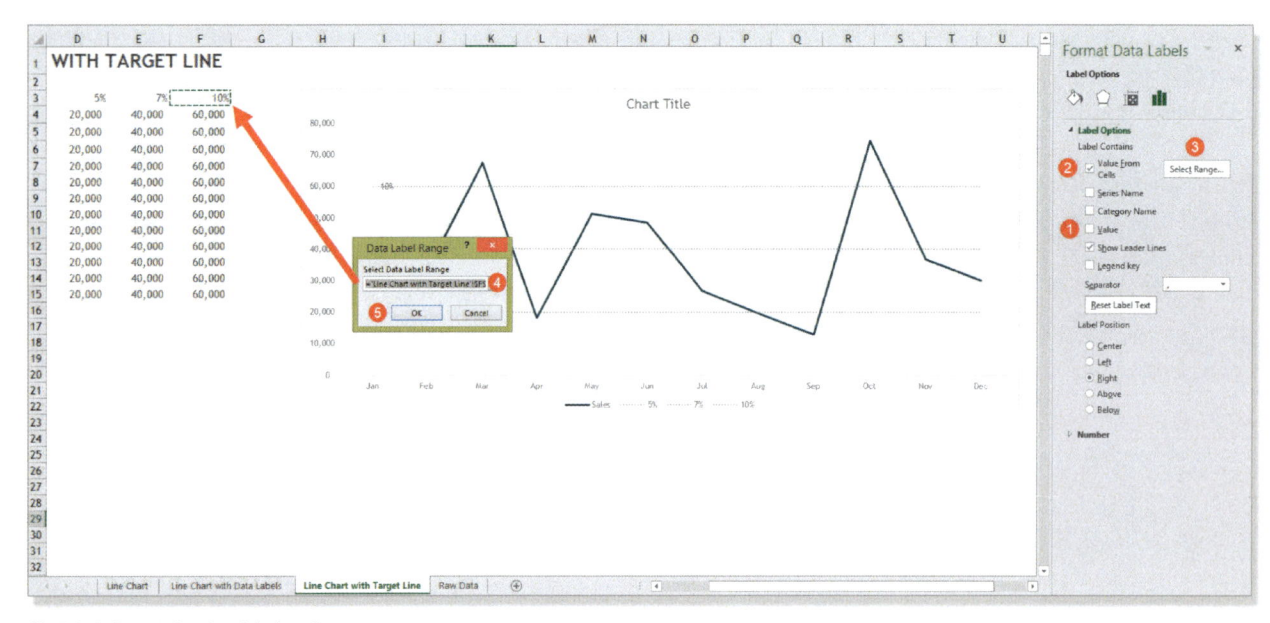

Fig 5.214: Customize data label options.

Step 4: Now let's add some data labels to our target lines. Select your first target line and open the **Chart Elements** flyout menu. Click the right-facing arrow next to **Data Labels** to open its flyout menu, and choose **More Options** to open the **Format Data Labels** pane. We don't need multiple data labels with the same value, so we'll need to go to town in the **Label Options** tab. First, under **Label Contains**, deselect **Values**. Next, select **Value From Cells**, then click the **Select Range** button to the right. In the **Data Label Range** dialog, set the range to the column header for your target **(Fig 5.214)**.

Step 5: We don't need the leader line that tethers the data label to where it first appeared along your

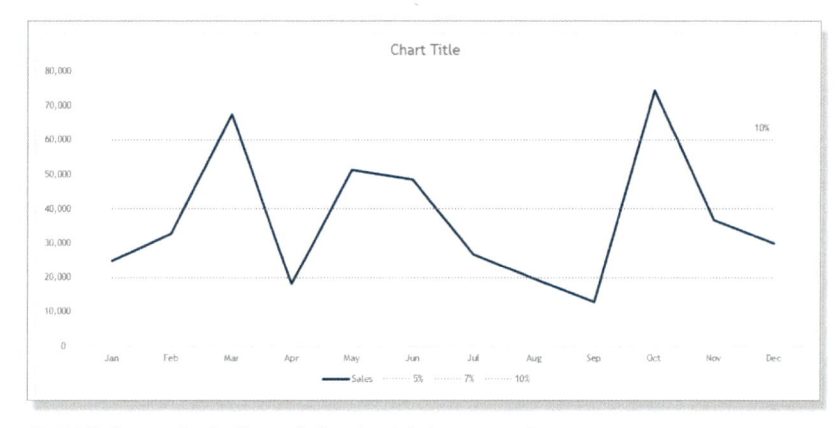

Fig 5.215: Remove leader line and align data label to your preference.

line, so deselect **Show Leader Lines**. Now you can click-and-drag your data label wherever you want without anyone being the wiser. I set **Label Position** to **Above**, then Shift-dragged it to the right side of my line to move it along a horizontal plane. If you only have one target line, you can just drag it without holding the Shift key. I also right aligned the text in my data label so they align regardless of how many characters are in the data label **(Fig 5.215)**. Data labels are center aligned by default.

Warning: Once you move the data label your Above setting will disappear. It's fine. Keep calm and carry on.

CHAPTER 5 · CREATE CHARTS

Step 6: Rinse and repeat with your other data labels. My eyes trip over items in a group that aren't properly aligned in a chart—so I go to admittedly rather extreme lengths to align my data labels. It would be more optimal if Excel allowed you to select each of your data labels and align them in the **Format Data Labels** pane. But that's not possible, so I drag out a thin, vertical line and use the **Align to Shape** option to sidle my data labels up to it. Delete the line when you're finished **(Fig 5.216)**.

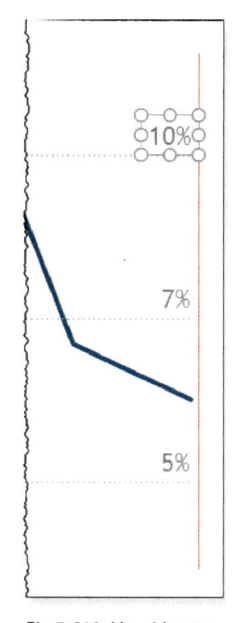

Fig 5.216: Yes, I know there's a clever term for people who go to such extreme measures to align their data labels.

Step 7: Now that we have data labels, we don't need the legend, so you can delete it.

Step 8: When I have months or days of the week, I give my horizontal axis some breathing room by specifying the interval between the labels. To access this setting, select the horizontal axis to open to the **Format Axis** pane. Navigate to **Labels ▶ Interval between labels ▶ Specify interval unit**. I set it to 2 **(Fig 5.217)**. You may wonder why Excel is treating my axis as text (if you have an eagle eye). It's because, in the sample data, months came in hard-coded as abbreviations, so Excel treats them as text.

Note: If I had been working with actual dates, I would have aggregated them into months using a pivot table and used custom number formatting (mmm) to get abbreviated months. Then I would have date-specific formatting options and would have set **Major Units** to 2.

💬 LEARN MORE See the Align Strategically section of the Design with Elegance chapter to learn how to align objects in your workbook.

Fig 5.217: Thin axis by specifying interval units.

211

CHAPTER 5 • CREATE CHARTS

> 💬 **LEARN MORE** See the Thin Out Axes section of the Design with Elegance chapter to learn how to reduce clutter in your axes.

Step 9: Clean up your chart to taste. I cleaned mine by taking the following steps:

- left aligned chart title
- made title more descriptive
- bumped up title font size

> 💬 **LEARN MORE** See the Design with Elegance chapter to learn how to perform these refinements.

Add a Trendline

It can be helpful to add a trendline to your chart, especially with line charts where sometimes the trend can be obfuscated if the data is erratic. To add one, right-click on the data series and choose **Add Trendline** from the contextual menu (or select it from the **Chart Elements** flyout menu).

The linear trendline is the easiest to wrap your mind around, but you should investigate the other options. The polynomial trendline is infinitely more powerful in demonstrating fluctuations in trends, especially with larger datasets, as you'll see in the Scatter Plot tutorial. A linear trendline indicates that social shares are negatively correlated with word count (meaning visitors were less inclined to share longer articles). But when I enabled a polynomial trendline, you see that at around 1,500 words social shares increased slightly. With that in mind, you may want to experiment with both before jumping to conclusions. Learn more about trendlines from the **office.com** website: **bit.ly/mdsepc-trendlines**.

Example File
line-chart.xlsx

Templates
If you'll use these charts again, save them as chart templates. Or download the templates included in the chart templates folder:

- Line chart: mds-line.crtx
- Sizzle line chart: mds-line-sizzle.crtx
- Line chart with target line: mds-line-target.crtx

Note: Before creating the line with target line chart template, I removed two of the target lines and removed the data labels. Without taking these steps, the template was unusable because it was too customized.

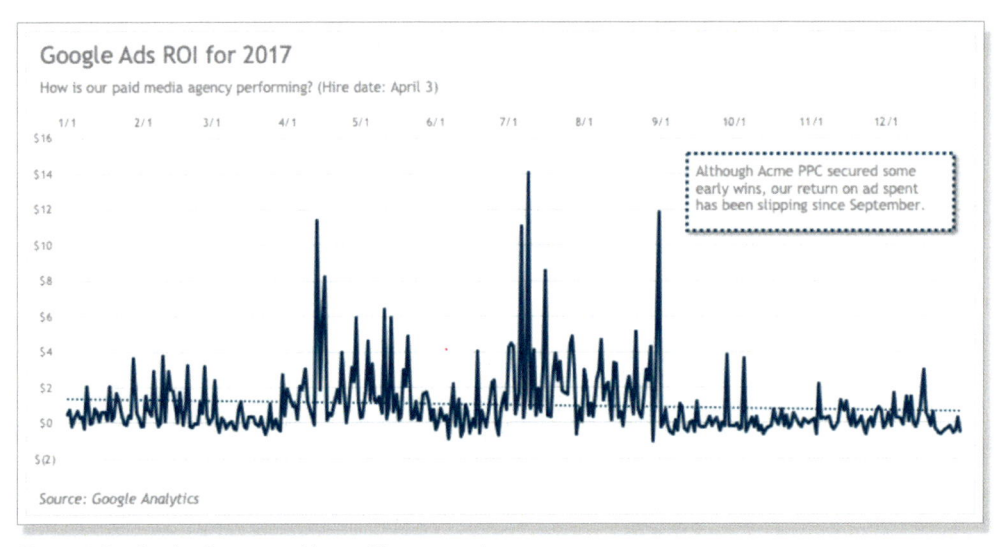

Fig 5.218: Excel makes it easy to add a trendline to your chart.

CHAPTER 5 • CREATE CHARTS

BAND CHART

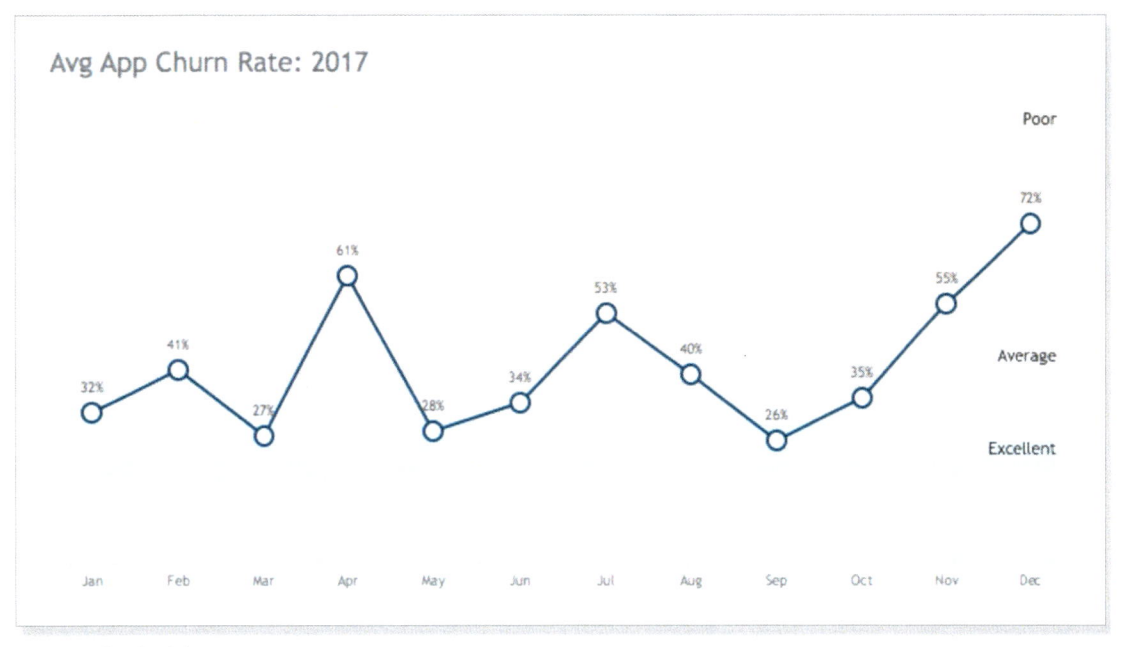

Avg App Churn Rate: 2017

Fig 5.219: Our final chart.

AKA
You may also see band charts called:
+ band-in-line charts
+ KPI line charts

Uses
Band charts are best used to:
+ plot data with confidence levels (in the case of confidence bands)
+ plot trend data that contains a target range
+ visualize data from customer satisfaction surveys

Tips
+ In my experience, band charts are used more with percentages than whole numbers. Target lines tend to be used more with whole numbers. (See the Add a Target Line variation of the Line Chart tutorial.)
+ Avoid band color(s) that are too obtrusive. The emphasis should be on the data; the band(s) are there to provide context.
+ If you're using bands to indicate levels of satisfaction, avoid colors that could be misinterpreted (e.g., red).
+ The more bands you want, the more metrics you'll need in your dataset.
+ Consider limiting your band colors to tints of your theme colors.

Category / Metric
1 category / 3+ metrics

Tutorial
Hundred-Foot View
We'll use a combination chart to corral our data. The app churn data will be set to a line chart, and the levels will be set to a stacked column chart that will add up to 100%. To create the bands, we'll set the gap width of our columns to 0%, causing them to touch. This will give the appearance of bands.

213

CHAPTER 5 • CREATE CHARTS

Step 1: Before we get started, we need to structure our data for the job. Since I'm creating bands to set thresholds for three levels of app churn rate (Excellent, Average, and Poor), my percentages needed to add up to 100%. Enter the value you want in the first row, then double-click the bottom-right corner of the cell to send it down the column. (Excel's default auto-fill option is copy, and it's faster than dragging.) Rinse and repeat for any additional columns **(Fig 5.220)**.

	Avg Churn Rate	Excellent	Average	Poor
Jan	32%	30%	20%	50%
Feb	41%	30%	20%	50%
Mar	27%	30%	20%	50%
Apr	61%	30%	20%	50%
May	28%	30%	20%	50%
Jun	34%	30%	20%	50%
Jul	53%	30%	20%	50%
Aug	40%	30%	20%	50%
Sep	26%	30%	20%	50%
Oct	35%	30%	20%	50%
Nov	55%	30%	20%	50%
Dec	72%	30%	20%	50%

Fig 5.220: Structure your data to cause it to stretch across your chart.

Step 2: Select any cell inside your dataset and navigate to **Insert ▶ Charts ▶ Insert Combo Chart (Fig 5.221)**. This initial chart is not for the faint of heart.

Step 3: Let's begin the cleanup effort by changing the levels (Excellent, Average, and Poor) to a stacked column chart. Right-click the chart and choose **Change Chart Type**. In the **Change Chart Type** dialog, set the data series that will define your bands to **Stacked Column** and your main data series (or multiple series) to **Line (Fig 5.222)**.

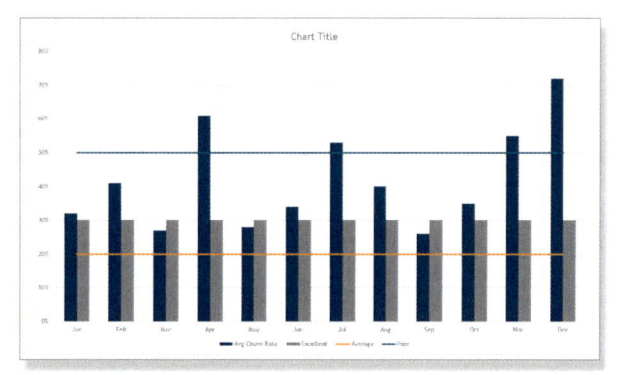

Fig 5.221: Our initial chart.

Step 4: Select your first data series and press Ctrl-1 to open the **Format Data Series** pane. Open the **Fill & Line** tab and set **Fill Color** to a lighter color. I also recommend bumping up **Transparency**. I set mine to 70% **(Fig 5.223)**. Rinse and repeat with the remaining data series in your stacked columns.

Step 5: I don't want the gridlines to cramp my style, so I deleted them **(Fig 5.224)**.

Step 6: Now we'll merge the columns to create the impression of bands. To start, select one of the data series, which will open the **Format Data Series** pane. Open the **Series Options** tab, if it doesn't open automatically, and set **Gap Width** to 0% **(Fig 5.225)**. Setting it for one will adjust it for all the series. What's good for the goose….

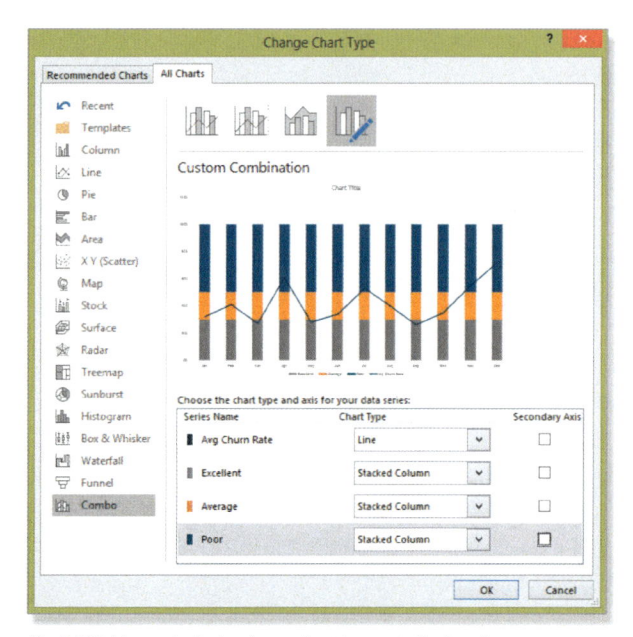

Fig 5.222: Use a stacked column chart to create the bands.

214

CHAPTER 5 • CREATE CHARTS

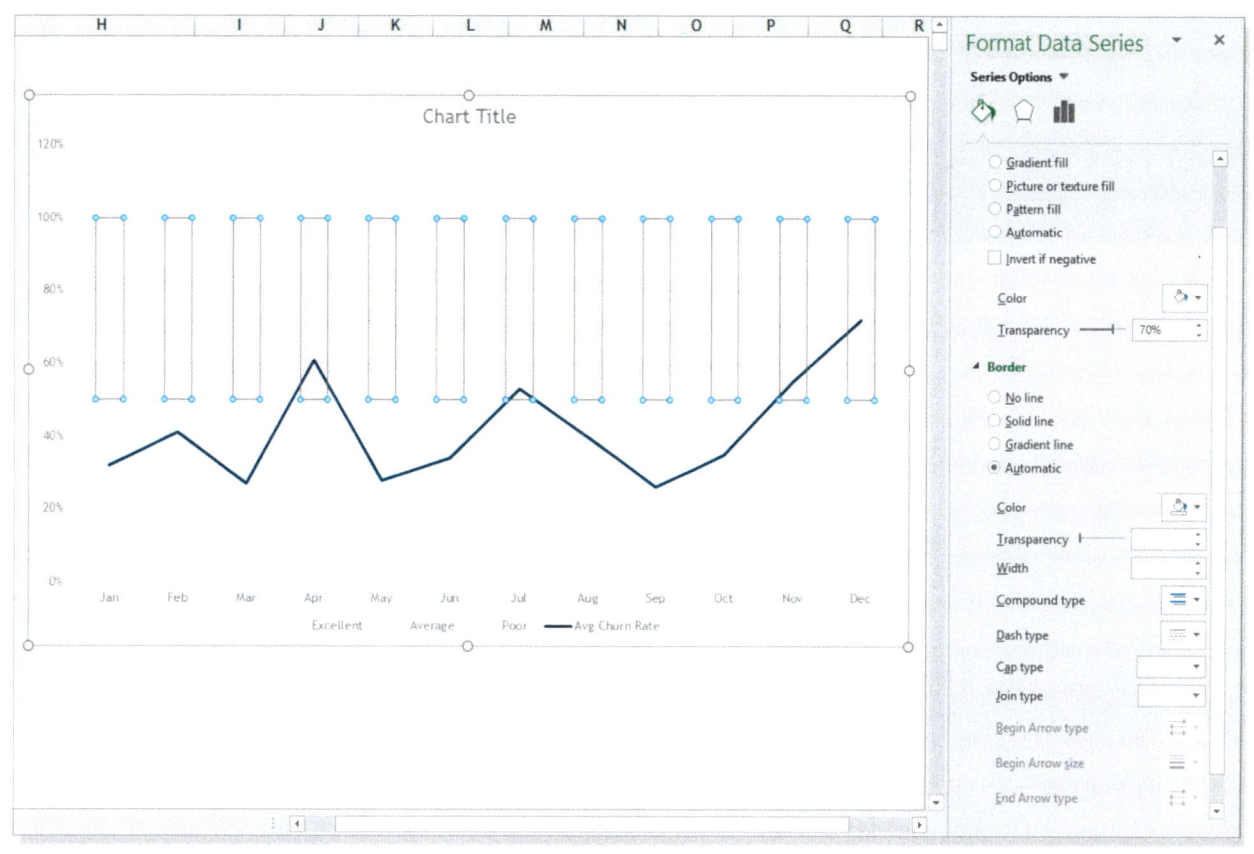

Fig 5.223: Adjust the fill and transparency of your data series to dial back the intensity.

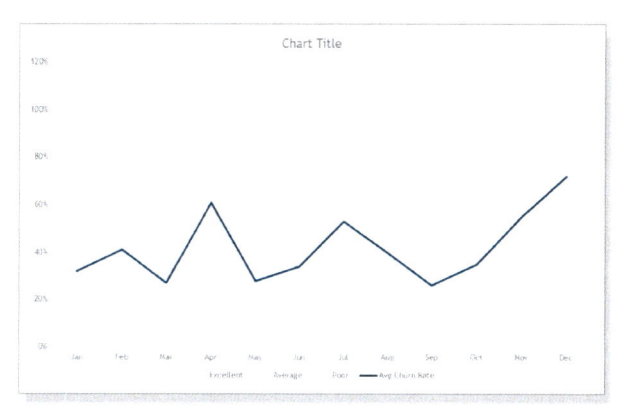

Fig 5.224: Buhbai, gridlines!

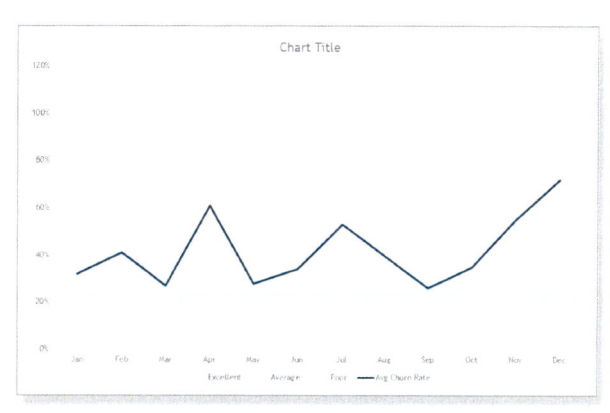

Fig 5.225: Set Gap Width to 0% to mash your bands together.

CHAPTER 5 · CREATE CHARTS

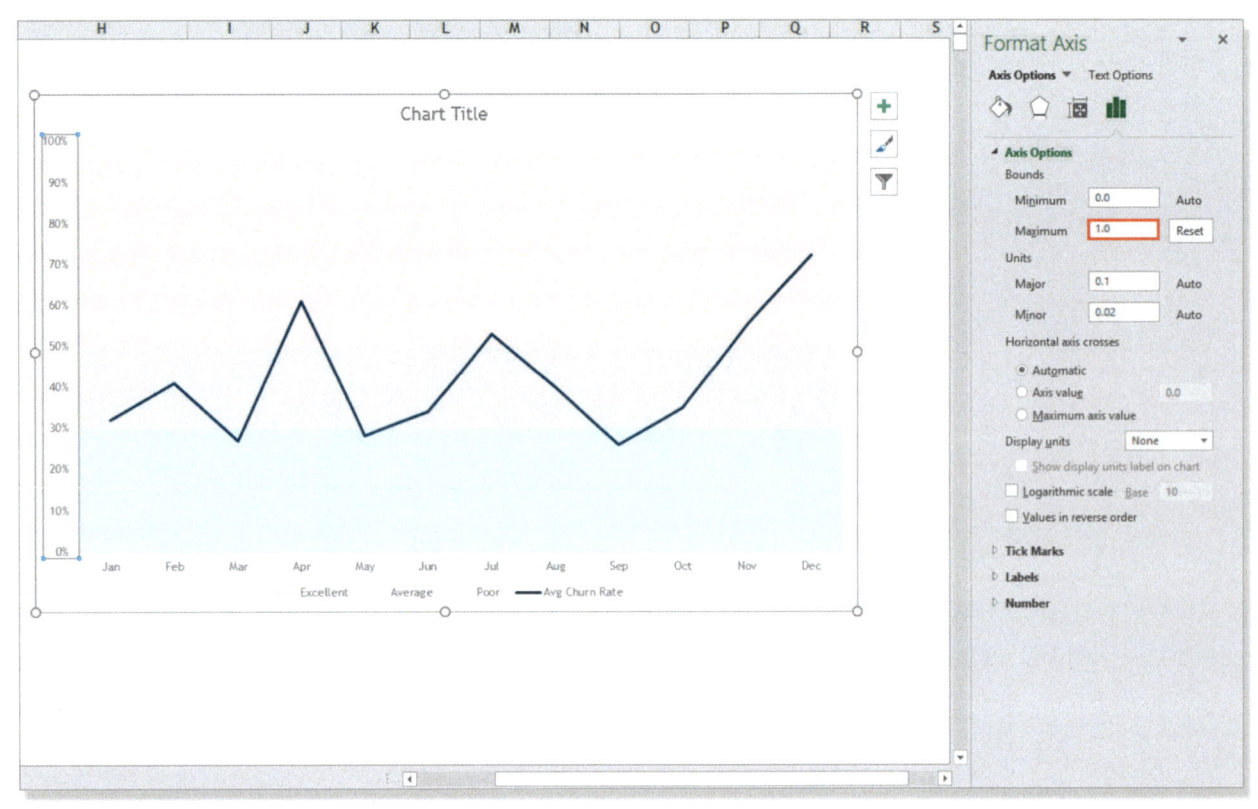

Fig 5.226: Set Maximum Bounds to your max value.

Step 7: We want the bands to fill our chart, so we need to set the max value of the vertical axis to 100% (or whatever your upper limit is). Select the axis to open the **Format Axis** pane. In the **Axis Options** tab, under **Bounds**, set **Maximum** to 1 **(Fig 5.226)**.

Step 8: Although I normally don't care for how busy data labels are in a line chart, it really doesn't work to have gridlines in a band chart, especially if the demarcations between the bands don't line up with the gridlines. To add data labels to your chart, select the line and open the **Chart Elements** fly-out menu. Next, click the right-facing arrow next to **Data Labels** and choose **More Options** to open the **Format Data Labels** pane. Open the **Text Options** tab and set **Label Position** to **Above (Fig 5.227)**.

Step 9: Delete the vertical axis, then drag the sizing handle on the left side of the **Plot Area** to fill the space **(Fig 5.228)**.

Step 10: To soften the blow of adding all those data labels, I added markers and formatted them following the technique I laid out in steps 4 – 6 of the Add Sizzle variation of the Line Chart tutorial. If you have room in your chart, you can embed your data labels in your markers **(Fig 5.229)**.

Step 11: You could use data labels to indicate the different levels, but aligning text boxes is easier so I went with that option (**Insert ▸ Text ▸ Text Box**). In the **Format Shape** pane, under the **Fill & Line** tab, I set **Fill** to **No fill** and **Line** to **No line.** Then I right aligned the text in my text box. When I finished formatting one, I copied and pasted it and switched out the text **(Fig 5.230)**.

CHAPTER 5 • CREATE CHARTS

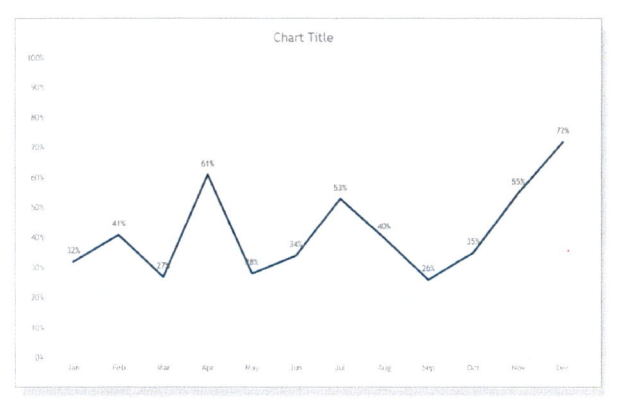

Fig 5.227: Add data labels so you can delete your gridlines.

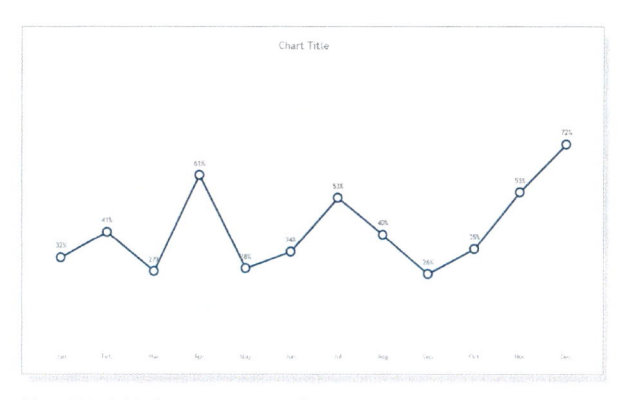

Fig 5.229: Add pizzazz to your markers.

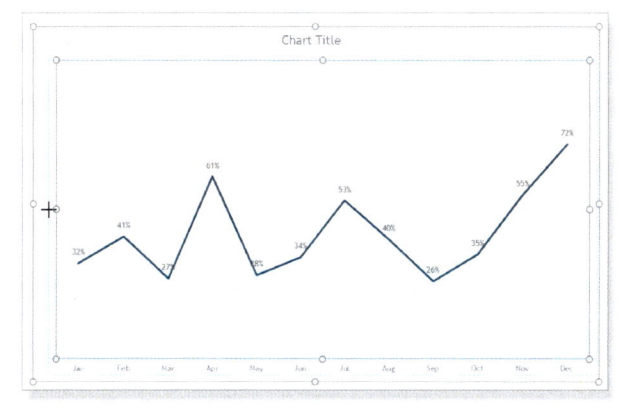

Fig 5.228: We have no need for the vertical axis.

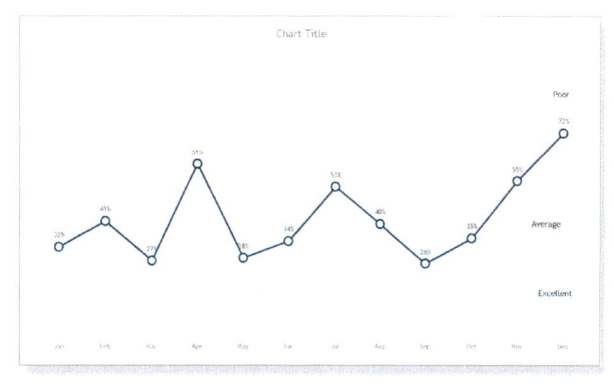

Fig 5.230: Create text boxes to serve as labels for the bands.

Step 12: To align the text boxes, select one, then press-and-hold the Shift or Ctrl key and select the others. Navigate to **Drawing Tools ▶ Format ▶ Arrange ▶ Align ▶ Align Right**. I used a rectangle to line everything up **(Fig 5.231)**.

LEARN MORE See the Align Strategically section of the Design with Elegance chapter to learn how to align objects in your workbook.

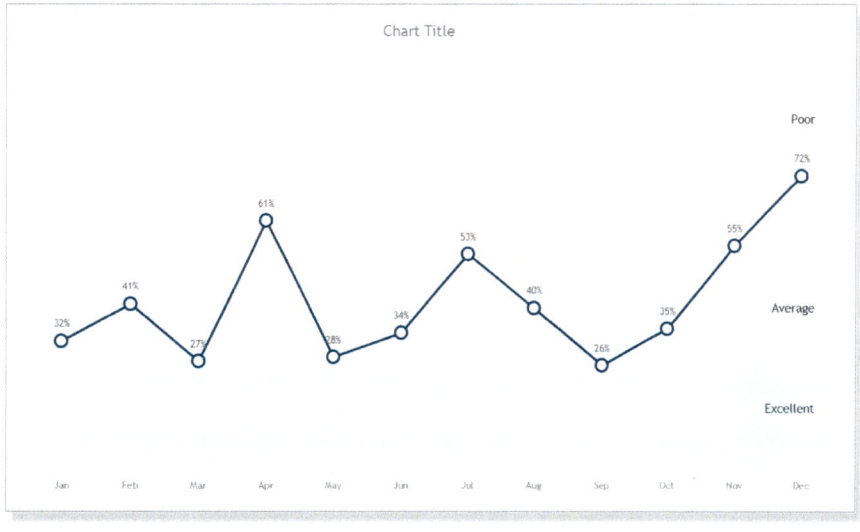

Fig 5.231: Align your text boxes.

217

CHAPTER 5 · CREATE CHARTS

Step 13: Clean up your chart to taste. I cleaned mine by taking the following steps:

- made title more descriptive
- left aligned chart title
- bumped up title font size

> 💬 **LEARN MORE** See the Design with Elegance chapter to learn how to perform these refinements.

Credit Shout out to the Chandoo blog for the inspiration to hack a stacked column chart to create the bands: **bit.ly/mdsepc-band-camp**.

Variation
Add Single Band
If you just need one band, set **Fill** for the other data series to **No fill (Fig 5.232)**.

Example File
band-chart.xlsx

Templates
If you'll use these charts again, save them as chart templates. Or download the templates included in the chart templates folder:

- Multiple bands: mds-band-multiple.crtx
- Single band: mds-band-single.crtx

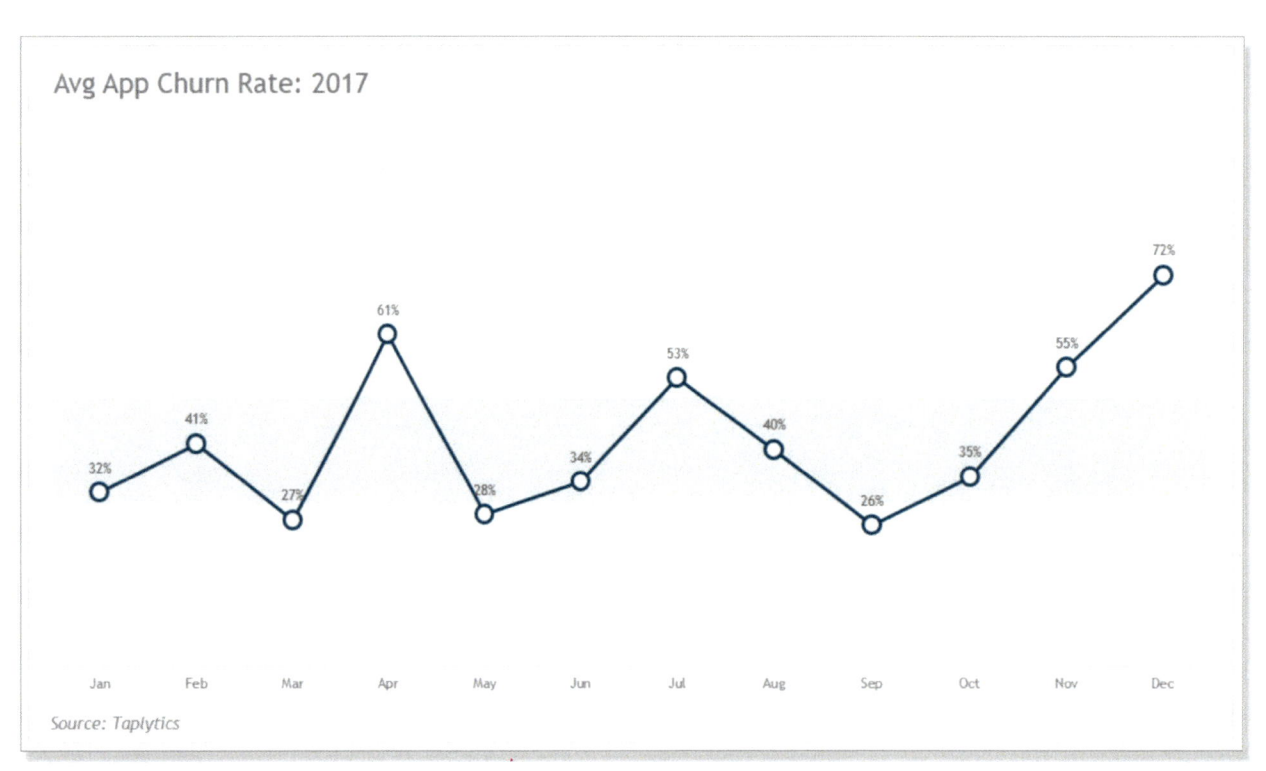

Fig 5.232: Remove fill from extraneous data series.

CHAPTER 5 • CREATE CHARTS

AREA CHART

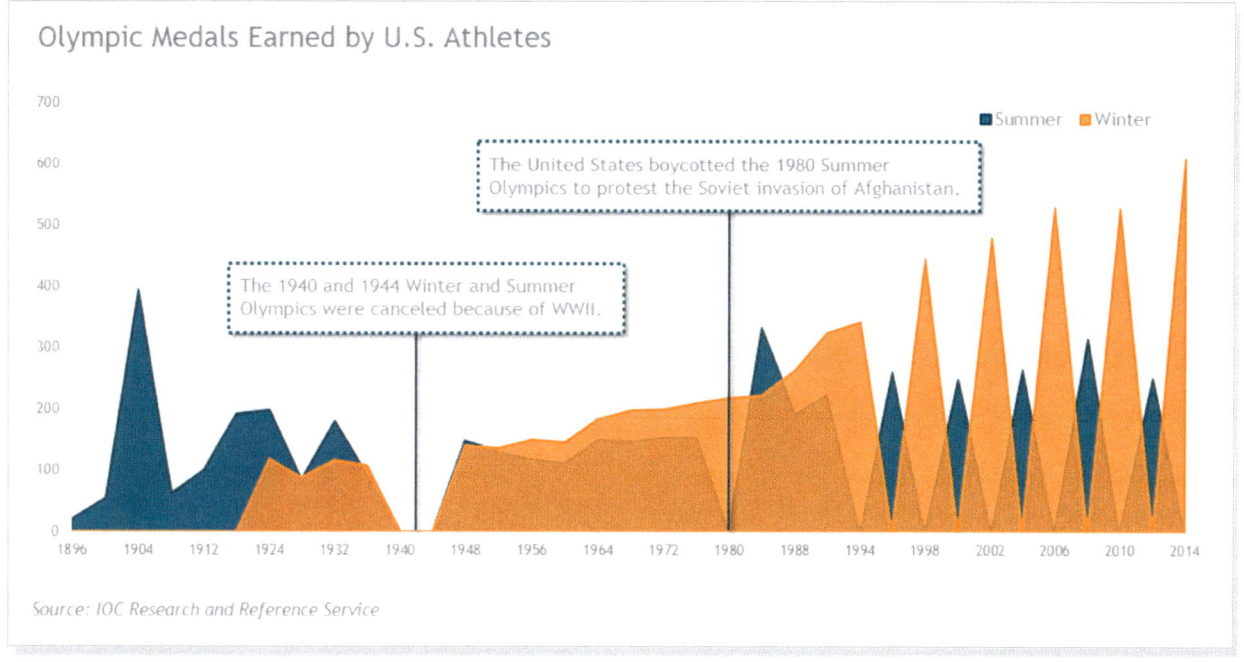

Fig 5.233: Our final chart.

AKA
You may also see area charts called mountain charts. (Clever. Verrry clever.)

Uses
Area charts are best used to:

+ represent volume
+ plot trend data
+ compare trends
+ compare continuous variables, such as distance, revenue, incidences of an event (e.g., disease)

Tips
+ Data can easily be obfuscated in area charts when you display more than one data series. If you have more than two series, you may want to consider using a line chart instead.
+ To reduce the risk of hiding data points, you can dial up the fill transparency.

+ You always need to decide how to deal with blank cells in your data, (i.e., as 0s or gaps). I recommend treating them as 0s—as you can see in the chart for this tutorial—because this will maintain the mountain-like appearance of your area chart. You can access this option in any chart via the Select Data Source dialog **(Fig 5.234)**.

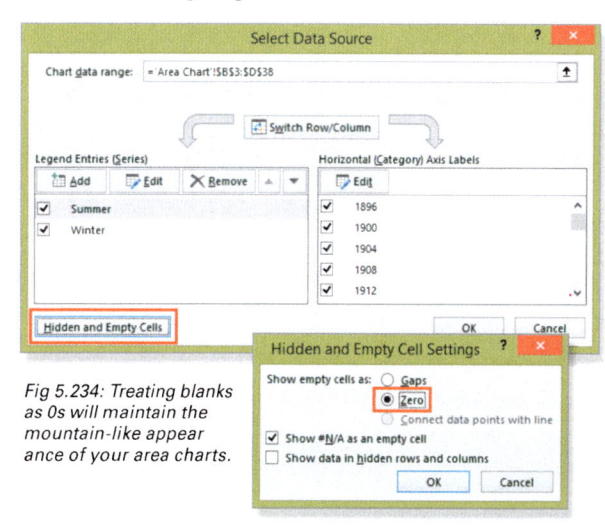

Fig 5.234: Treating blanks as 0s will maintain the mountain-like appearance of your area charts.

CHAPTER 5 • CREATE CHARTS

> **💬 LEARN MORE** See the Handling Blanks section of the Massage Your Data chapter to learn how to chart blanks cells.

- If you only have one category, you can plot mutliple metrics (within reason). If you have two categories—as we do in the example dataset—you can only plot one metric.

- Adding a border to your data series can add clarity and crispness, especially if you increase the fill transparency. If you only have one data series and want to add a border without tweaking the transparency settings, you can just use a slightly darker version of your fill color.

- Avoid using the area chart for non-continuous data. With fairly rare exceptions (such as plotting distances), this will mean restricting your use of an area chart to datasets that contain a time element.

- If you're using tabular data, you may need to use a pivot table to transform your data into a crosstab format. I've included a Pivot Table worksheet in the downloadable workbook to show you how I set mine up. Click anywhere in the pivot table to open the PivotTable Fields pane. From there you can easily see how I constructed it.

Category / Metric
1–2 categories / 1+ metrics

Tutorial
Step 1: Select any cell inside your dataset and navigate to **Insert ▸ Charts ▸ Insert Line or Area Chart ▸ 2-D Area ▸ Area (Fig 5.235)**.

Step 2: If you have more than one data series, I recommend increasing the transparency of the fill and contrasting that with a border that has no transparency applied to it. To do this, select one of your data series, and press Ctrl-1 to open the **Format Data Series** pane. Open the **Fill & Line** tab and go to town **(Fig 5.236)**. I used the following settings:

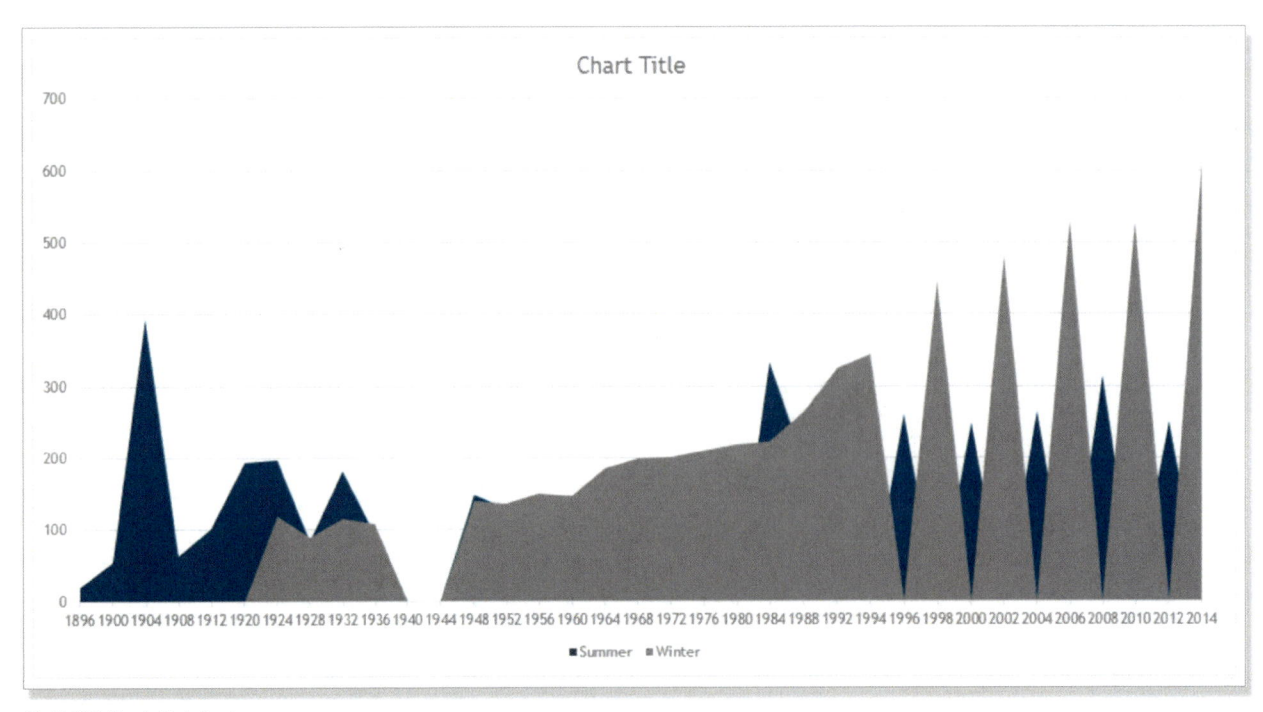

Fig 5.235: Our initial chart.

CHAPTER 5 · CREATE CHARTS

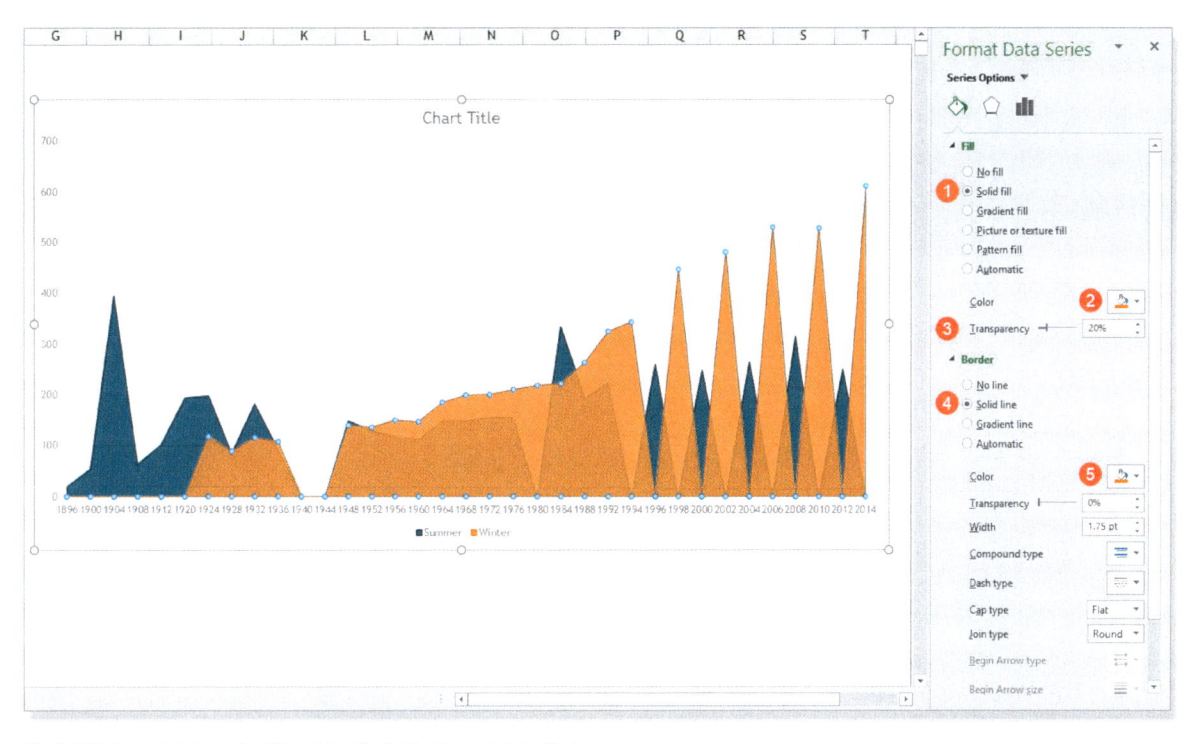

Fig 5.236: Format data series fill and border in the Format Data Series pane.

- **Fill**: Solid Fill
- **Fill Color**: blue/orange
- **Transparency**: 20%
- **Border**: Solid line
- **Outline color**: same as fill
- **Width**: 1.75 pt

Step 3: Clean up your chart to taste. I cleaned mine by taking the following steps:

- made title more descriptive
- left aligned chart title
- bumped up title font size
- added annotations for data source and explanations of anomalies
- bumped up transparency of gridlines
- moved legend
- thinned horizontal axis

LEARN MORE See the Design with Elegance chapter to learn how to perform these refinements.

Example File
area-chart.xlsx

Warning: Excel may reset Fill to Automatic when you open the file. This is a bug I swatted at but couldn't kill. If you change Fill back to Solid fill, you'll need to reapply the transparency. So. Annoying.

Template
If you'll use this chart again, save it as a chart template. Or download the template included in the chart templates folder: mds-area.crtx.

BASELINE DELTA CHART

Uses

Baseline delta charts are best used to:

- plot data with negative values, e.g., temperatures, return on investment (ROI), profit/loss over time (if combined in one column since this chart type uses only one metric)
- compare data across discrete categories
- plot trend data

Tips

- I don't normally recommend using a column chart for continuous data; it's better suited for aggregated data (e.g., by month) or categorical data. But Excel only gives you the option to use a different color for negative values with column and bar charts. You don't have this option with line or area charts. I hope that someday Microsoft will give us this feature for area charts.
- At the time of writing there appears to be a bug associated with the Inverted Fill Color option. Each time I reopen the file, the last column with a negative value is given a white fill and black line **(Fig 5.238)**.
- If your heart is set on an area chart, you can try the hack that Jon Peltier details in this post: **bit.ly/mdsepc-baseline-area**. It didn't pass the elegance sniff test for this book because of how stiff it looks, but it's probably your only choice, shy of using a macro. Or Tableau. #zing

- A baseline delta chart is just a diverging bar chart with one metric.
- You'll need to move your horizontal axis labels out of the way with this chart.

Category / Metric

1 category / 1 metric

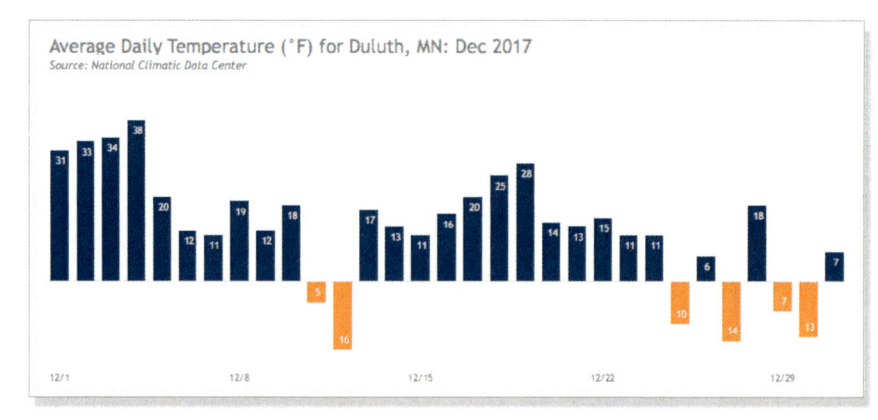

Fig 5.237: Our final chart.

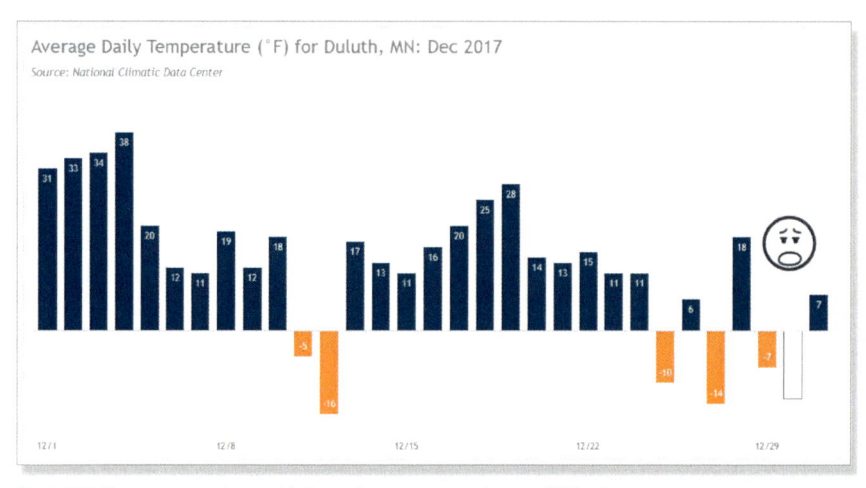

Fig 5.238: You may experience this bug when you set an inverted fill color.

222

CHAPTER 5 • CREATE CHARTS

Tutorial

Step 1: Select any cell inside your dataset and navigate to **Insert** ▸ **Charts** ▸ **Insert Column or Bar Chart** ▸ **2-D Column** ▸ **Clustered Column (Fig 5.239)**.

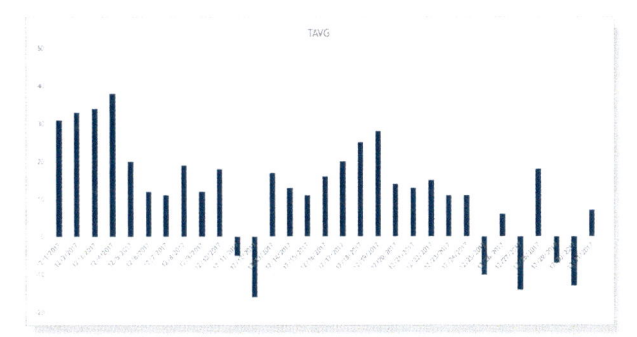

Fig 5.239: Our initial chart.

Fig 5.240: You will want to relocate your axis if you have negative values.

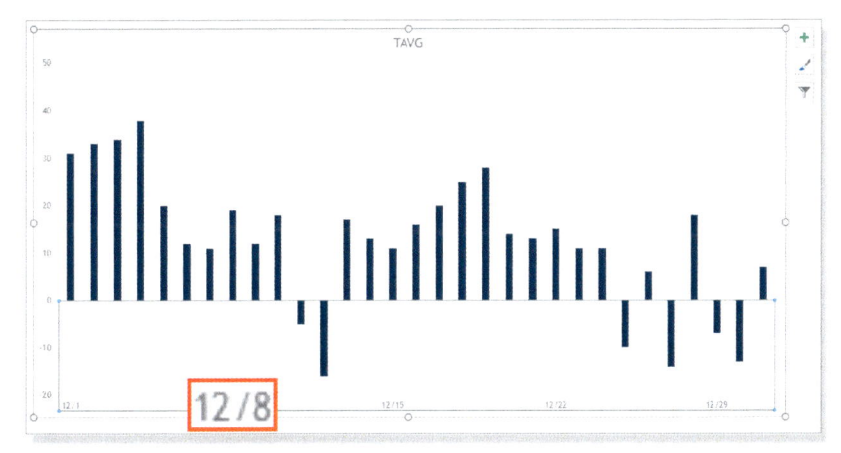

Fig 5.241: Thin out an unnecessarily busy axis.

Step 2: We need to deal with that horizontal axis before we can get anywhere. Select it and press Ctrl-1 to open the **Format Axis** pane. In the **Axis Options** tab, under **Labels**, set **Label Position** to anywhere but **Next to Axis**. I chose **Low** to move it due south **(Fig 5.240)**.

Step 3: Let's thin out that rat's nest of an axis. In the **Axis Options** section of the **Axis Options** tab (because who even has time to create unique labels?), I set **Major Units** to 7 days. If you want to customize the number formatting, hop down to the **Number** section of the **Axis Options** tab. I set mine to m/d to reduce noise **(Fig 5.241)**.

> 💬 **LEARN MORE** See the Thin Out Axes section of the Design with Elegance chapter to learn how to reduce clutter in your axes and the Customize Number Formatting section to learn how to finesse your number formatting.

Step 4: I like to bring values into my column charts via data labels whenever feasible, but it usually requires fattening them up by reducing **Gap Width**. I set mine to 41%. Oddly specific, I know.

> 💬 **LEARN MORE** See the Adjust Chart Series Spacing section of the Design with Elegance chapter to learn how to customize the space between columns.

223

CHAPTER 5 • CREATE CHARTS

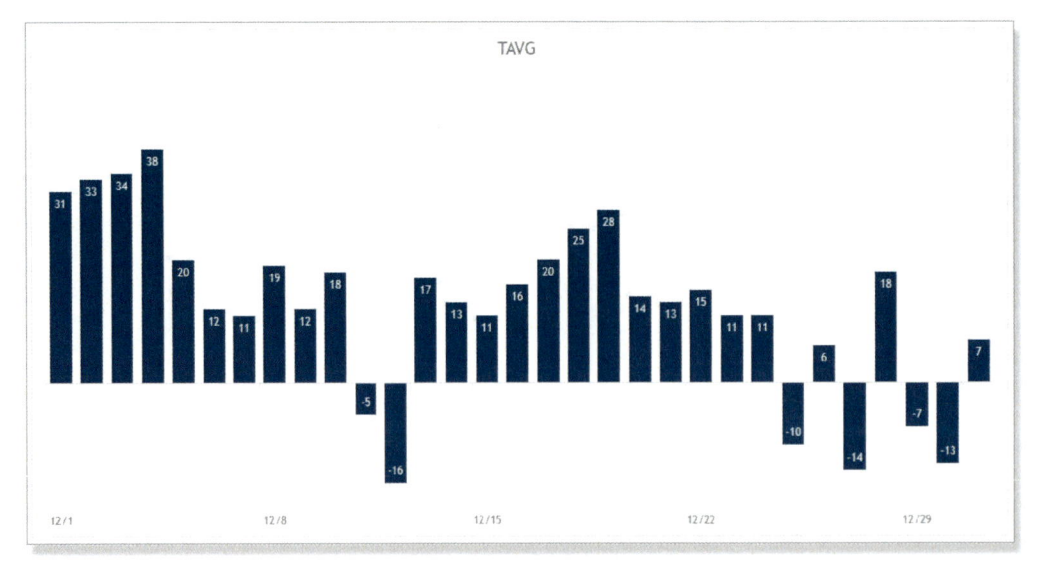

Fig 5.242: Go Spartan on your chart to minimize clutter and add finesse.

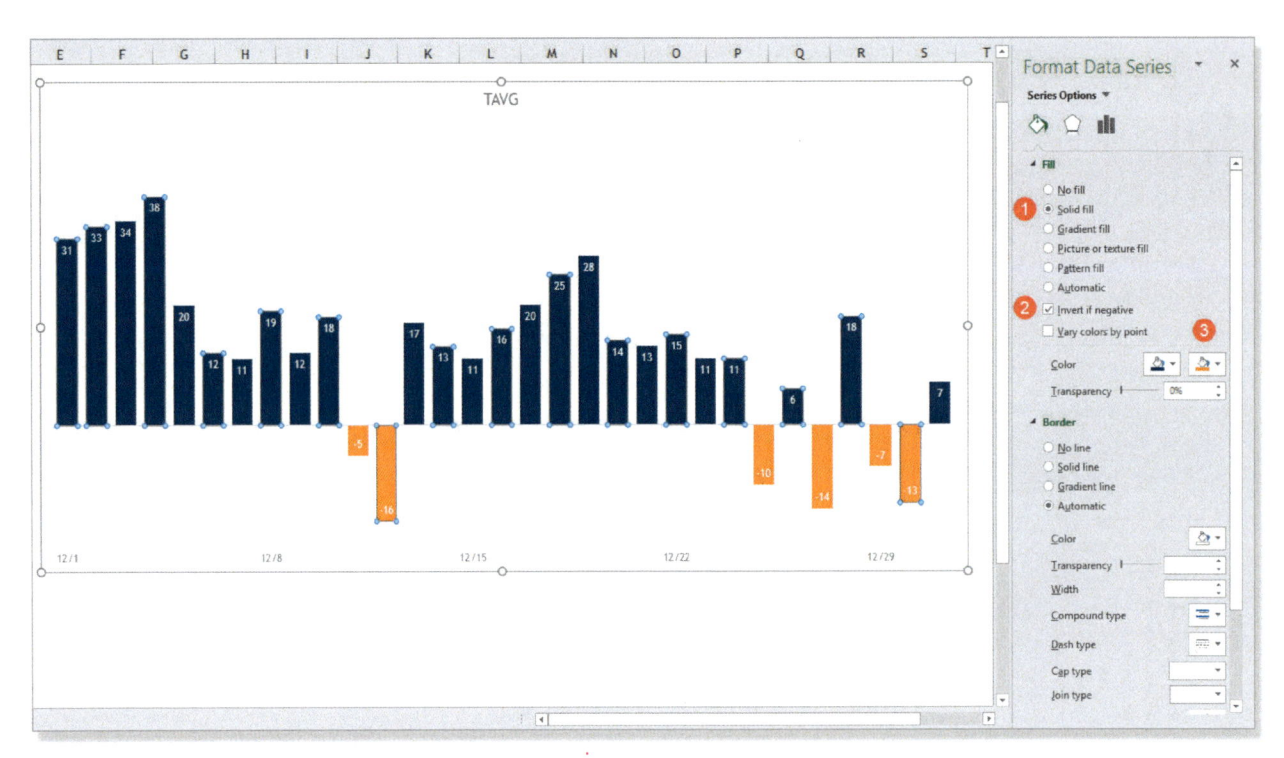

Fig 5.243: Set columns that represent negative colors to a different fill color.

Step 5: Now let's bring our values into our chart. Select one of the columns to select the entire series. From the **Chart Elements** flyout menu, click the right-facing arrow next to **Data Labels** to open the flyout menu and select **Inside End (Fig 5.242)**.

💬 **LEARN MORE** See the Add Data Labels section of the Design with Elegance chapter to learn how to add data labels.

CHAPTER 5 • CREATE CHARTS

Step 6: Now for the fun part: Let's change the color of the columns that represent negative values. First, select a column to select the data series. Then, in the **Format Data Series** pane, under the **Fill & Line** tab, set **Fill** to **Solid fill**. Then select **Invert if negative**. Now you can choose your color from the **Inverted Fill Color** drop-down menu that magically appears **(Fig 5.243)**.

Warning: At the time of writing, Excel sets **Fill** to **Automatic** by default. You have the option to select the **Invert if negative** setting while your series is set to **Automatic**; however, you must set **Fill** to **Solid fill** to trigger the **Inverted Fill Color** option **(Fig 5.244)**. It's kind of like an Easter egg in Excel…that went undiscovered in the backyard and was left to rot.

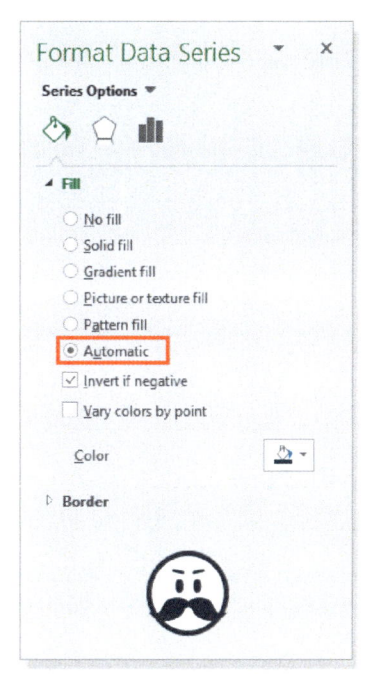

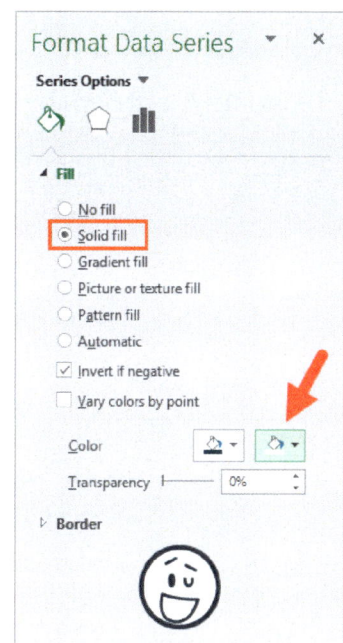

Fig 5.244: I hope this saves you some of the suffering I experienced!

Step 7: Clean up your chart to taste. I cleaned mine by taking the following steps:

+ made title more descriptive
+ left aligned chart title
+ bumped up title font size
+ added annotation for data source

LEARN MORE See the Design with Elegance chapter to learn how to perform these refinements.

Note: I ordered the weather data for Duluth from the National Climatic Data Center for Dec 2017 to get data with negative values but took some artistic license to distribute the negative values. The original, unaltered data is included in the Raw Data worksheet of the workbook. #transparency

Example File
baseline-delta-chart.xlsx

Template
If you'll use this chart again, save it as a chart template. Or download the template included in the chart templates folder: mds-baseline-delta.crtx.

Warning: When I tested the chart template, the last negative bar came through with the funky fill, as expected, because of the aforementioned bug, but the text box with the source information didn't come through either. Ymmv.

CHAPTER 5 · CREATE CHARTS

STACKED AREA CHART

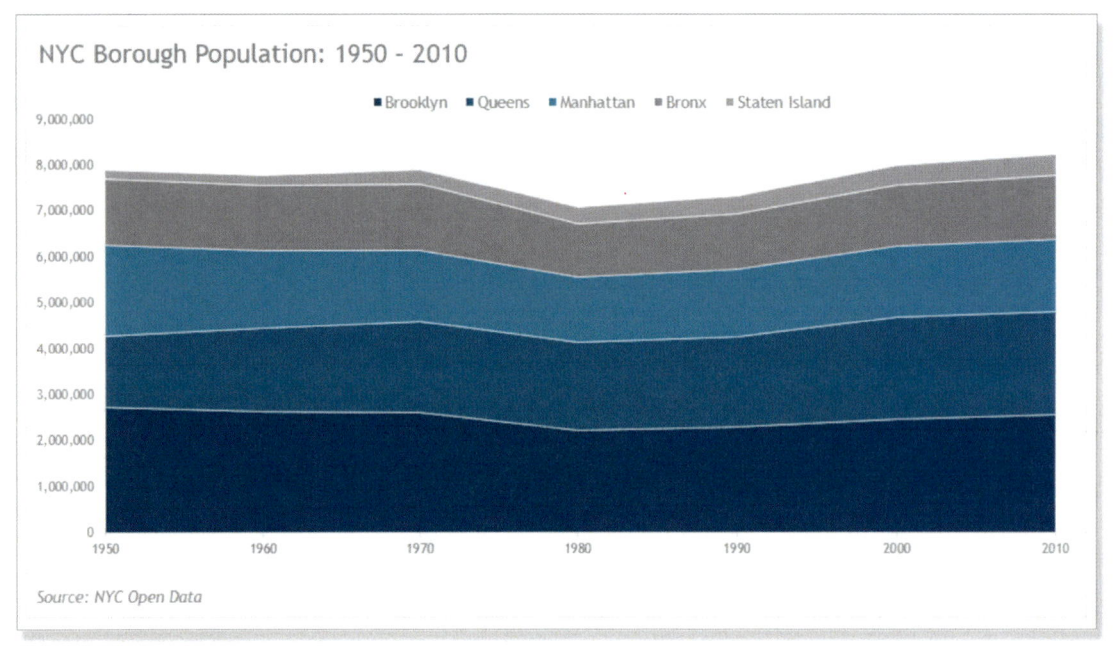

Fig 5.245: Our final chart.

Uses

Stacked area charts are best used to:

- demonstrate how each data series contributes to the whole
- plot data over time
- plot trend data

Tips

- Be careful with your choice of color with stacked area charts. Microsoft's default colors are jarring. You should always use your branded colors. If you need to use additional colors, it's better to use tints of your branded colors.
- If you use tints, I recommend putting the lighter tints at the top of the stack.

> 💬 **LEARN MORE** See the Stick to Branded Colors section of the Design with Elegance chapter to learn about the importance of restricting your design choices to your branded colors.

- Another option to adding sizzle to your data series is to increase the transparency of your data series, then add a border of that color at 100%. I only recommend using this option if you're confident your chart will stay against a white background.
- Sorting your data series is important with stacked area charts. I recommend placing your larger data series on the bottom of the stack, with the smallest series at the top. I usually employ a total column or row to determine this stacking order.
- If a data series starts or ends in the middle of the chart, it can be helpful to add an annotation with an explanation because readers' eyes will be drawn toward these unnatural ledges in your visualization. Better to meet the distraction head on with an explanation like, "We started running Facebook ads in April" or "Product X was recalled and removed from the shelves, causing the drop-off in revenue."

CHAPTER 5 • CREATE CHARTS

- If you find that your area chart's edges are too jagged and you're plotting along a timeline, you may want to consider aggregating the time element (e.g., aggregate days to weeks or months) to reduce the noise. Aggregated time elements tend to work better to express trends, whereas more granular timelines are better at highlighting the impact of disparate events (e.g., a day your website was down or sales on Black Friday).

- Gridlines are almost always necessary with area charts, but I always increase the transparency and/or change them to a lighter color to reduce their intensity.

- If you use markers, keep them as subtle as possible to reduce the potential for distraction. If you really need them, you may want to consider using a column chart instead.

- You can add a straight vertical line that intersects the line(s) to note a significant event (such as the redesign of a website or date a test began). If you add one, I recommend keeping the line thin and making it dotted or broken to keep it from becoming too cumbersome. You can also use a data label to add an annotation, which will connect it to the line. You will lose that connection with a text box.

- You always need to decide how to deal with blank cells in your data (i.e., as 0s or gaps). I recommend treating them as 0s—as you will see in the chart for the tutorial—because this will maintain the mountain-like appearance of your area chart. You can access this option in the Select Data Source dialog **(Fig 5.246)**.

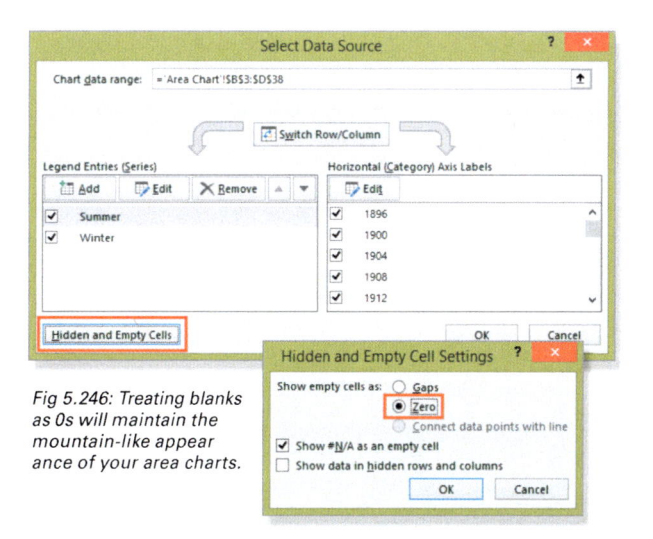

Category / Metric
2 categories / 1 metric

Tutorial
Step 1: When creating a stacked area chart, I recommend adding a Total column to your data. If you're anything like me, when you're sitting and looking at your dataset, it can be confusing to know whether to add totals to your rows or columns. It helps to take a step back and think about the role these totals will play. To wit, we'll be using them to sort our data series (e.g., the bridges), so we will need to calculate totals by bridge.

Tip: An easy way to calculate the total of a column or row is to select the cell adjacent to your column or row and press Alt-=. This will drop a SUM function into the cell using the contiguous cells that contain numerical data. Then drag the fill handle of that cell to apply the formula to your other columns or rows **(Fig 5.247)**.

Step 2: To sort your data series, right-click any cell

Fig 5.247: Adding a total column gives you the ability to sort your series.

in your Total column and choose **Sort ▶ Sort Largest to Smallest**.

Step 3: Click-and-drag to select all columns except the Total column, then navigate to **Insert ▶ Charts ▶ Insert Line or Area Chart ▶ 2-D Area ▶ Stacked Area (Fig 5.248)**.

Fig 5.246: Treating blanks as 0s will maintain the mountain-like appearance of your area charts.

227

CHAPTER 5 • CREATE CHARTS

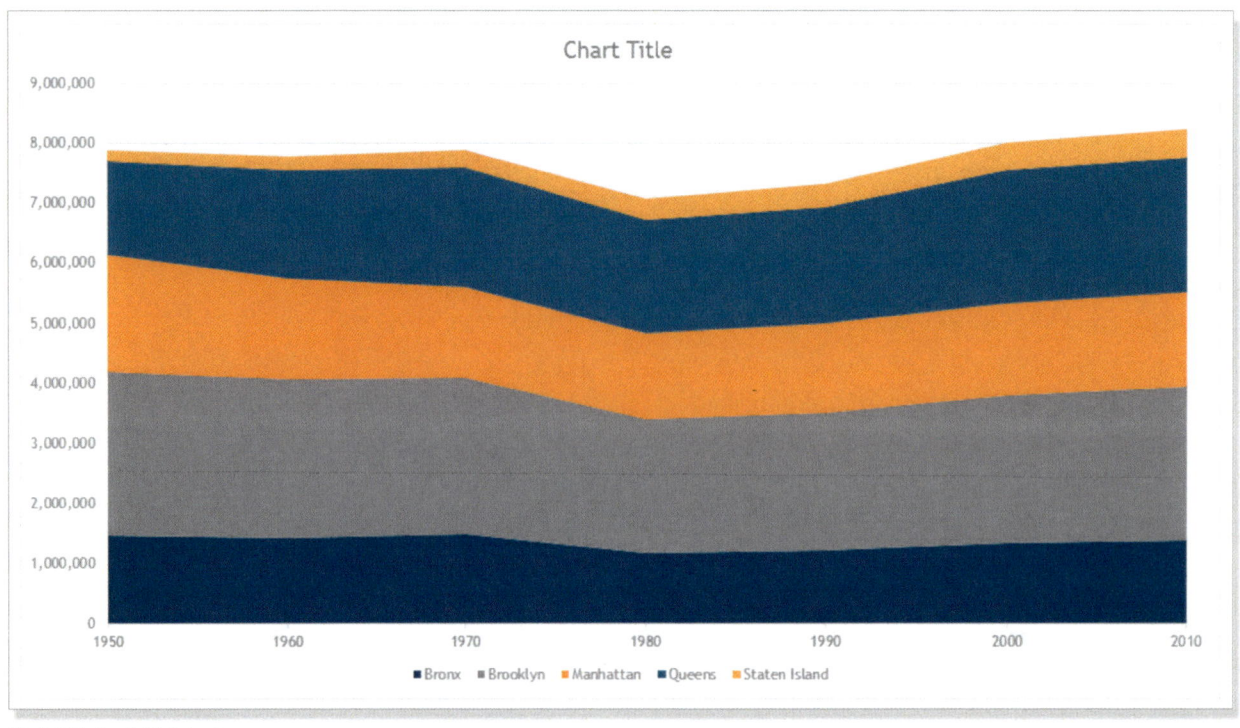

Fig 5.248: Our initial chart.

Step 4: Clean up your chart to taste. I cleaned mine by taking the following steps:

- made title more descriptive
- left aligned chart title
- bumped up title font size
- added annotation for data source
- removed gridlines
- bumped up legend font size
- moved legend
- added border to data series
- added a subtle border to the Plot Area to keep the top axis value from appearing as though it is floating

LEARN MORE See the Design with Elegance chapter to learn how to perform these refinements.

Example File
stacked-area-chart.xlsx

Template
If you'll use this chart again, save it as a chart template. Or download the template included in the chart templates folder: Stacked area: mds-stacked-area.crtx.

COMPOSITION

PIE CHART

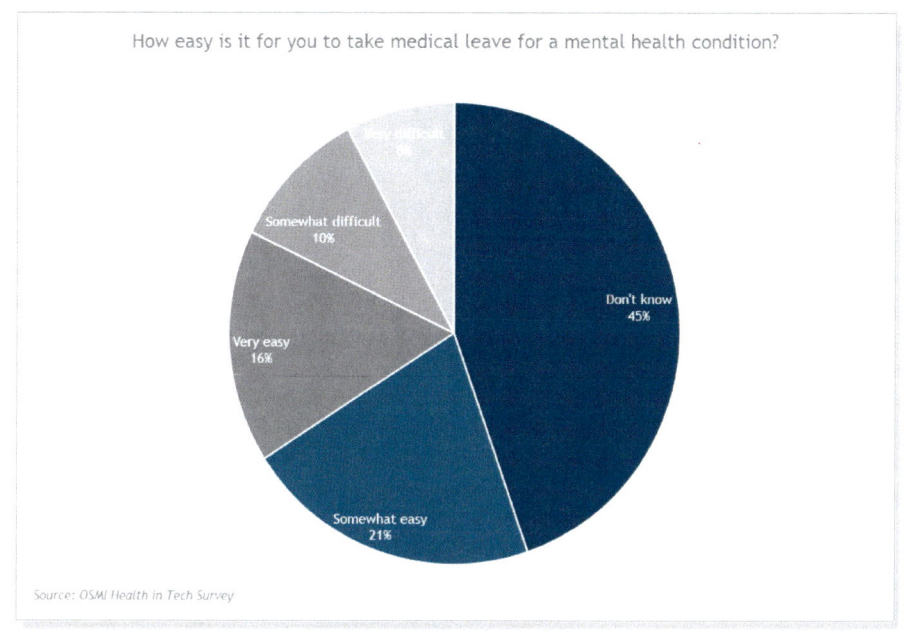

Fig 5.249: Our final chart.

AKA
You may also see pie charts called:

- dessert charts
- candy chart (donut chart only)

Uses
Pie charts are best used to:

- offend the intelligence of your viewers (lol j/k…sort of)
- plot small datasets
- demonstrate how each data series contributes to the whole
- visualize data from a single snapshot in time

Tips

- Pie charts are more difficult to compare than bar or column charts, especially if your slices are similar in size. If you have data series with similar values, I recommend using a bar or column chart.

- I don't recommend using pie charts for a data series with more than five values. Again, bars and columns work better.

- Sort your data to put your pie slices in descending order. Unsorted pie charts are difficult to interpret and look amateur.

- If your pie slices are big enough, bringing labels into the slices also saves the reader from having to consult the legend. Or you can add those data labels outside the slices. Then you can delete the legend.

- Instead of using different colors, you could use different tints of one of your branded colors.

- Pie of Pie and Bar of Pie charts are very difficult to interpret. You should (for the love of all things data) strongly consider using bar or column charts before employing either of these ill-conceived chart types.

- If you're using tabular data, the easiest way to tally values for a category is with a pivot table. In fact, if you pull a category that contains text, Excel will automatically aggregate by count. I've included a Pivot Table worksheet in the downloadable workbook to show you how I set mine up.

Category / Metric
0 categories / 1 metric

Tutorial

Step 1: Sort your data by right-clicking in any cell of your value column and selecting **Sort ▸ Sort Largest to Smallest**. Next, select any cell inside your dataset and navigate to **Insert ▸ Charts ▸ Insert Pie or Doughnut Chart ▸ 2-D Pie ▸ Pie (Fig 5.250)**.

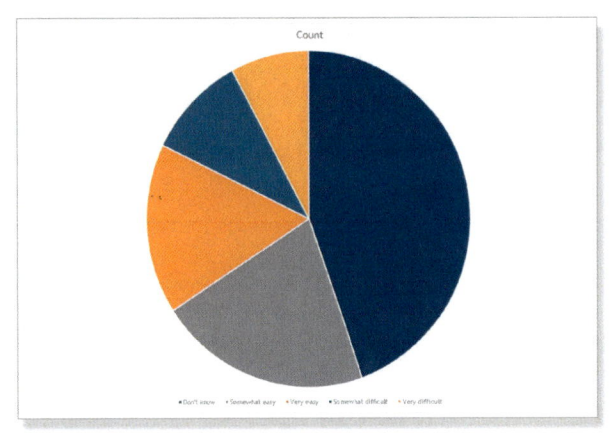

Fig 5.250: Our initial chart.

Step 2: I love that Excel now adds a white border to your pie slices (1.5 pt wide). If you don't want them, you can remove them. Select any data point in your chart (i.e., slice of the pie), then press Ctrl-1 to open the **Format Data Series** pane. Under the **Fill & Line** tab, set **Border** to **No line (Fig 5.251)**.

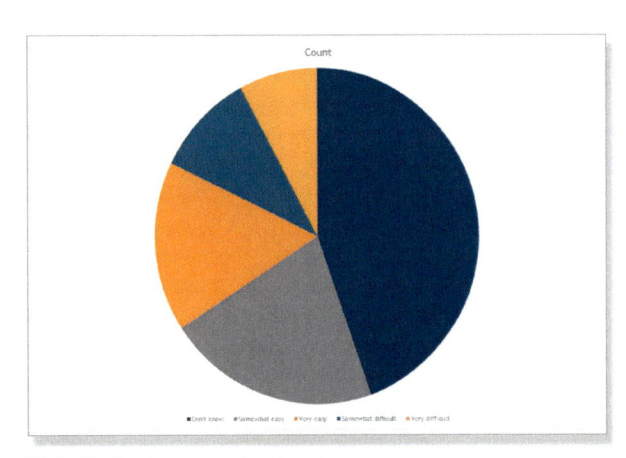

Fig 5.251: Good bye, sexy borders. :(

Step 3: I recommend bringing your data inside your chart. To add data labels, select your chart to open the **Chart Elements** flyout menu. Click the right-facing arrow next to **Data Labels** to open the flyout menu and select **More Options** to open the **Format Data Labels** pane. In the **Label Options** tab, select what you want included in your labels. I pulled in **Category Name** and **Value** and set **Separator** to **New Line**. Then I set **Font Color** to white and bumped up **Font Size** under **Home ▸ Font**. Now you can delete the legend **(Fig 5.252)**.

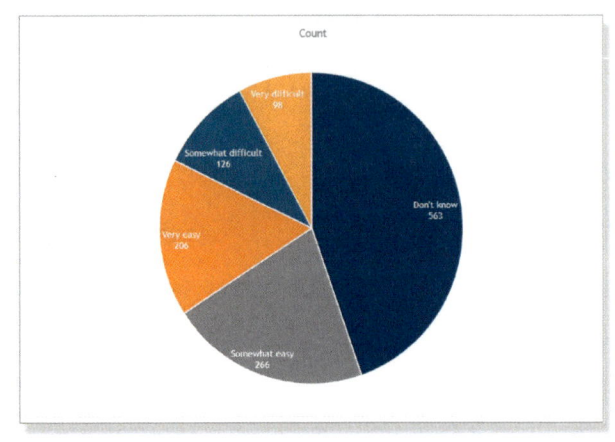

Fig 5.252: I'm a big fan of bringing data into a pie chart.

Step 4: Clean up your chart to taste. I cleaned mine by taking the following steps:

- made title more descriptive
- bumped up title font size
- added annotation for data source
- customized fill colors of pie slices

💬 LEARN MORE See the Design with Elegance chapter to learn how to perform these refinements.

Variations

Accent with Color

You can draw attention to one data point by applying a different color to it. It is, of course, essential that you bring values into the chart if you do this.

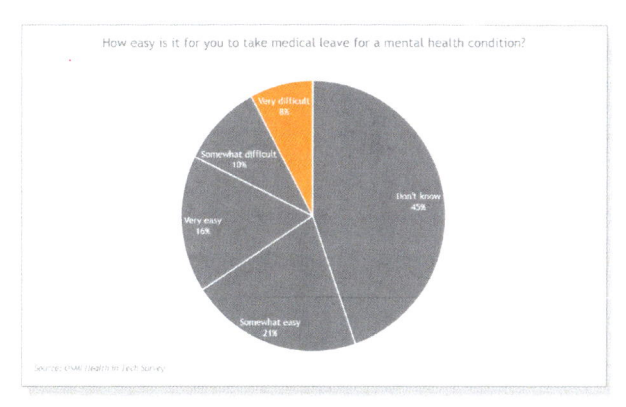

Fig 5.253: Use an accent color to draw attention to one data point.

Exploding Pie Chart

You can also draw attention to a particular data point by pulling one piece of the pie out a bit. To give it a nudge, select the chart, then click the data point you want to explode to isolate it. Click-and-drag it to pull it out from the center.

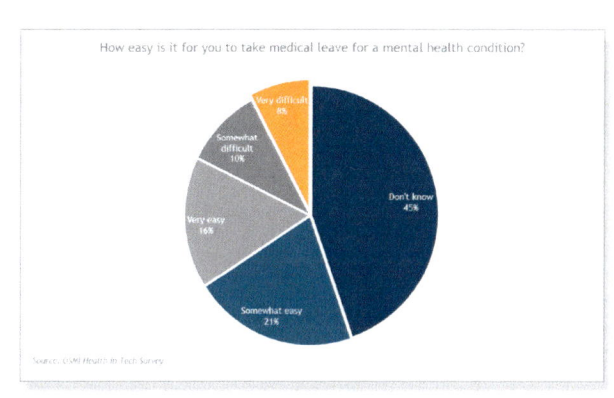

Fig 5.254: Our final chart.

Donut Chart

A fun alternative to a pie chart is a donut chart. To create one, select any cell inside your dataset and navigate to **Insert ▶ Charts ▶ Insert Pie or Doughnut Chart ▶ Doughnut ▶ Doughnut**.

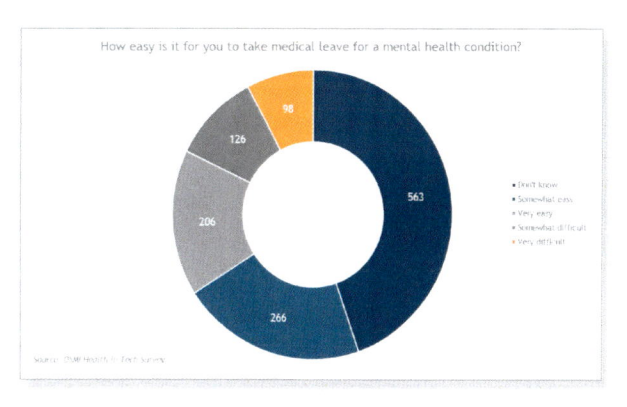

Fig 5.255: Our final chart.

Exploding Donut Chart

Same instructions as the exploding pie, just with a donut chart.

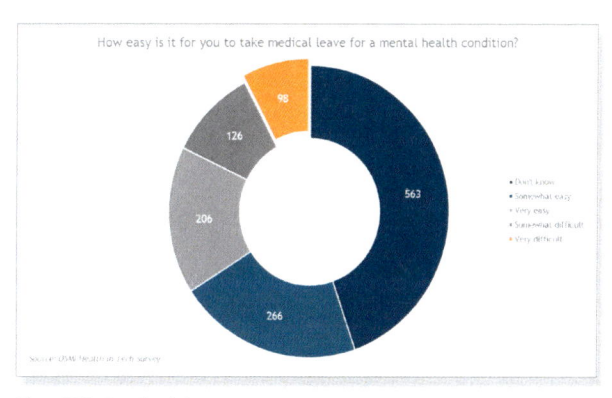

Fig 5.256: Our final chart.

Example File

pie-chart.xlsx

Templates

If you'll use these charts again, save them as chart templates. Or download the templates included in the chart templates folder:

- Pie chart: mds-pie.crtx
- Pie chart with accent color: mds-pie-accent.crtx
- Exploding pie chart: mds-pie-exploding.crtx
- Donut chart: mds-donut.crtx
- Exploding donut chart: mds-donut-exploding.crtx

CHAPTER 5 • CREATE CHARTS

100% STACKED BAR CHART

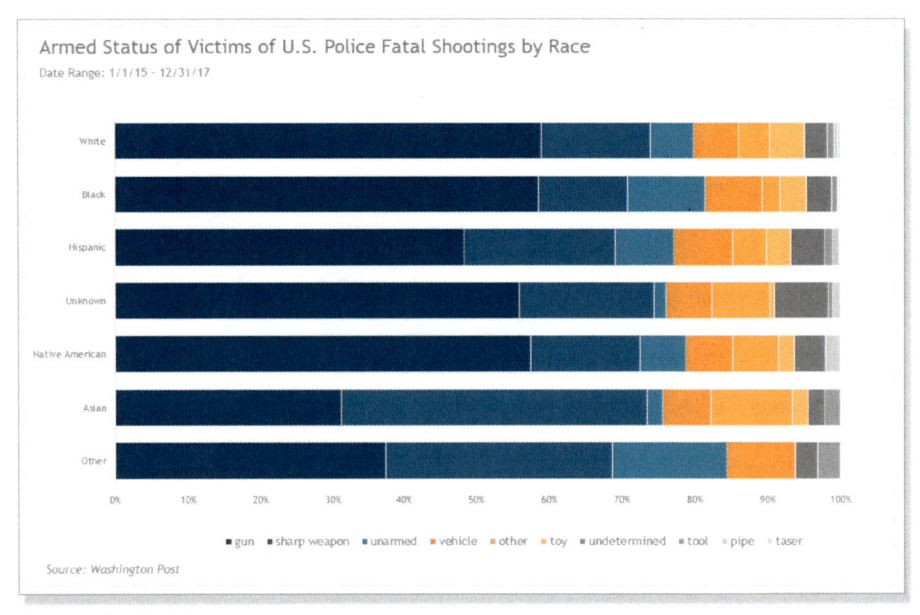

Fig 5.257: Our final chart.

AKA
You may also see 100% stacked bar charts called ratio bar (or column) charts.

Uses
100% stacked bar charts are best used to:

- demonstrate how each data series contributes to the whole
- plot data over time (column only)
- visualize rating scales, such as Likert scales

Tips
- 100% stacked bar/column charts can be powerful tools, but keep in mind that the only data series that can be easily compared are the bookends because they have baselines (the ends of the bars).
- It can be powerful to make all the data series gray except the one you want to highlight (e.g., % of expenses from a problematic department in the organization, % of complaints by product category, % of comments for a particular blog category).
- Stacked bars can be easier for some to process than stacked columns, especially if you have more than three category elements. This is because it's easier to read across than in an upward direction.
- Stick to your branded colors. If you have more data series than branded colors, consider using tints of your branded colors.

> **LEARN MORE** See the Stick to Branded Colors section of the Design with Elegance chapter to learn about the importance of restricting your design choices to your branded colors.

- Consider adding a white border to your data series to accentuate the bounds of each data series and to add visual interest.
- If you're using tabular data, you may need to use a pivot table to transform your data into a cross-tab format. I've included a Pivot Table worksheet in the downloadable workbook to show you how I set mine up. Click anywhere in the pivot table to open the PivotTable Fields pane. From there you can easily see how I constructed it.

232

CHAPTER 5 • CREATE CHARTS

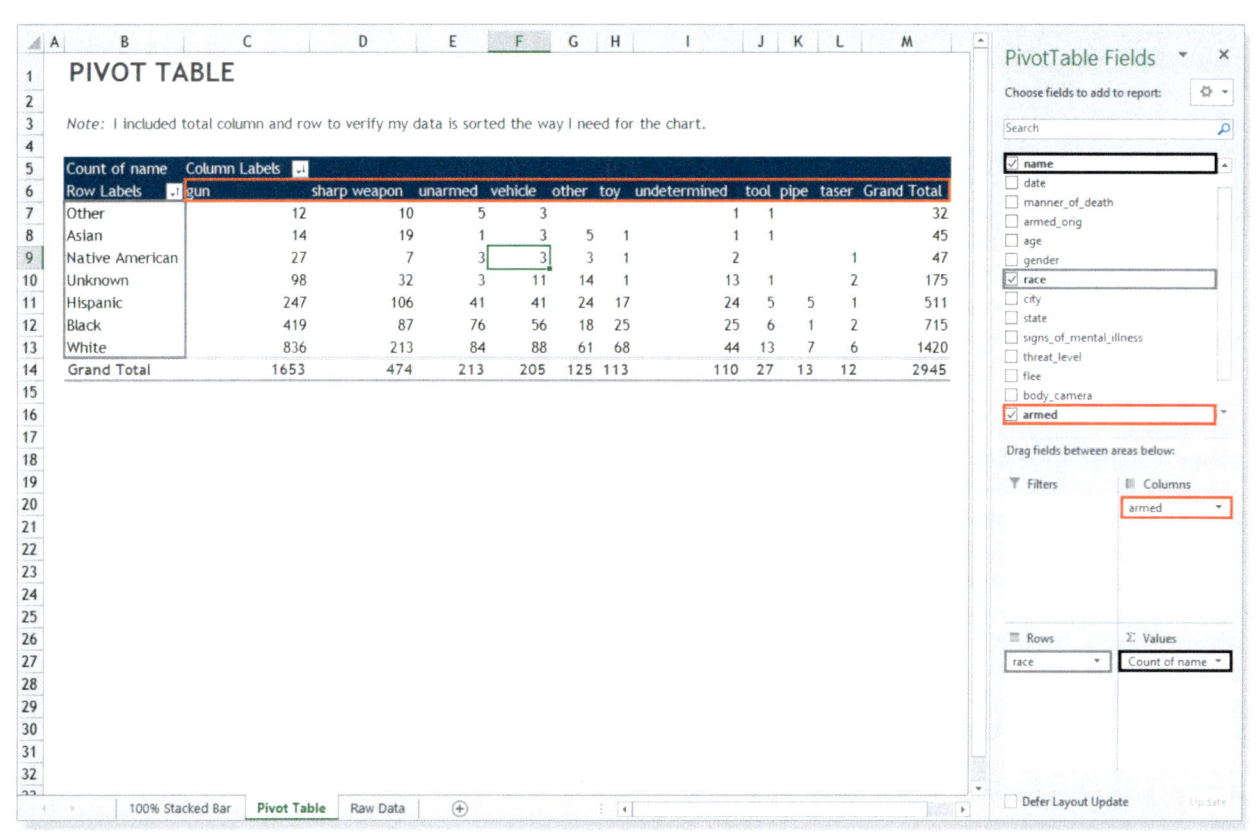

Fig 5.258: Prepare your data using a pivot table.

Category / Metric

2 categories / 1 metric

Tutorial

Step 1: This chart requires quite a bit of massaging of your data before you can jump in—unless you like chaotic, unsorted charts. Since I need my data in a crosstab format—and data is rarely exported in this format—you'll probably need to use a pivot table to corral it. In the example chart, I pulled armed into the **Columns** field and race into **Rows**. Then I pulled name into **Values** and summarized by count. This gave me the count of deaths by counting the names of people who were killed by police officers **(Fig 5.258)**.

> **LEARN MORE** See the Pivot Tables section of the Massage Your Data chapter to learn how to rock pivot tables.

Step 2: Next, we need to sort both our columns and rows. First, enable totals for both rows and columns by selecting any cell in your pivot table, then navigating to **PivotTable Tools ▸ Design ▸ Layout ▸ Grand Totals** and select **On for Rows and Columns**.

Step 3: To get the races with the largest number of deaths at the top of our chart, we'll need to sort our rows in ascending order. This is because we're creating a bar chart. If we were creating a 100% stacked column chart, we would sort in descending order. The easiest way to sort, in my opinion, is to right-click on any cell in the total column and select **Sort ▸ Sort Smallest to Largest**.

233

CHAPTER 5 • CREATE CHARTS

Count of name	Column									
Row Labels	gun	sharp weapon	unarmed	vehicle	other	toy	undetermined	tool	pipe	taser	Grand Total
Other	12	10	5	3			1	1			32
Asian	14	19	1	3	5	1	1	1			45
Native American	27	7	3	3	3	1	2			1	47
Unknown	98	32	3	11	14	1	13	1		2	175
Hispanic	247	106	41	41	24	17	24	5	5	1	511
Black	419	87	76	56	18	25	25	6	1	2	715
White	836	213	84	88	61	68	44	13	7	6	1,420
Grand Total	1,653	474	213	205	125	113	110	27	13	12	2,945

Fig 5.259: Column totals are sorted in descending order and row totals in ascending order.

	gun	sharp weapon	unarmed	vehicle	other	toy	undetermi	tool	pipe	taser
Other	12	10	5	3			1	1		
Asian	14	19	1	3	5	1	1	1		
Native America	27	7	3	3	3	1	2			1
Unknown	98	32	3	11	14	1	13	1		2
Hispanic	247	106	41	41	24	17	24	5	5	1
Black	419	87	76	56	18	25	25	6	1	2
White	836	213	84	88	61	68	44	13	7	6

Fig 5.260: We no longer need the pivot table.

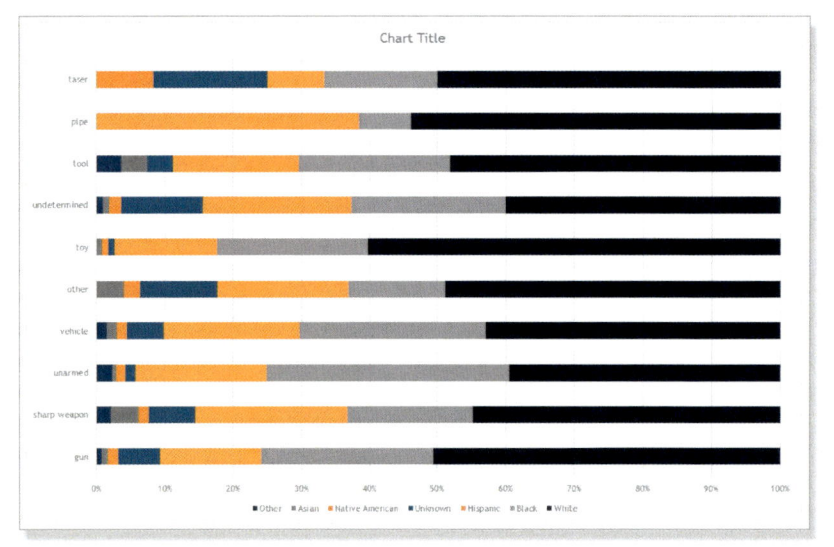

Fig 5.261: Our initial chart.

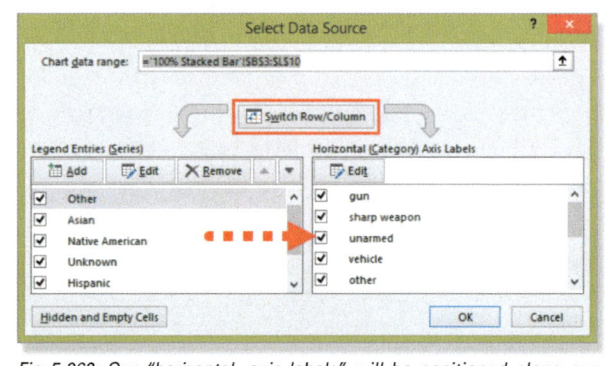

Fig 5.262: Our "horizontal...axis labels" will be positioned along our vertical axis. #jesustakethewheel

Step 4: We also want to sort our columns in descending order to get the "weapons" subjects were most frequently holding at the front of the bar for each race. To do this, right-click on any cell in your total row and select **Sort ▶ Sort Largest to Smallest** (Fig 5.259).

Note: I place weapons in quotation marks because, if you look at the Raw Data worksheet of the example workbook, you'll see some of the items classified as weapons in police reports included objects like staplers, beer bottles, chairs, and even a pen in one instance. I'm not weighing in on the debate by doing this; I just want to be as transparent as possible, especially since I classified weapons into new categories to avoid having 75 data series (the total number of unique weapons included in the original dataset).

Step 5: Copy-and-paste your massaged data from the pivot table as values from the **Paste Options** button **(Fig 5.260)**.

Step 6: Select any cell inside your dataset and navigate to **Insert ▶ Charts ▶ Insert Column or Bar Chart ▶ 2-D Bar ▶ 100% Stacked Bar (Fig 5.261)**.

Step 7: Notice a problem? This is a good example dataset to illustrate that, regardless of how you set up your pivot table, Excel will take the category with the least number of values and use it for the

234

CHAPTER 5 • CREATE CHARTS

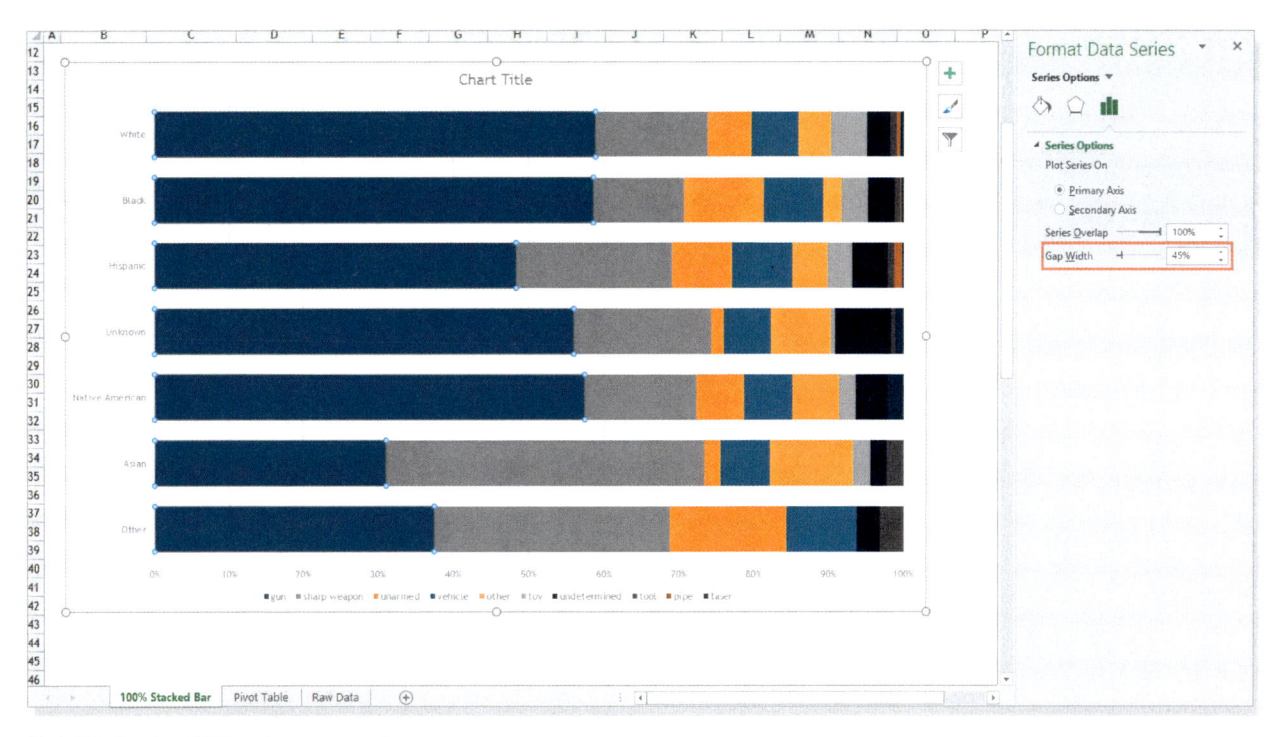

Fig 5.263: Use Gap Width to fatten up your bars.

data series. In my case, that was the race category. But I want my data series to be created from the weapons and to use race as my axis labels. If this happens to you, just right-click inside your chart and choose **Select Data** to open the **Select Data Source** dialog. Then click the **Switch Row/Column** button to make your categories trade places. This will result in 10 data series (weapons) divided into seven category elements (race), if you're following along with the example file **(Fig 5.262)**.

Step 8: I don't like all the white space, so I chose to fatten my bars up a bit. I set **Gap Width** to 45% **(Fig 5.263)**.

> **LEARN MORE** See the Adjust Chart Series Spacing section of the Design with Elegance chapter to learn how to customize the space between columns.

Step 9: I'm also not wild about the layout of colors. (I know, I just want the world.) If I'm dealing with a chart with this many data series (remember, we have 10), instead of introducing colors that don't align with my branding, I prefer to use tints of my (or my client's) branded colors **(Fig 5.264)**.

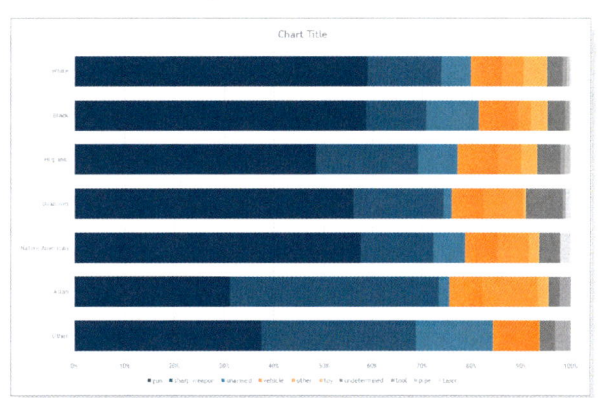

Fig 5.264: You're not beholden to use the colors Excel serves up.

> **LEARN MORE** See the Stick to Branded Colors section of the Design with Elegance chapter to learn about the importance of restricting your design choices to your branded colors.

235

Step 10: Clean up your chart to taste. I cleaned mine by taking the following steps:

- made title more descriptive
- left aligned chart title
- bumped up title font size
- added annotation for data source
- removed gridlines
- bumped up legend font size
- added border to data series

> **LEARN MORE** See the Design with Elegance chapter to learn how to perform these refinements.

Variation
Add Data Labels
If you have a relatively small dataset and you want to surface the values of the individual data series, you can easily add data labels to your series. I had to reduce the range of the example file **(Fig 5.265)**.

> **LEARN MORE** See the Add Data Labels section of the Design with Elegance chapter to learn how to add data labels.

Example File
100%-stacked-bar.xlsx

Template
If you'll use this chart again, save it as a chart template. Or download the templates included in the chart templates folder:

- mds-100%-stacked-bar.crtx
- mds-100%-stacked-bar-labels.crtx

Your chance of being able to "set it and forget it" with this template is low, but with some modifications, you'll have a much more elegant chart than you'll get using Excel's default settings.

Note: If you have a larger dataset, you may need to switch the rows and columns. (See step 7.) Also, you will need to add borders to any additional data series, if you have more than six. Finally, you will need to customize the colors of your additional series because Excel allows you to set six colors in your theme. And I know you're using a theme. Right? *glares*

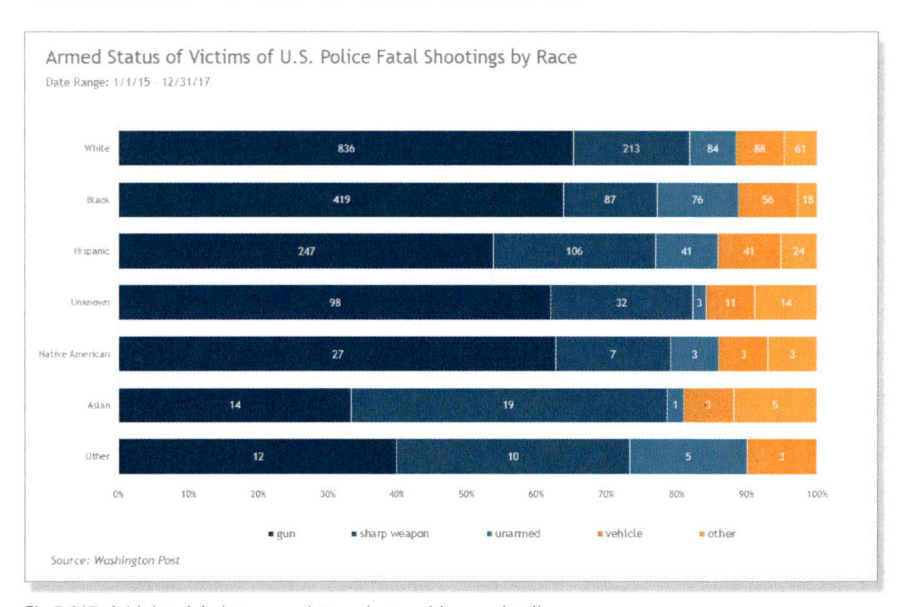

Fig 5.265: Add data labels to your data series to add more detail.

CHAPTER 5 • CREATE CHARTS

100% STACKED AREA CHART

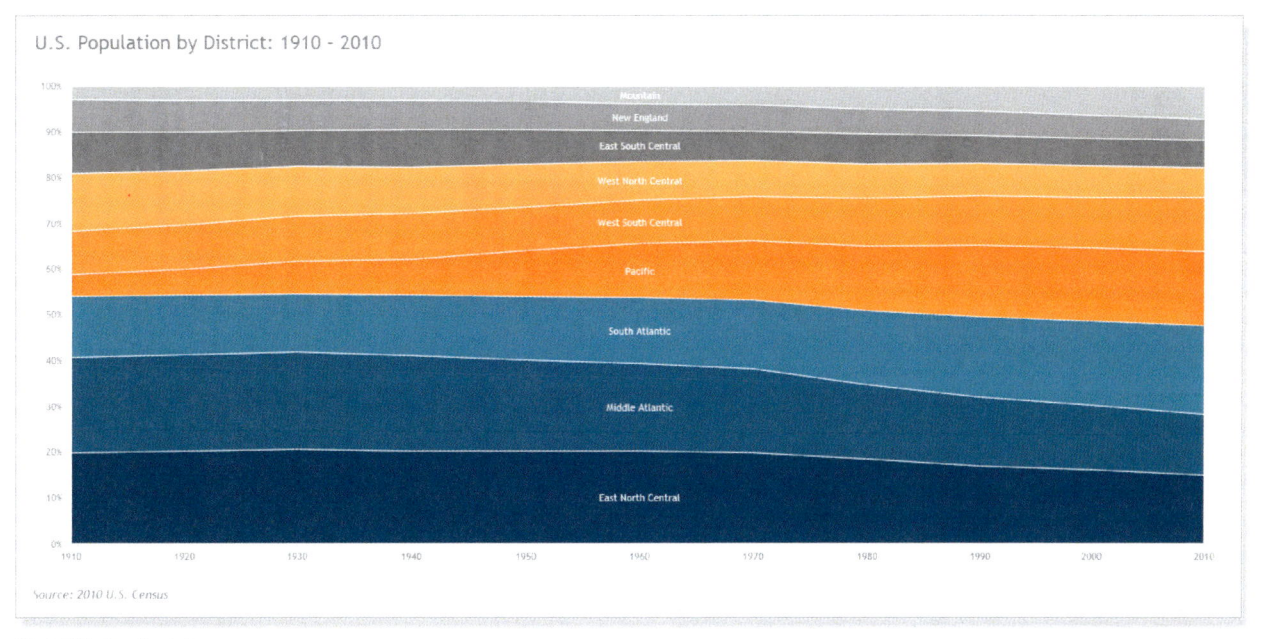

Fig 5.266: Our final chart.

AKA
You may also see 100% stacked area charts called ratio area charts.

Uses
100% stacked area charts are best used to:

- demonstrate how each data series contributes to the whole
- plot data over time
- plot trend data
- compare trends

Tips

- 100% stacked area charts can be powerful tools, but keep in mind that the only data series that can be easily compared are the bookends because they have baselines (the floor and ceiling of the Plot Area).
- It can be powerful to make all the data series gray except the one you want to highlight (e.g., % of expenses from a problematic department in the organization, % of complaints by product category, % of comments for a particular blog category).
- Stick to your branded colors. If you have more data series than branded colors, you may want to use tints of your branded colors.

LEARN MORE See the Stick to Branded Colors section of the Design with Elegance chapter to learn about the importance of restricting your design choices to your branded colors.

- Consider adding a white border to your data series to accentuate the bounds of each series and to add visual interest.

Category / Metric
2 categories / 1 metric

Tutorial
Step 1: The data source I used for this chart came already pivoted—meaning the category was already divided into columns—so we don't

237

need to start with a pivot table. But if the dataset you work with doesn't come pivoted (which is more standard and a better practice), you'll need to follow steps 1–5 of the 100% Stacked Bar Chart tutorial. However, your sort order will depend on the data you're charting and the story you're telling with it.

Step 2: Select any cell inside your dataset and navigate to **Insert ▸ Charts ▸ Insert Column or Bar Chart ▸ Insert Line or Area chart ▸ 2-D Area ▸ 100% Stacked Area (Fig 5.267)**.

Step 3: Just as we did in step 9 of the 100% Stacked Bar Chart tutorial, I modified the layout of colors in my chart using tints of my branded colors. It's my preference to put similar colors together, but you do you!

> **LEARN MORE** See the Stick to Branded Colors section of the Design with Elegance chapter to learn about the importance of restricting your design choices to your branded colors.

Step 4: Clean up your chart to taste. I cleaned mine by taking the following steps:

- made title more descriptive
- left aligned chart title
- bumped up title font size
- added annotation for data source
- removed gridlines
- added data labels
- removed legend
- added border to data series

> **LEARN MORE** See the Design with Elegance chapter to learn how to perform these refinements.

Example File
100%-stacked-area.xlsx

Fig 5.267: Our initial chart.

Template

If you'll use this chart again, save it as a chart template. Or download the template included in the chart templates folder: mds-100%-stacked-area.crtx.

If you have more data series than the chart I created the template from, you'll need to modify the additional series to match the ones included in the template.

Fix Data Labels

By default, Excel applies **Value** to your data labels, so you'll see the value for each data point in a data label. Pretty overwhelming. The template used **Series Name**. If you want to use **Series Name** for your remaining data series, select one of the data labels to select all of them for that series. In the **Format Data Labels** pane that should open automatically, open the **Label Options** tab, then deselect **Values** and select **Series Name**. Do that for each of your data series.

When you're finished switching out the data labels, select one of your formatted data labels, then double-click the **Format Painter** under **Home ▶ Clipboard**. This will allow you to apply the formatting of this data label to multiple labels (sans the number formatting). When you're finished copying the formatting over to the rest of your data labels, click the **Format Painter** again or press the Esc key.

Apply Border

You'll also need to add a border to your additional data series.

> **LEARN MORE** See the Add Border to Touching Data Series section of the Design with Elegance chapter to learn how to add borders to your data series.

TREEMAP

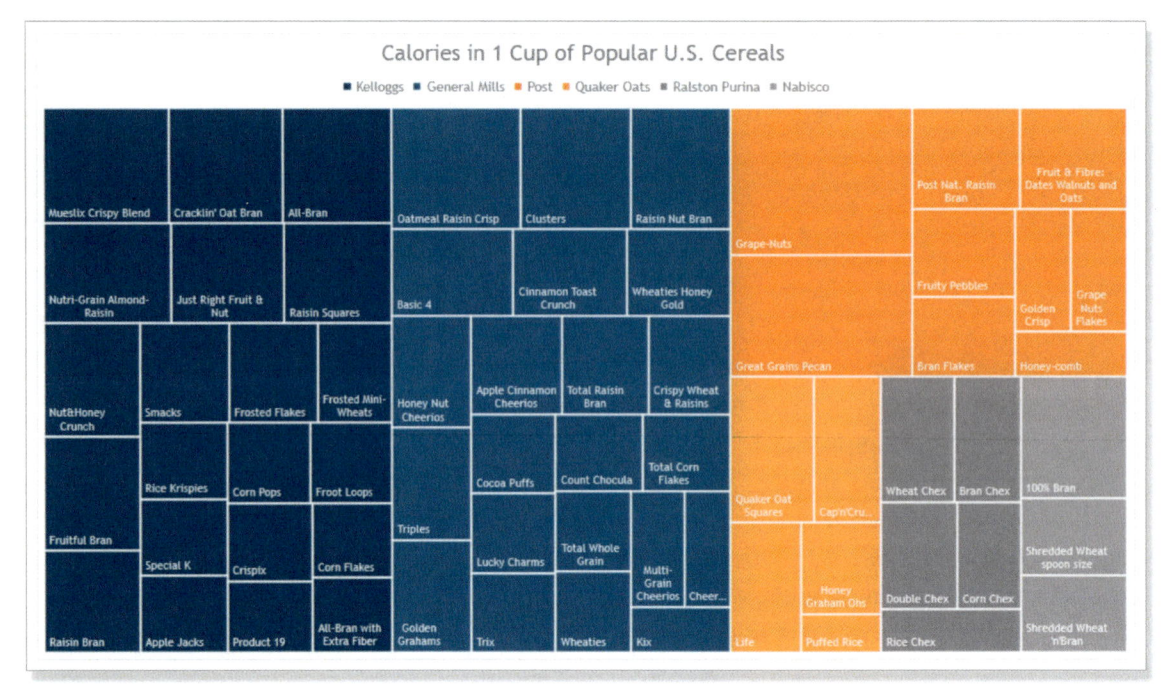

Fig 5.268: Our final chart.

AKA
You may also see treemaps called nested area charts.

Uses
Treemaps are best used to:

- provide a hierarchical view of data using nested rectangles
- highlight the relationship of a root node to its child nodes
- compare parts of a whole within the hierarchy
- condense hundreds, if not thousands, of data points in one chart

Tips
- All the charts new to Excel 2016 (Histogram, Waterfall, Pareto, Box & Whisker, Treemap, and Sunburst) tend to be a bit volatile and cause Excel to crash much more than other charts so save often.
- At the time of writing, you can't move the chart title, link the title to a cell (to create a dynamic title), adjust the Plot Area, or save chart templates with any of the charts new to Excel 2016. And if you add a text box (e.g., to add a data source) and move your chart, the text box doesn't move with the chart. (You will need to group the chart and text box if you want to move or scale them together.)
- One disappointment with treemaps in Excel is you can only chart one metric. Other tools, like Tableau and even Google Analytics, use a heatmap to indicate a second metric.
- If you have more than two categories, a sunburst chart is a better choice because they radiate out.
- You can assign one color for your treemap or a discrete color to each top category. I prefer the latter.
- Sorting your data is critical to effectively structuring a treemap. Your treemap will look amateur and be more difficult to interpret if it's not sorted. It's like bed head for data.

Category / Metric
2+ categories / 1 metric

CHAPTER 5 • CREATE CHARTS

Tutorial

Step 1: First, you will want to create a pivot table from your data (**Insert ▸ Tables ▸ PivotTable**). Set **Report Layout** to **Tabular** (**PivotTable Tools ▸ Design ▸ Layout ▸ Report Layout ▸ Show in Tabular Form**). I also got rid of the decimal places in the calories column, as well as the aggregation type in the heading (Sum of). Finally, I turned off **Grand Totals** from the **Layout** group (**Grand Totals ▸ Off for Rows and Columns**), leaving me with a table format that's ready to be massaged **(Fig 5.269)**.

💬 **LEARN MORE** See the Pivot Tables section of the Massage Your Data chapter to learn how to rock pivot tables.

Step 2: Next, we'll need manufacturer to be repeated in every cell in that column, which you can set under the **Report Layout** menu by selecting **Repeat All Item Labels (Fig 5.270)**.

Step 3: To sort your dataset by your top category (*mfr* in the example dataset), you'll need to add subtotals to your pivot table, if Excel didn't automatically add them. With any cell in your pivot table selected, select **Show all Subtotals at Bottom of Group** from the **Subtotals** menu.

💬 **LEARN MORE** See the Pivot Tables section of the Massage Your Data chapter to learn how to rock pivot tables.

Step 4: To get your top category in the upper-left corner of the treemap, sort your pivot table in descending order by subtotal. To sort, right-click on any subtotal value in the calories column and choose **Sort ▸ Sort Largest to Smallest** from the contextual menu **(Fig 5.271)**.

Fig 5.269: Set your pivot table up for success.

Fig 5.270: We can't have blank cells in our final output.

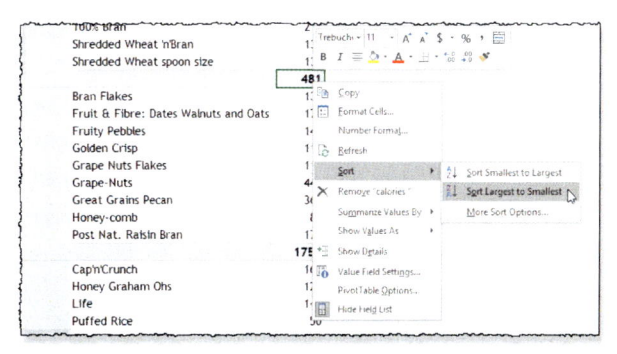

Fig 5.271: Use subtotals in your pivot table to sort your categories.

241

Step 5: To produce a finessed treemap, you also need your subcategory to be sorted in descending order. We can sort by right-clicking anywhere inside your metric column (except a subtotal cell) and choosing **Sort ▸ Sort Largest to Smallest** from the contextual menu. This will sort the data in your top-level category in descending order. Now remove subtotals from your pivot table by selecting **Do Not Show Subtotals** from the **Subtotals** menu. They've served their purpose. Copy-and-paste your massaged data from the pivot table as values from the **Paste Options** button (**Paste Values ▸ Values & Number Formatting**).

Step 6: Select any cell inside your dataset and navigate to **Insert ▸ Charts ▸ Insert Hierarchy Chart ▸ Treemap ▸ Treemap (Fig 5.272)**.

Step 7: I don't like the way the manufacturer labels blend with the cereal labels. There's just not enough of a difference in the formatting to make them stand out. A possible remedy is to use banners for the top category. To enable this, select anywhere inside the **Plot Area** and press Ctrl-1 to open the **Format Data Series** pane. Open the **Series Options** tab and set **Label Options** to **Banner (Fig 5.273)**. If this works for your data, great! I wanted it to work for mine, but Excel used one of my theme colors, which then blended with my data point that

Fig 5.272: Our initial chart.

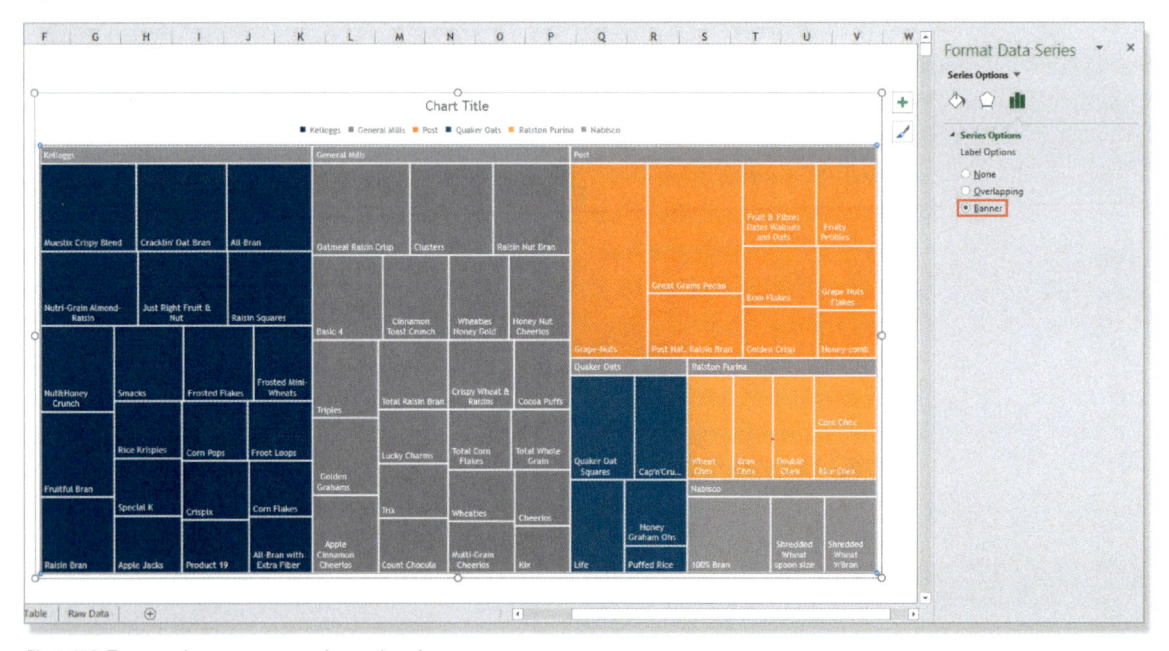

Fig 5.273: Treemap banners are a swing and a miss.

CHAPTER 5 · CREATE CHARTS

Fig 5.274: You can modify the fill colors of your data points.

modify the colors of your data points. In the **Fill & Line** tab, set **Fill Color** to one of your branded colors. Rinse and repeat with each of your data points **(Fig 5.274)**. If I run out of branded colors and need to use tints, I prefer to group similar tints. You'll need to figure out what style works for you.

Tip: If you're careful not to click on a data label, you will be able to select each data point without leaving Excel's isolation mode (my term, not Microsoft's).

used the same color. And you can't modify these banners. It's better to turn them off and rely on the legend than to endure poorly formatted labels, in my opinion. In the end, I chose **None**. This means I can't delete my legend, which I'm a huge fan of doing. *sigh*

Step 8: Select a data point by clicking the chart once and then the data point to isolate it. If you're successful, the other rectangles will fade into the background. This should open the **Format Data Point** pane. (If it doesn't, press Ctrl-1 to open it.) You can now

Step 9: If you want to bring values into your treemap (provided you have the room to support them), you can easily add data labels. Select the chart and open the **Chart Elements** flyout menu in the upper-right corner. Click the right-facing arrow next to **Data Labels** and choose **More Data Label Options** from the flyout menu to open the **Format Data Labels** pane. Under the **Label Options** tab, select **Value** to add values to the data points. I also set **Separator** to **(New Line) (Fig 5.275)**. Then I noped on out of there (in other words, removed the values, in adult speak) because the chart gave me vertigo.

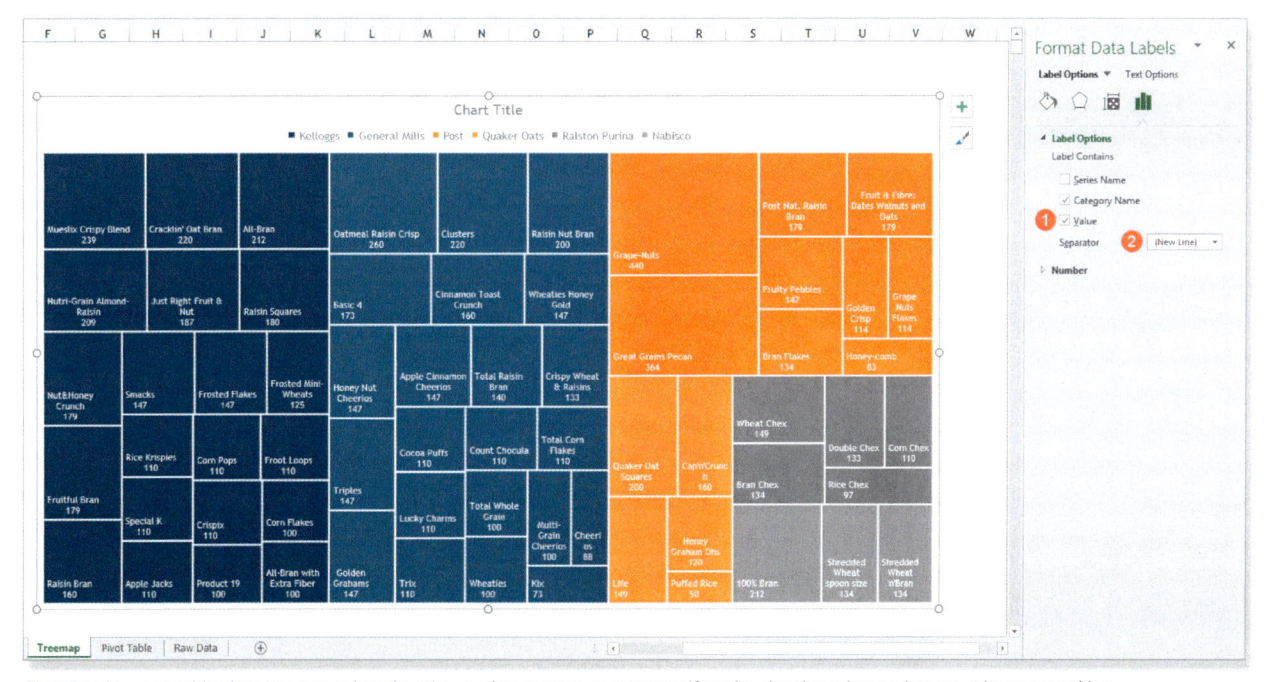

Fig 5.275: You can add values to your points, but they make your treemap messy if you're charting a larger dataset. #doesnotsparkjoy

243

CHAPTER 5 • CREATE CHARTS

Step 10: At the time of writing, Excel doesn't allow you to move the title. I tried everything: dragging, copying/pasting, changing the settings in the **Title Options** tab of the **Format Chart Title** pane, unplugging and plugging my computer back in, creating an effigy of a tree and sacrificing acorns to it…all to no effect. You can't even replace it with a text box. You can try, but Excel doesn't provide sizing handles around the Plot Area to move it down. This means you also can't add an annotation with your source info either. Thanks, Obama.

Step 11: Clean up your chart to taste. I cleaned mine by taking the following steps:

- made title more descriptive
- bumped up title font size
- bumped up legend font size

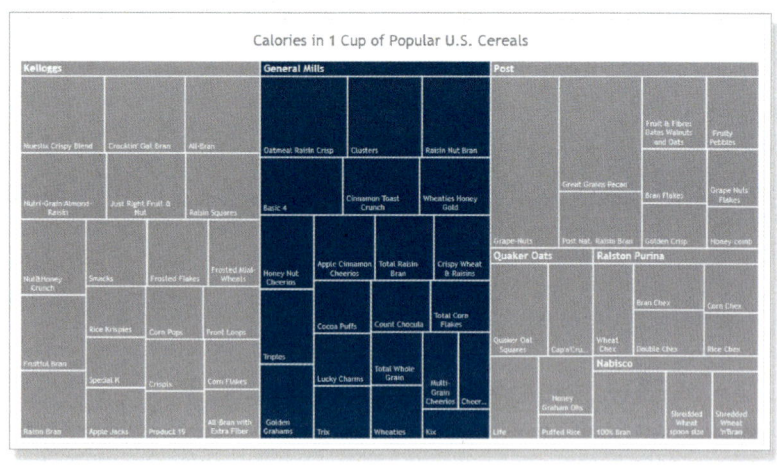

Fig 5.276: Our final chart.

Variation

Highlight with Color

You can use color to highlight a particular data point (e.g., all the General Mills cereals) by graying out the other data points and adding one of your darker colors to the one you want to spotlight.

Step 1: Select a data point by clicking the chart once to select the chart and then the data point to isolate it. If you're successful, the other rectangles will fade into the background. Press Ctrl-1 to open the **Format Data Point** pane. Open the **Fill & Line** tab and set **Fill**

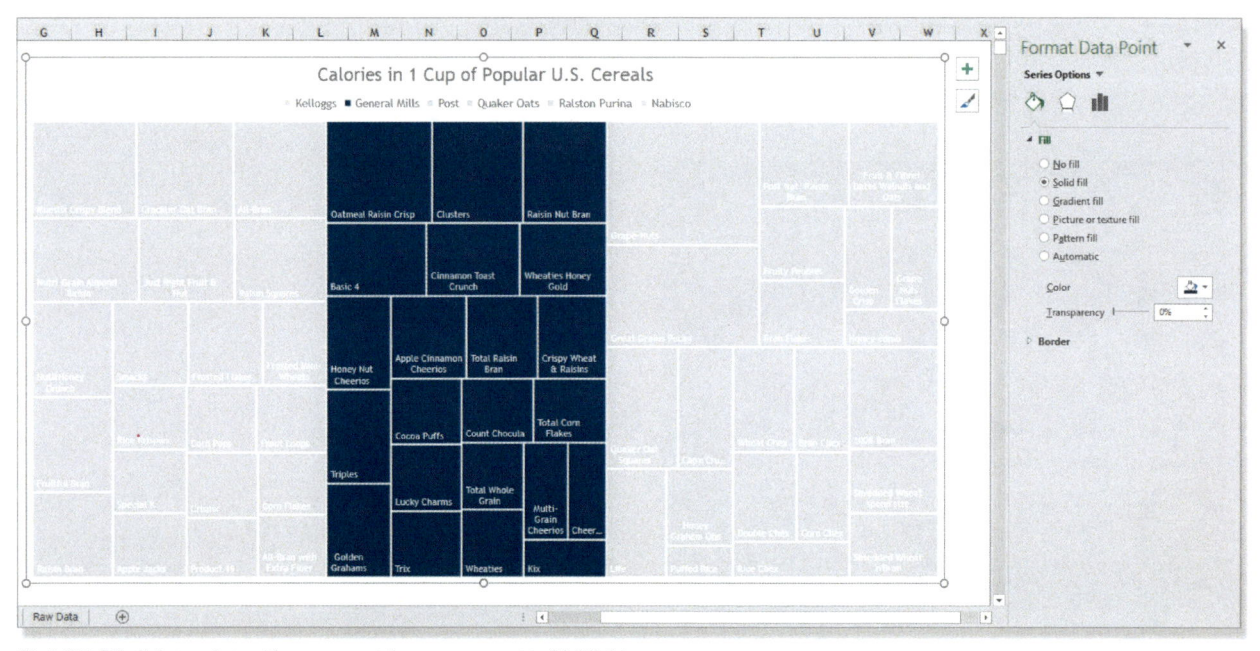

Fig 5.277: Fill all data points with gray except the one you want to highlight.

244

Color to a light gray. Rinse and repeat with each of the data points you want to deemphasize. Then assign the point you want to emphasize one of your branded colors **(Fig 5.277)**.

Step 2: Since most of our data points are now the same color, our legend is about as useless as a peephole on a glass door. Consequently, I enabled banners **(Fig 5.278)**. To do this, select the **Plot Area**, which should open the **Format Data Series** pane. In the **Series Options** tab, set **Label Options** to **Banner**. To make them stand out more, I increased the font size and boldfaced them. To apply formatting to an individual banner label, select it twice to isolate it (but don't double-click).

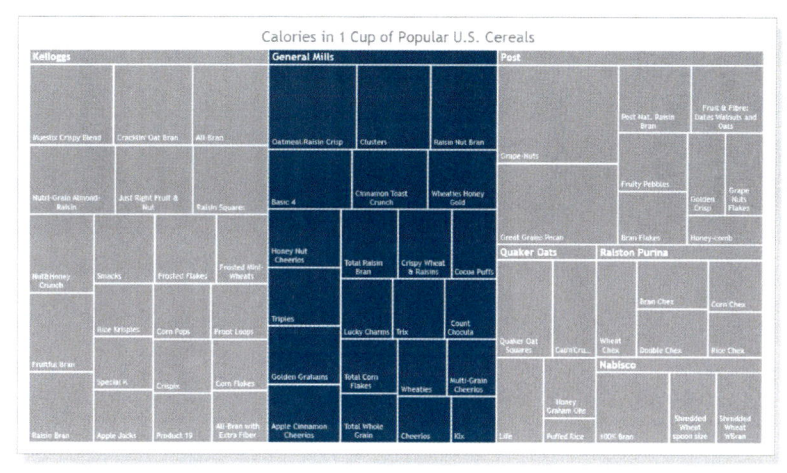

Fig 5.278: Banners are a good idea with when you gray out most of your chart.

Step 3: Our chart title looks like it's wedged awkwardly between the top border of our Chart Area and Plot Area. As mentioned in the tips above, you can't adjust the Plot Area of a treemap, at least not at the time of writing. After grumbling to myself a while, I faked white space by entering a line break above and below the title and setting my font size to 8 pt **(Fig 5.279)**.

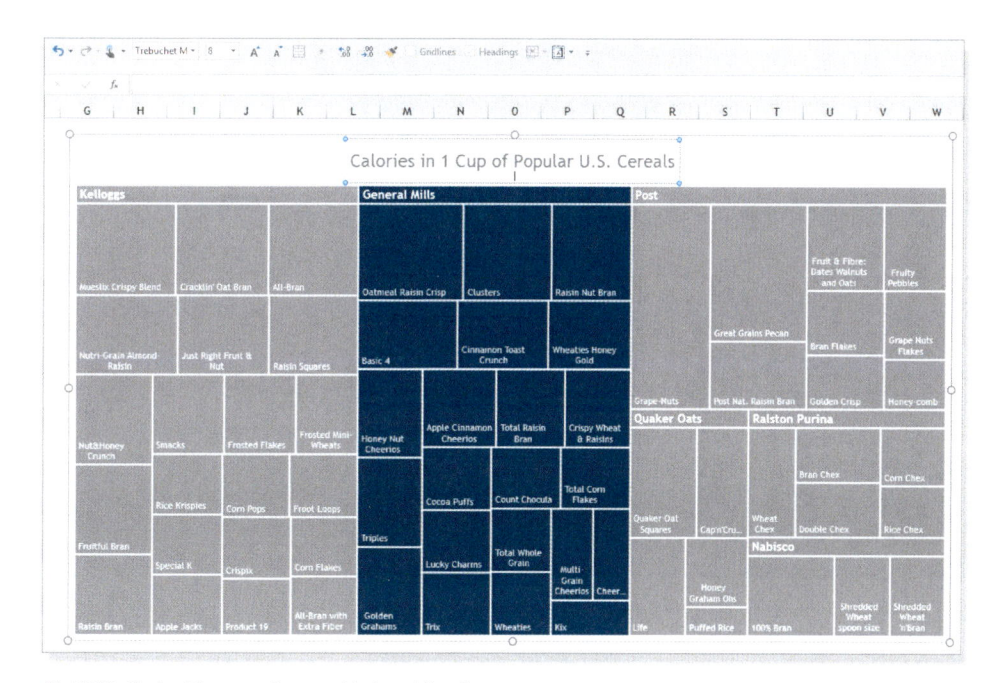

Fig 5.279: Hack white space in your title by adding line spaces.

Example File

treemap.xlsx

Template

At the time of writing, you can't save a treemap as a chart template. You'll need to create it from scratch or save it in a file where you keep all your charts that won't save as templates.

CHAPTER 5 • CREATE CHARTS

SUNBURST CHART

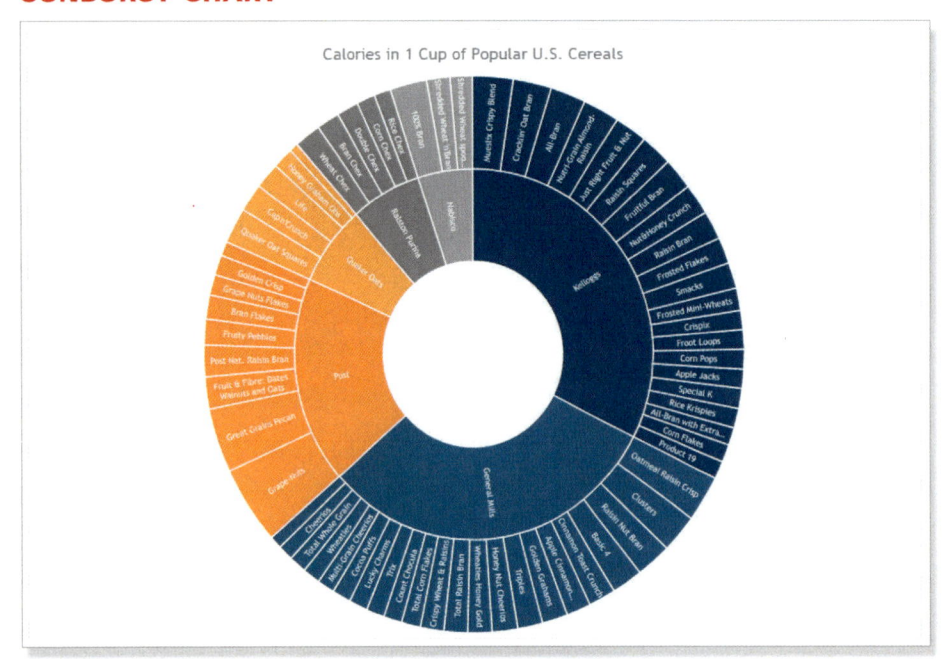

Fig 5.280: Our final chart.

AKA
You may also see sunburst charts called:

- belt charts
- multilevel pie charts
- radial treemaps
- ring charts

Uses
Sunburst charts are best used to:

- visualize hierarchical data
- demonstrate how each data point contributes to the whole
- highlight the relationship of a root node to its child nodes

Tips

- All the charts new to Excel 2016 (Histogram, Waterfall, Pareto, Box & Whisker, Treemap, and Sunburst) tend to be a bit volatile and cause Excel to crash much more than other charts so save often.

- At the time of writing, you can't move the chart title, link the title to a cell (to create a dynamic title), adjust the Plot Area, or save chart templates with any of the charts new to Excel 2016. And if you add a text box (e.g., to add a data source) and move your chart, the text box doesn't move with the chart. (You will need to group the chart and text box if you want to move or scale them together.)

- The most efficient way to get your data sorted properly is with a pivot table. I've included a Pivot Table worksheet in the downloadable workbook to show you how I set mine up. Click anywhere in the pivot table to open the PivotTable Fields pane. From there you can easily see how I constructed it.

- Just as with the treemap, you can use one color for your sunburst chart or assign a discrete color to each data point.

- Arcs are more difficult for our brains to interpret than rectangles, especially among data points that are similar in scope. Consequently, if you want to emphasize the values of your hierarchical data,

246

CHAPTER 5 · CREATE CHARTS

a treemap might be a better choice. If you want to emphasize the relationships among your data points, a sunburst chart might be more effective.

* If you want to emphasize one category element more than the others, you can gray out the others and only apply color to that it.

Category / Metric
2+ categories / 1 metric

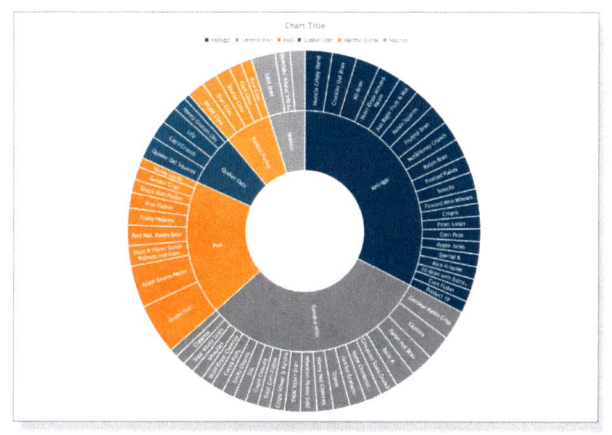

Fig 5.281: Our initial chart.

Tutorial

Step 1: To prepare your data to plot, follow steps 1 – 4 in the Treemap tutorial.

Step 2: Select any cell inside your dataset and navigate to **Insert ▶ Charts ▶ Insert Hierarchy Chart ▶ Sunburst ▶ Sunburst (Fig 5.281)**.

Step 3: You can modify the colors of your subcategories by selecting the chart and then clicking anywhere inside a data point to isolate it. Press Ctrl-1 to open the **Format Data Point** pane. Under the **Fill & Line** tab, set **Fill Color** to one of your branded colors **(Fig 5.282)**. Rinse and repeat with all your top-level categories. If I run out of branded colors and need to use tints, I prefer to group similar tints. You'll need to figure out what style works for you.

Tip: If you're careful not to click on a data label, you will be able to easily select each data point without leaving Excel's isolation mode (my term, not Microsoft's).

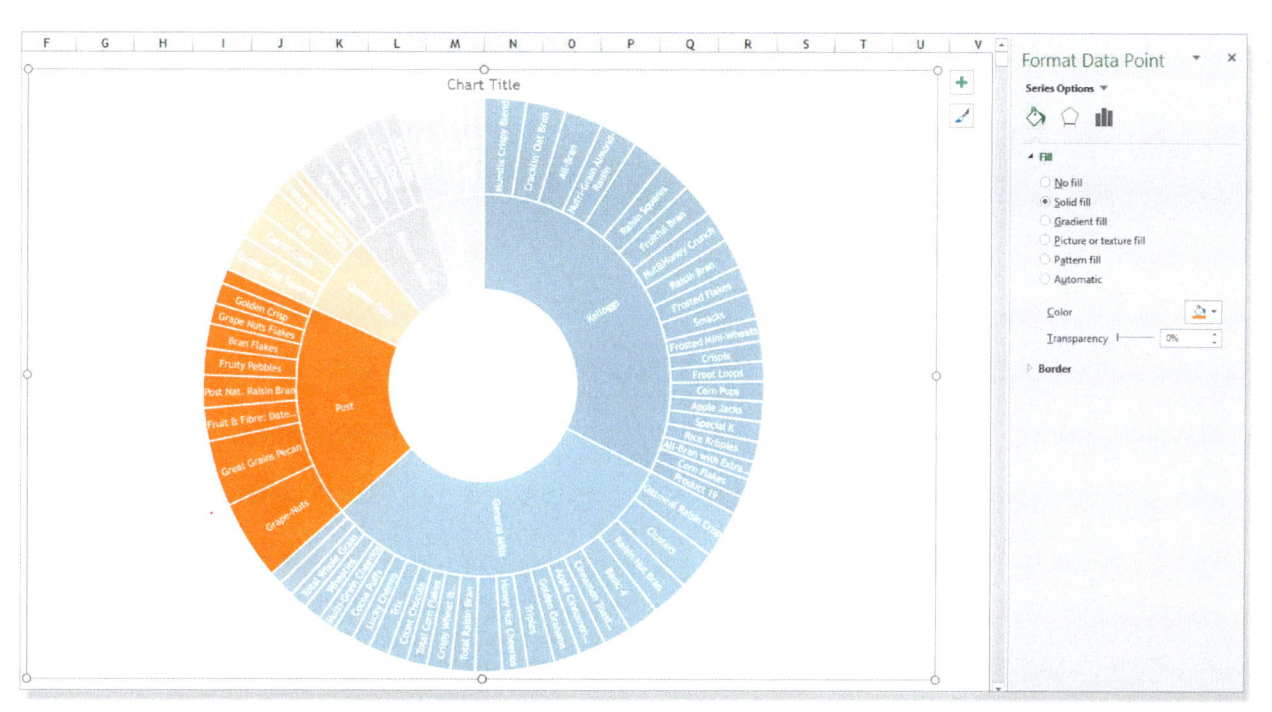

Fig 5.282: You can modify the fill colors of your categories.

247

CHAPTER 5 • CREATE CHARTS

Step 4: Clean up your chart to taste. I cleaned mine by taking the following steps:

- made title more descriptive
- bumped up title font size
- added white space to chart title with line breaks

💬 LEARN MORE See the Design with Elegance chapter to learn how to perform these refinements.

Variation

Highlight with Color

You can use color to highlight a particular data point (e.g., all the Post cereals) by graying out the other data points and adding one of your darker colors to the data point you want to spotlight. You can do this by taking the steps laid out in the Highlight with Color variation of the Treemap tutorial.

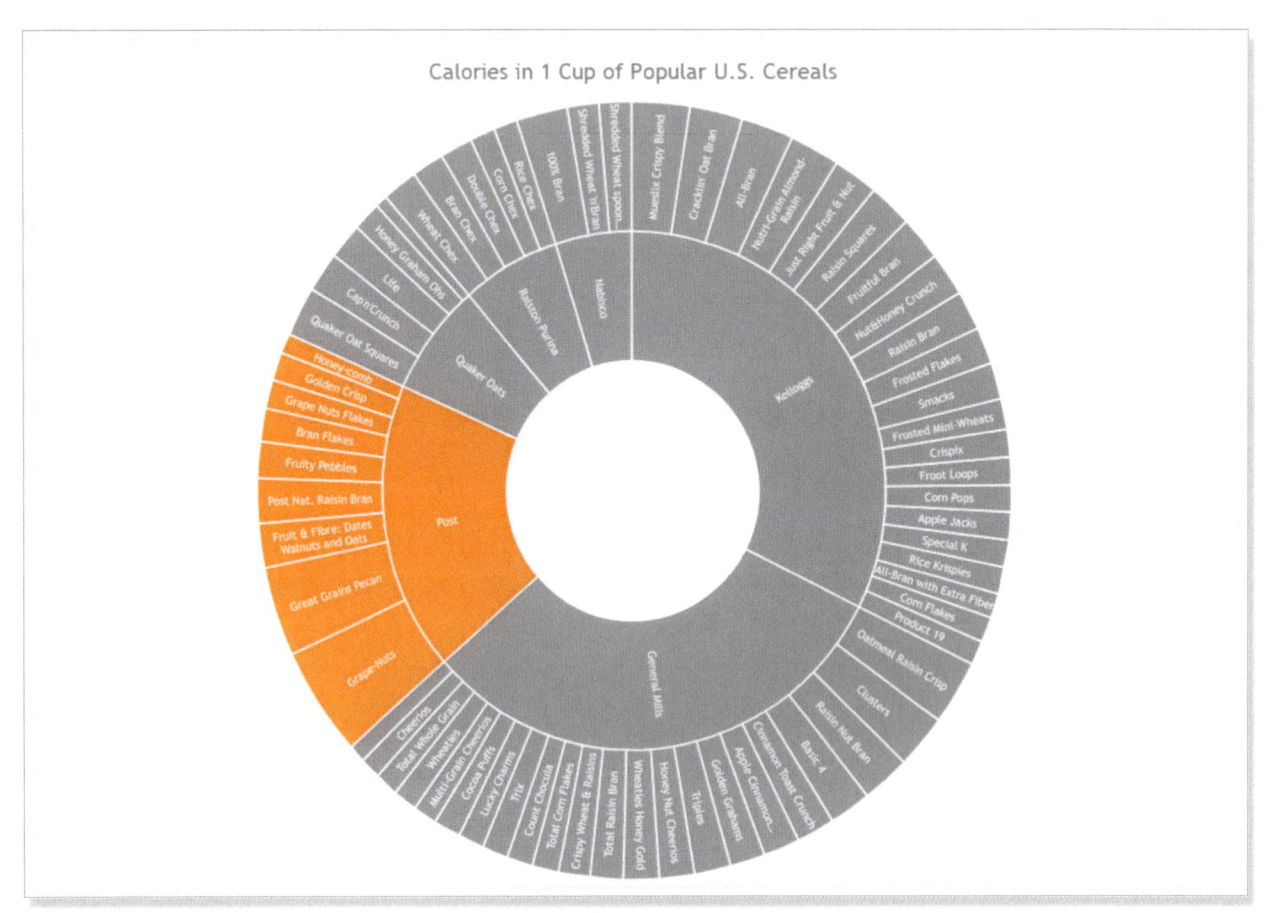

Fig 5.283: Our final chart.

Example File

sunburst-chart.xlsx

Template

At the time of writing, you can't save a sunburst chart as a chart template. You'll need to create it from scratch or save it in a file where you keep all your charts that won't save as templates.

DISTRIBUTION

BOX & WHISKER CHART

First, a personal note about box plots. Before I started writing this book, I spent 50+ hours learning the programming language R. I took a total of eight courses because I have a penchant for thoroughness (if you can't tell). One interesting thing I noticed, across instructors and courses, was that most of them would pull their data into a box plot before doing anything else with it. They used box plots to see how data was distributed, as well as how each of the elements was skewed, if there were outliers, how much the data deviated from the median (indicating how similar or dissimilar the data was), etc. That gave me a whole new appreciation for box plots.

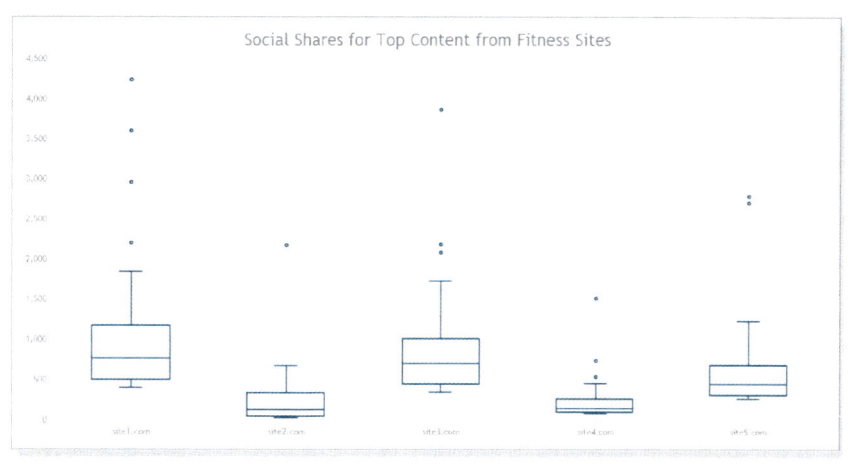

Fig 5.284: Our final chart.

I realized that, as an analyst who deals mostly with marketing data, I'm too quick to aggregate my data without looking at things like skew, kurtosis, deviation, median, and outliers. I've also gained a greater understanding of when median is a better metric than mean. For example, if you have a dataset with outliers that carry the threat of skewing the mean, median is a better metric.

Case in point: In April 2018, Walmart's CEO, Doug McMillon, came under scrutiny for raking in $22.8 million in earnings in fiscal 2018, which ended Jan 31, 2018. CNN's headline read, "Walmart's CEO earns 1,188 times as much as the company's median worker." Why didn't CNN use the mean? Very simply because McMillon's salary would have skewed the mean value.

A box plot is used to graphically depict groups of data through their quartiles. It's derived from five values, as you can see in the diagram I created to demystify them **(Fig 5.285)**.

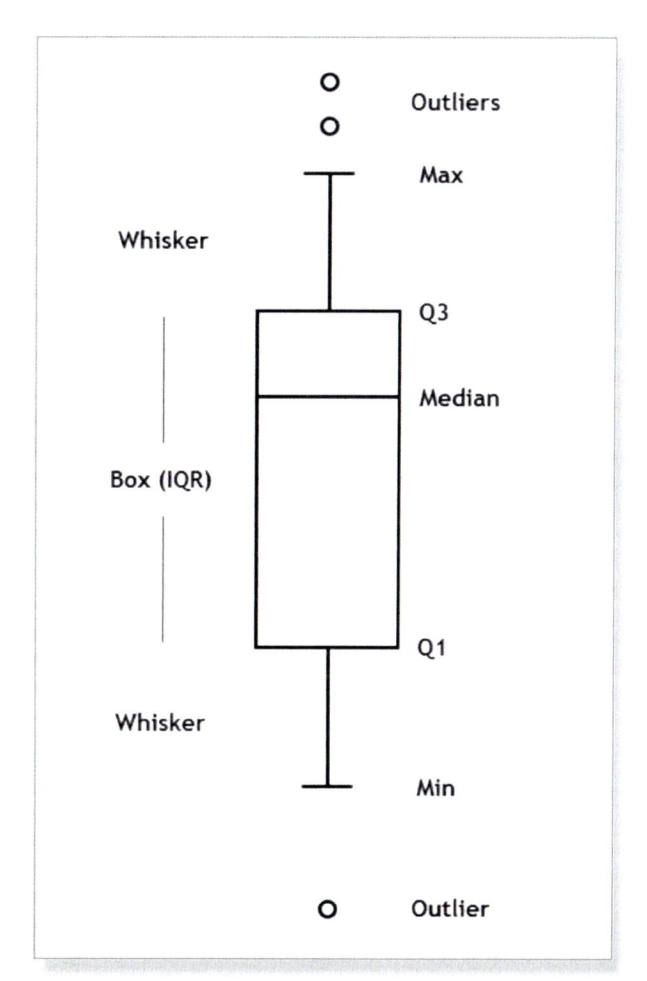

Fig 5.285: A box plot is worth figuring out.

- Min: the smallest value in the dataset (outliers excepted)
- Q1: 25% of the data points fall below this point
- Median: the middle number in a range of numbers (or the average of two middle numbers)
- Q3: 75% of the data points fall below this point
- Max: the largest value in the dataset (outliers excepted)
- Box: the interquartile range (IQR) measures the spread of the data
- Outlier(s): values that fall outside Q1 − (1.5×IQR) or Q3 + (1.5×IQR)

AKA

You may also see box & whisker charts called:
- box diagrams
- box plots
- five-number summary charts

Uses

Box & whisker charts are best used to:
- quickly visualize the distribution of data in datasets with or without a category
- visualize outliers that can distort measures of central tendency in datasets (such as mean or median)
- view skew—or the asymmetry of the probability distribution—in datasets
- compare multiple datasets from independent sources that are related to each other in some way, such as test scores among schools or classrooms, data from before and after an experiment, or data from disparate audiences (such as race, gender, or age)

Tips

- All the charts new to Excel 2016 (Histogram, Waterfall, Pareto, Box & Whisker, Treemap, and Sunburst) tend to be a bit volatile and cause Excel to crash much more than other charts so save often.
- At the time of writing, you can't move the chart title, link the title to a cell (to create a dynamic title), adjust the Plot Area, or save chart templates with any of the charts new to Excel 2016. And if you add a text box (e.g., to add a data source) and move your chart, the text box doesn't move with the chart. (You will need to group the chart and text box if you want to move or scale them together.)
- Know the jargon. If the median is closer to the lower quartile (in other words, mean > median) it has a positive skew; if it's closer to the upper quartile (in other words, median > mean) it has a negative skew; if it's in the center of the box, it's considered a normal (or symmetric) distribution. (You don't even want to know how I remember this…….. Okay, fine. I'll tell you. I think, *Only with data is it better to be mean*. Get it? If the mean is greater, that's good. Ymmv.)
- Consider removing outliers before calculating the mean. If in doubt, toss it out. Not really. I'm just all amped up from that last bullet point. If you do this, make sure you document your dirty deed with an annotation.
- It's often a good idea to provide a secondary analysis of outliers.
- Keeping all the data points in the box can make it visually overwhelming.
- I will refer to them as box & whisker charts in this chapter because that's the nomenclature Excel uses (*sigh*), but the cool kids just call them box plots. If you call them anything else on the playground, you might lose your lunch money.

Category / Metric

0–1 categories / 1 metric

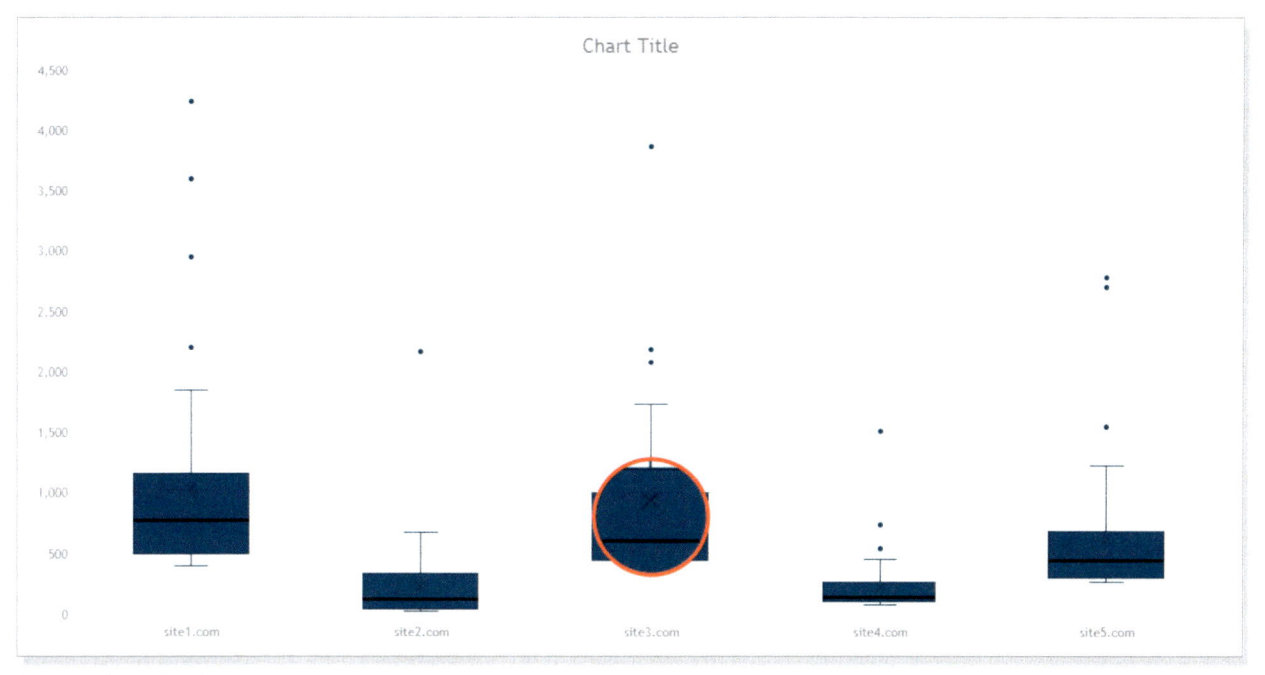

Fig 5.286: Our initial chart.

Tutorial

Step 1: Because we're dealing with frequencies, you want to have a column with your category (optional) and a column with your values. I pulled the top 50 socially shared pages from five popular fitness sites for my box plot. Select any cell inside your dataset and navigate to **Insert ▸ Charts ▸ Insert Statistic Chart ▸ Box and Whisker ▸ Box and Whisker (Fig 5.286)**.

Note: I zoomed in on a mean marker because they can be difficult to see.

Segue: Right from the start we can see a few things from our data:

- Each of the boxes has a positive skew because the median line is lower than the mean marker. Even if we didn't have the mean markers in our charts, we would know that they are all skewed because the median line in each box plot is closer to the lower quartile.

- site1.com has received the most social shares on average, as evidenced by the position of its mean marker compared to the other sites'.

- site1.com also has the greatest amount of deviation in the number of shares among its posts, as evidenced by the height of its box.

- site4.com has the most consistency in the number of social shares its posts receive, as evidenced by having the shortest box.

- Outliers in this context are posts that experienced some virality (or bombed, if there had been any outliers hanging out under the boardwalk). I would want to investigate the content to see if we could replicate that magic.

- There's more deviation in site3.com's and site5.com's 4th quartiles than their 1st, 2nd and 3rd quartiles combined. This may suggest that some of their top content for that month was on the cusp of going viral. If I were an analyst for either of these sites (which, as a fitness freak, would be hella fun!), I'd want to look at the content in that quartile to see if there's some way they could squeeze more from future articles, like the ones that wanted to go viral but just couldn't seem to break away from the pack.

CHAPTER 5 • CREATE CHARTS

* Whoever is responsible for site1.com's social efforts deserves a raise. Four viral posts in a month is fantastic.

Note: We're operating under the assumption that the sample we used is representative of all posts, which would need to be tested to withstand scientific rigor.

Okay, back to our regularly scheduled program….

Step 2: You know by now how I feel about gridlines. So icky. Since we're mostly using box plots to see patterns and distributions, I'll delete them.

Step 3: I like to remove the fill from the boxes to make it easier to see the median line and to reduce the ink in the chart **(Fig 5.287)**. To remove the fill, select a box to select all of them, then press Ctrl-1 to open the **Format Data Series** pane. In the **Fill & Line** tab set **Fill** to **No fill**. If you'd like your borders to be thicker, set **Border** to **Solid line** and bump up **Width**. I set mine to 1.25 pt.

Step 4: If you like the mean marks, by all means (heh), keep them in. I don't think they add much value, so I remove them. If you want to remove yours, select a box to select your entire data series and open the **Format Data Series** pane. In the **Series Options** tab deselect **Show mean markers**.

Step 5: When I'm analyzing the data, I sometimes like to take a peek at the individual data points **(Fig 5.288)**. It's a stop gap measure because I don't keep them in the final presentation. If you want a more granular picture of your data, you could create a dot plot. If you want to see them, select **Show inner points** from the **Series Options** tab of the **Format Data Series** pane. Unless they tell a compelling story though, give them a pink slip afterwards.

💬 **LEARN MORE** See the Dot Plot tutorial to learn how to create dot plots.

Step 6: Clean up your chart to taste. I cleaned mine by taking the following steps:

* made title more descriptive
* bumped up title font size
* added white space to chart title with line breaks

Example File
box-plot.xlsx

Template
At the time of writing, you can't save a box & whisker plot as a chart template. You'll need to create it from scratch or save it in a file where you keep all your charts that won't save as templates.

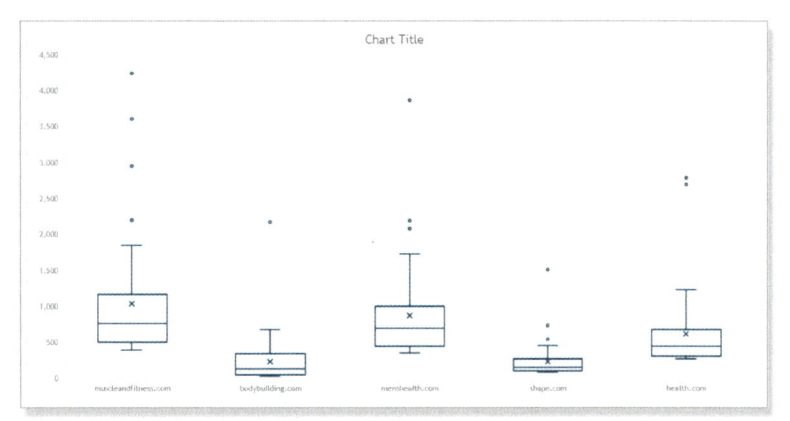

Fig 5.287: Remove fill and bump up the border width to clean up your boxes.

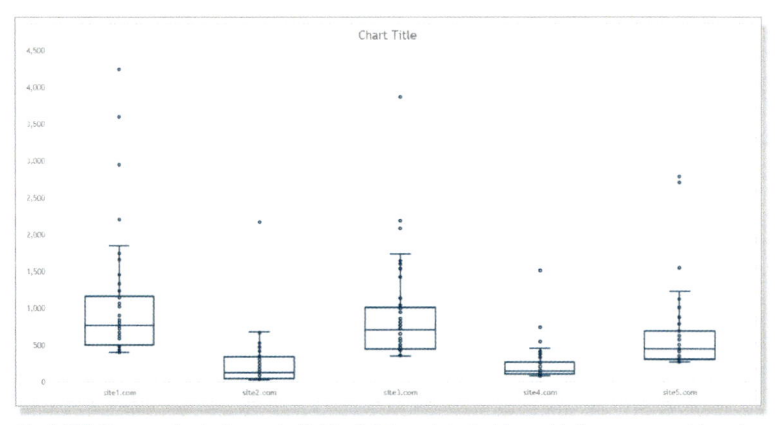

Fig 5.288: You can look at your individual data points, but I wouldn't recommend keeping them for the final presentation.

HISTOGRAM

Uses

Histograms are best used to:

- display frequencies within a distribution
- visualize datasets with continuous values
- visualize datasets with a variable that can be displayed in individual histograms, such as a dataset with multiple years of the same data
- plot datasets that can be easily binned
- demonstrate the spread of data in a dataset
- plot quality control performance metrics

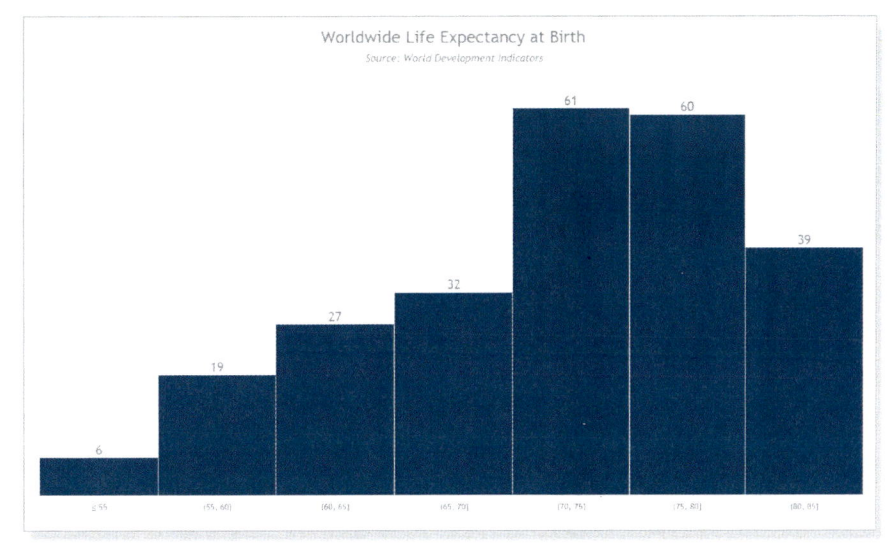

Fig 5.289: Our final chart.

Tips

- All the charts new to Excel 2016 (Histogram, Waterfall, Pareto, Box & Whisker, Treemap, and Sunburst) tend to be a bit volatile and cause Excel to crash much more than other charts so save often.
- At the time of writing, you can't move the chart title, link the title to a cell (to create a dynamic title), adjust the Plot Area, or save chart templates with any of the charts new to Excel 2016. And if you add a text box (e.g., to add a data source) and move your chart, the text box doesn't move with the chart. (You will need to group the chart and text box if you want to move or scale them together.)
- A histogram shaped like a bell is normally distributed. You can find a simple explanation of what the different histogram shapes (or distributions) mean on the **statisticshowto.com** website: **bit.ly/mdsepc-distributions**.
- Don't add space in between your bins. Doing this will make your histogram look more like a column chart, which measures aggregated data, not frequencies.
- You need a category (e.g., country) to create the distribution of your metric (e.g., life expectancy), but Excel only lets you plot the metric.
- Be careful to not make bins so small, your insights are lost in bins that are too granular.
- A histogram may also be normalized to display relative frequencies.
- If you find that you have a number of small bins to the left or right of your histogram, you can use Excel's Underflow and Overflow bin options.
- I try to keep my bin ranges as clean as possible. Making your ranges unnecessarily granular will just reduce comprehension.
- If you add data labels, you may want to place them outside the bins. At the time of writing, if you choose Inside End, there's no top margin applied to these text boxes, so they look like they're trying to cram all their life possessions into a tiny, two-bedroom apartment in New York. From what I've read.
- Unfortunately, Excel doesn't give you a way to

CHAPTER 5 • CREATE CHARTS

overlay your histogram with a normal distribution curve. I hope this functionality will be added in a future version.

+ If you want to draw special attention to one of your bins, you can gray out the others.

Category / Metric
0 categories / 1 metric

Tutorial
Step 1: Select any cell inside your dataset and navigate to **Insert** ▶ **Charts** ▶ **Insert Statistic Chart** ▶ **Histogram** ▶ **Histogram (Fig 5.290)**.

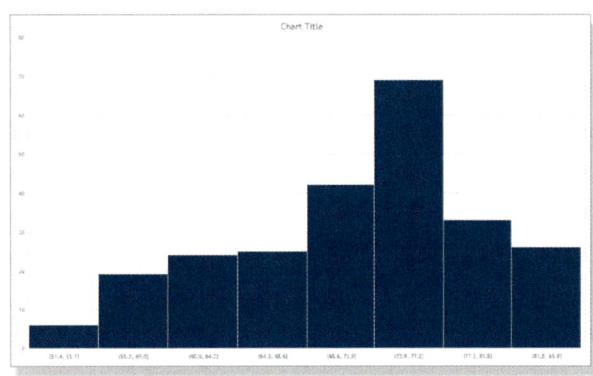

Fig 5.290: Our initial chart.

Step 2: As I warned in the tips, Excel's default bins can be unnecessarily granular. To clean them up, select your horizontal axis and press Ctrl-1 to open the **Format Axis** pane. Open the **Axis Options** tab and set **Bins** to **Bin width**. Then set **Bin width** to something that will be less granular. Excel set mine to 4.3. I changed it to 5 **(Fig 5.291)**. If you want to modify the number of bins, play with this setting until you get your desired results.

Step 3: Those bin widths are pretty funky. That's because Excel starts counting from the smallest value in your dataset. My smallest value was 51.4, so Excel started incrementing by 5 from that value. You can clean your axis up by using an underflow bin. If you'd like to set one, select **Underflow bin** and enter a value that makes sense for your dataset. I set mine to 55. Then Excel labeled my first bin <= 55 **(Fig 5.292)**.

Warning: If you work on a Mac too, you'll need to access these bin settings from the **Format Data Series** pane.

Step 4: I don't like unnecessary decimal points, so I removed them from the horizontal axis. To remove yours, select the axis to open the **Format Axis** pane,

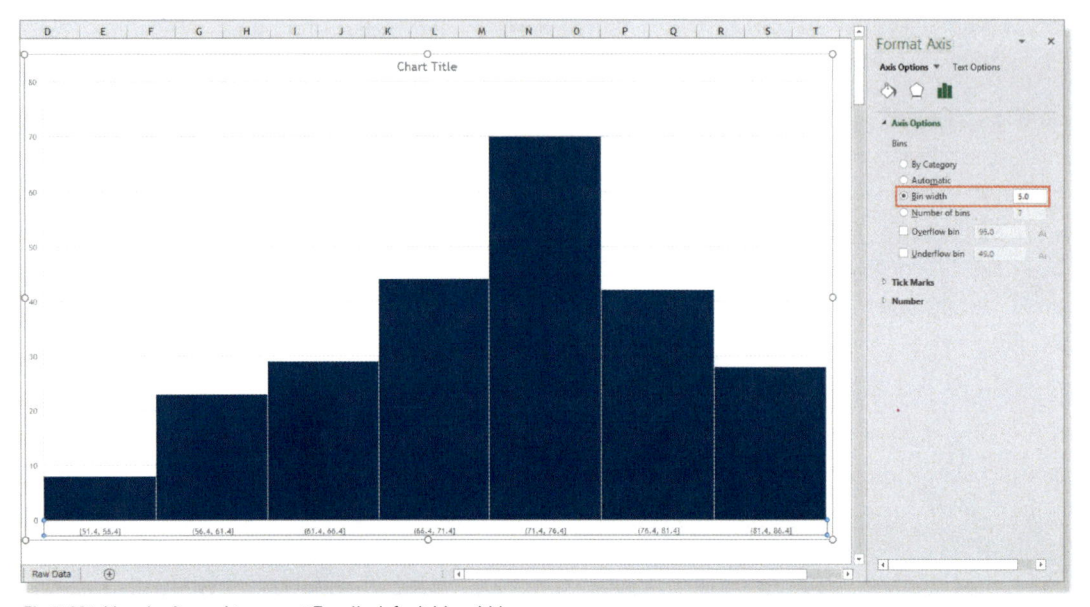

Fig 5.291: You don't need to accept Excel's default bin widths.

254

then open the **Axis Options** tab. Under the **Number** section, set **Decimal places** to 0 **(Fig 5.293)**.

Warning: At the time of writing, Excel went rabid and crashed twice while testing this feature, so you may want to save your file before touching this setting.

Step 5: Clean up your chart to taste. I cleaned mine by taking the following steps:

+ made title more descriptive
+ left aligned chart title
+ bumped up title font size
+ added annotation for data source
+ removed gridlines
+ added data labels

Example File
histogram.xlsx

Template
At the time of writing, you can't save a histogram as a chart template. You'll need to create it from scratch or save it in a file where you keep all your charts that won't save as templates.

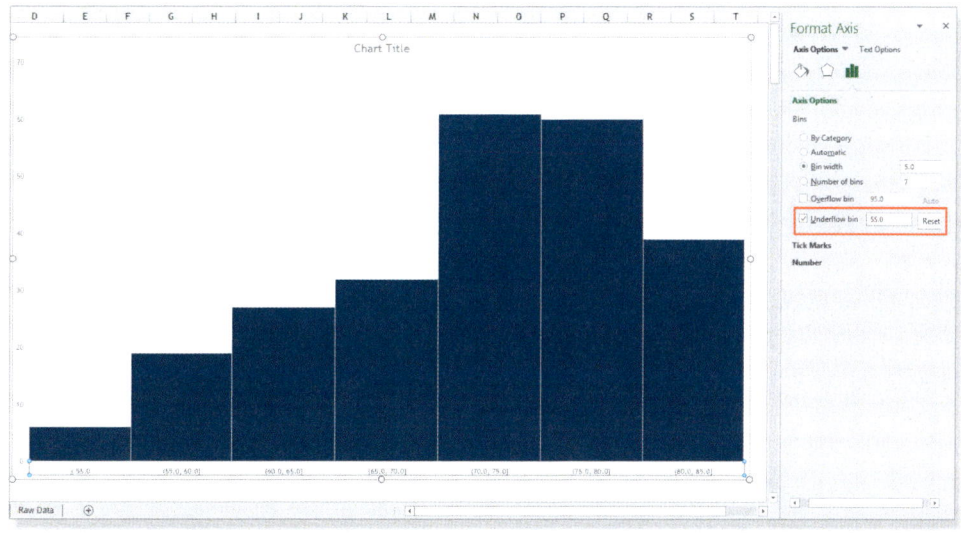

Fig 5.292: Set Underflow bin to clean up your axis.

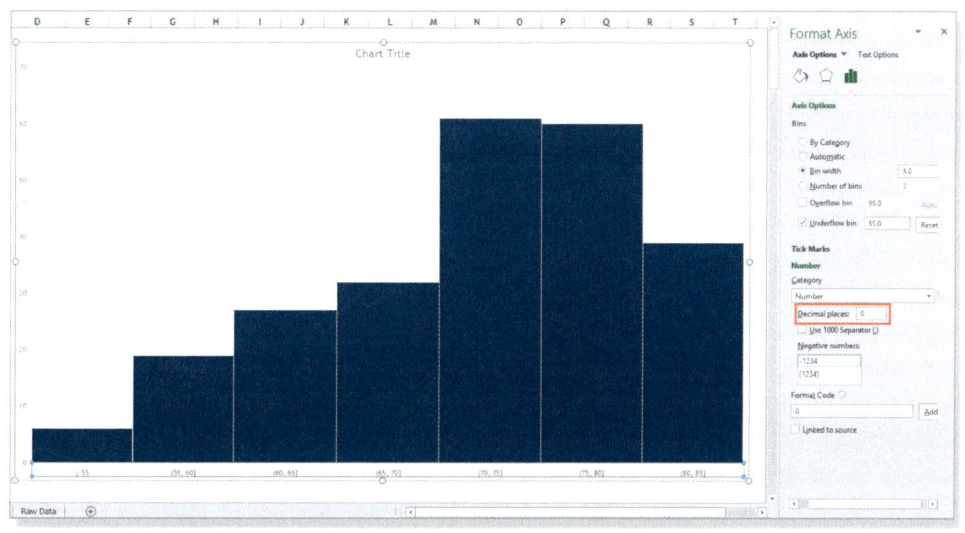

Fig 5.293: If you're working with whole numbers, you may want to remove decimals.

CHAPTER 5 • CREATE CHARTS

PARETO CHART

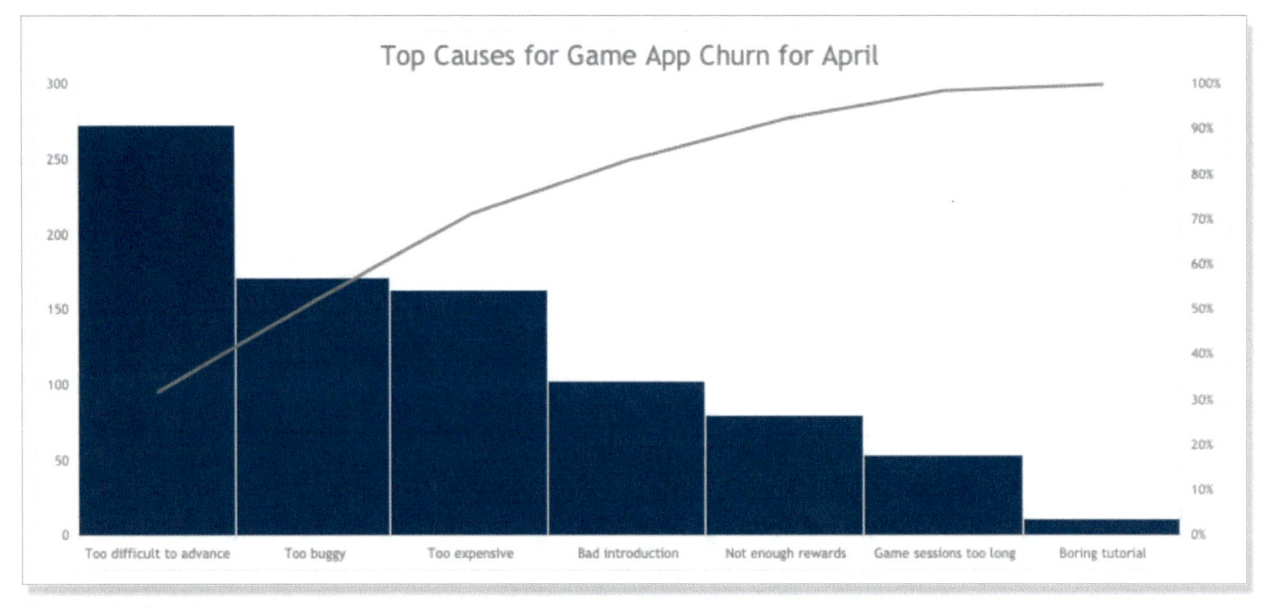

Top Causes for Game App Churn for April chart

Fig 5.294: Our final chart.

Uses
Pareto charts are best used to:
- analyze the occurrences of problems in a process
- identify steps in a process that have the greatest impact

Tips
- All the charts new to Excel 2016 (Histogram, Waterfall, Pareto, Box & Whisker, Treemap, and Sunburst) tend to be a bit volatile and cause Excel to crash much more than other charts so save often.
- At the time of writing, you can't move the chart title, link the title to a cell (to create a dynamic title), adjust the Plot Area, or save chart templates with any of the charts new to Excel 2016. And if you add a text box (e.g., to add a data source) and move your chart, the text box doesn't move with the chart. (You will need to group the chart and text box if you want to move or scale them together.)
- Think of a Pareto chart as a histogram sorted in descending order with a line chart that arcs from 0% to 100% as it stretches across your chart.
- You can't add markers to the line chart. I wanted to add data labels to markers so I could delete both axes, but there's no marker option in the Format Pareto Line pane. So I left both axes in the chart, even though I used data labels for the histogram. Keeping only one of them would be like seeing only one of the creepy hallway twins from *The Shining*.
- You can't thin out either of the vertical axes. A 10% increment is a bit noisy for my taste. Hopefully Microsoft will correct this in a future version.
- Determine the time period that is most appropriate for your analysis (e.g., day, week, month, year), and indicate it in your chart.
- Pareto charts are already difficult to interpret for the layperson, so it's imperative to keep the chart as clean as possible.
- Keep axis labels as short as possible.
- It's better to position labels at an angle than at 90° (completely sideways).
- Although you can format columns with discrete colors, it is distracting without adding value.

CHAPTER 5 · CREATE CHARTS

Category / Metric
1 category / 1 metric

Tutorial
Hundred-Foot View
A Pareto chart combines a histogram and a line chart. The bins are sorted in descending order, and the line represents the cumulative percentage of the total number of occurrences. For this reason, a Pareto chart has two axes. The line starts at 0% and arches toward 100%, although Excel doesn't make it extend to the bottom-left corner of the chart for aesthetic reasons.

Step 1: Select any cell inside your dataset and navigate to **Insert ▶ Charts ▶ Insert Statistic Chart ▶ Histogram ▶ Pareto (Fig 5.295)**.

Step 2: Clean up your chart to taste. I cleaned mine by taking the following steps:

- made title more descriptive
- bumped up title font size
- lightened the gridlines

💬 LEARN MORE See the Design with Elegance chapter to learn how to perform these refinements.

Example File
pareto-chart.xlsx

Template
At the time of writing, you can't save a Pareto chart as a template. You'll need to create it from scratch or save it in a file where you keep all your charts that won't save as templates.

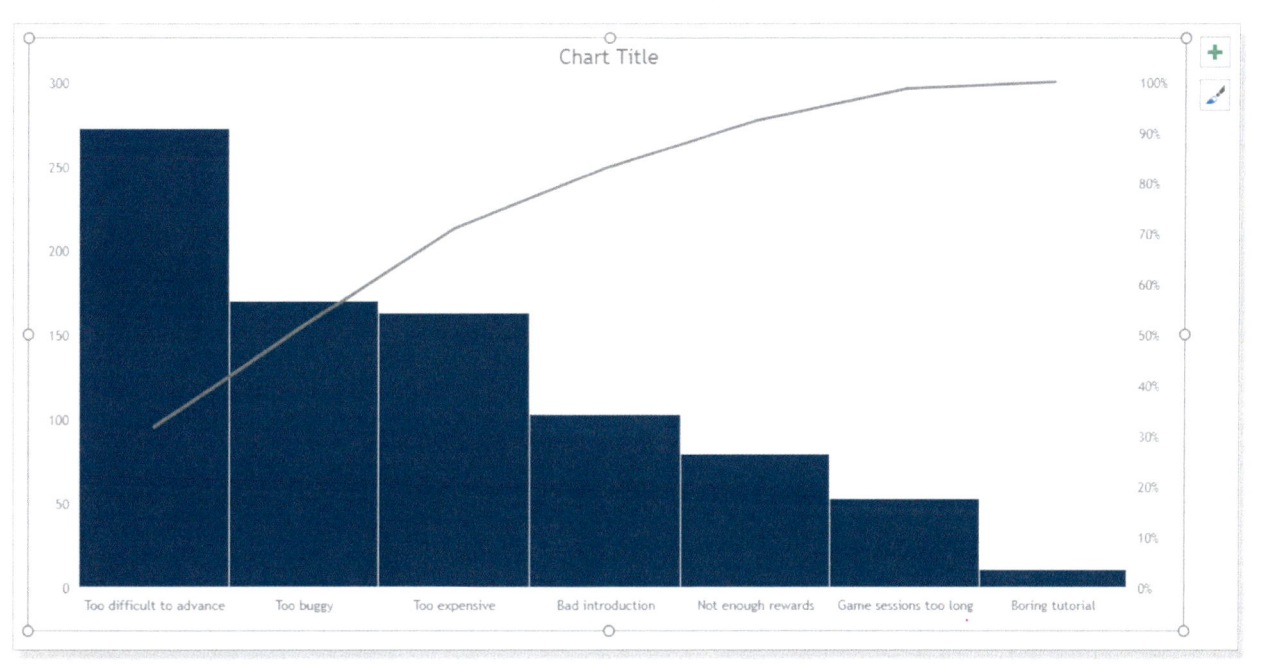

Fig 5.295: Our initial chart.

CHAPTER 5 • CREATE CHARTS

STOCK CHART

High Low Close for GOOG
Nov 2017

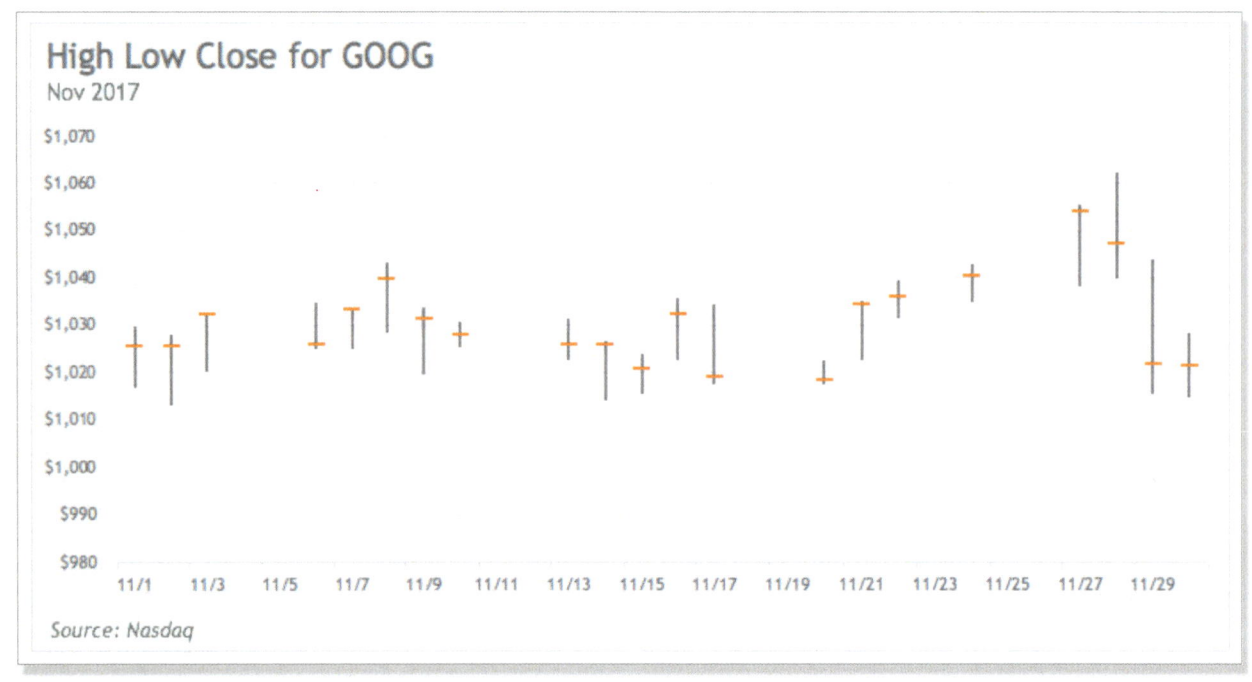

Source: Nasdaq

Fig 5.296: Our final chart.

AKA

You may also see stock charts called:

- candlestick charts
- Japanese candlestick charts
- Open-High-Low-Close charts

Uses

Stock charts are best used to:

- represent trading patterns over short periods of time
- visualize price movements of a security, stock, bond, derivative, currency, or commodity
- aid in making decisions in trading
- demonstrate fluctuations in any metric, not just financial

Learn more from the Candlestick Chart page on Wikipedia: **bit.ly/mdsepc-candlestick**.

Tips

- Each candlestick typically represents one day of trading.

- Although the Open-High-Low-Close chart is visually similar to the box & whisker chart, it serves a very different purpose.

- Stock charts must be formatted in a very specific order. Date must be the first column in your dataset. Although the metrics listed below are not all required, if they're included, they must be in this order (from left to right in your spreadsheet):

  - Volume
  - Open
  - High
  - Low
  - Close

258

CHAPTER 5 • CREATE CHARTS

- The color of the box represents whether prices opened higher or lower than the close. A stock that closes at a price lower than it opened is typically indicated by a box with a fill, whereas a stock that closes at a price higher than it opened is indicated with a box with no fill. The solid color helps to really maximize that sinking feeling you experience when your stock betrays you **(Fig 5.302)**.

- Excel's default lines and markers for the High-Low-Close chart are very thin. If your audience is over 40, it may be difficult for them to see the close markers. I like to thicken everything up.

- The vertical (y) axis in stock charts will often not start at 0. This is one of those [very few] times you want to accentuate slight differences in a chart.

Category / Metric
1 category / 3–5 metrics

Tutorial
Note: All the stock charts will share step 1, so the rest of the tutorials in this section will start with step 2 to save repetition.

Step 1: If your table includes only what you'll be charting, select any cell inside your dataset and navigate to **Insert ▸ Charts ▸ Insert Waterfall or Stock Chart ▸ Stock**, then choose an appropriate stock chart. If your table includes more than what you'll be charting (as the example dataset does), click-and-drag over the columns you want to include, provided your range is contiguous. If your range is non-contiguous, press-and-hold the Ctrl key while you drag.

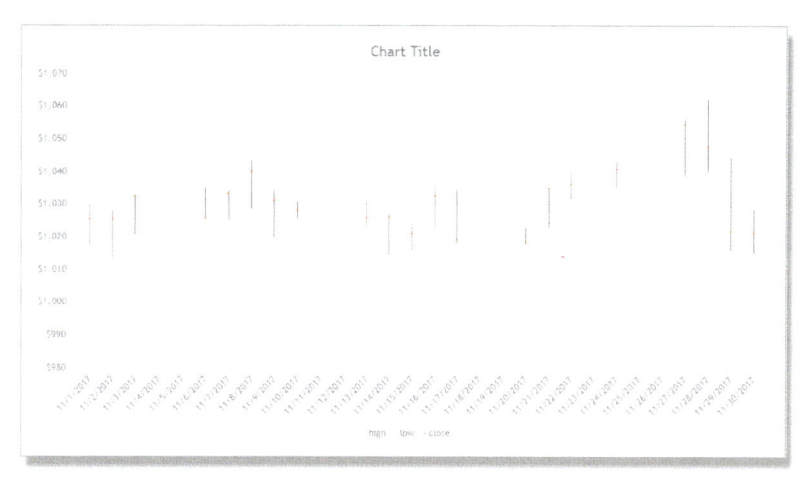

Fig 5.297: Our initial chart.

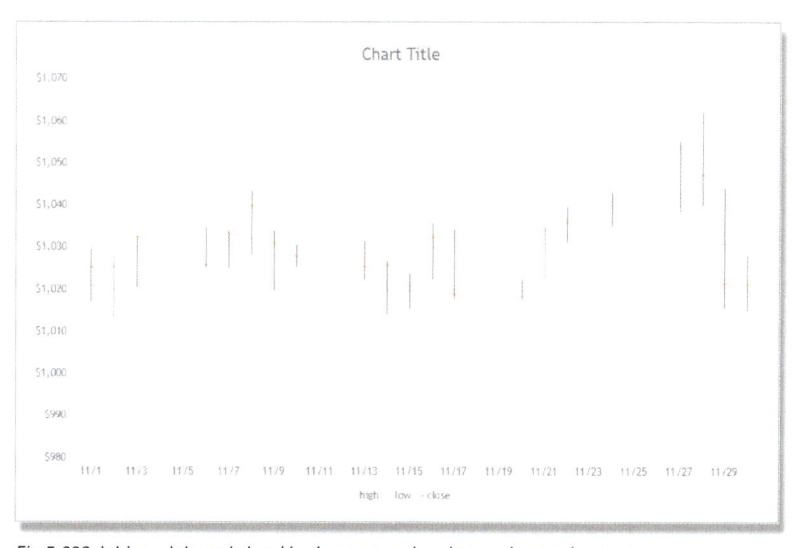

Fig 5.298: I thinned the axis by skipping every other date and removing year.

Note: You may also find it housed under **Insert Waterfall**, **Funnel**, **Stock**, **Surface** or **Radar Chart**.

High-Low-Close
Step 2: Choose **High-Low-Close** from the **Stock** menu of options **(Fig 5.297)**.

Step 3: The first thing we need to do is thin out that horizontal axis. I'll remove the year and skip every other date **(Fig 5.298)**. Trust me: no one will think the even dates went MIA.

CHAPTER 5 • CREATE CHARTS

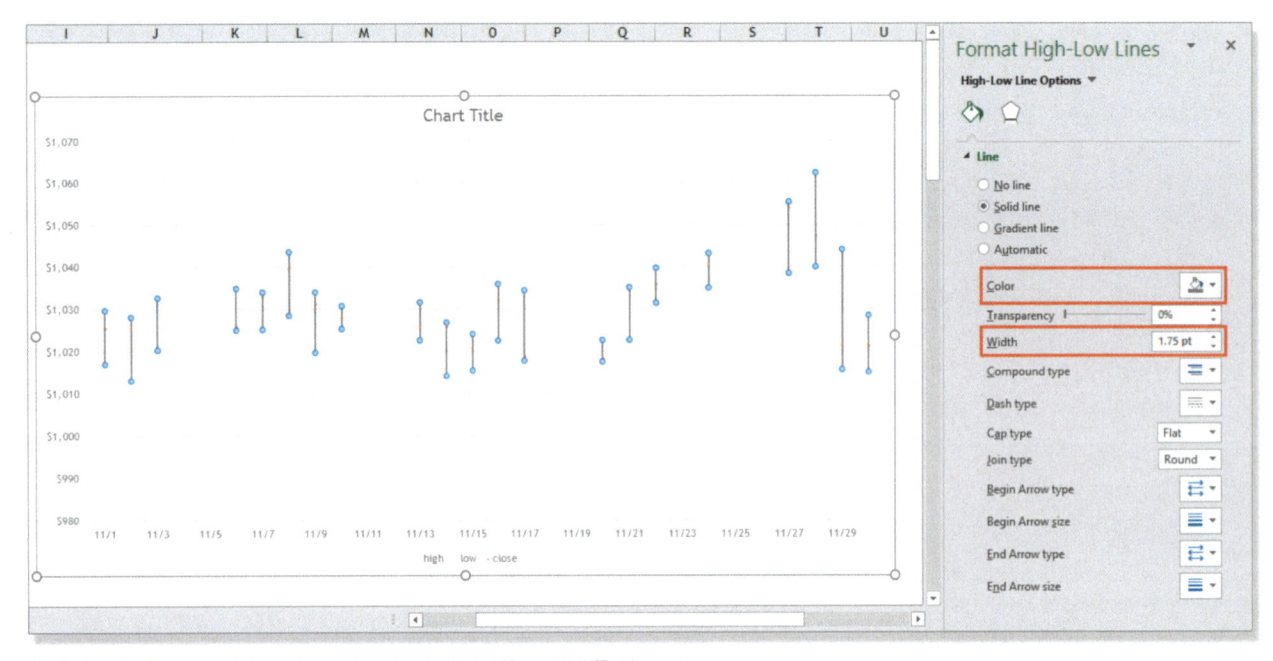

Fig 5.299: Don't hesitate to thicken up your high-low lines if they're difficult to see.

> 💬 **LEARN MORE** See the Thin Out Axes section of the Design with Elegance chapter to learn how to reduce clutter in your axes.

Step 4: The next order of business is putting some fat on the data series. This isn't a Victoria's Secret Fashion Show. To modify the width of the lines, select one to select the entire data series and press Ctrl-1 to open the **Format High-Low Lines** pane.

I set **Outline color** to gray and **Width** to 1.75 pt **(Fig 5.299)**.

Step 5: If you thicken your lines, you'll need to do the same to your markers, or they will get swallowed up by the lines. Select one to select all of them, which will open the **Format Data Series** pane. Open the **Fill & Line** tab and select **Marker** at the top of the tab. Under **Marker Options**, I bumped **Size** up to 9 **(Fig 5.300)**.

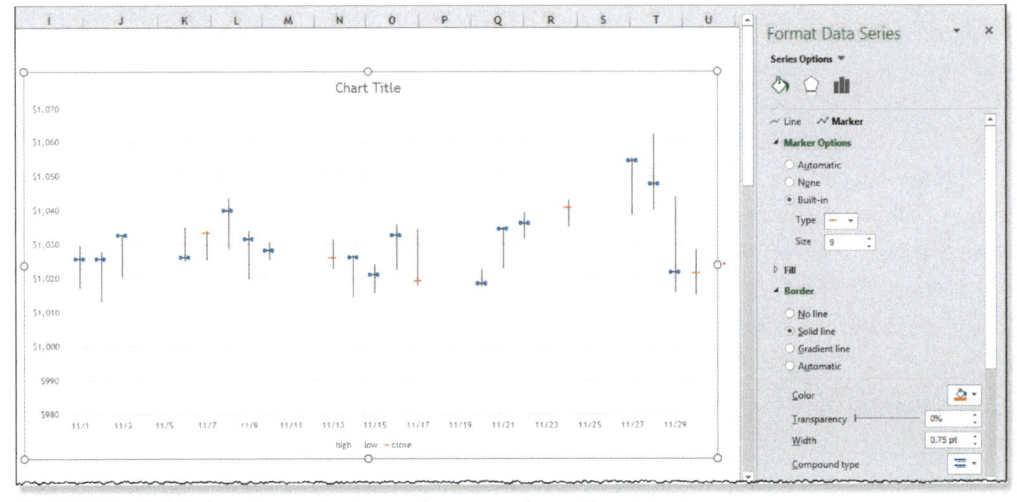

Fig 5.300: Excel's default markers are hard to see, but you can format them to taste.

Step 6: Clean up your chart to taste. I cleaned mine by taking the following steps:

- made title more descriptive
- left aligned chart title
- bumped up title font size
- added annotation for data source
- added subtitle
- bumped up transparency of gridlines
- removed legend

💬 **LEARN MORE** See the Design with Elegance chapter to learn how to perform these refinements.

Open-High-Low-Close

The OHLC chart is pretty intuitive: If the stock closed at a higher price than it opened (yay!), it's typically displayed as hollow. If it closed at a lower price than it opened at (boo!), it's displayed as filled **(Fig 5.302)**.

Step 2: Choose **Open-High-Low-Close** from the **Stock** menu of options **(Fig 5.303)**.

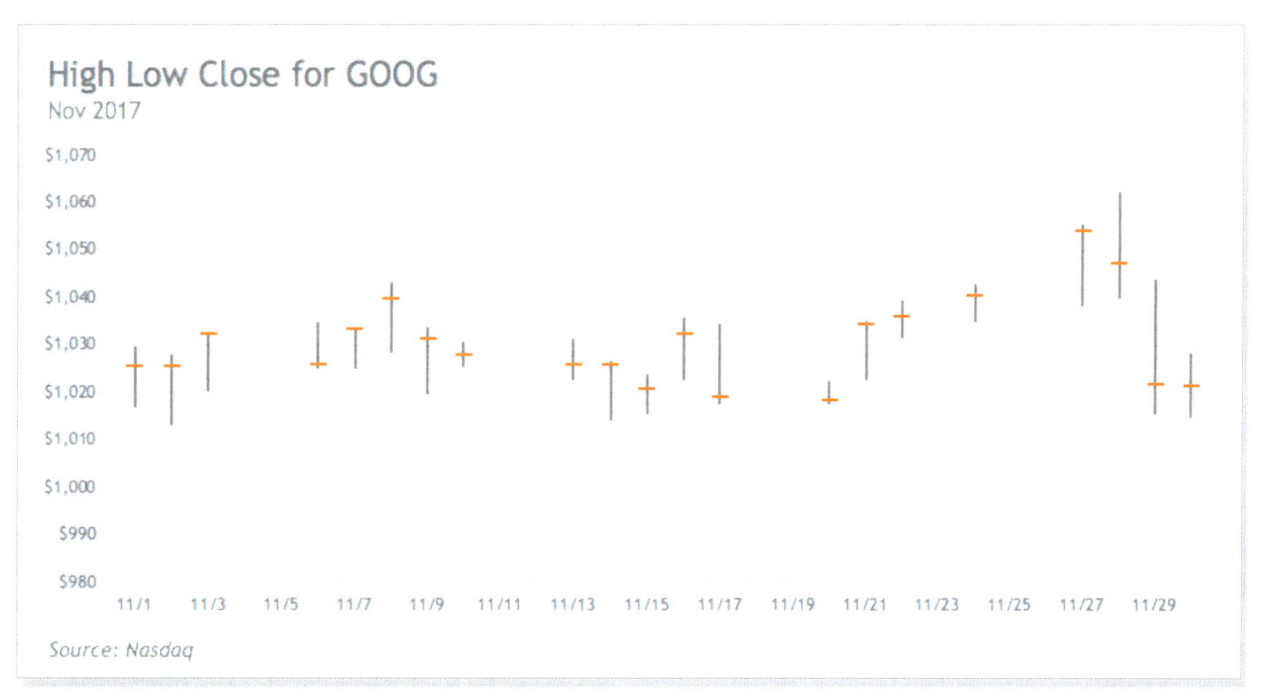

Fig 5.301: Our final chart.

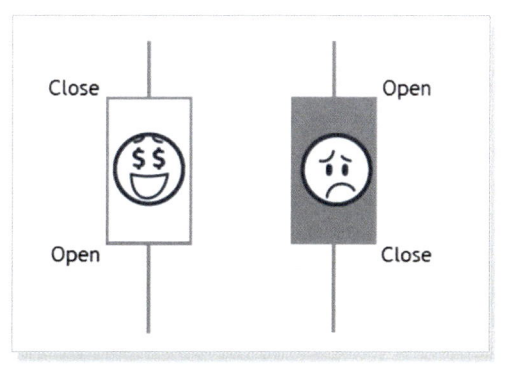

Fig 5.302: How to read an OHLC chart

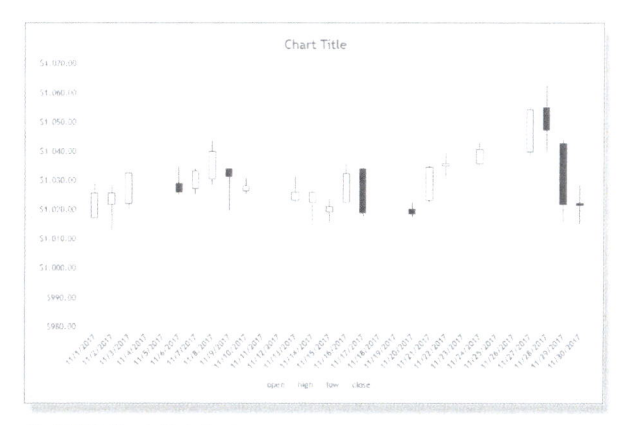

Fig 5.303: Our initial chart.

261

CHAPTER 5 • CREATE CHARTS

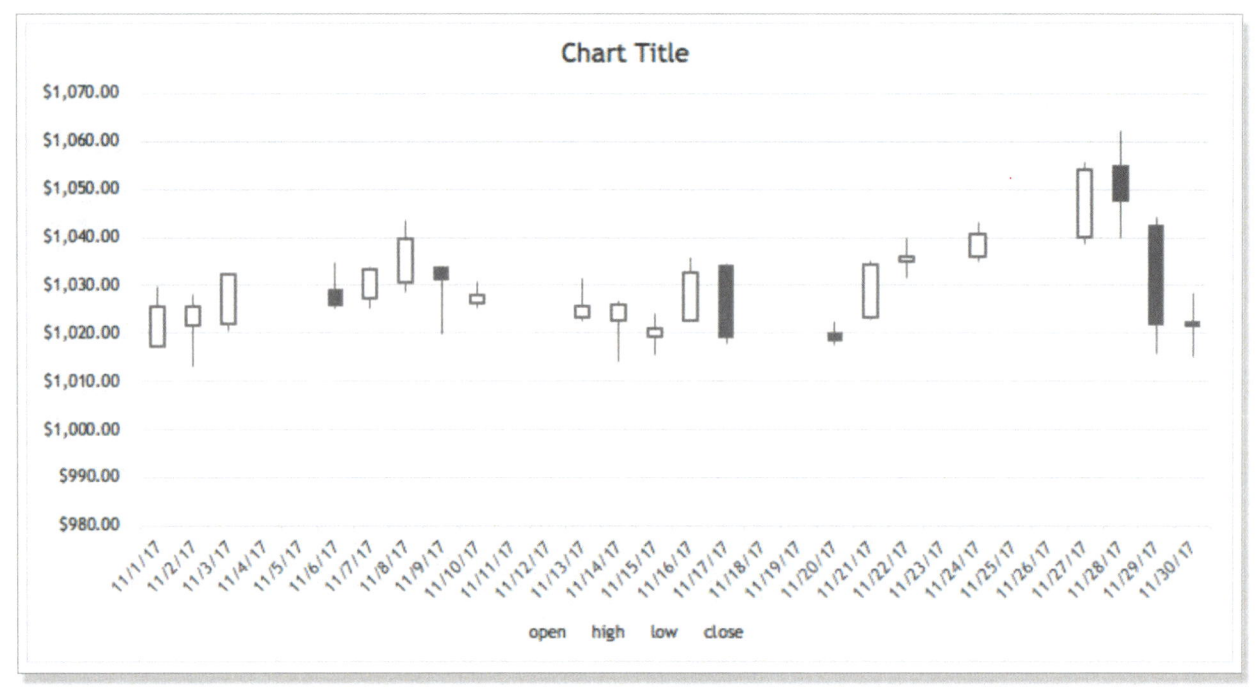

Fig 5.304: I like to thicken my up bar borders to make them pop and visually offset the sucker punch that is the down bars.

Step 3: I like to thicken the bar borders for the up (or hollow) bars. This is a sheer preference issue. If you're picking up what I'm putting down, you can thicken yours by first selecting one of the up bars to select the entire data series. This will open the **Format Up Bars** pane. Open the **Fill & Line** tab and dial up **Width**. I set mine to 1.25 pt **(Fig 5.304)**.

Step 4: Clean up your chart to taste. I cleaned mine by taking the following steps:

+ made title more descriptive
+ left aligned chart title
+ bumped up title font size
+ added subtitle
+ added annotation for data source
+ bumped up transparency of gridlines
+ removed legend
+ customized number formatting of vertical axis
+ made high-low lines thicker (see the High-Low-Close tutorial)

LEARN MORE See the Design with Elegance chapter to learn how to perform these refinements.

262

Volume-High-Low-Close

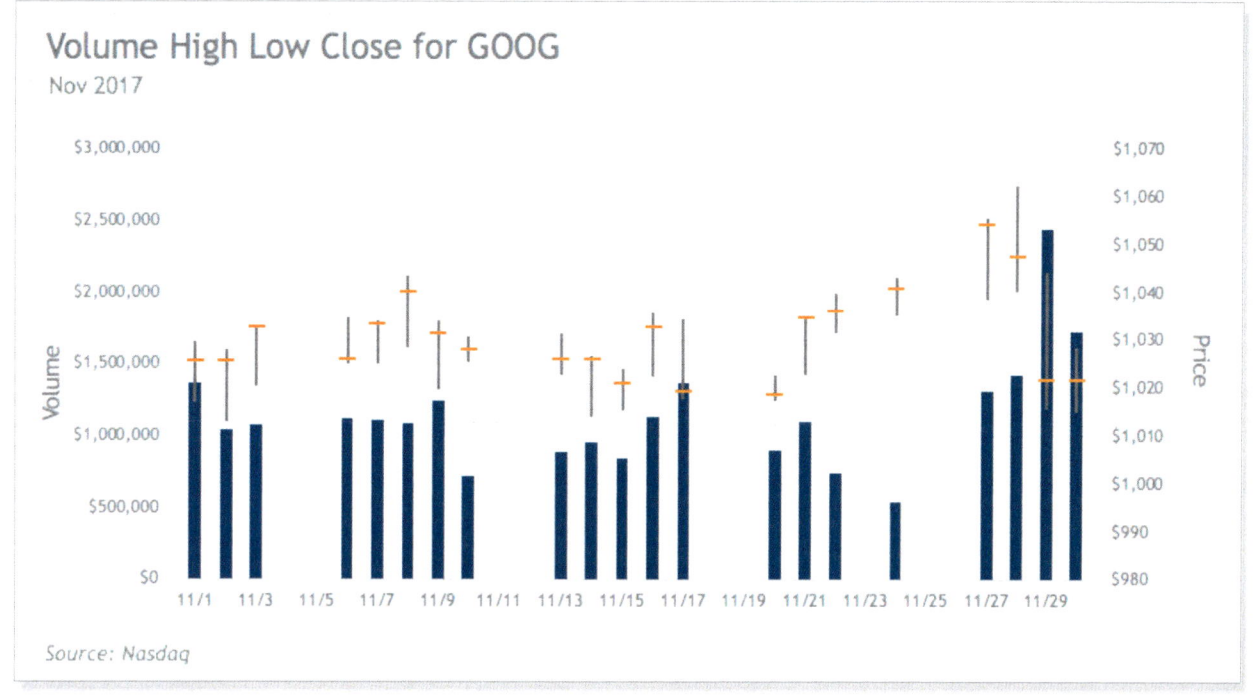

Fig 5.305: Our final chart.

Step 2: Choose **Volume-High-Low-Close** from the **Stock** menu of options **(Fig 5.306)**.

Step 3: Clean up your chart to taste. I cleaned mine by taking the following steps:

- made title more descriptive
- left aligned chart title
- bumped up title font size
- added subtitle
- added annotation for data source
- bumped up transparency of gridlines
- removed legend
- added vertical axis titles
- rotated axis title on secondary axis
- customized number formatting of vertical axis
- made high-low lines thicker (see the High-Low-Close tutorial)

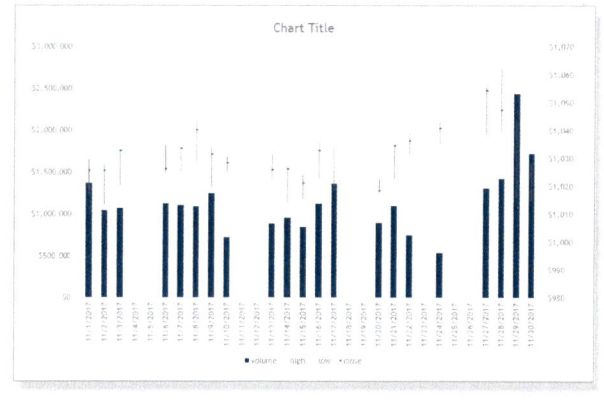

Fig 5.306: Our Initial chart.

LEARN MORE See the Design with Elegance chapter to learn how to perform these refinements.

263

Volume-Open-High-Low-Close

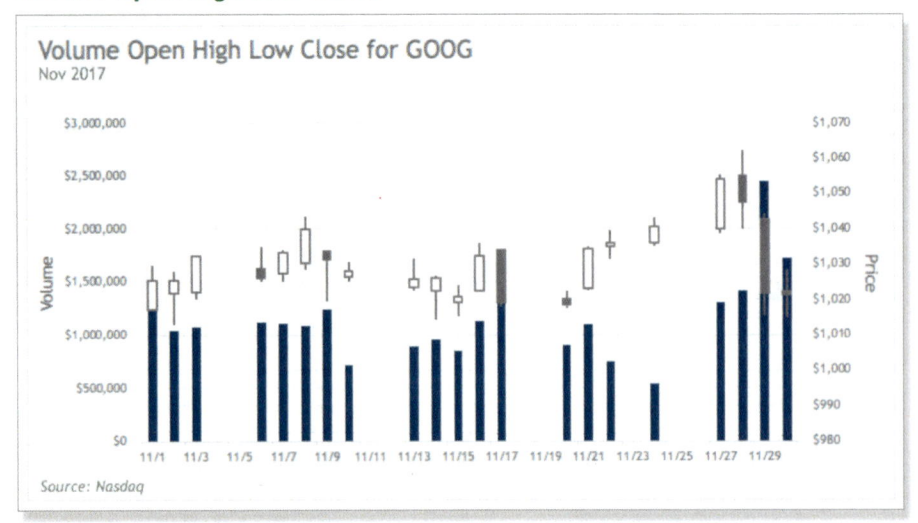

Fig 5.307: Our final chart.

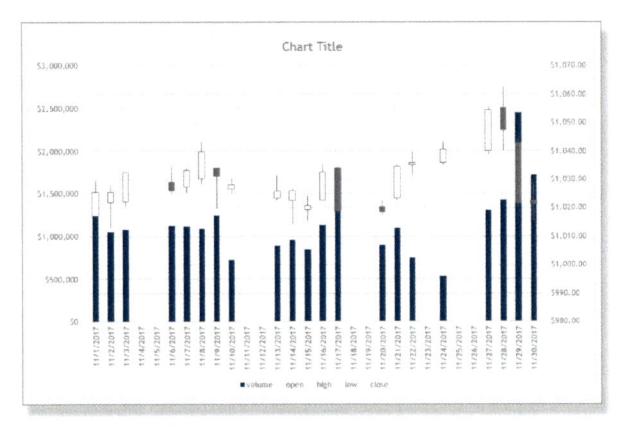

Fig 5.308: Our initial chart.

Step 2: Choose **Volume-Open-High-Low-Close** from the **Stock** menu of options **(Fig 5.308).**

Step 3: Clean up your chart to taste. I cleaned mine by taking the following steps:

- made title more descriptive
- left aligned chart title
- bumped up title font size
- added subtitle
- added annotation for data source
- bumped up transparency of gridlines
- removed legend
- added vertical axis titles
- rotated axis title on secondary axis
- customized number formatting of vertical axis
- made high-low lines thicker (see the High-Low-Close tutorial)
- made up bar borders thicker (see the High-Low-Close tutorial)

💬 **LEARN MORE** See the Design with Elegance chapter to learn how to perform these refinements.

Example File
stock-chart.xlsx

Templates
If you'll use these charts again, save them as chart templates. Or download the templates included in the chart templates folder:

- HLC: mds-stock-hlc.crtx
- OHLC: mds-stock-ohlc.crtx
- VHLC: mds-stock-vhlc.crtx
- VOHLC: mds-stock-vohlc.crtx

CHAPTER 5 · CREATE CHARTS

RELATIONSHIP

SCATTER PLOT

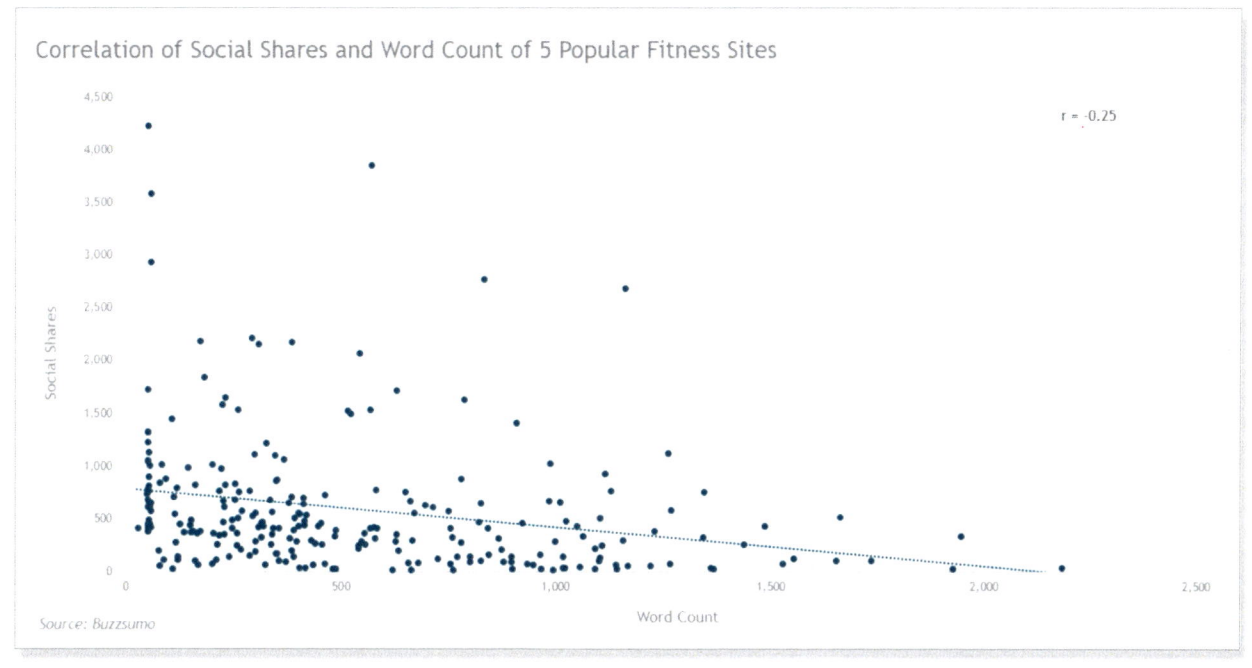

Correlation of Social Shares and Word Count of 5 Popular Fitness Sites

r = -0.25

(Y-axis: Social Shares, 0 to 4,500; X-axis: Word Count, 0 to 2,500)

Source: Buzzsumo

Fig 5.309: Our final chart.

AKA

You may also see scatter plots called:

- scatter charts
- scatter diagrams
- scatter graphs
- scattergrams
- XY charts

Uses

Scatter plots are best used to:

- plot two variables to find out if there's a correlation between the two
- test the impact of one continuous variable (an independent variable under the control of the experimenter) on another (dependent variable) in a study
- assess quality control
- find outliers in a dataset

265

Tips

- The concept of correlation is the most essential to grasp when working with scatter plots—or bubble charts for that matter. Below is a simple guide to help you determine, at a glance, whether the two metrics you're plotting against each other are positively correlated (as the variable on the x axis increases so does the variable on the y axis), negatively correlated (as the variable on the x axis increases the variable on the y axis decreases), or not correlated—or so weakly correlated that the appearance of correlation could be the result of chance **(Fig 5.310)**.

- One important thing to keep in mind with scatter plots is that correlation doesn't necessarily suggest causation. Finding a correlation between two factors is often a jumping-off point for more analysis. For example, in the downloadable dataset I used for the tutorial, I wanted to see if there was a correlation between word count and social shares for five popular fitness sites. With my test, longer posts were shared less often. This doesn't necessarily mean that these sites should restrict their word count to improve social sharing. Other factors could be impacting that negative correlation. For example, they might need to chunk longer articles better, add anchors on the page to allow visitors to jump to a section, reduce page load time, add more images...you get the idea. Or their longer articles might appeal to a more technical audience who might not be as active on social media—or might be less willing to share resources they think might give them a competitive edge in their industry.

- It's often a good idea to include a trendline—also called a regression line or line of best fit—any time you create a scatter plot to easily see if there is a positive correlation, negative correlation, or no correlation.

- Excel offers you the option to include the R^2 value with your trendline. I prefer to use the correlation coefficient—technically, the Pearson product-moment correlation coefficient—with my trendlines (indicated by a lowercase r). Both quantify the amount of correlation between two variables, but r ranges between -1 and 1, with -1 representing a perfect negative correlation, 1 representing a perfect positive correlation, and 0 representing no correlation. Because R^2 is squared, it's expressed as a positive value, which I find less intuitive. If you want to add r to your chart, you'll need to do it manually with a data label or text box. I used a text box in the example file because it gave me more control over the placement of this annotation.

- Although Excel offers you the ability to connect your individual data points with smooth or straight lines, these are very rarely seen out in the wild,

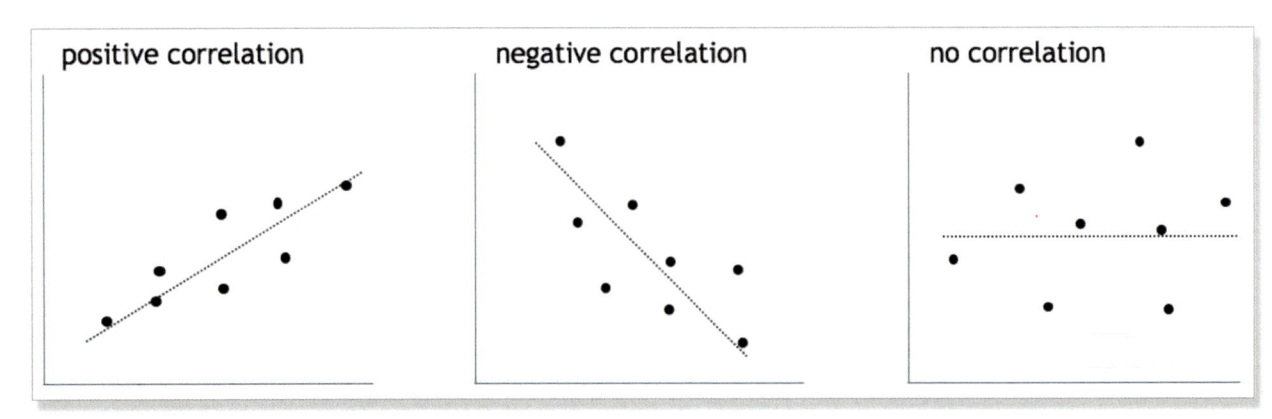

Fig 5.310: Correlation is critical to understand when working with scatter plots.

CHAPTER 5 • CREATE CHARTS

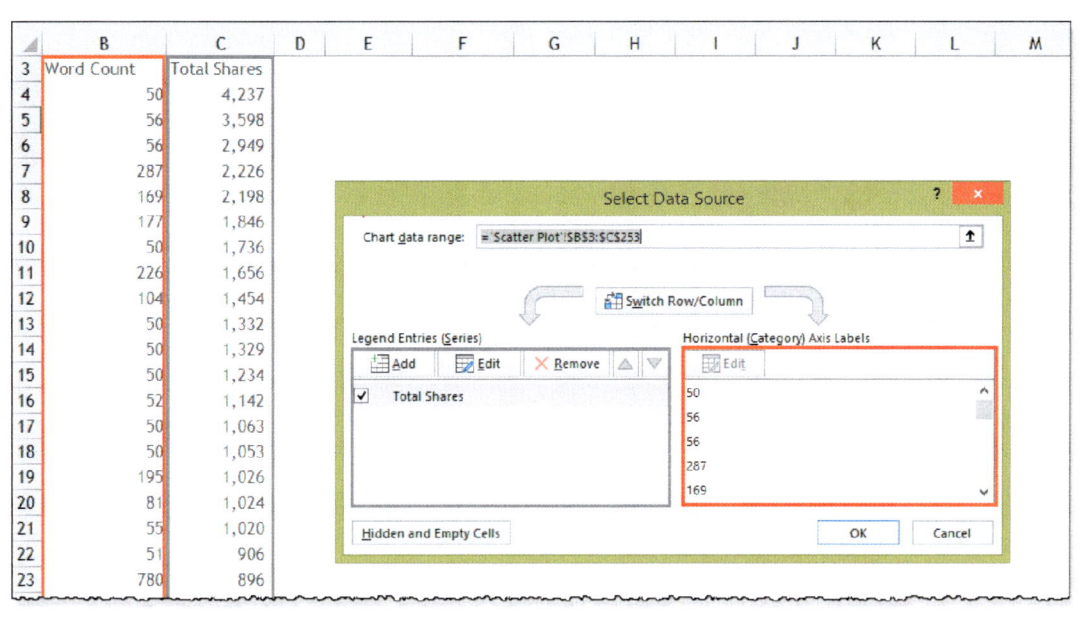

Fig 5.311: Double check your x and y values before manually assigning axis titles.

although we use them in the Dumbbell Plot tutorial. #hackfoolery

- If your dots are densely populated, increasing the transparency of your markers can help compensate for the overlap.

- I like to customize formatting of the trendline to make it more elegant, as you'll see in the tutorial.

- In your dataset, the metric that you're testing or analyzing (usually referred to as a variable in statistics) should be in the left column so that it will be plotted in the x axis. In our example dataset, that's word count. The variable that you're observing for impact should be in the right column so that it will be plotted in the y axis. In our example dataset, that's social shares. I wanted to see if word count impacted social sharing.

- At the time of writing, Excel doesn't add axis titles by default with your scatter plots, which I think is an oversight. Even if you add them through the Chart Elements flyout menu, Excel will add two titles labeled Axis Title. Thanks, Excel.

Very helpful. To check to see which of your variables is in each axis, right-click anywhere in your chart and choose Select Data from the contextual menu to open the Select Data Source dialog **(Fig 5.311)**. You kind of need to do some mental gymnastics to figure out which one is which, but I look in the Horizontal Axis Labels field—*horizontal* being the key phrase. Whatever's in there is plotted in my x axis. Elementary, my dear Watson! . Or you can just think of it as whatever is in your left column will show up in the right field of your Select Data Source dialog and vice versa. Yes, I'm the queen of mnemonic devices. SAT survival skills FTW!

- If you want to map a third variable (MOAR DATAS!), you can do that with a bubble chart.

💬 **LEARN MORE** See the Bubble Chart tutorial to learn how to create a bubble chart.

267

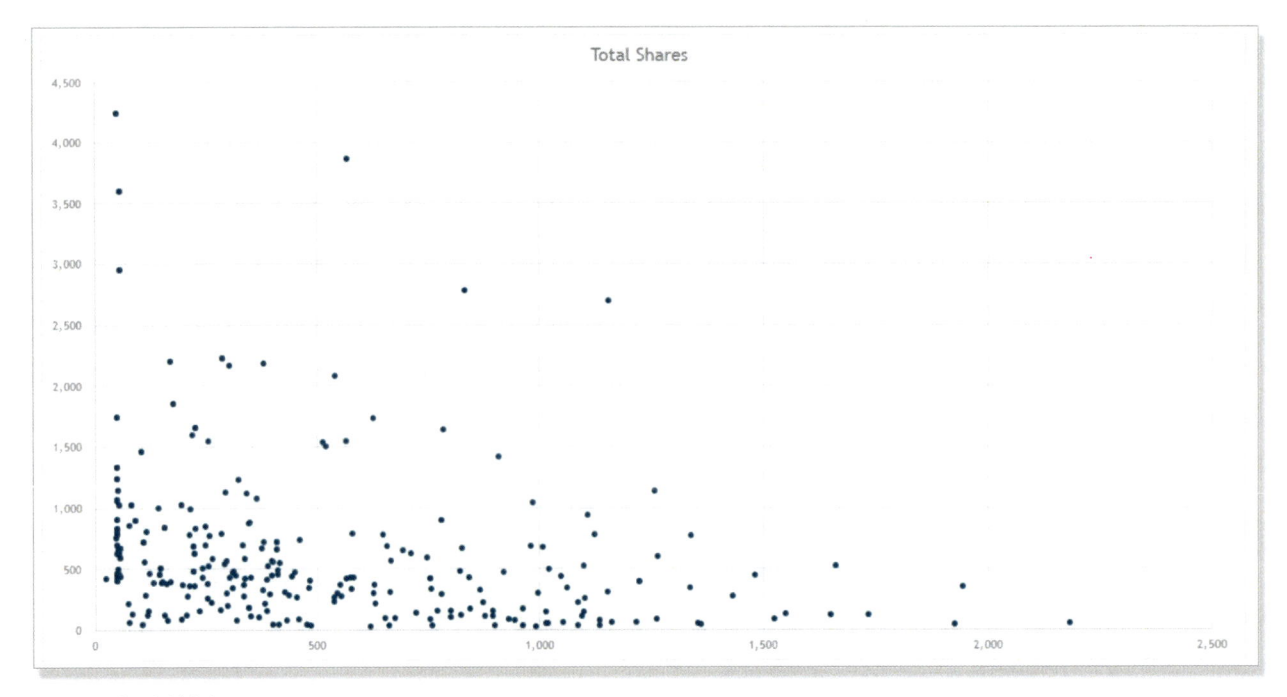

Fig 5.312: Our initial chart.

Category / Metric
0 categories / 2 metrics

Note: I show you how to include a category in the Variation section, but it's a hack. By default, Excel only supports metrics in scatter plots. For this reason, I don't include it in the category count.

Tutorial
Step 1: Select any cell inside your dataset and navigate to **Insert ▸ Charts ▸ Insert Scatter (X, Y) or Bubble Chart ▸ Scatter ▸ Scatter (Fig 5.312)**.

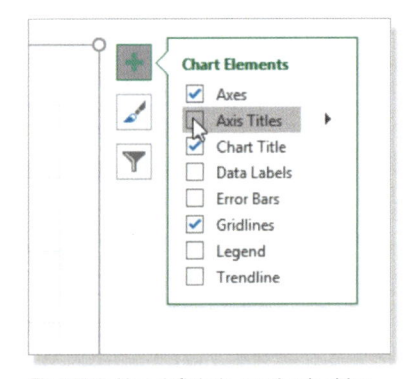

Fig 5.313: You definitely need axis titles with scatter plots!

Step 2: We need to add axis titles. Select the chart and open the **Chart Elements** flyout menu in the upper-right corner. Select **Axis Titles (Fig 5.313)**. Change the titles to ones that make sense with your data, following the guidance in the tips. I also bumped up **Font Size** to 12 pt.

Step 3: The label position of the horizontal axis isn't terrible. Depending on my mood and the time I need to put toward obsessing over augmentations the average person will never notice, I usually tweak it. But the positioning of the axis title for the vertical axis is pretty tragic. To remedy

CHAPTER 5 • CREATE CHARTS

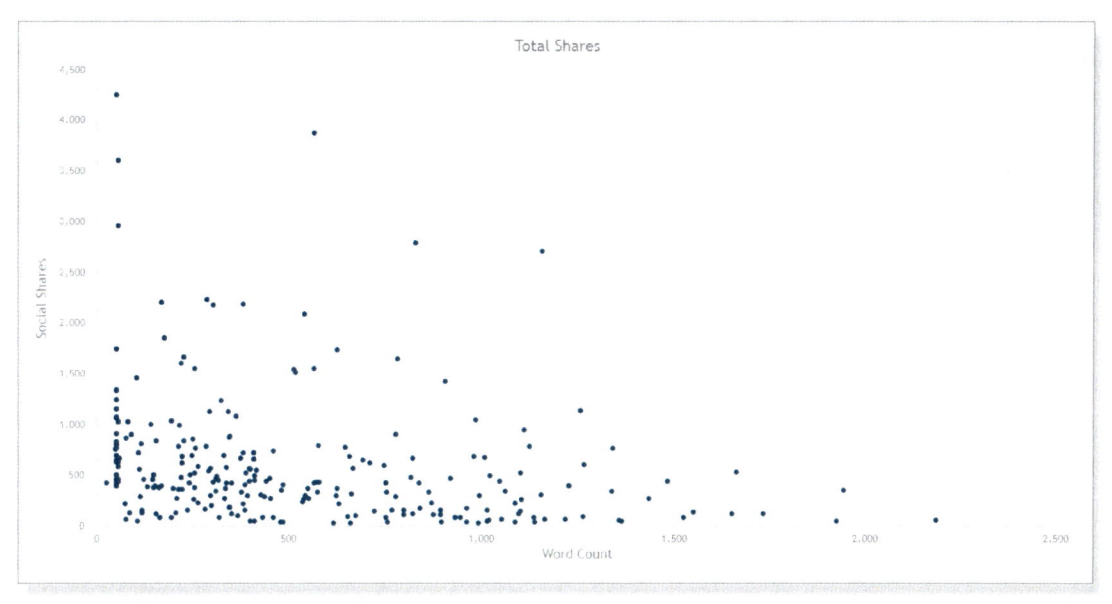

Fig 5.314: Whoa, clingy vertical axis title...Boundaries.

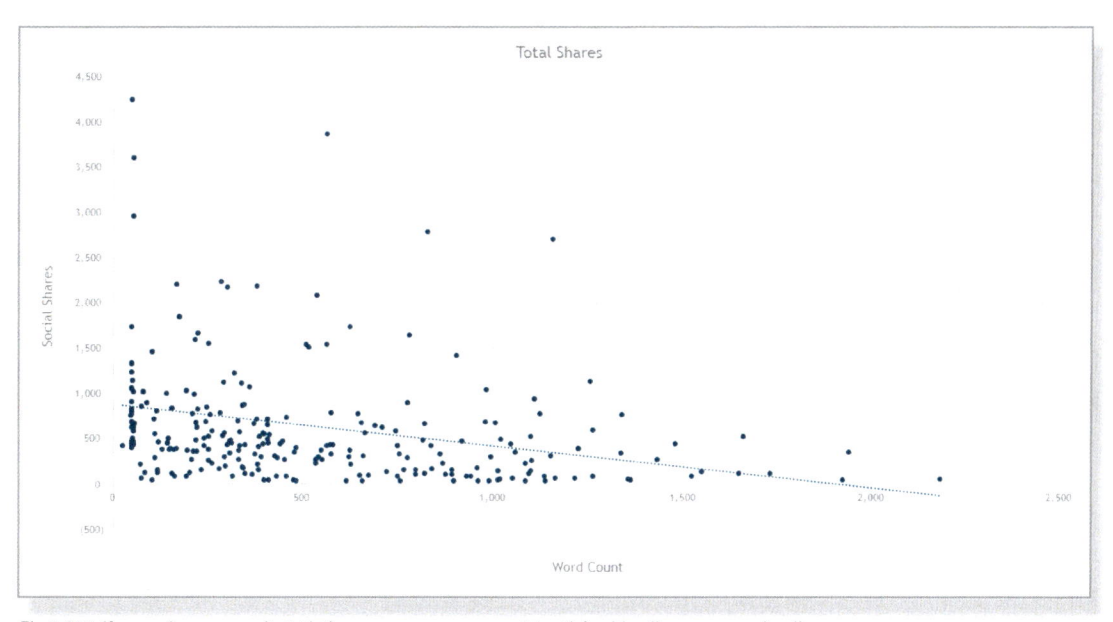

Fig 5.315: If your viewers aren't statistics savvy, you may want to stick with a linear regression line.

this, I move the sizing handle on the left border of the **Plot Area** to the right to give it some breathing space and then nudge the axis title a bit to the left (because it moves with the **Plot Area**). Shift-dragging it will prevent it from becoming uncentered with your axis **(Fig 5.314)**.

Step 4: If you want to add a trendline (and you usually do), select your chart and open the **Chart Elements** flyout menu. Click the right-facing arrow next to **Trendline** to open the flyout menu and choose **More Options** to open the **Format Trendlines** pane. Excel will give you a linear trendline by default **(Fig 5.315)**.

269

CHAPTER 5 • CREATE CHARTS

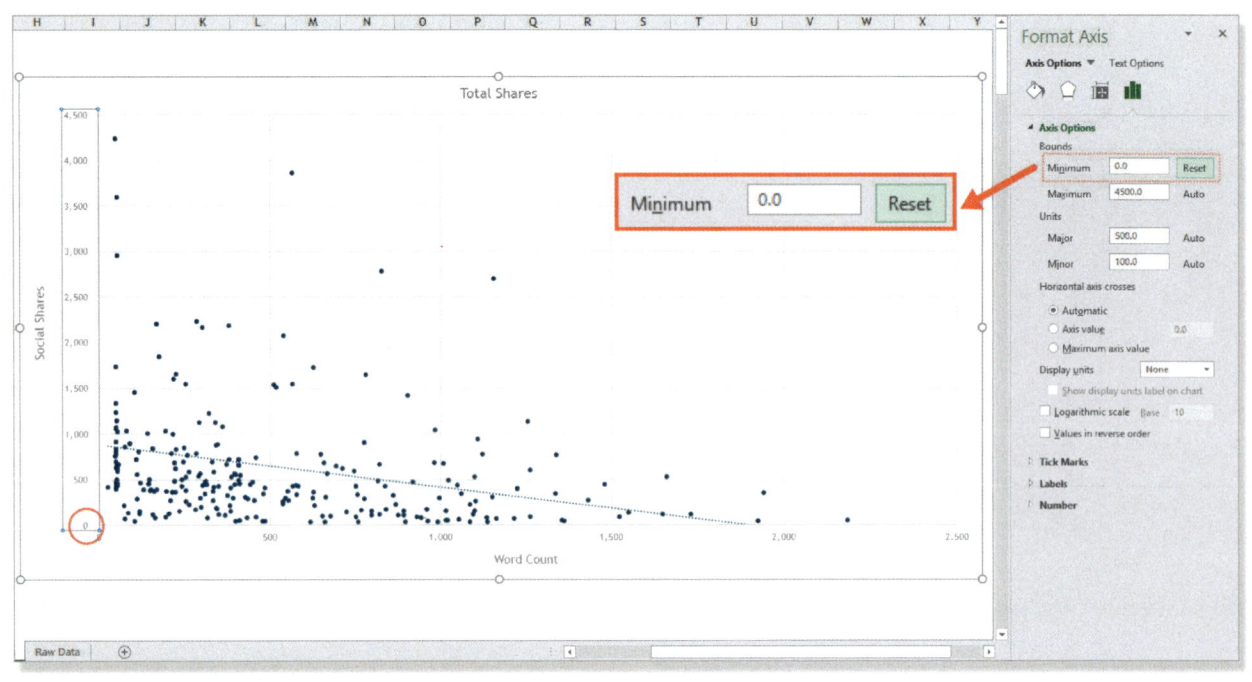

Fig 5.316: Set Minimum Bounds to 0 to get rid of negative numbers in your vertical axis.

Step 5: One thing I don't like about the linear regression line in the example file is that it dips below sea level, causing the y axis to add negative numbers. I don't want anything in my visualizations that could cause viewers to stumble. To wit, someone could look at that and understandably think to themselves, *Wait. How can you have fewer than 0 social shares? No bueno.* If it bugs you as much as it does me, you can offset that by manually setting the minimum value of your vertical axis to 0. To do this, select your vertical axis to open the **Format Axis** pane. In the **Axis Options** tab, under **Bounds,** set **Minimum** to 0 **(Fig 5.316)**.

Step 6: As I mentioned in the tips, Excel gives you the ability to add R²—known as the coefficient of determination—to your regression line. But I'm a princess and prefer to add the correlation coefficient (r) because I feel like the range of -1 to 1 is more intuitive and universally recognized. If you are partial to R², you can easily add it. Just select the trendline to open the **Format Trendline** pane and select **Display R-squared value on chart** in the southernmost point of the **Trendline Options** tab. However, if you want to add the correlation coefficient to your chart, you'll need to calculate it in a cell in your worksheet using the CORREL function, then pull it into the chart using a data label or text box. I used a text box **(Fig 5.317)**.

> 💬 **LEARN MORE** See the Add Annotations section of the Design with Elegance chapter to learn how to create text boxes or the Add Data Labels section to learn how to add data labels.

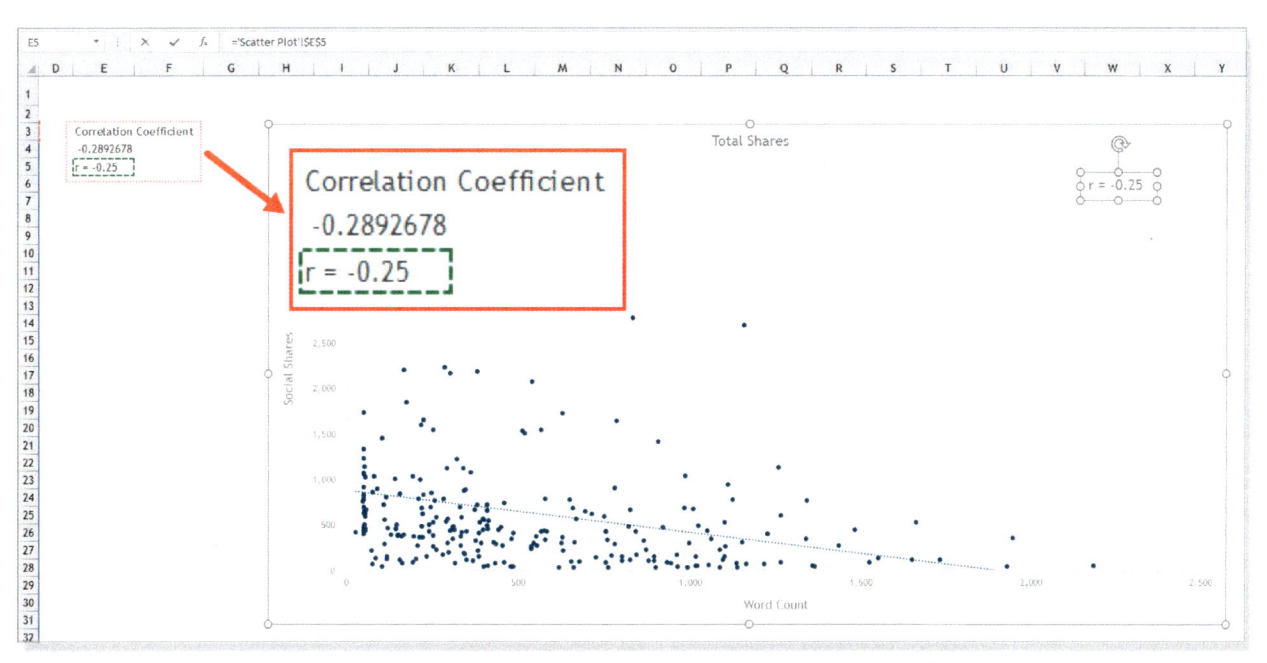

Fig 5.317: It's obvious there's a negative correlation between word count and social shares; adding r just quantifies it.

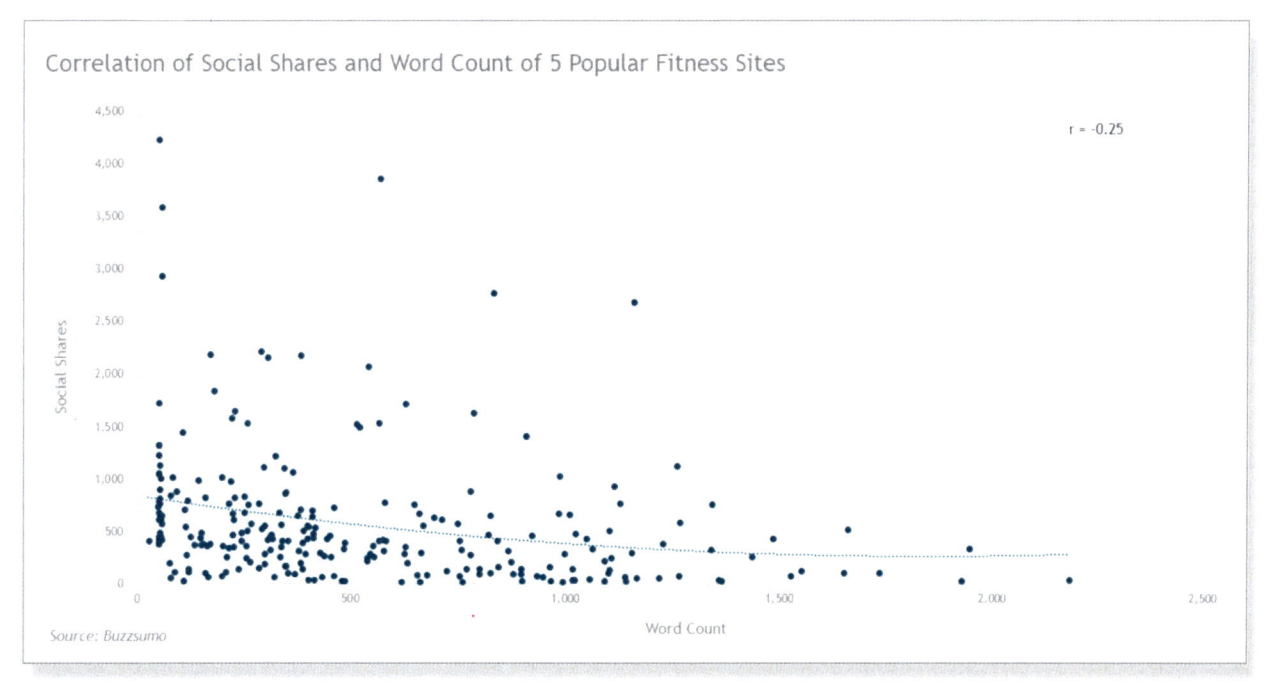

Fig 5.318 : Our final chart.

CHAPTER 5 · CREATE CHARTS

Step 7: Clean up your chart to taste. I cleaned mine by taking the following steps:

- made title more descriptive
- left aligned chart title
- bumped up title font size
- added annotation for data source
- removed gridlines

> 💬 **LEARN MORE** See the Design with Elegance chapter to learn how to perform these refinements.

Variations
Apply Polynomial Trendline
In many cases—especially with smaller datasets—a linear regression line will suffice. But if you have a larger dataset, the **Polynomial** option (under the **Trendline Options** tab of the **Format Trendline** pane) can be helpful in visualizing fluctuations. The **Order** setting will determine how many hill/valley pairs you'll get: An order of 2 will give you one, 3 will give you two, etc. You can see with the example dataset that social shares are generally lower with posts with higher word counts until you get to about 1,500 words. Then there's a slight uptick in social shares.

Add a Category
Excel doesn't let you add a category to your scatter plot by default, but have Excel's limitations stopped us at any point over the course of this book? No. So why let them now? Turns out you can hack a scatter plot to add segmentation to your chart **(Fig 5.319)**. If you understand how the Select Data Source dialog works and are comfortable putzing around in there, the hack is pretty straightforward.

> 💬 **LEARN MORE** See the Hack Excel's Charting Engine section of the Design with Elegance chapter to learn how to rock the Select Data Source dialog.

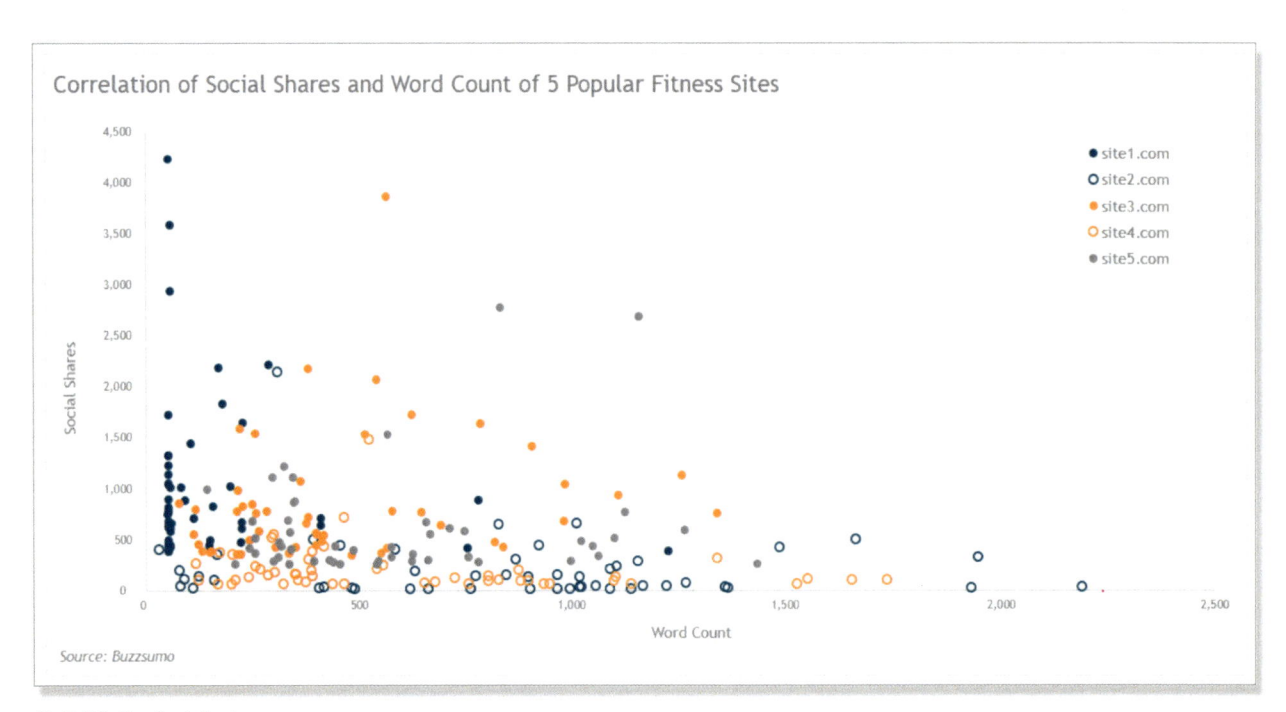

Fig 5.319: Our final chart.

CHAPTER 5 • CREATE CHARTS

site1.com	55	455
site1.com	145	453
site1.com	56	434
site1.com	50	433
site1.com	754	420
site1.com	50	397
site1.com	50	396
site1.com	1,226	393
site2.com	304	2,166
site2.com	1,009	672
site2.com	825	662
site2.com	1,661	523
site2.com	389	521
site2.com	921	466
site2.com	451	466
site2.com	1,483	440

Fig 5.320: I applied a pattern fill to every other category.

Step 1: Add a category column to your dataset. Alternatively, you could create a table for each of your category/series pairs. But that could get unwieldy pretty fast. I prefer to work with one dataset. To make it easy to spot your individual category elements, you can apply a subtle pattern fill to every other one. That's what I did in the example data **(Fig 5.320)**.

Step 2: Select your first category element's x and y values (site1.com) by selecting the first value and Shift-clicking the value in the bottom-right corner of its data range. Don't select the column headers or category names, just the cells with data. (Alternatively, you can click-and-drag over the data points for your first category element to select.) Next, navigate to **Insert ▸ Charts ▸ Insert Scatter (X, Y) or Bubble Chart ▸ Scatter ▸ Scatter** (Fig 5.321).

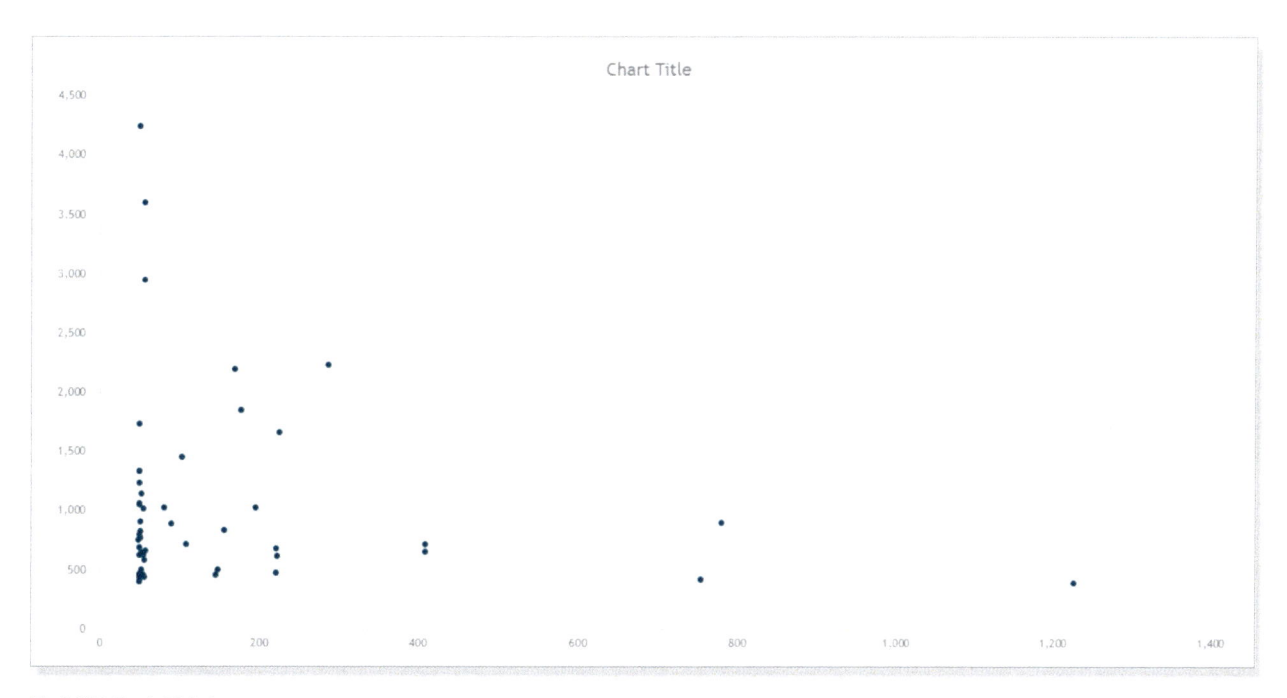

Fig 5.321: Our initial chart.

273

CHAPTER 5 • CREATE CHARTS

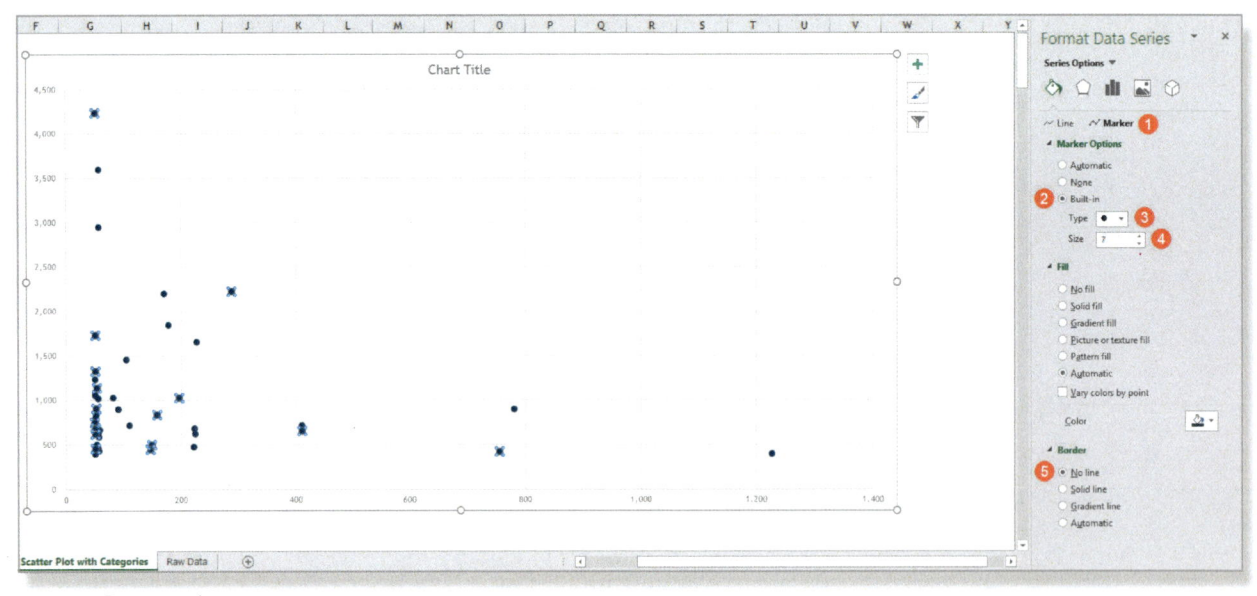

Fig 5.322: Format markers to taste.

Step 3: Let's start by formatting our markers. Select one to select the entire data series, then press Ctrl-1 to open the **Format Data Series** dialog. Open the **Fill & Line** tab and click **Marker** at the top of the tab **(Fig 5.322)**. I used the following settings:

- **Marker Options**: Built-in
- **Type**: Circle
- **Size**: 7
- **Border**: No line

Step 4: We now need to circle back and add our category to our series. We'll do this by pointing the **Series Name** field to one of our labels. To get started, right-click on your chart and choose **Select Data** from the contextual menu. In the **Select Data Source** dialog, let Excel know where it can find the name **(Fig 5.323)**. Alternatively, you could type your category name into this field. I prefer the flexibility of pointing to it. This way, if you ever change the name of a

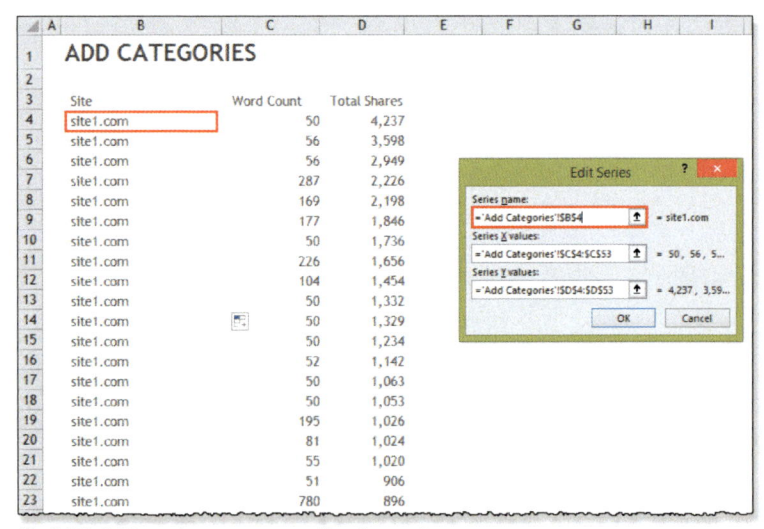

Fig 5.323: Add a series name to your first data series to manually build a legend as we add more series.

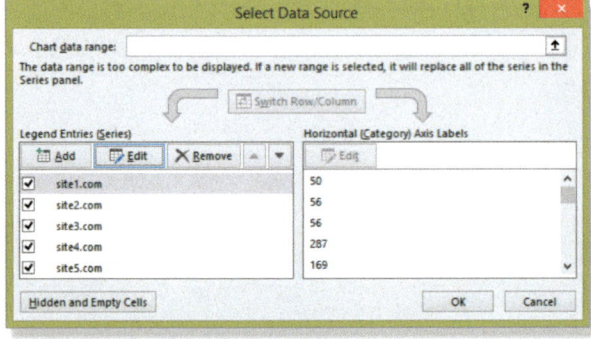

Fig 5.324: Presto! Take THAT, Excel, for not letting us create categories by default!

274

CHAPTER 5 · CREATE CHARTS

data series, your chart will update automatically. Then click **OK** to close the **Edit Series** dialog.

Step 5: You can build your additional data series right inside the **Select Data Source** dialog **(Fig 5.324)**. It's easy!

💬 **LEARN MORE**   See the Hack Excel's Charting Engine section of the Design with Elegance chapter to learn how to add new data series.

Step 6: Our chart needs a legend, and we need to add it manually since we stitched our chart together manually (like a BEAST). You can add it from the **Chart Elements** flyout menu **(Fig 5.325)**.

Step 7: Now you can edit the markers for each of your series and format them to taste.

Warning: I've seen charts use different marker types. This can junk up your visualization fast, especially because a 7 pt dot might appear larger than a 7 pt triangle. Also, if you use lighter tints of your brand colors—as I did in this book so that I constrain all my visualizations to the same limitations as an organization with only three brand colors—it will be difficult to differentiate your colors. What I chose to do—to maximize impact without cheapening the visualization—was use hollowed-out versions of two of my colors by using only a border. I accomplished this by adding a circle marker with a line and no fill **(Fig 5.326)**.

Tip: If you have trouble finding a data series, you can navigate to **Chart Tools ▸ Format ▸ Current Selection** and select it from the drop-down menu.

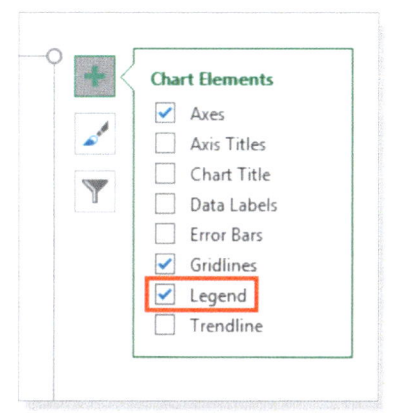

Fig 5.325: You need to add a legend to your chart manually since you went a little rogue to create this chart.

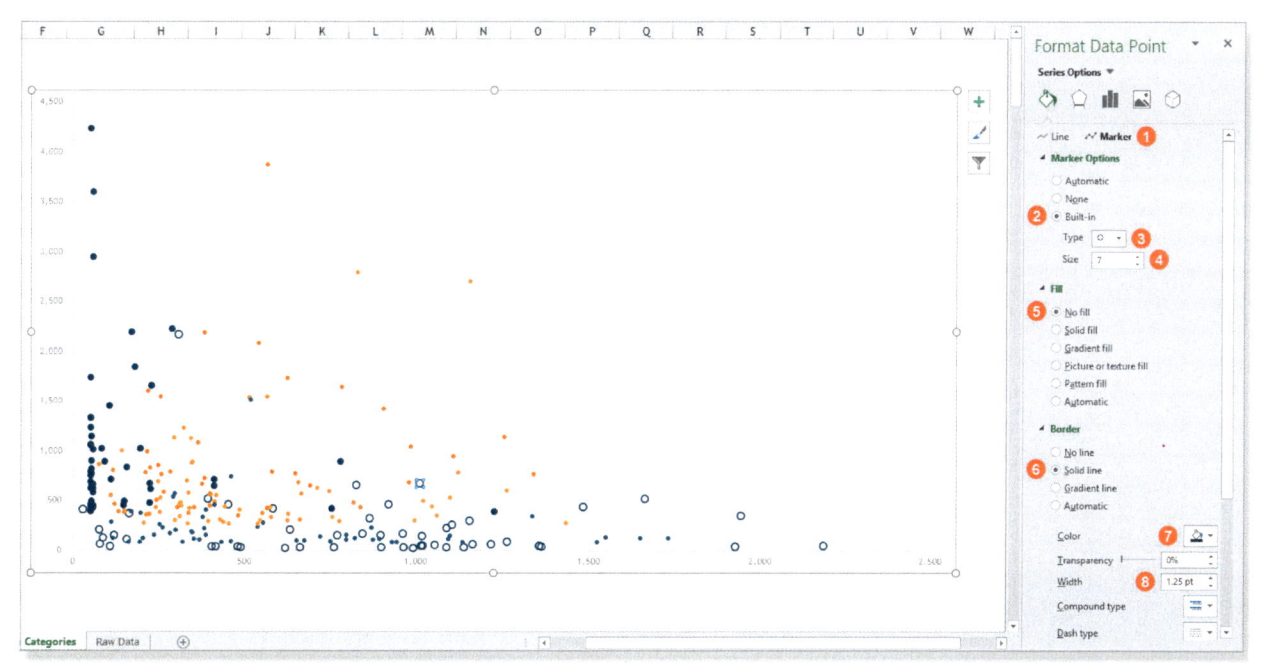

Fig 5.326: Use hollow markers if you run out of colors.

CHAPTER 5 • CREATE CHARTS

Step 8: Clean up your chart to taste. I cleaned mine by taking the following steps:

- made title more descriptive
- left aligned chart title
- bumped up title font size
- added annotation for data source
- removed gridlines
- bumped up legend font size
- moved legend
- added axis titles

> 💬 **LEARN MORE** See the Design with Elegance chapter to learn how to perform these refinements.

Isolate Trendlines

If you add a category to your scatter plots, you have the perfect opportunity to see if a data series (or multiple series) is skewing overall trends by isolating the trendlines.

To do this, apply trendlines to each of your data series and remove the formatting from the markers.

When I did this with the example dataset, I could see that site1.com's data skewed the aggregated trendline. For site3.com and site5.com, there was surprisingly a positive correlation between word count and social shares. That finding begs further analysis, and a marketing analyst worth his weight in gold would do well to dig into that data to unearth more insights.

Example File
scatter-plot.xlsx

Template
The scatter plot won't save as a template. It saves, but the scatter plot it spawns is unrecognizable. You'll need to create it from scratch or save it in a file where you keep all your charts that won't save as templates.

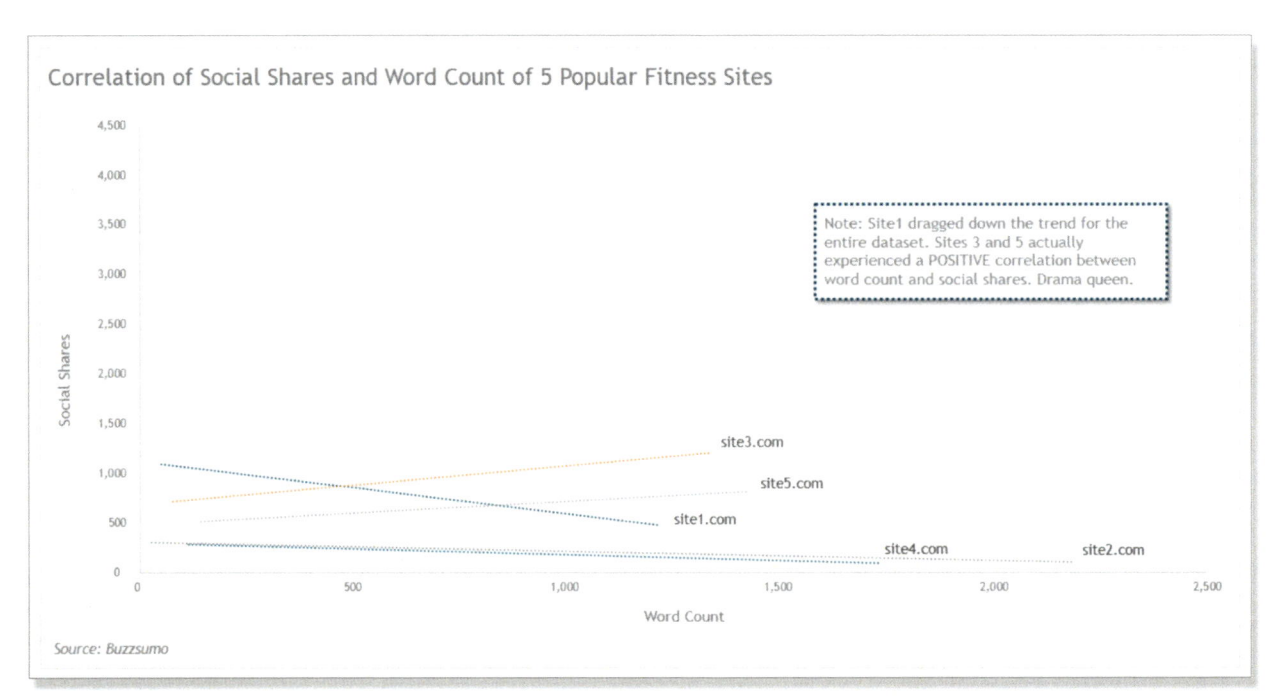

Fig 5.327: Our final chart.

CHAPTER 5 • CREATE CHARTS

BUBBLE CHART

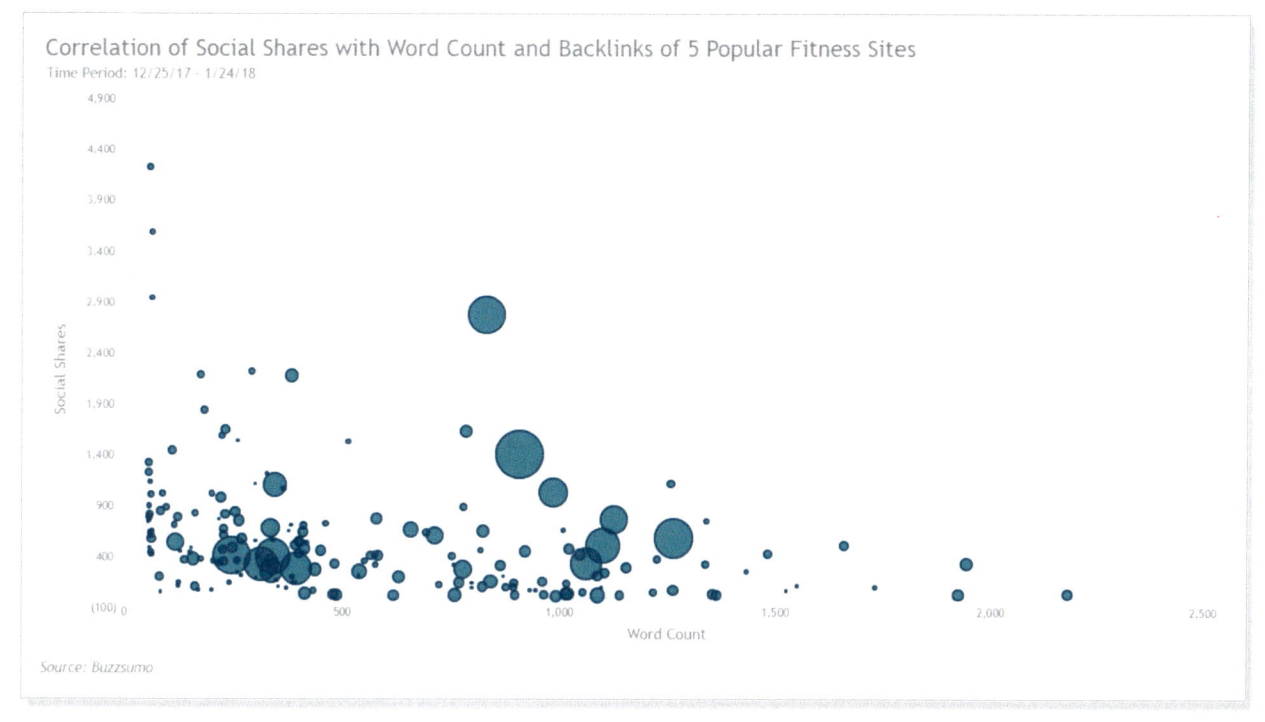

Fig 5.328: Our final chart.

Uses
Bubble charts are best used to:

- add a third metric to a scatter plot, with the third metric determining the size of the bubbles
- plot three continuous variables (metrics) to find out if there's a correlation between the three
- find outliers in a dataset
- assess quality control

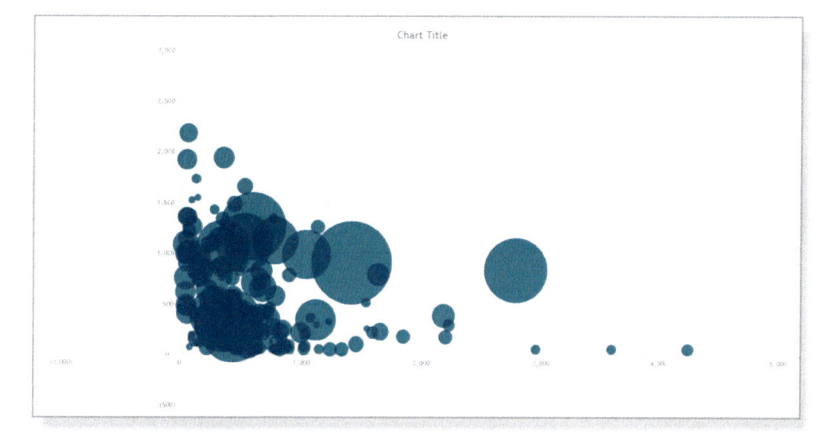

Fig 5.329: Our initial chart.

Tips
All the tips in the previous Scatter Plot tutorial apply to the bubble chart, sans those regarding trendlines. You don't usually see trendlines on bubble charts. It would be kind of like wearing flip flops with a tux.

Category / Metric
0 categories / 3 metrics

Tutorial
Step 1: Select only the data in your dataset (no headings) and navigate to **Insert ▸ Charts ▸ Insert Scatter (X, Y) or Bubble Chart ▸ Bubble ▸ Bubble (Fig 5.329)**.

277

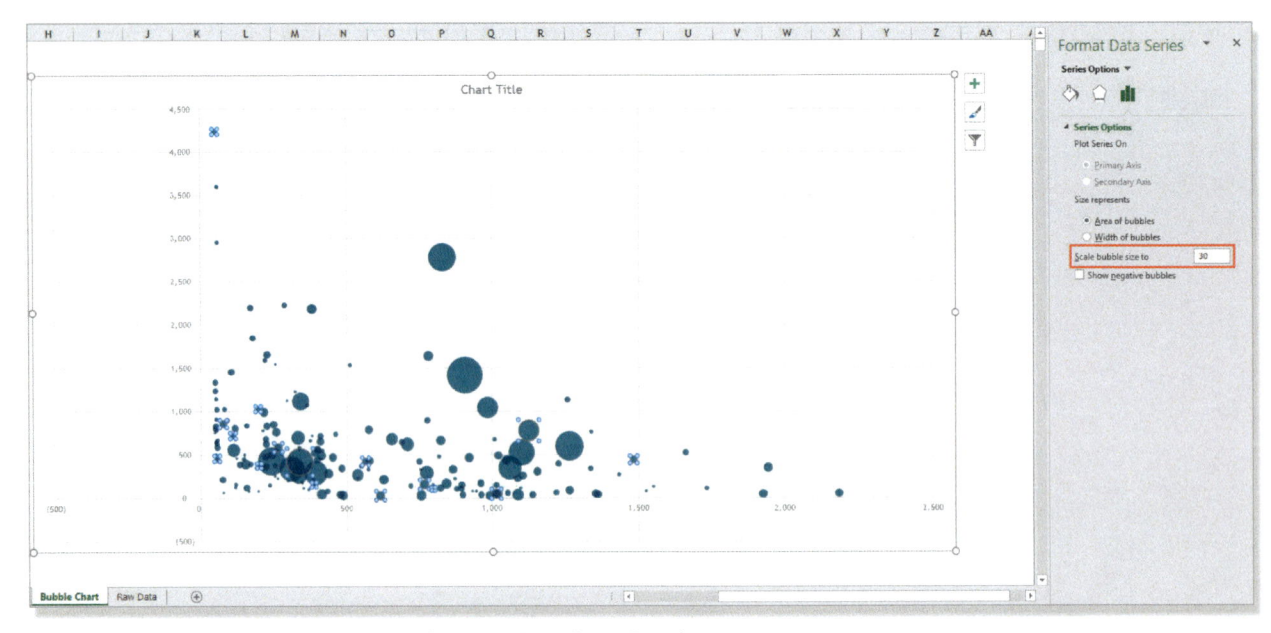

Fig 5.330: Adjust the scale of your bubbles to make them universally smaller or larger.

Step 2: These bubbles are a bit large. We need to scale them back. To access the scale option, select a bubble to select the entire series, then press Ctrl-1 to open the **Format Data Series** pane. In the **Series Options** tab, keep **Size represents** set to **Area of bubbles**. Then set **Scale bubble size to** to whatever works for your dataset. I set mine to 30 **(Fig 5.330)**.

Step 3: I normally zero out both axes. With my dataset, however, I needed to set the **Minimum Bounds** of the vertical axis to -100 so that my bubbles wouldn't be cut off **(Fig 5.331)**. I still set the **Minimum Bounds** of my horizontal axis to 0 though. I also reformatted negative values to use a minus sign instead of parentheses using the code below:

```
_ (* #,##0 _ ); _ (* -#,##0; _ (* "0"?? _ ); _ (@ _ )
```

> 💬 **LEARN MORE** See the Thin Out Axes section of the Design with Elegance chapter to learn how to reduce clutter in your axes and the Customize Number Formatting section to learn how to finesse your number formatting.

Step 4: Excel adds transparency to the bubbles by default. If you want to adjust the transparency, select your data series to open the **Format Data Series** pane. Open the **Fill & Line** tab and set **Fill** to **Solid fill**, which will surface the **Transparency** slider. Adjust the transparency by adjusting the slider or entering a value. I set mine to 30%.

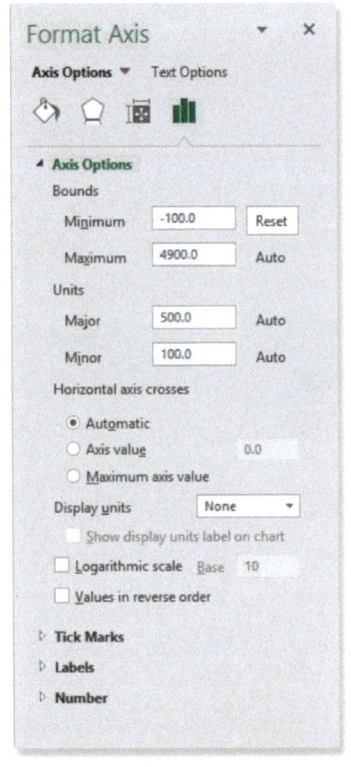

Fig 5.331: Modify the Minimum Bounds setting if your bubbles get cut off.

CHAPTER 5 • CREATE CHARTS

Step 5: If you apply transparency to your bubbles, you may want to assign a border to help them pop a bit. I use the same color as the fill (but without transparency, of course). I set **Width** to 1.25 pt.

Step 6: Clean up your chart to taste. I cleaned mine by taking the following steps:

- made title more descriptive
- left aligned chart title
- bumped up title font size
- added subtitle
- added annotation for data source
- removed gridlines
- added axis titles

💬 **LEARN MORE** See the Design with Elegance chapter to learn how to perform these refinements.

Variation
Add a Category
You can add a category to your bubble chart using the same technique demonstrated in the Add a Category variation of the Scatter Plot tutorial **(Fig 5.332)**.

Example File
bubble-chart.xlsx

Template
The bubble chart won't save as a template. You'll need to create it from scratch or save it in a file where you keep all your charts that won't save as templates.

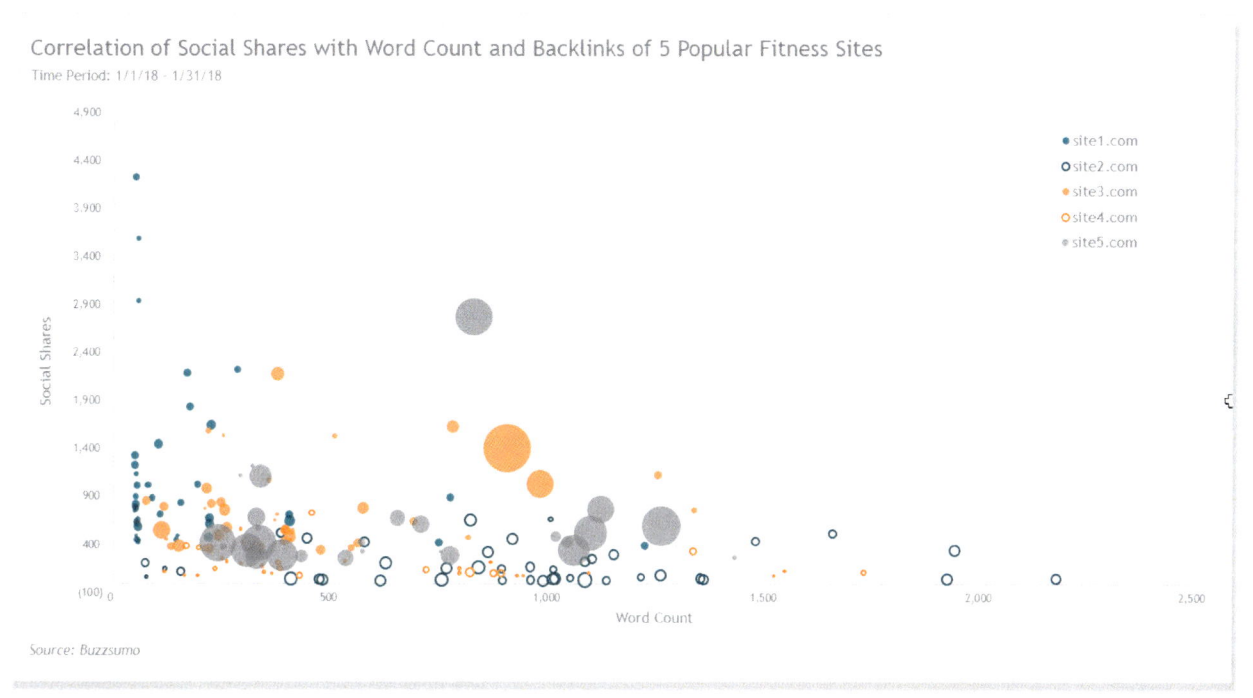

Fig 5.332: Our final chart.

279

CHAPTER 5 • CREATE CHARTS

HEATMAP

AKA
You may also see heatmaps called:
- density tables
- grid matrix heatmaps
- heat tables

Uses
Heatmaps are best used to:
- compare relative values of data in a matrix using color
- visualize data you might normally represent with a scatter plot
- highlight variance across two variables in a dataset
- identify high traffic times for websites, retail stores, call centers, databases, hospitals, etc.
- monitor quality control issues

Tips
- A heatmap is just a table with conditional formatting applied.
- You could apply conditional formatting to the entire table, row by row, or column by column. But having a different scale for each row or column can make interpretation cumbersome.
- Be careful using red and green in heatmaps because red-green colorblindness is very common, especially among men.

> **LEARN MORE** See the Design for Colorblind Users section of the Design with Elegance chapter to learn how to design with accessibility in mind.

- Adding a white border to your cells can help add visual interest and clarify cell boundaries.
- Set your columns to equal widths and rows to equal heights. You can make columns the same width by selecting them and dragging one of the handles. When you let go, they will be the same width. Or you can use the Paste Special dialog.

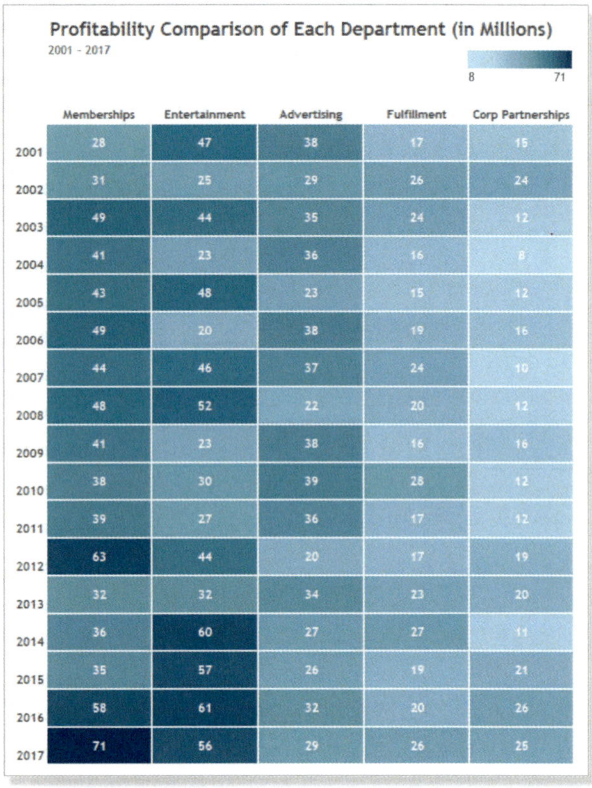

Fig 5.333: Our final chart.

> **LEARN MORE** See the Copy-and-Paste Tips section of the Design with Elegance chapter to learn cool ways to copy values and formatting.

- You can keep values in your heatmap or remove them, depending on whether you want to communicate general trends or specific values.
- Excel won't generate a legend for a heatmap, but you can create one manually using a rectangle filled with a gradient, which I recommend.
- If you're using tabular data, you may need to use a pivot table to transform your data into a cross-tab format. I've included a Pivot Table worksheet in the downloadable workbook to show you how I set mine up. Click anywhere in the pivot table to open the PivotTable Fields pane. From there you can easily see how I constructed it.

280

CHAPTER 5 • CREATE CHARTS

Category / Metric

2 categories / 1 metric

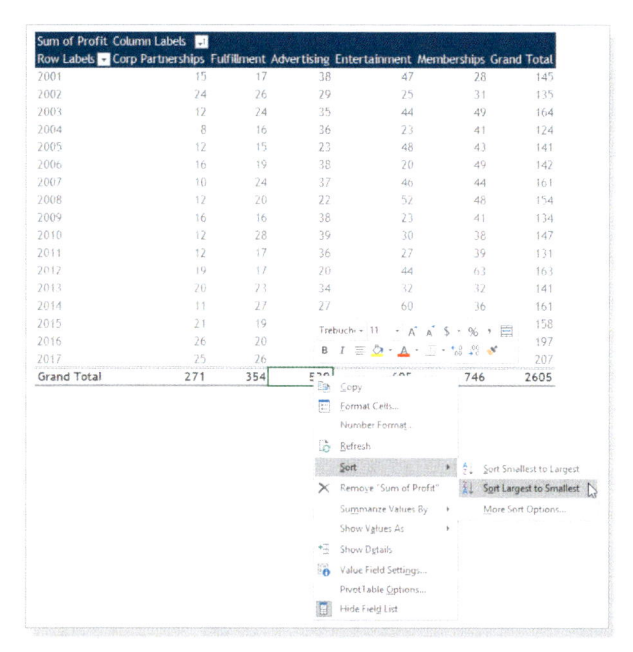

Fig 5.334: Sort your columns using the Grand Total row.

Tutorial

Step 1: The first thing you need to do is put your data in crosstab format using a pivot table. I sorted my table using the column totals. To do this, right-click any cell in the Grand Total row. From the contextual menu choose **Sort** ▶ **Sort Largest to Smallest (Fig 5.334)**. Copy-and-paste your massaged data from the pivot table (sans the total row/column) as values from the **Paste Options** button (**Paste Values** ▶ **Values & Number Formatting**).

Step 2: Select your data cells, then navigate to **Home** ▶ **Styles** ▶ **Conditional Formatting** ▶ **Color Scales** ▶ **More Rules**. With **Select a Rule Type** set to **Format all cells based on their values**, keep **Format Style** set to **2-color scale**. Set **Minimum Color** and **Maximum Color (Fig 5.335)**. You want to make sure there's enough of a difference between your minimum and maximum colors for it to pop in your heatmap. I prefer to use a tint of one of my branded colors for the minimum color, but it's totally legal to use disparate colors, as long as they're in your theme. Click **OK** to close the **New Formatting Rule** dialog **(Fig 5.336)**.

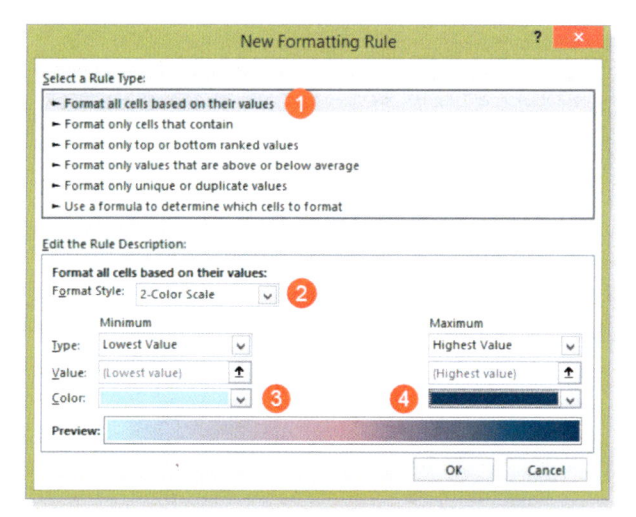

Fig 5.335: Use color scales as the basis for your heatmap.

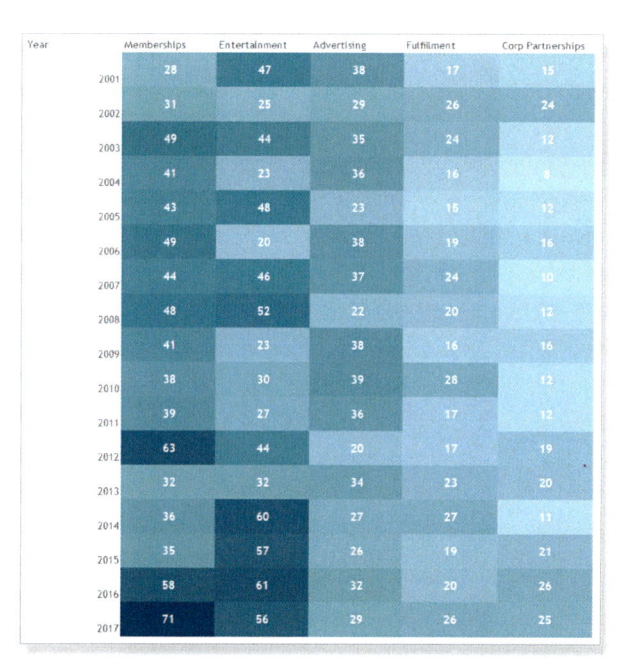

Fig 5.336: Our heatmap is starting to take shape.

281

Step 3: Next, we'll add white borders to our cells. This is optional, but it adds finesse to a heatmap. First, select all the data cells, then press Ctrl-1 to open the **Format Cells** dialog. Select the **Border** tab to access border settings. Under **Line**, set **Style**, then **Color** **(Fig 5.337)**. The tricky thing is, because we're adding white borders, you won't be able to see if you've selected the borders or not, so you might feel like you're flying blind. (Because you are.) Next, you'll need to set the borders by either clicking the lines inside the **Border** field or by selecting from the **Presets**.

I selected both the **Outline** and **Inside** presets. Click OK to close the **Format Cells** dialog **(Fig 5.338)**.

Step 4: Now we want to center our data inside each of the cells, including our headings and Year column. To do this, select your entire table and pop back into the **Format Cells** dialog, then select the **Alignment** tab. Set both **Horizontal** and **Vertical** to **Center** **(Fig 5.339)**. I also reduced the width of my Year column and removed its heading because I am prone to tweak.

Fig 5.337: Very unhelpful preview of your white borders.

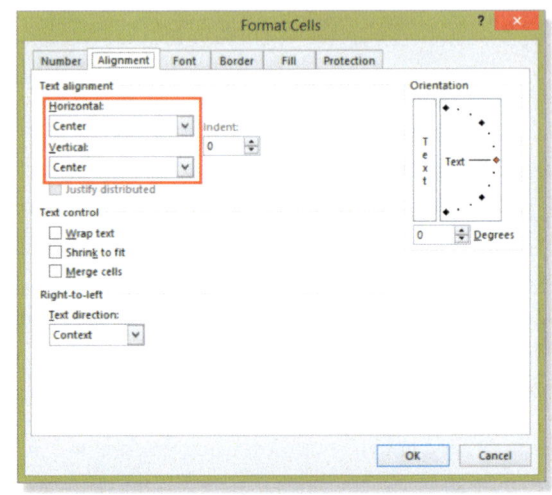

Fig 5.339: Center text in cells in the Format Cells dialog.

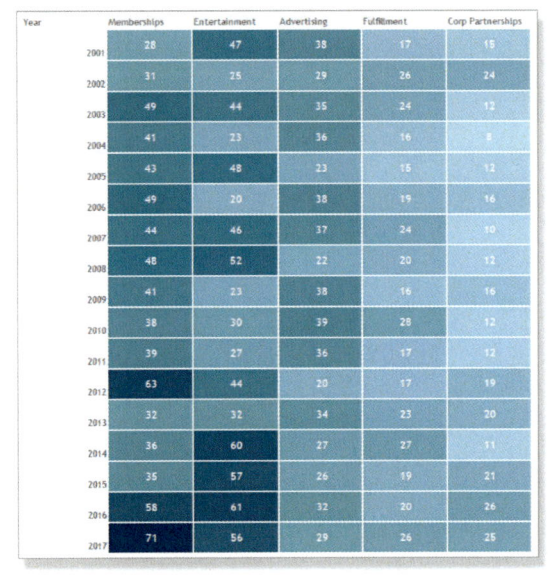

Fig 5.338: Add a white border to your heatmap for finesse.

Fig 5.340: Format your heatmap's text to taste.

CHAPTER 5 • CREATE CHARTS

Step 5: To make the font pop more, I styled it. If you want to format your heatmap's font, open the **Fonts** tab. I set **Font style** to **Bold** and **Color** to white. Click **OK** to close the **Format Cells** dialog when you're finished formatting **(Fig 5.340)**.

Step 6: Add an intuitive title above your heatmap, as well as a subtitle, if your data calls for it **(Fig 5.341)**. I like to add one earlier in the process than I do with standard charts because it helps me stay focused on the KPIs the heatmap is designed to illustrate.

Fig 5.341: Add an intuitive title for your heatmap.

Step 7: You should really have a legend to support your heatmap. To create one, drag out a rectangle (**Insert ▶ Illustrations ▶ Shapes ▶ Rectangles ▶ Rectangle**). Then press Ctrl-1 to open the **Format Shape** pane. Open the **Fill & Line** tab to access your formatting options **(Fig 5.342)**. I used the following settings for my rectangle:

+ **Fill**: Gradient fill
+ **Type**: Linear
+ **Direction**: Linear Right
+ **Gradient stops**
  Color: same colors you used for the min/max color

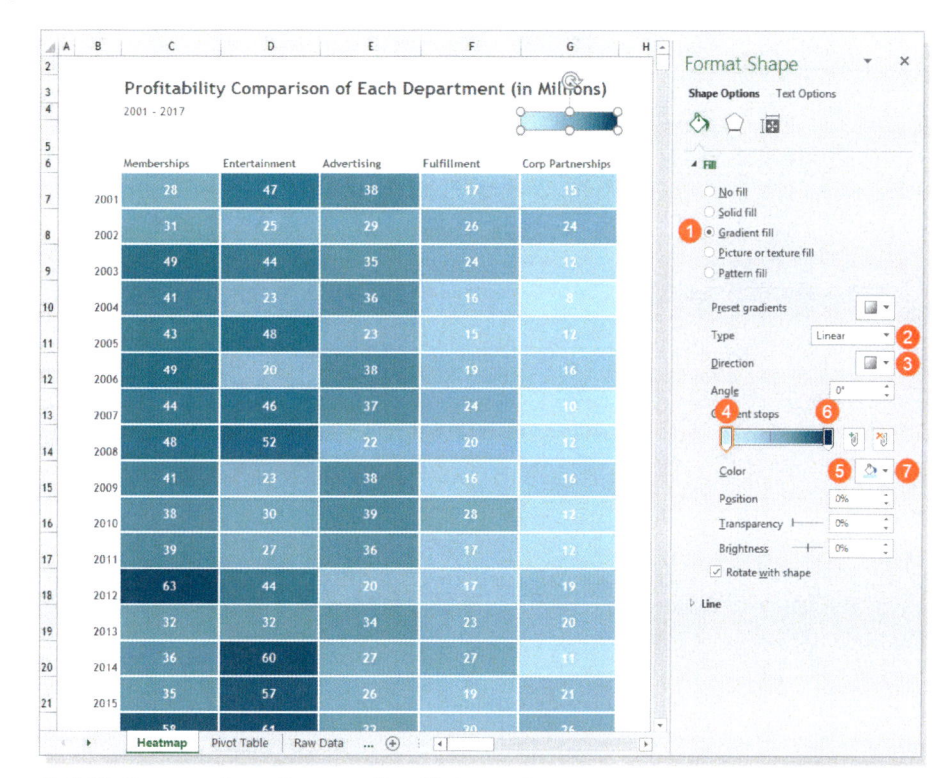

Fig 5.342: Create your legend from a gradient-filled rectangle.

283

CHAPTER 5 • CREATE CHARTS

Fig 5.343: Format your legend text box and line it up with your gradient-filled rectangle.

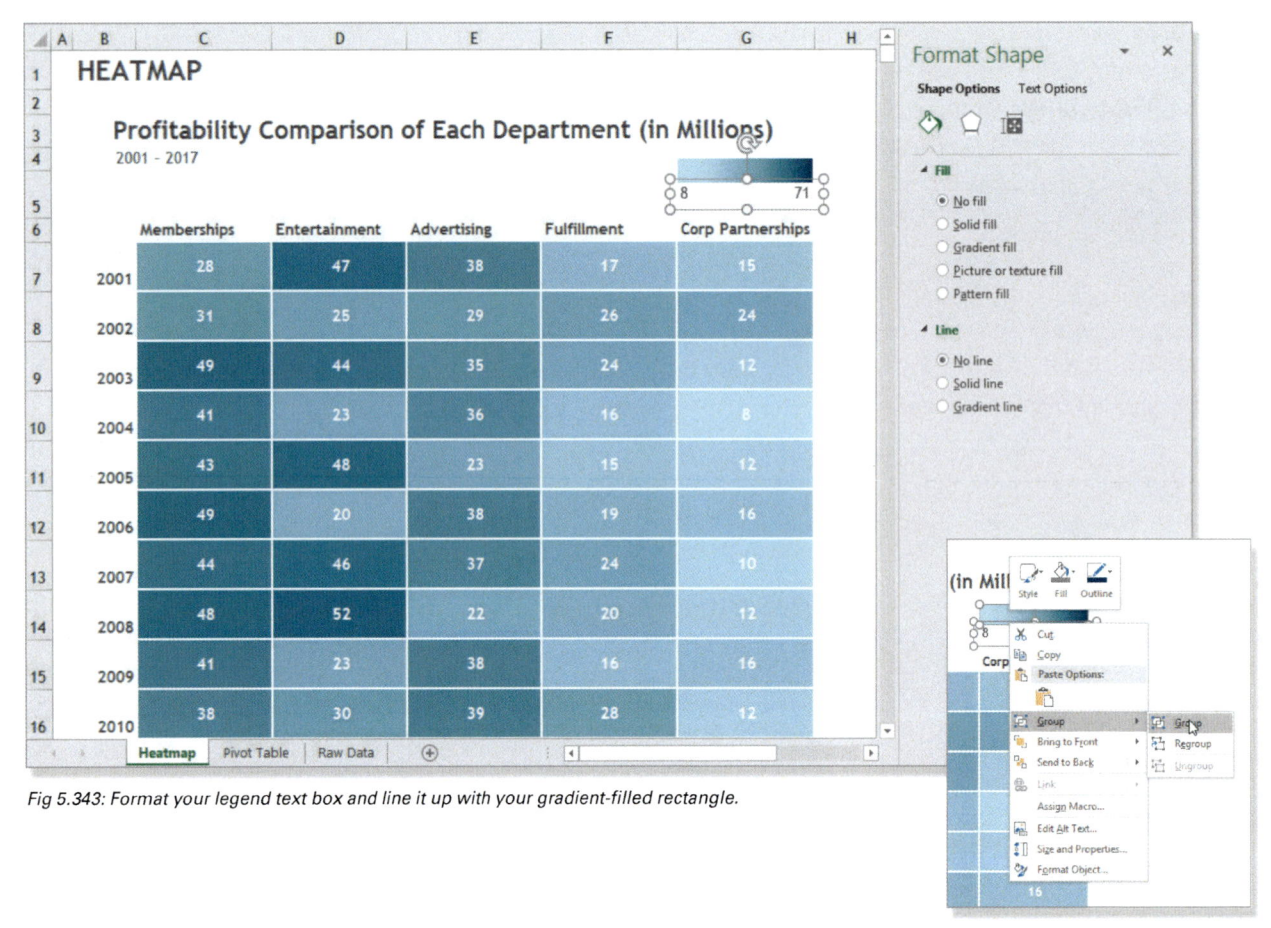

Fig 5.344: Group all legend objects, so you can move them together.

Step 8: Finally, enter your low and high values into a text box. (Or you can use two text boxes, if you want to give them their own rooms.) Select the text box(es), then press Ctrl-1 to open the **Format Shape** pane. Set **Fill** to **No fill** and **Line** to **No line (Fig 5.343)**.

> 💬 **LEARN MORE** See the Add Annotations section of the Design with Elegance chapter to learn how to create text boxes.

Step 9: Group your values text box(es) with your rectangle so that you can move them together. To start, select both elements while holding down the Shift or Ctrl key, then right-click and choose **Group ▸ Group** from the contextual menu **(Fig 5.344)**.

Example File
heatmap.xlsx

Template
Since heatmaps aren't charts they can't be saved as a template. But you can use the Format Painter to apply the formatting from one heatmap to another dataset.

> 💬 **LEARN MORE** See the Paint Formats section of the Design with Elegance chapter to learn how to format using the Format Painter.

284

CHAPTER 5 • CREATE CHARTS

RADAR CHART

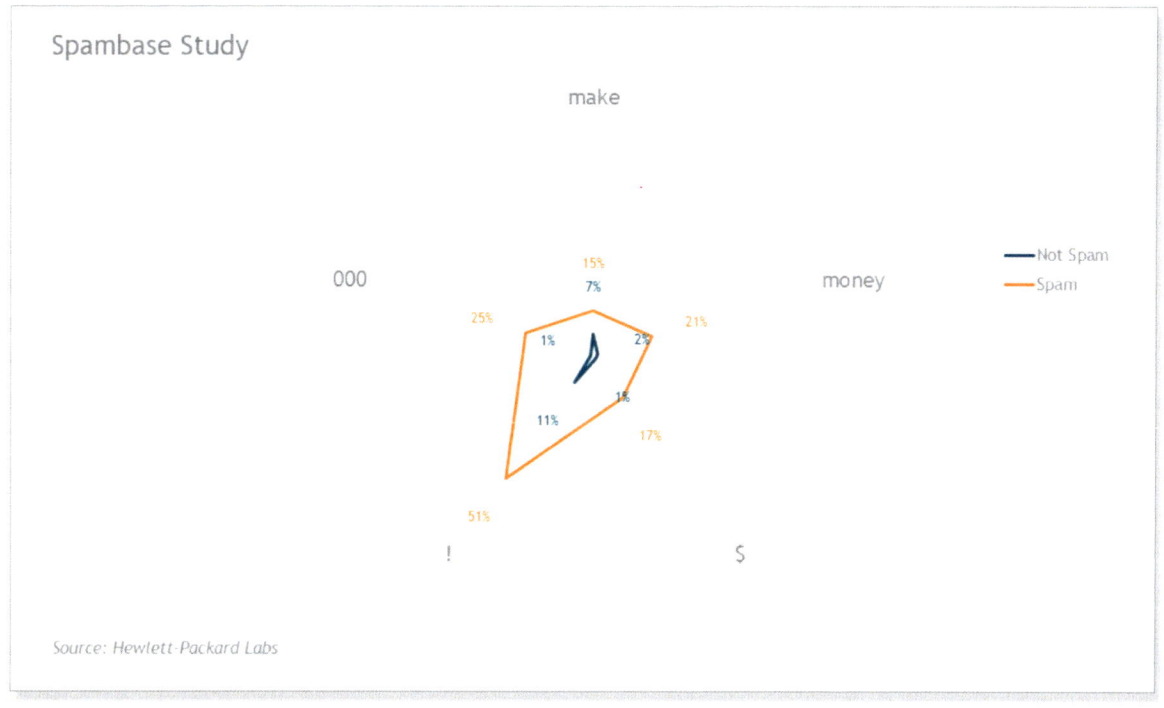

Spambase Study

make · 000 · money · $ · ! · 51% · 11% · 17% · 1% · 2% · 1% · 21% · 7% · 15% · 25%

— Not Spam
— Spam

Source: Hewlett-Packard Labs

Fig 5.345: Our final chart.

AKA

You may also see radar charts called:

+ circular charts
+ cobweb charts
+ irregular polygons
+ kiviat diagrams
+ polar charts
+ polar plots
+ spider charts
+ star charts
+ star plots
+ web charts

Uses

Radar charts are best used to:

+ visualize a small dataset with a category
+ display clusters in data
+ compare performance of populations (e.g., athletes, students, employees, airlines)
+ spot outliers in datasets
+ highlight commonalities and differences in data

Tips

+ You don't want to use radar charts with datasets that have too many data series. Labels start overlapping, and the clustering gets difficult to interpret.
+ Consider using data labels so you can fire your gridlines.
+ Any hub-and-spoke chart is processed by our brains like a circle, which is harder to interpret than rectangles.
+ If you only have one category, you can plot multiple metrics, but if you have two categories (as with the example chart), you can only plot one metric.
+ If you're using tabular data, you may need to use a pivot table to transform your data into a crosstab format. I've included a Pivot

285

Table worksheet in the downloadable workbook to show you how I set mine up. Click anywhere in the pivot table to open the PivotTable Fields pane.

Category / Metric
1–2 categories / 1+ metrics

Tutorial

Step 1: My dataset has a category with two elements (*spam* and *not spam*) and a metric, so I needed to configure my data into a crosstab format using a pivot table **(Fig 5.346)**. The data in the example file comes from a sample of a study performed by researchers, which aimed to analyze the frequency of a number of keywords and symbols in emails that were classified as spam and those that weren't. My pivot table aggregates a percentage of instances of the words money and make, as well as the symbols !, $, and 000, so I summarized by average instead of sum **(Fig 5.347)**. I copied and pasted my pivot table into another worksheet and cleaned it up.

Step 2: Select any cell inside your dataset and navigate to **Insert ▸ Charts ▸ Insert Surface or Radar Chart ▸ Radar ▸ Radar** (**Fig 5.348**).

Step 3: If you want to change the formatting of either of your data series (as I did), press Ctrl-1 to open the **Format Data Series** pane, then open the **Fill & Line** tab. Set **Line** to **Solid line** and **Outline color** to the color of your choice.

Step 4: With the data series still selected, we'll add data labels so that we can delete all those obnoxiously busy gridlines—unless, of course, you're into spider webs. From the **Chart Elements** flyout menu, select **Data Labels**. Rinse and repeat for each data series in your chart. Then delete the gridlines **(Fig 5.349)**. That feeling is called data catharsis. Cherish it. (I didn't delete the axis for a reason that will be revealed in step 6. #foreshadowing)

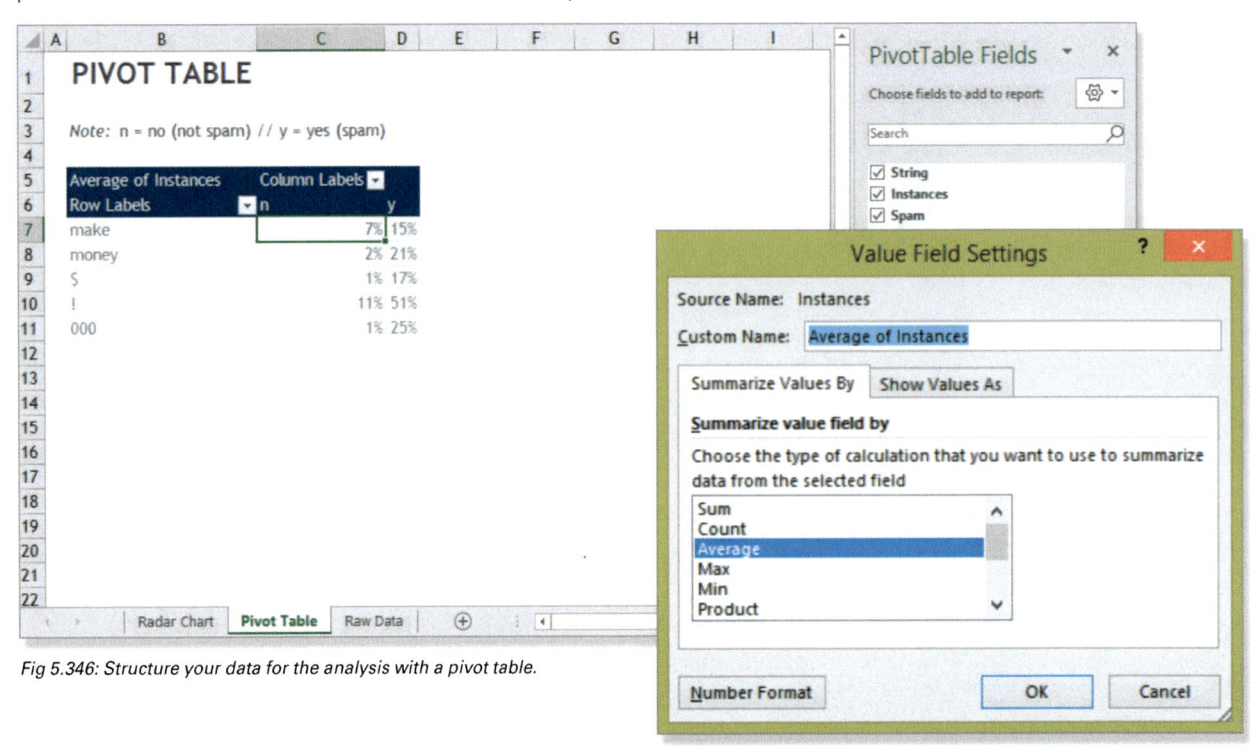

Fig 5.346: Structure your data for the analysis with a pivot table.

Fig 5.347: Summarize by average instead of sum for percentages.

CHAPTER 5 • CREATE CHARTS

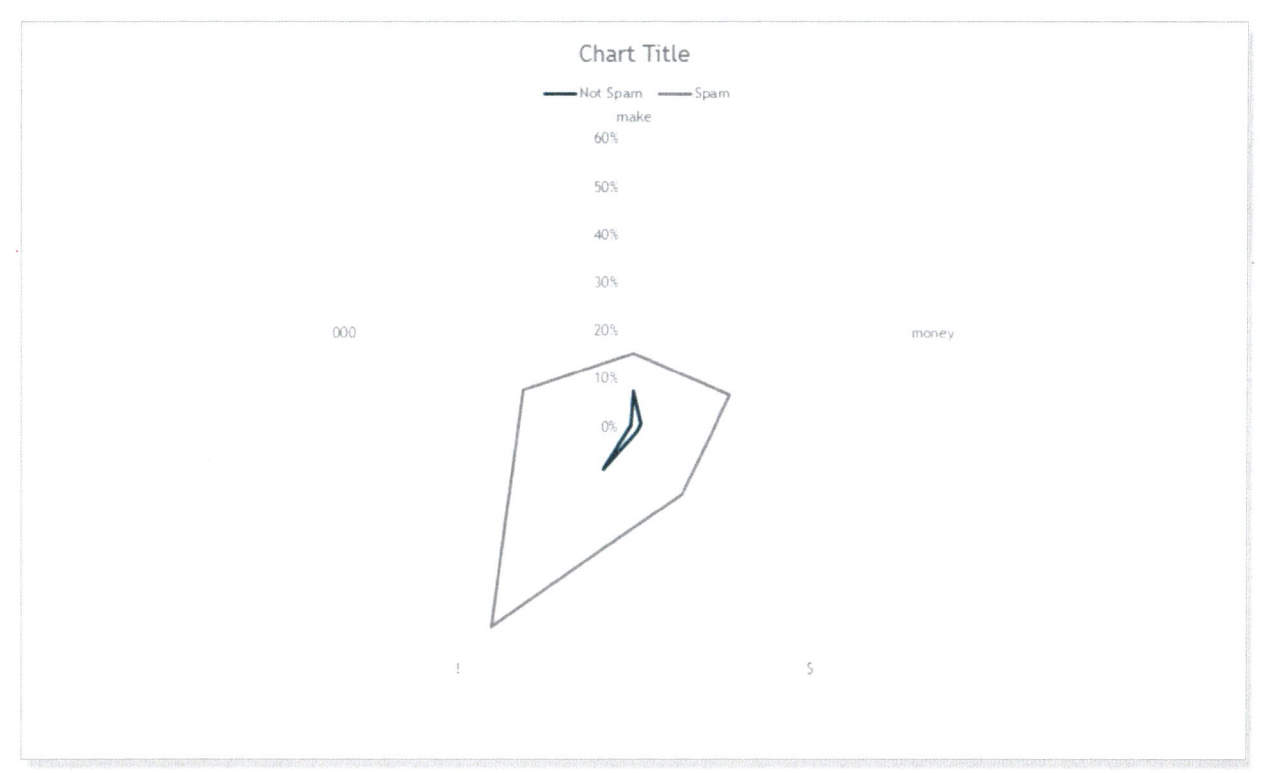

Fig 5.348: Our initial chart.

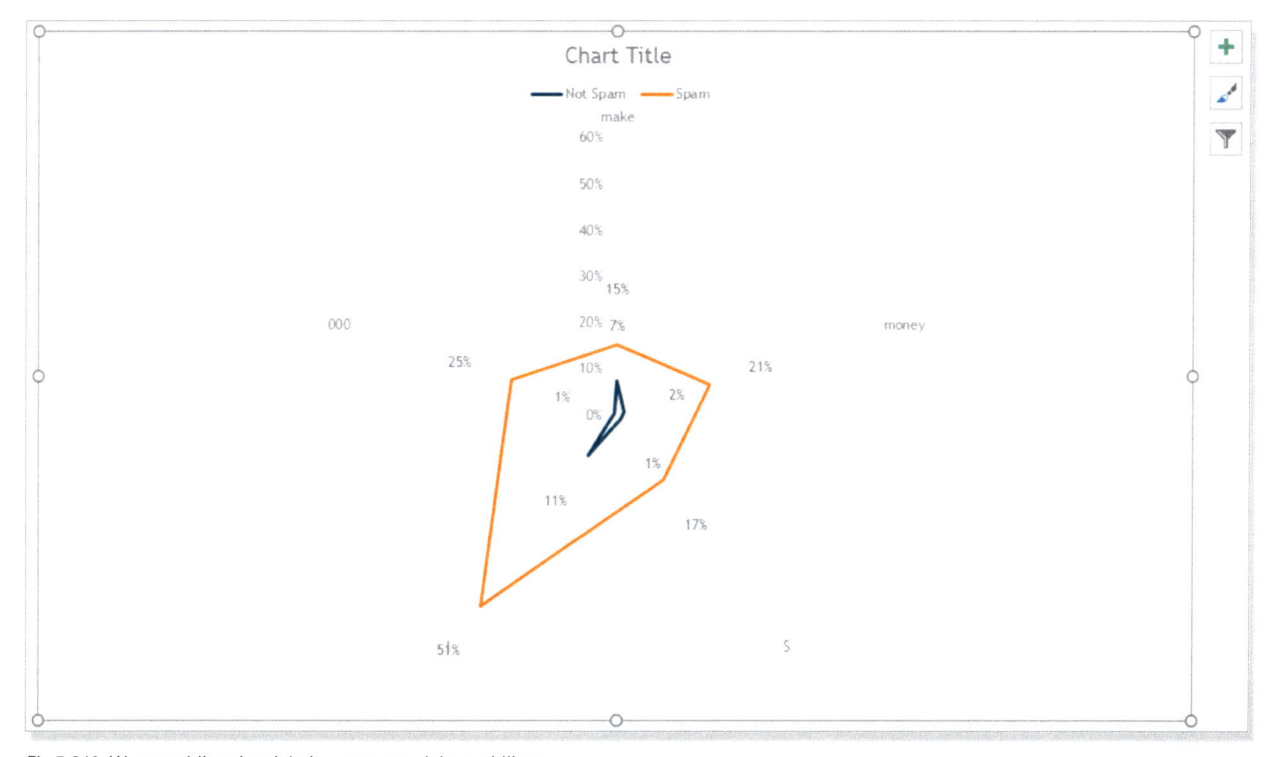

Fig 5.349: We are adding data labels, so we can delete gridlines.

287

CHAPTER 5 • CREATE CHARTS

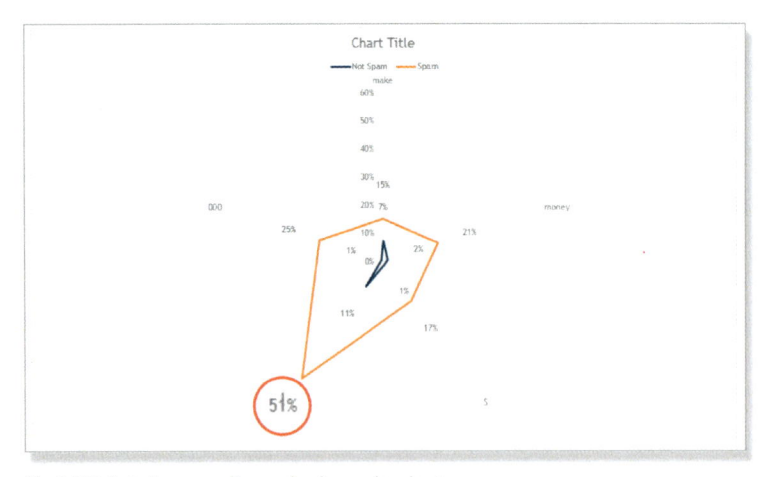

Fig 5.350: Labels can easily overlap in a radar chart.

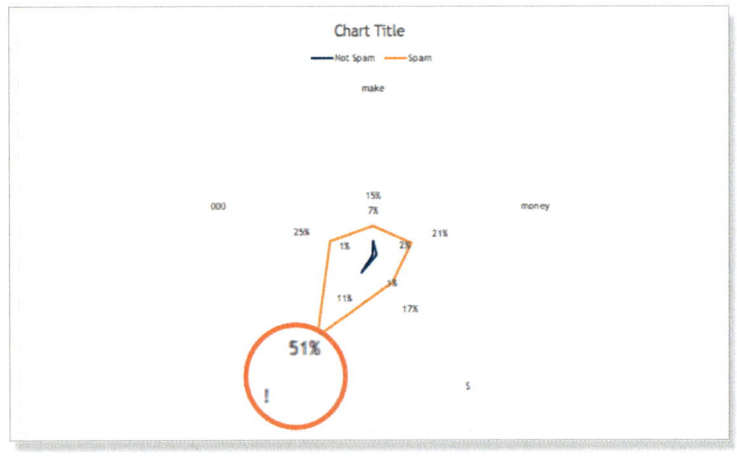

Fig 5.351: Our series labels no longer need to share a double bed.

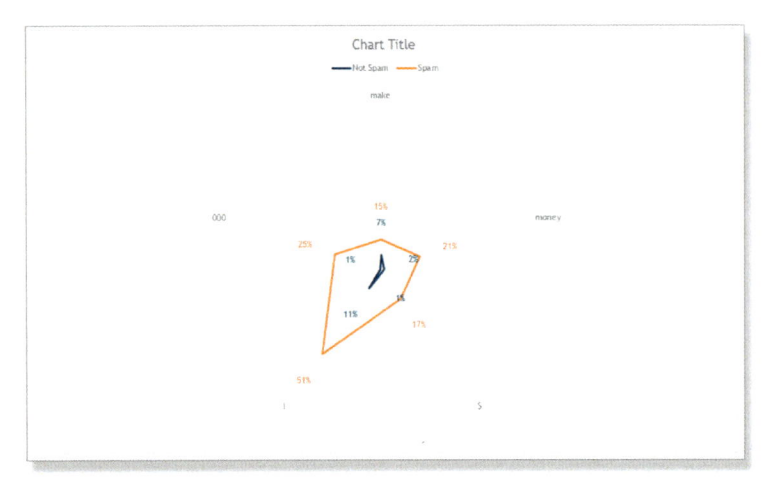

Fig 5.352: If your data series are close together, you may want to make the data series and labels matchy matchy.

Step 5: One risk with the radar chart is that you may have a data label that bumps up against or overlaps with one of your category labels, as I did with the chart from the example data **(Fig 5.350)**.

Step 6: The way I chose to resolve this was to adjust the maximum label of the y axis. If you run into this issue and need to adjust yours, select the axis to open the **Format Axis** pane. In the **Axis Options** tab, set **Maximum Bounds** to a value that gives your labels breathing room. I bumped mine up from .6 to .8. Once this issue is resolved, you can delete the axis **(Fig 5.351)**. You may also need to make your **Plot Area** smaller, if the category label at the top of the chart gets a little too cozy with your chart title.

Step 7: To bolster the association between the data series and data labels, I'll make the data labels the same color as the series **(Fig 5.352)**. This is purely optional. The data series in my dataset were pretty close together. If you opt to customize your data labels, select one of the data labels to select all for that data series. In the **Format Data Labels** pane, select **Text Options.** In the **Text Fill & Outline** tab, under **Text Fill**, set **Fill Color** to match your data series. Rinse and repeat for any other data series in your dataset.

CHAPTER 5 • CREATE CHARTS

Step 8: There's a lot going on in this chart, so you may want to increase the font size of your category labels. I increased mine to 14 pt by selecting them and navigating to **Home ▶ Font ▶ Font Size (Fig 5.353)**.

Step 9: The legend looks odd at the top of the chart. You can move your legend by selecting it to open the **Format Legend** pane. In the **Legend Options** tab, I set **Legend Position** to **Right** and selected **Show the legend without overlapping the chart (Fig 5.354)**. This didn't impact my chart, but it could with future datasets when I save my chart as a template. Then I bumped up **Font Size** to 11 pt.

Step 10: Clean up your chart to taste. I cleaned mine by taking the following steps:

- made title more descriptive
- left aligned chart title
- bumped up title font size
- added annotation for data source

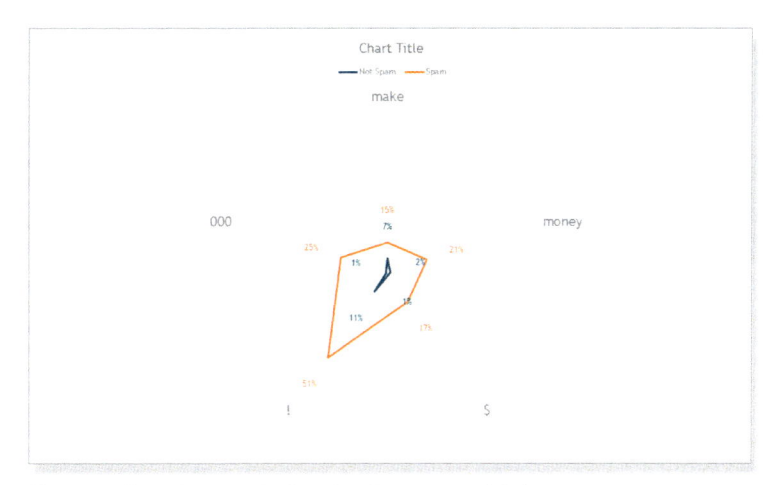

Fig 5.353: I like to bump up the font size of my category labels.

💬 **LEARN MORE** See the Design with Elegance chapter to learn how to perform these refinements.

Example File
radar-chart.xlsx

Template
If you'll use this chart again, save it as a chart template. Or download the template included in the chart templates folder: mds-radar.crtx.

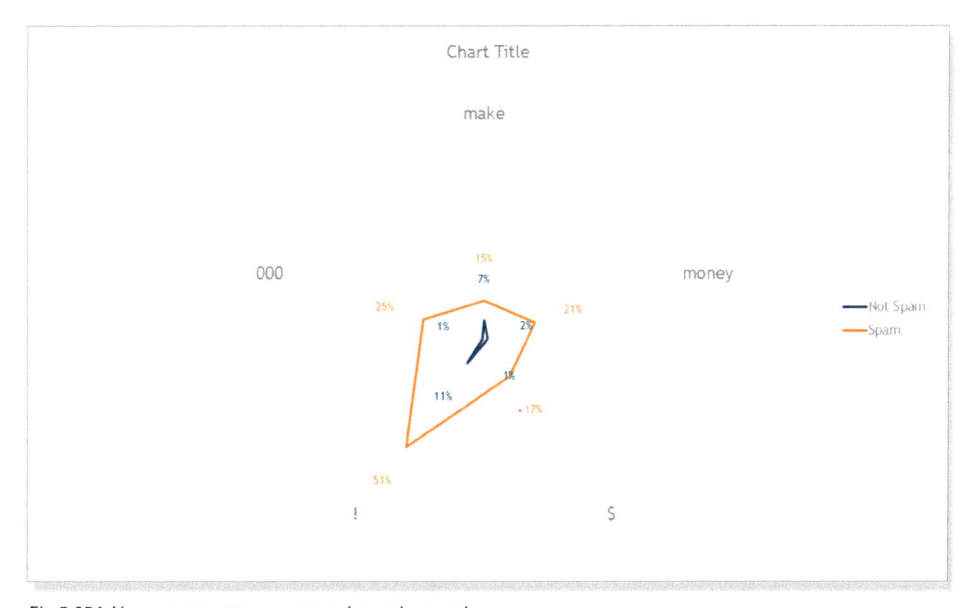

Fig 5.354: You may want to move your legend around.

289

CHAPTER 5 • CREATE CHARTS

PROGRESS
GAUGE CHART

AKA
You may also see gauge charts called speedometer charts.

Uses
Gauge charts are best used to:

- visualize the differential between a goal and actual values
- express a story using a familiar metaphor
- display performance metrics against a goal metric, such as sales, fundraising, weight loss, etc.

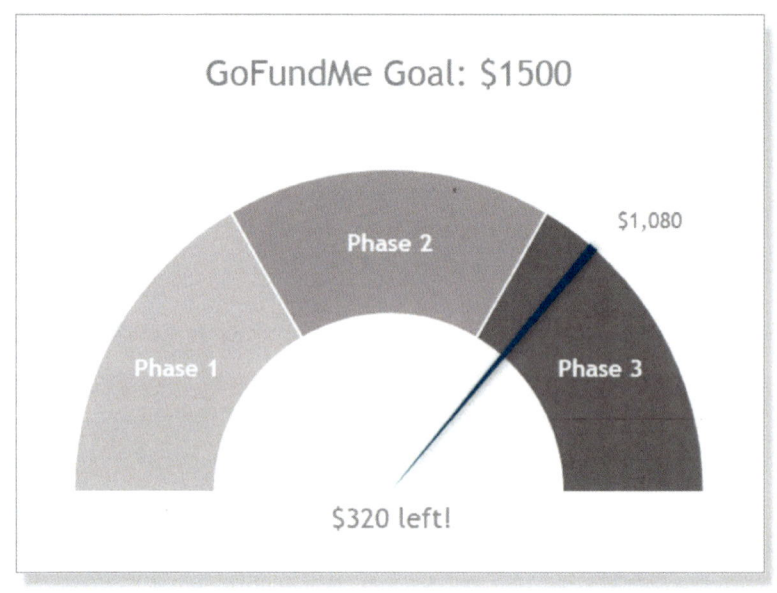

Fig 5.355: Our final chart.

Tips

- Be careful not to throw your branding out the window with gauge charts.
- You don't need 3-D effects, like drop shadows and bevels. They can cheapen your dashboard, in my opinion.
- I'm a professed minimalist, but less is more with gauge charts, which don't get a lot of respect because of how unwieldy they quickly become. If a gauge chart ignites flashbacks of the days of Clippy popping up to ask me repetitive questions, my eyes will roll into the back of my head.
- In the same way you wouldn't have axis labels set to increment every 5 units when your maximum value is 100, be careful to not add too many labels to your gauge. Unnecessary pixels cause anxiety.
- It's easy to get tangled up in the data and math. Laying out your raw data in an organized fashion will go a long way toward making future gauge charts easy to create.
- I provide additional guidance on how I laid out my data in the optional downloadable workbook. It gets way in the weeds for people who like structures and systems (*raises hand*). You don't need it to create a gauge chart. It's an over-the-top effort to ensure no one loses as much sleep as I did, trying to figure out the logic behind this whacky, hacky chart.
- As simple as the metaphor of a gauge is, this is the most complicated chart in the book. I went to great lengths to provide as many tips as possible to help flatten the learning curve.

Category / Metric
2 categories / 2 metrics

Note: The assignment of categories and metrics can be a little confusing for the gauge chart because of the helper tables. But since this chart combines a donut and pie chart—and each uses one category and one metric, our gauge chart uses two of each.

Tutorial

Hundred-Foot View

I had the hardest time wrapping my mind around the logic behind the gauge chart. I watched and read an insane number of tutorials and was frustrated by the fact that they all dealt with numbers less than 360 (the number of degrees in a circle). Since I wanted a tutorial that wasn't limited by this restriction, I really had to wrestle through the logic.

Also, every tutorial I drilled through stated what to do without explaining why. Or maybe I was too dense to understand their erudite strategies.

Either way, my goal here is to explain the strategy in a way that's easy to understand!

Let's start with a diagram **(Fig 5.356)**.

donut chart. The sole purpose of the pie chart is to give us a pointer that moves clockwise as your data updates. The pie chart will technically contain four values:

1. A value equal to our goal amount (which will make up the lower half of the pie chart and will be also be invisible)
2. The amount raised
3. The amount yet to raise
4. The width of our pointer

The first three will be hidden; only the pointer "slice" will be visible. Each of these hidden values will be used to position the pointer so that it dynamically updates as new donations come in for our GoFundMe campaign.

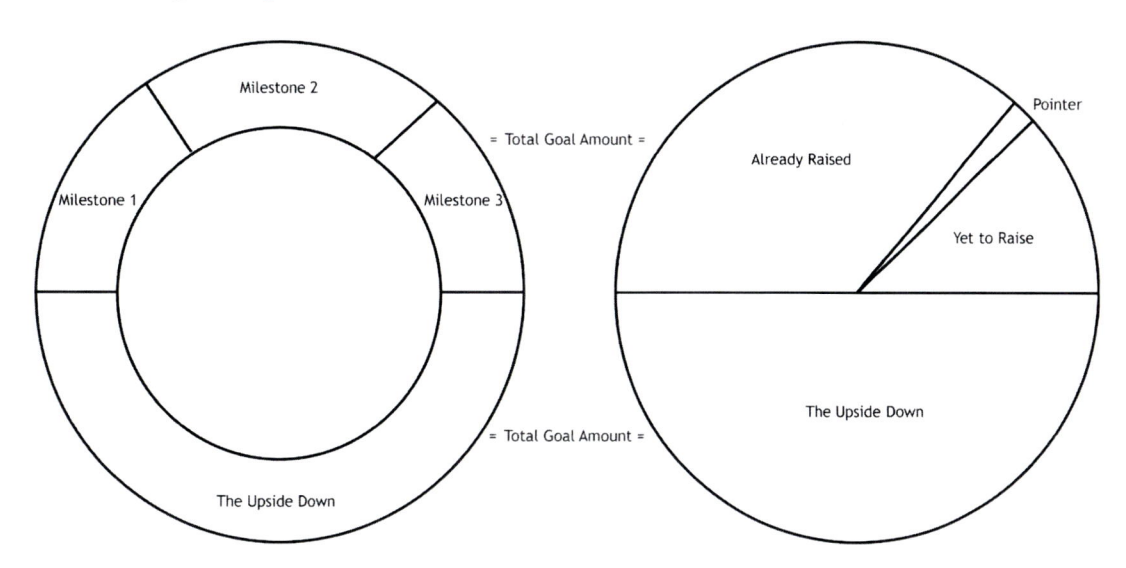

Fig 5.356: The gauge chart is tricky. Hopefully this diagram demystifies it a bit.

We'll first build a donut chart. Its sole purpose is to house our milestones. The bottom part of the donut chart will be invisible (no fill or border). Think of it as dialing in to a conference call in a nice top with pajama bottoms. No one will see them so no harm no foul.

Then we'll essentially superimpose a pie chart onto the

Since you don't want a pie chart sitting on top of your donut chart, we'll add a new data series to our donut chart via the Select Data Source dialog, then change the chart type to Pie Chart—effectively creating a super freaky combination chart.

Alright, enough navel gazing. It's time to get to work!

CHAPTER 5 · CREATE CHARTS

💬 **LEARN MORE** See the Hack Excel's Charting Engine section of the Design with Elegance chapter to learn how to rock the Select Data Source dialog.

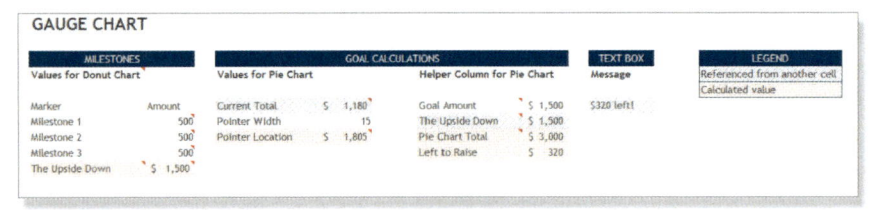

Fig 5.357: Organizing your data will make building your chart much easier.

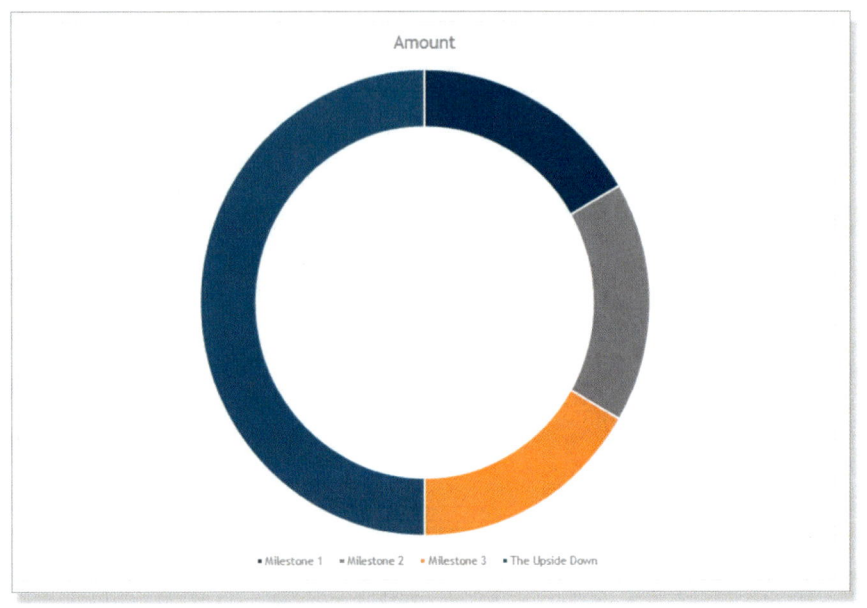

Fig 5.358: Our initial chart.

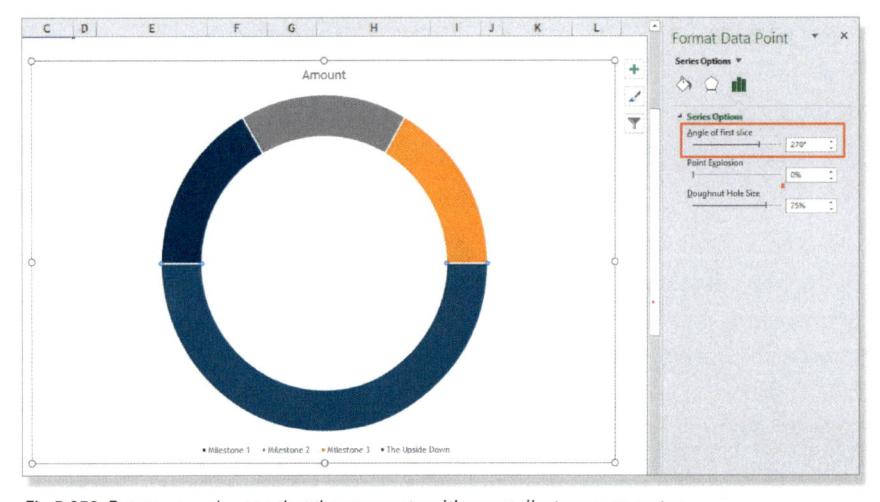

Fig 5.359: Rotate your chart so that the segments with your milestones are on top.

Step 1: Before you can start building your chart, you'll need to arrange the data. If you download the example file, every cell with a triangle in the upper-right corner has a comment with pretty exhaustive explanations. Of course, when you're adding a gauge chart to a dashboard, the data underpinning it should be in another worksheet **(Fig 5.357)**.

Step 2: Select any cell in the table that contains your milestones and navigate to **Insert ▸ Charts ▸ Insert Pie or Doughnut Chart ▸ Doughnut ▸ Doughnut (Fig 5.358)**.

Step 3: Select a slice of the donut chart and press Ctrl-1 to open the **Format Data Series** pane. The first thing we need to do is rotate our donut chart on its axis. We want the half with the three sections to sit in the northern hemisphere and the Upside Down (any *Stranger Things* fans out there?) to be in the southern hemisphere. To rotate it, open the **Series Options** tab and set **Angle of first slice** to 270° **(Fig 5.359)**. I'd love to let you think I just remember that, but the reality is I just drop 90° into that field and increment by 90° until it's positioned correctly. There's no "lefty loosey, righty tighty" mnemonic device for gauge charts. Sorry.

CHAPTER 5 · CREATE CHARTS

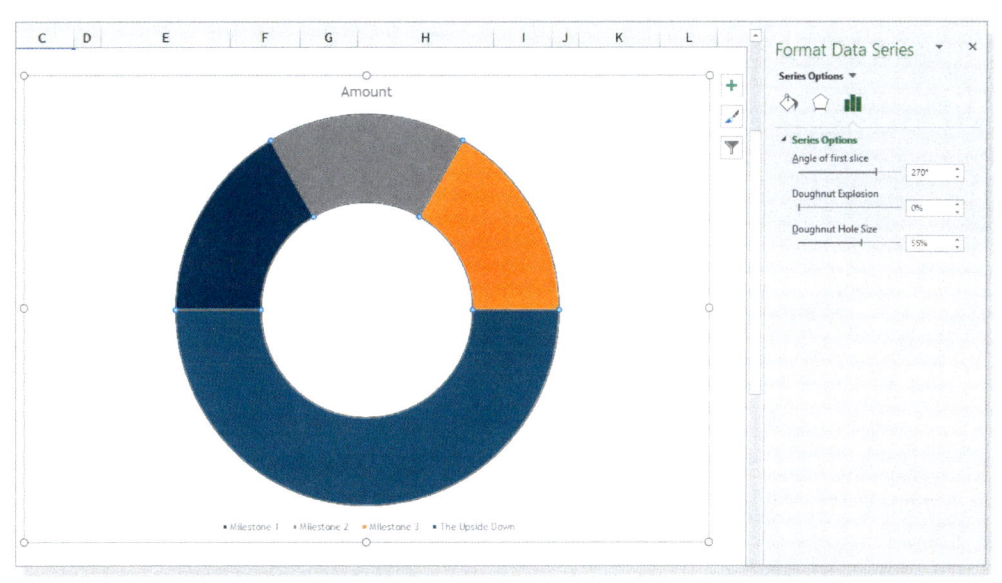

Fig 5.360: Adjust Doughnut Hole Size to thicken up the donut.

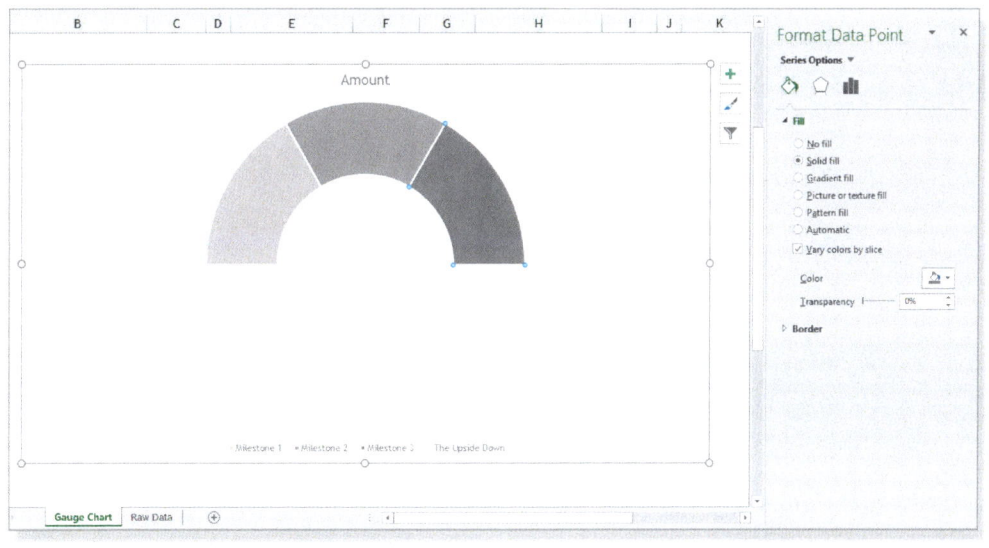

Fig 5.361: Set the fills of your data points.

Step 4: If you want your gauge chart to be thicker, reduce **Doughnut Hole Size**, also available from the **Series Options** tab. I set mine to 55% **(Fig 5.360)**.

Step 5: Select one of the slices of your donut chart, which will open the **Format Data Point** pane. Open the **Fill & Line** tab and set **Fill** to **Solid fill**. Also, select the **Vary colors by slice** option. Set **Color**. Rinse and repeat with each of your milestone slices (data points). I used the gray colors from my theme to minimize distraction and direct users' eyes to the pointer. Next, remove the fill from the bottom half of the donut by setting **Fill** to **No fill** and **Border** to **No line (Fig 5.361)**.

Step 6: We don't need the legend, so it can go the way of its people.

293

CHAPTER 5 • CREATE CHARTS

Step 7: Now we'll add another data series, which will come in as another donut chart, but then we'll switch it to a pie chart in the next step. It sounds complicated, but we're just changing it to get a pointer that's the right shape. Creating a pointer from a donut chart would be weird. Like a water pistol that shoots…

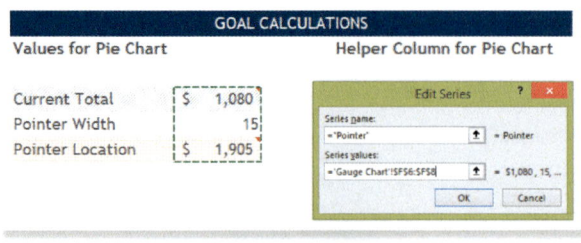

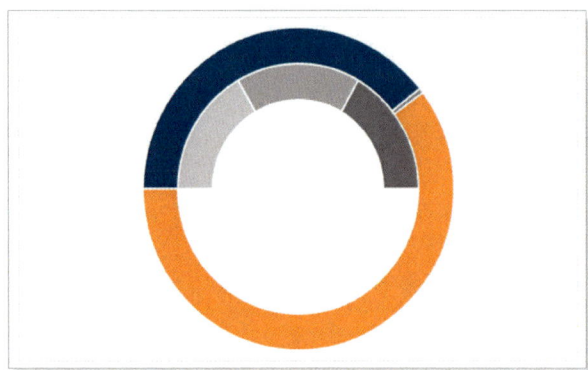

Fig 5.362: Add the data series we'll use to build the pointer.

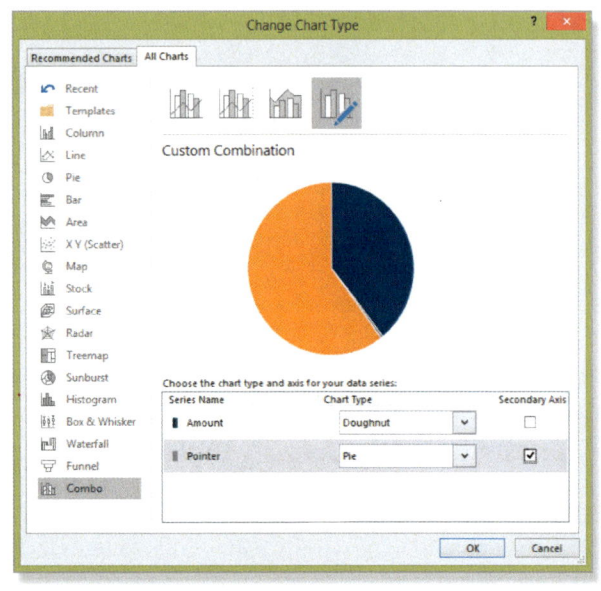

Fig 5.363: Change your chart type to pie.

jelly! To get started, right-click on the chart and choose **Select Data** from the contextual menu to open the **Select Data Source** dialog. Then click the **Add** button in the **Legend Entries (Series)** field. In the **Edit Series** dialog, point to the cell containing your series name or type it into the **Series name** field. I typed in "Pointer." Then, click-and-drag over the values in your data table. For the example workbook that's in cells F6:F9 **(Fig 5.362)**. When you're finished tinkering, click **OK** to close the dialog.

Step 8: Now we'll change our second donut chart into a pie chart. First, right-click on the series and choose **Change Series Chart Type** from the contextual menu, which will open the **Change Chart Type** dialog. Set your new series to **Pie** and select the **Secondary Axis** checkbox **(Fig 5.363)**. If you don't put it on a secondary axis, it'll hide behind your donut chart, and changing the order of the series in the **Select Data Source** dialog won't fix it. (I tried…for the sake of science.) Click **OK** to close the **Change Chart Type** dialog.

Step 9: Select a slice of the pie to open the **Format Data Series** pane. Just as we did with the donut chart, set **Angle of first slice** in the **Series Options** tab to the same value you applied to your donut chart. I set it to 270° in the example workbook **(Fig 5.364)**.

Step 10: Switch over to the **Fill & Line** tab. Set **Fill** for each of the data series points except the pointer to **No fill** and **Border** to **No line**. Then set **Fill** for the pointer slice to a color of your choice.

Step 11: With the pointer slice still selected, open the **Series Options** tab. Set **Point of Explosion** to a value that will bump your slice a bit outside the boundaries of the pie chart. I set mine to 3% **(Fig 5.365)**. This step is optional, but it makes your pointer look a little more realistic, in my opinion.

294

CHAPTER 5 · CREATE CHARTS

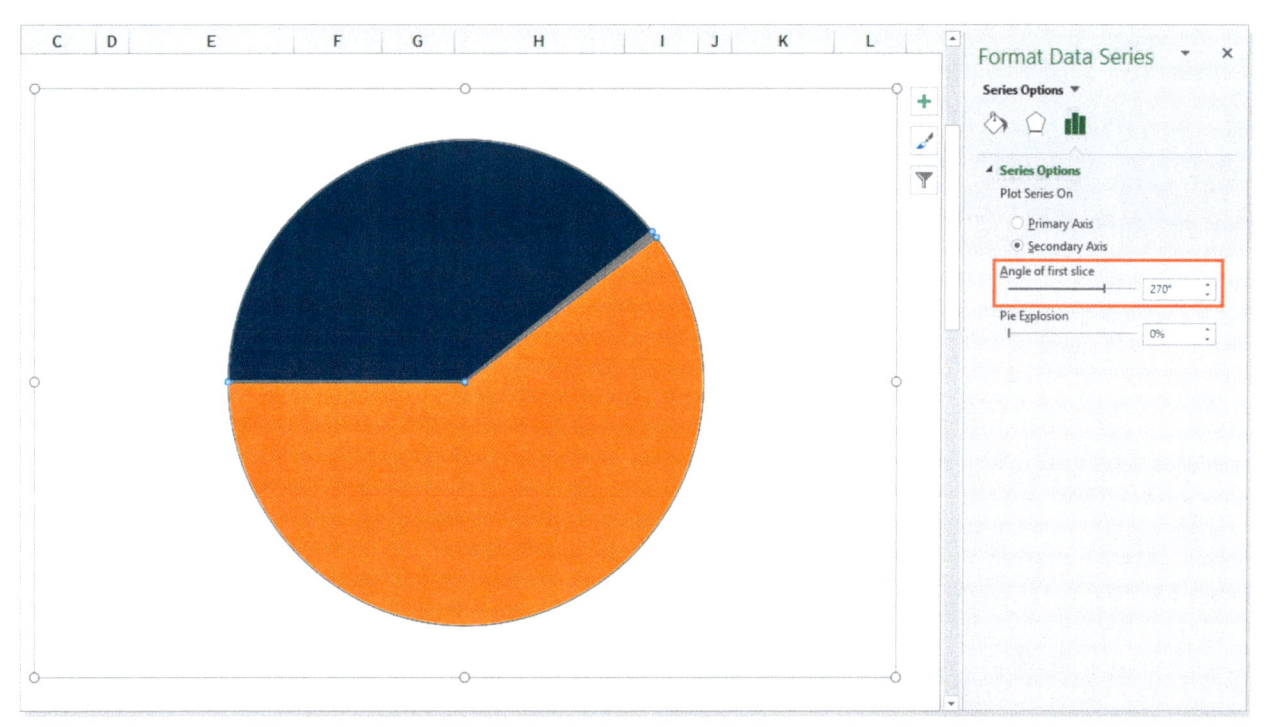

Fig 5.364: Set the Angle of first slice to the same angle you used for the donut chart.

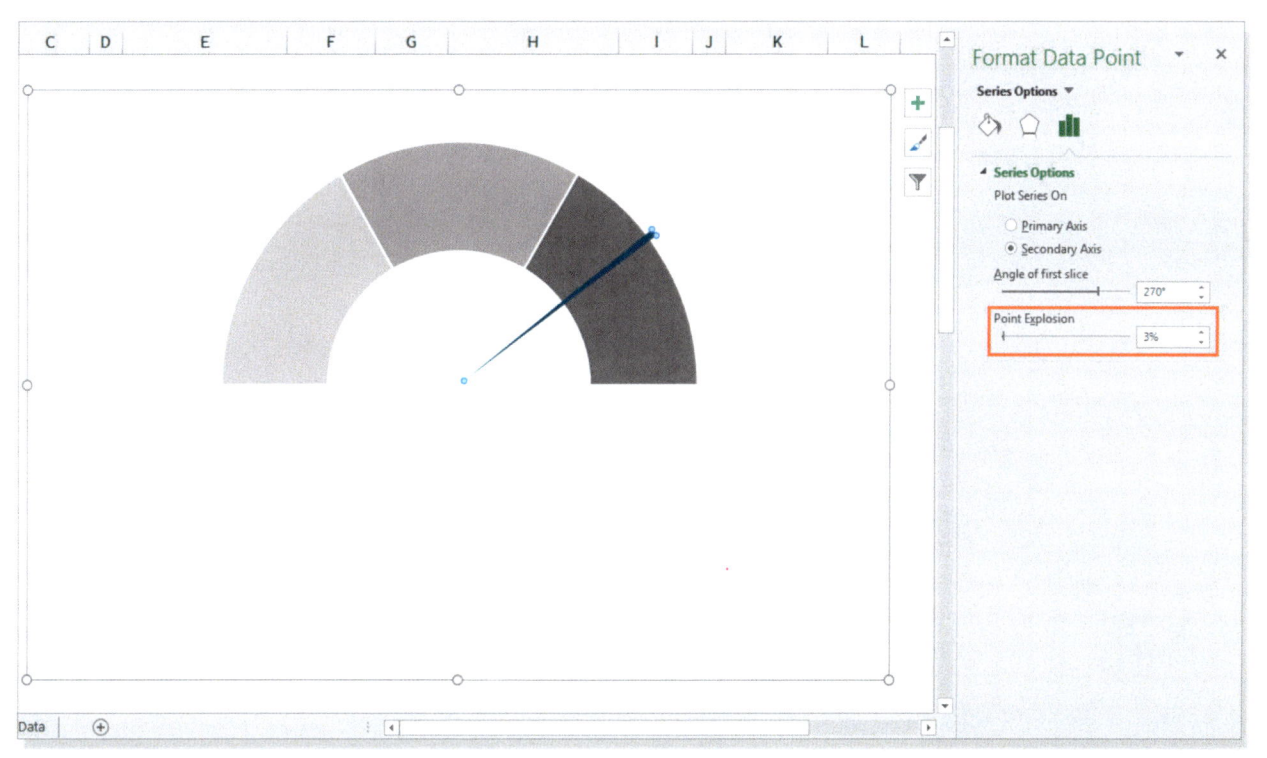

Fig 5.365: Set Point of Explosion to a small value to make your pointer look more realistic.

295

Step 12: If you want to give your pointer a little depth, you can add a slight drop shadow to it. With it still selected, switch to the **Effects** tab. I added the **Offset: Bottom Right** preset (first option under **Outer**) but then dialed **Distance** back to 1 pt **(Fig 5.366)**.

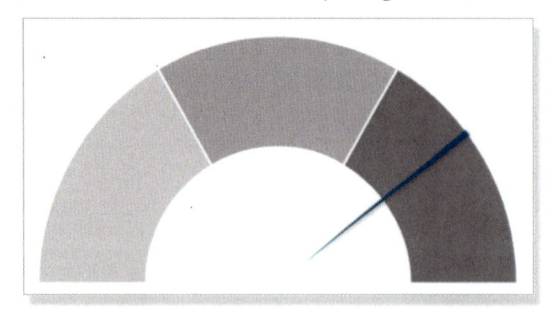

Fig 5.366: You can add a touch of drop shadow to your pointer to give it depth.

Step 13: Next, we'll add a data label to our pointer. Select your pointer slice, then open the **Chart Elements** flyout menu and click the right-facing arrow next to **Data Labels**. Choose **More Options** from the flyout menu to open the **Format Data Labels** pane, then open the **Label Options** tab. For some reason, you don't have

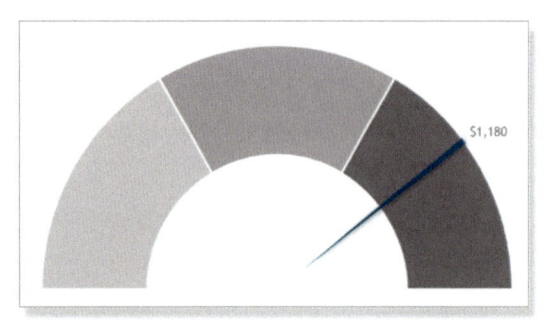

Fig 5.367: Add a data label to your pointer.

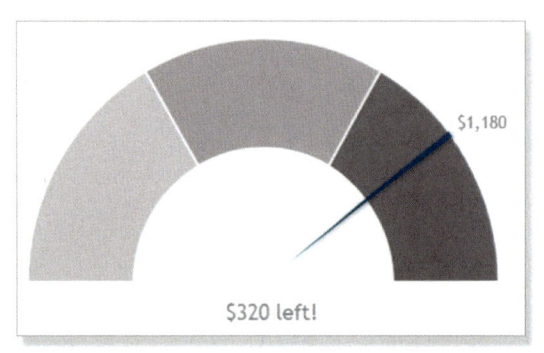

Fig 5.368: You can add a message to your chart with a text box.

the **Value From Cells** option we used in our Bar Chart, Column Chart, and Line Chart tutorials. But we can do the same thing manually. With the data label still selected, enter an = sign into the **Formula Bar**, then point to the cell that contains your current total (F6 in the example workbook) and press Enter to apply your formula. Format the text to taste **(Fig 5.367)**.

Step 14: You will want to test your chart to make sure it updates properly. I tested mine by adding a new value to my raw dataset, as if someone had just donated. Then I temporarily removed some of the data to make sure the needle moves in both directions.

Step 15: This is purely optional, but you can also add an annotation of some kind to your gauge chart. I wanted to pull in the amount remaining in our fundraiser. But you can't combine text and a cell reference in a text box, so you need to combine them using a TEXT function in a cell, then pull that cell into your text box **(Fig 5.368)**.

> **LEARN MORE** See the Add Annotations section of the Design with Elegance chapter to learn how to create text boxes.

Step 16: Finally, add a chart title to your chart via the **Chart Elements** flyout menu, format to taste, and you're good to go!

Credit: Shout out to the Carpe Diem YouTube channel for the inspiration to use the Pie Explosion feature to get the needle to pop: **bit.ly/mdsepc-gauge**.

Example File
gauge-chart.xlsx

Template
Because this chart is so highly customized and needs to be built from the ground up, Excel won't let you save it as a chart template. You'll need to create it from scratch or save it in a file where you keep all your charts that won't save as templates.

WATERFALL CHART

AKA
You may also see waterfall charts called:

- bridge charts
- flying bricks charts
- Mario charts

Uses
Waterfall charts are best used to:

- shed light on how an initial value is affected by a series of positive and/or negative values

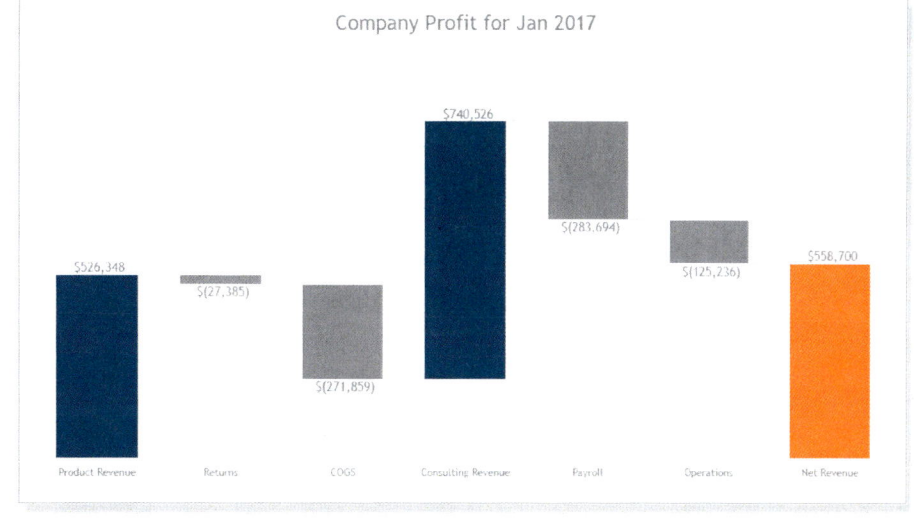

Fig 5.369: Our final chart.

- analyze any process, workflow, or funnel that includes increments and decrements in value
- manage or audit contract acquisitions and losses in industries like marketing, legal, and construction
- visualize any process that uses a running total

Tips
- All the charts new to Excel 2016 (Histogram, Waterfall, Pareto, Box & Whisker, Treemap, and Sunburst) tend to be a bit volatile and cause Excel to crash much more than other charts so save often.
- At the time of writing, you can't move the chart title, link the title to a cell (to create a dynamic title), adjust the Plot Area, or save chart templates with any of the charts new to Excel 2016. And if you add a text box (e.g., to add a data source) and move your chart, the text box doesn't move with the chart. (You will need to group the chart and text box if you want to move or scale them together.)
- Usually the initial and the final value bookends are represented by whole columns, whereas the intermediate values are characterized by columns that appear to be floating.

- You want to use one fill color to denote positive values, another for negative values, and yet another for the final value.
- Values can be measured against categories (as in the example workbook) or a time series.
- Be careful using red and green in waterfall charts because red-green colorblindness is very common, especially among men.

💬 **LEARN MORE** See the Design for Colorblind Users section of the Design with Elegance chapter to learn how to design with accessibility in mind.

- Under normal circumstances, I'm a huge fan of bringing data labels inside your bars to give your visualization cleaner edges. But the data labels for the new charts in Excel 2016, at the time of writing, offer limited formatting options. There's very little top margin, causing them to bump up against the edges of your visualization. (It reminds me of Mr. Incredible huddled up in a near-fetal position in his car. But that might just be me….) Anyway, if you manually reposition them, you will need to manually update them any time you update your chart.

CHAPTER 5 · CREATE CHARTS

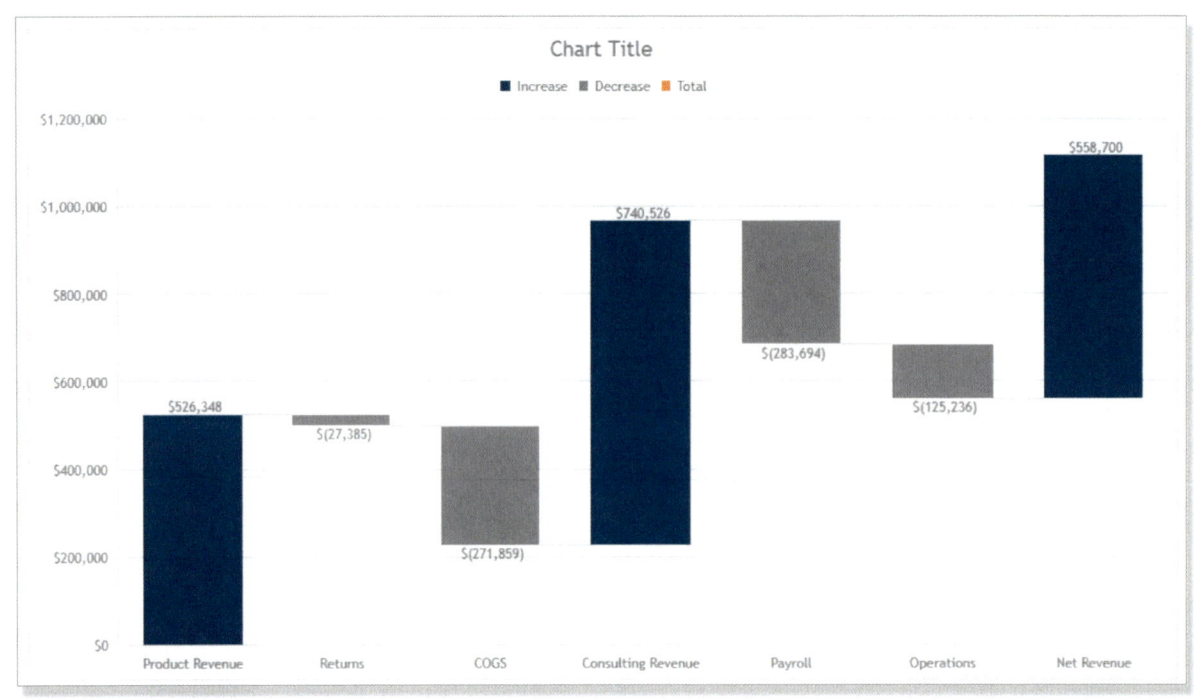

Fig 5.370: Our initial chart.

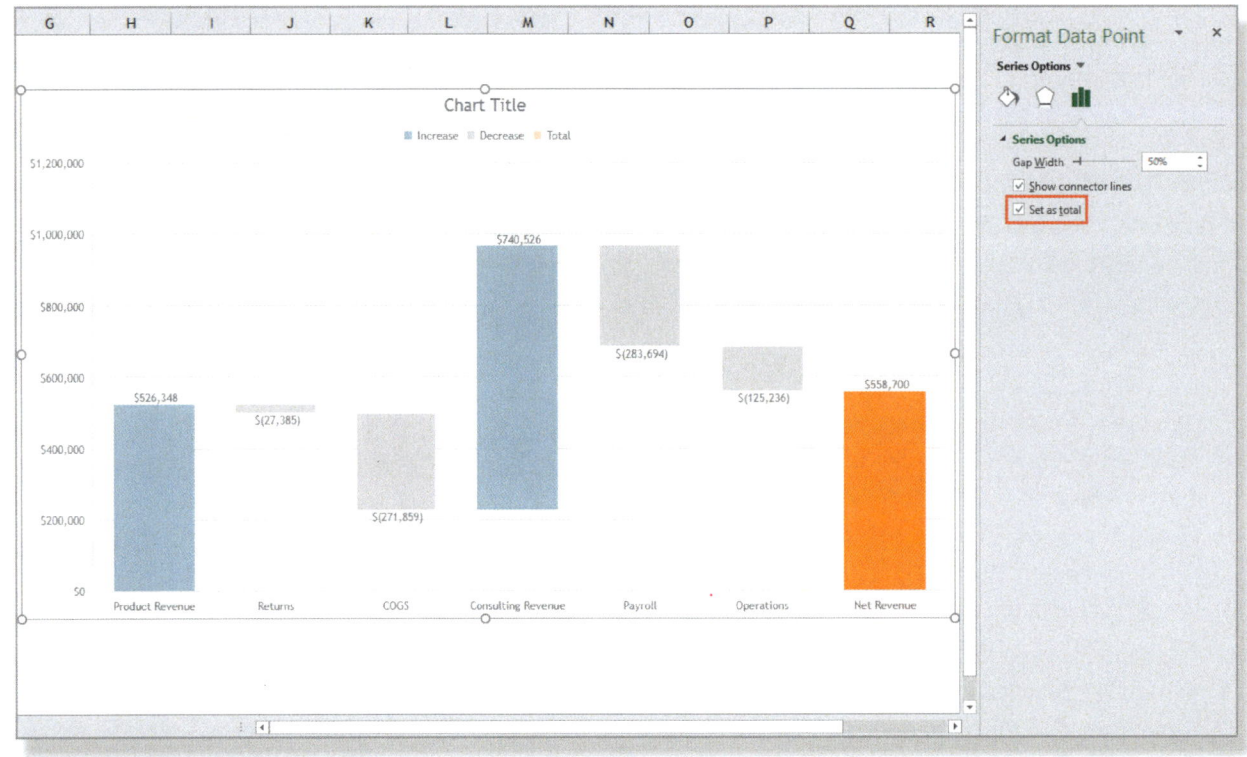

Fig 5.371: Set last bar as your net total value.

Hopefully the Excel team will iron out some of these bugs in a future update.

- Avoid long horizontal axis labels so they don't get flipped on their sides (one of my biggest data viz pet peeves).

Category / Metric
0 categories / 4+ metrics

Tutorial
Step 1: Select any cell inside your dataset and navigate to **Insert** ▸ **Charts** ▸ **Insert Waterfall or Stock Chart** ▸ **Waterfall** ▸ **Waterfall (Fig 5.370)**.

Note: You may also find it housed under **Insert Waterfall, Funnel, Stock, Surface or Radar Chart**.

Step 2: You need to tell Excel which value your net total is; Excel doesn't assume it's the last row in your table. To reassign your last bar, select that data point (i.e., floating brick), press Ctrl-1 to open the **Format Data Point** pane, and select **Set as total** in the **Series Options** tab **(Fig 5.371)**. Excel will assign it a different color from your theme.

Step 3: Clean up your chart to taste. I cleaned mine by taking the following steps:

- made title more descriptive
- bumped up title font size
- removed gridlines
- removed legend
- removed vertical axis

💬 **LEARN MORE** See the Design with Elegance chapter to learn how to perform these refinements.

Example File
waterfall-chart.xlsx

Template
At the time of writing, you can't save a waterfall chart as a chart template. You'll need to create it from scratch or save it in a file where you keep all your charts that won't save as templates.

CHAPTER 5 • CREATE CHARTS

MAPS

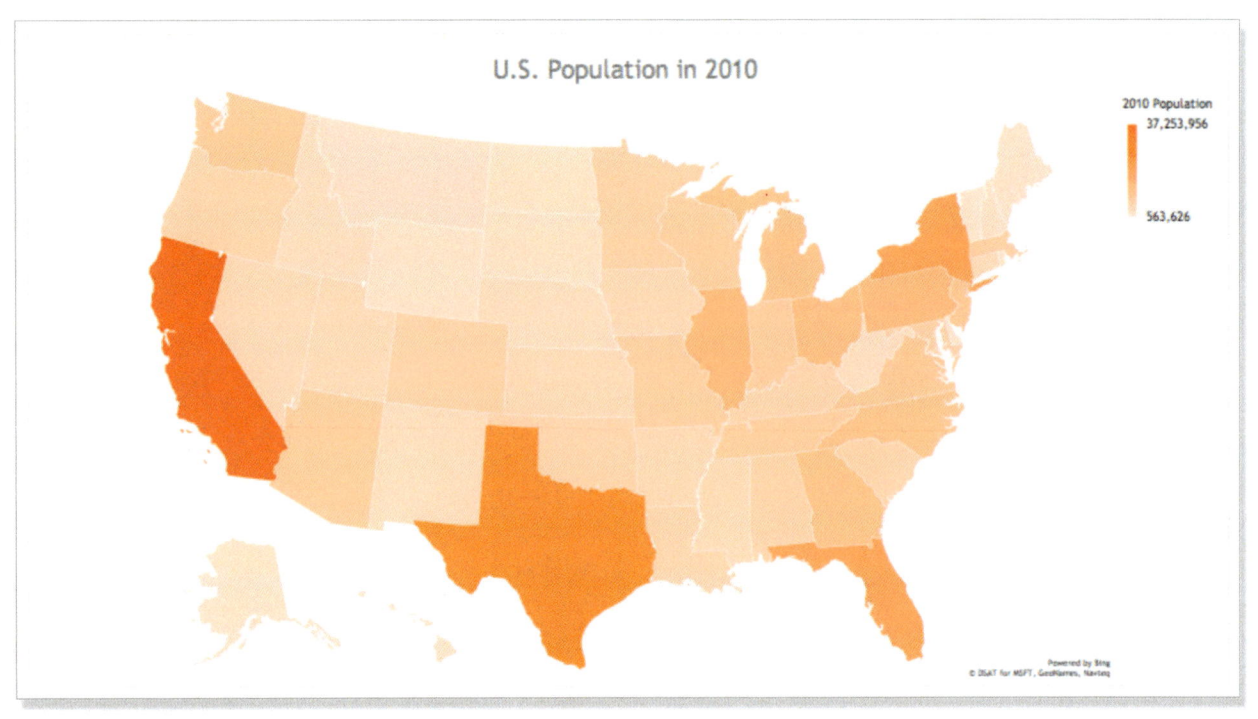

Fig 5.372: Our final chart.

Note: Maps are available to 365 subscribers only.

AKA
You may also see maps called:
- choropleths
- filled maps
- geo charts

Uses
- tether data to geographic boundaries
- generate heatmaps superimposed on maps

Tips
- At the time of writing, you can't move the chart title, link the title to a cell (to create a dynamic title), adjust the Plot Area, or save your map chart as a template. And if you add a text box (e.g., to add a data source) and move your chart, the text box doesn't move with the chart. (You will need to group the chart and text box if you want to move or scale them together.)
- If you don't see the Map option under the Insert menu, follow the instructions on the Microsoft website to enable this feature: **bit.ly/mdsepc-maps**.
- You need to enable Excel to use Bing to generate your map.
- Map charts in Excel automatically detect geographical data, such as countries, regions, states, provinces, counties, and postal codes.
- If you're creating 3-D maps, you can feed Excel longitude and latitude values.
- Maps can display both values and categories.
- You might get geocoding errors if your dataset includes cities that exist in multiple states, such as Salem. You can reduce the risk of errors by providing more data, such as the state/region.

Category / Metric
1 category / 1 metric

CHAPTER 5 • CREATE CHARTS

Tutorials

Note: We'll cover two mapping strategies to give this chart type a little more breadth of coverage. You'll still want to play with Excel's settings and options. There's so much more you can do with map charts. Learn more from the **office.com** website: **bit.ly/mdsepc-map-chart**.

Option 1

Step 1: Select any cell inside your dataset and navigate to **Insert ▸ Charts ▸ Insert Map Chart ▸ Filled Map (Fig 5.373)**.

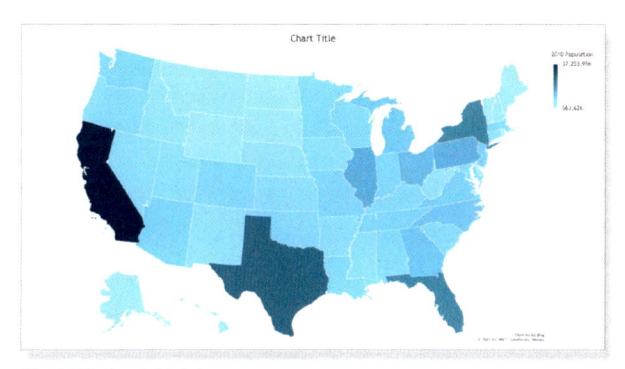

Fig 5.373: Our initial chart.

Step 2: I'm not a fan of the gradient Excel creates from my branded blue color. Something about it is reminiscent of the packaging for chlorine tablets. The silver lining to this cloud is my disdain gave me the perfect excuse to tinker with Excel's map formatting options. Getting started is a little tricky. You need to select an area of your map (not just the **Plot Area**) to open the **Format Data** pane. If you select the **Plot Area** outside the boundary of your map and press Ctrl-1 you'll open the **Format Plot Area** pane instead of the **Format Data Series** pane. If you think about it, it's no different from selecting a chart; however, with a map it's easy to think of the empty areas as ocean instead of white space. Next, open the **Series Options** tab if it didn't open automatically. Under **Series Color**, set **Minimum** and **Maximum** to your preferred colors (preferably different shades of the same color). Alternatively, you can use three colors, which will create a diverging color palette from your colors instead of the sequential color palette you get when you work with two colors. I liked the simplicity and elegance of using my branded orange **(Fig 5.374)**.

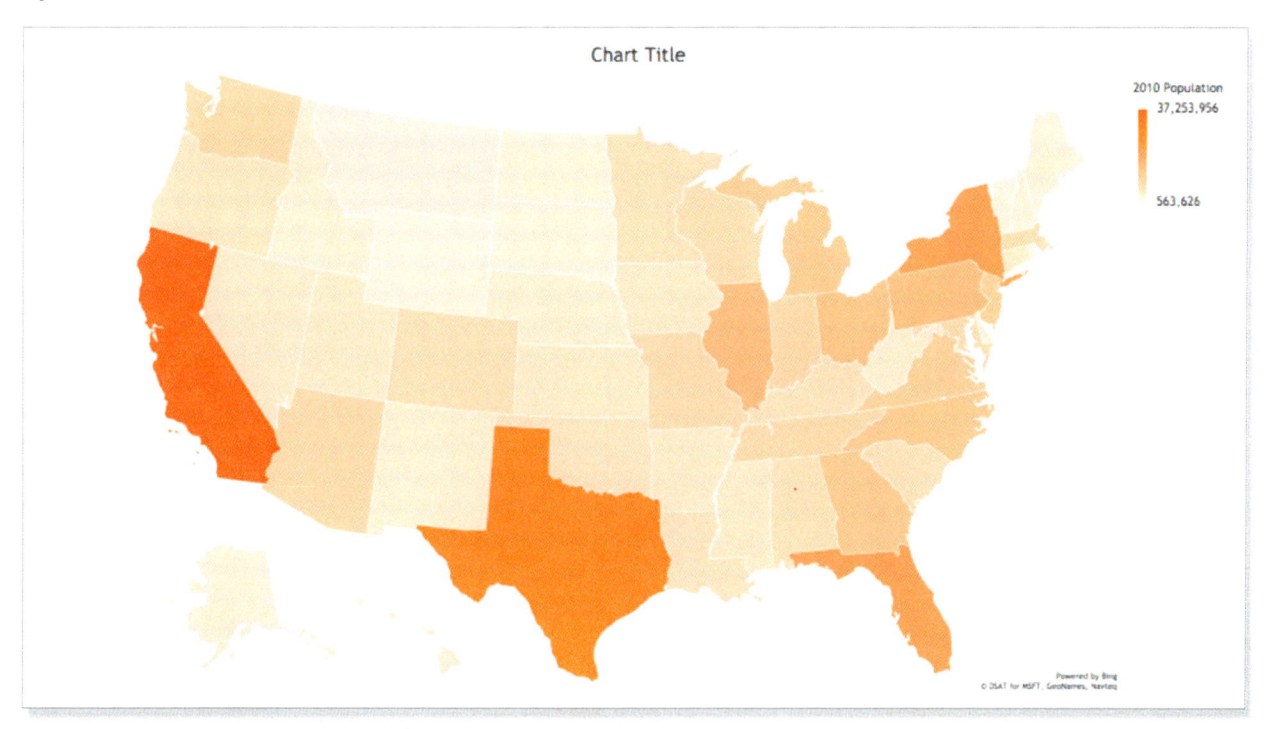

Fig 5.374: Set Minimum and Maximum Series Color for the gradient your map will use.

301

Step 3: While still in the **Series Options** tab, set **Map projection model** to a model of your choice. The default model Excel chose for my map was the **Albers projection**, which I kept **(Fig 5.375)**. But the **Mercator model** is also a very popular model.

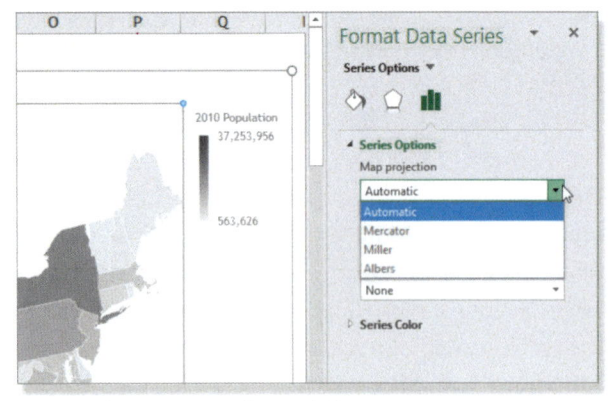

Fig 5.375: Select the projection model for your map.

Step 4: While still in the **Series Options** tab, select **Map area** for your map. Automatic will be the best option in most circumstances **(Fig 5.376)**. Your map area will update if you add a data point outside your current country/region, which is pretty slick.

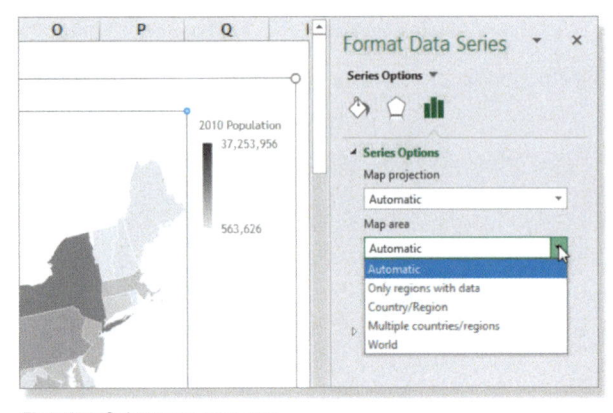

Fig 5.376: Select your map area.

Step 5: Next, set **Map labels**. Keep in mind, at the time of writing, Excel doesn't give you the ability to customize the fonts. It does at least use the font associated with your theme, but I generally wouldn't opt for black font. It's starker and less elegant, in my opinion, than a dark gray. I turned them on for the screenshot and then turned them back off **(Fig 5.377)**.

Tip: If you want control, forgo this option and add data labels instead. Adding **Category Name** will bring in your labels and **Value** will bring in the value (duh). Setting **Separator** to **(New Line)** is the most elegant option, in my opinion.

> **LEARN MORE** See the Add Data Labels section of the Design with Elegance chapter to learn how to add data labels.

Step 6: Clean up your chart to taste. I cleaned mine by taking the following steps:

- made title more descriptive
- bumped up title font size
- added white space to chart title with line breaks
- moved legend

> **LEARN MORE** See the Design with Elegance chapter to learn how to perform these refinements.

Option 2

I just wanted to include another map to demonstrate different formatting. With this map I used the following options:

- set **Map Projection** to Miller
- set **Map Area** to Only regions with data
- Set **Map labels** to Best fit only
- set **Series Color** to my darker orange and the **Lighter 60%** version of it that Excel automatically generates
- added data labels (**Category Name** and **Value**)
- added white space to chart title with line breaks

Example File

map-chart.xlsx

Template

At the time of writing, you can't save a map chart as a chart template. You'll need to create it from scratch or save it in a file where you keep all your charts that won't save as templates.

CHAPTER 5 · CREATE CHARTS

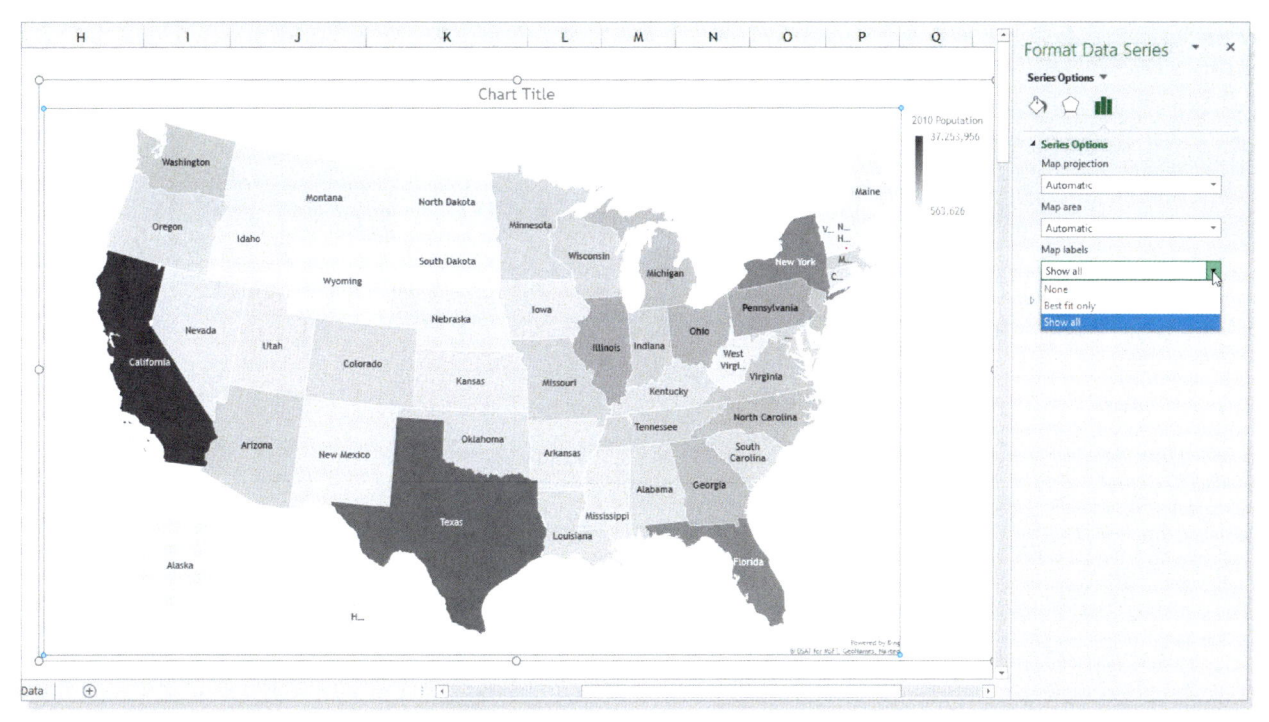

Fig 5.377: Select your map labels.

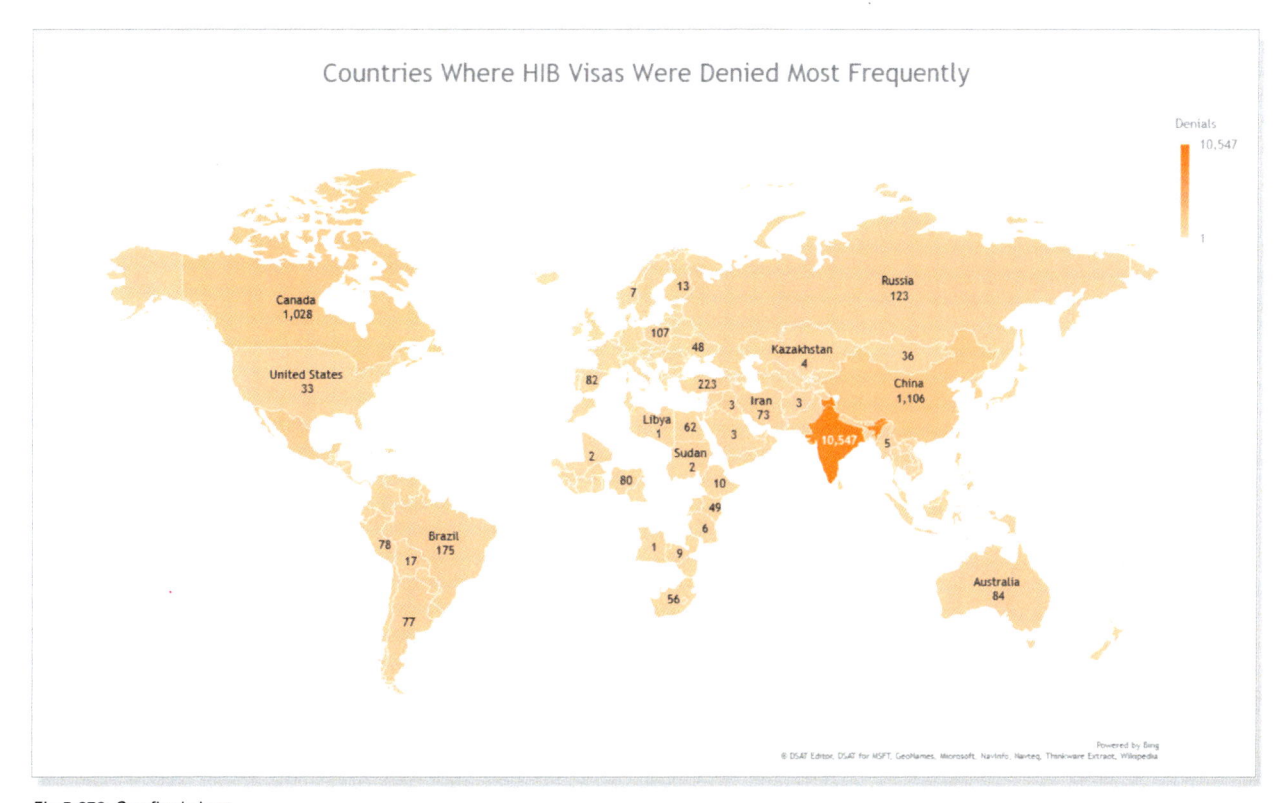

Fig 5.378: Our final chart.

303

CHAPTER 6

Parting Thoughts

GO STRUT YOUR STUFF

Okay, you are now ALL AMPED UP with some serious tools and strategies you can use to set yourself up for a new career or promotion—or just increased effectiveness in communicating with data in your current position. What to do now? If you're looking to use these skills to get a new job or promotion, I *highly* recommend either adding a portfolio to your website, if you have one, or creating a simple portfolio site on a free blogging platform like Tumblr. I helped my daughters, Destinee and Victoria, do this when they were in college. Now they don't need them, but Destinee kept hers on Tumblr (**destineecushing.tumblr.com**), and Victoria moved hers to **VictoriaCushing.com**. You can see an example of Victoria's Data Visualization category in her portfolio: **bit.ly/mdsepc-vc-data**.

She now works for Microsoft, and her data vis skills speak for themselves. But having a portfolio you can point to on your resume will give you a runway for your newfound data vis skills. And no one needs to know you used one of my chart template files, if you choose to go that direction!

JOIN IN THE FUN

I'd love to see the visualizations you create using this book! If you'd like to share your newest creation, simply use the #makingdatasexy hashtag on Twitter, Facebook, or the 'Gram.

You can also follow Making Data Sexy on social media:

Twitter: @makingdatasexy

Facebook: /makingdatasexy

Instagram: @makingdatasexy

And if you'd like to follow yours truly on Twitter, I'm at @AnnieCushing. I tweet data and marketing news and resources almost every weekday.

MORE RESOURCES

You can check out the resources I create regularly for marketers and website owners, such as my *Hundreds of Tools for Marketers* Google Doc, which breaks down hundreds of common marketing tasks and lists tools you can use to accomplish them. Creating excessively thorough Google Docs and templates (and now books) is a Catcher-in-the-Rye-level compulsion for me.

Other resources include:

Free

- Site Audit Checklist
- Campaign Tagging Tool
- Event Tracking Tools

Paid

- Site Audit Template
- Analytics Audit Template
- Dashboard Course
- Marketing Strategy Guide

Go to **annielytics.com/resources** to learn more.

SUGGESTIONS? CORRECTIONS?

Have a burning idea or suggestion for the book? This book wouldn't have been possible without the contribution of clever Excel aficionados pushing Excel to its limits. Feel free to email me yours at **books@makingdatasexy.com**. If I publish future copies of the book and incorporate your idea, I will happily give you credit.)

You can also follow any changes/errors for the book at **bit.ly/mdsem-errata**.

LIKE IT? REVIEW IT!

If you liked the book and it helped you in some way, I'd love for you to share your experience with others in an Amazon review. My hope is to help as many people learn how to make their data sexy as possible!